Protein Folding Handbook
Edited by J. Buchner and T. Kiefhaber

Further Titles of Interest

K. H. Nierhaus, D. N. Wilson (eds.)

Protein Biosynthesis and Ribosome Structure

ISBN 3-527-30638-2

R. J. Mayer, A. J. Ciechanover, M. Rechsteiner (eds.)

Protein Degradation

ISBN 3-527-30837-7 (Vol. 1)
ISBN 3-527-31130-0 (Vol. 2)

G. Cesareni, M. Gimona, M. Sudol, M. Yaffe (eds.)

Modular Protein Domains

ISBN 3-527-30813-X

S. Brakmann, A. Schwienhorst (eds.)

Evolutionary Methods in Biotechnology

ISBN 3-527-30799-0

Protein Folding Handbook

Edited by Johannes Buchner and Thomas Kiefhaber

WILEY-VCH Verlag GmbH & Co. KGaA

Editors

Prof. Dr. Johannes Buchner
Institut für Organische Chemie und
Biochemie
Technische Universität München
Lichtenbergstrasse 4
85747 Garching
Germany
johannes.buchner@ch.tum.de

Prof. Dr. Thomas Kiefhaber
Biozentrum der Universität Basel
Division of Biophysical Chemistry
Klingelbergstrasse 70
4056 Basel
Switzerland
t.kiefhaber@unibas.ch

Cover
Artwork by Prof. Erich Gohl, Regensburg

■ This book was carefully produced.
Nevertheless, authors, editors and publisher
do not warrant the information contained
therein to be free of errors. Readers are
advised to keep in mind that statements,
data, illustrations, procedural details or
other items may inadvertently be
inaccurate.

Library of Congress Card No. applied for
A catalogue record for this book is available
from the British Library.
Bibliographic information published by Die
Deutsche Bibliothek
Die Deutsche Bibliothek lists this
publication in the Deutsche
Nationalbibliografie; detailed bibliographic
data is available in the Internet at http://
dnb.ddb.de

© 2005 WILEY-VCH Verlag GmbH & Co.
KGaA, Weinheim
All rights reserved (including those of
translation in other languages). No part of
this book may be reproduced in any form –
by photoprinting, microfilm, or any other
means – nor transmitted or translated
into machine language without written
permission from the publishers. Registered
names, trademarks, etc. used in this book,
even when not specifically marked as such,
are not to be considered unprotected by law.

Printed in the Federal Republic of
Germany.
Printed on acid-free paper.

Typesetting   Asco Typesetters, Hong Kong
Printing   betz-druck gmbh, Darmstadt
Bookbinding   Litges & Dopf Buchbinderei
GmbH, Heppenheim

ISBN-13 978-3-527-30784-5
ISBN-10 3-527-30784-2

Contents

Part I, Volume 1

Preface *LVIII*

Contributors of Part I *LX*

I/1 Principles of Protein Stability and Design *1*

1 Early Days of Studying the Mechanism of Protein Folding *3*
Robert L. Baldwin
1.1   Introduction *3*
1.2   Two-state Folding *4*
1.3   Levinthal's Paradox *5*
1.4   The Domain as a Unit of Folding *6*
1.5   Detection of Folding Intermediates and Initial Work on the Kinetic Mechanism of Folding *7*
1.6   Two Unfolded Forms of RNase A and Explanation by Proline Isomerization *9*
1.7   Covalent Intermediates in the Coupled Processes of Disulfide Bond Formation and Folding *11*
1.8   Early Stages of Folding Detected by Antibodies and by Hydrogen Exchange *12*
1.9   Molten Globule Folding Intermediates *14*
1.10 Structures of Peptide Models for Folding Intermediates *15*
      Acknowledgments *16*
      References *16*

2 Spectroscopic Techniques to Study Protein Folding and Stability *22*
Franz Schmid
2.1   Introduction *22*
2.2   Absorbance *23*
2.2.1   Absorbance of Proteins *23*
2.2.2   Practical Considerations for the Measurement of Protein Absorbance *27*

2.2.3	Data Interpretation 29
2.3	Fluorescence 29
2.3.1	The Fluorescence of Proteins 30
2.3.2	Energy Transfer and Fluorescence Quenching in a Protein: Barnase 31
2.3.3	Protein Unfolding Monitored by Fluorescence 33
2.3.4	Environmental Effects on Tyrosine and Tryptophan Emission 36
2.3.5	Practical Considerations 37
2.4	Circular Dichroism 38
2.4.1	CD Spectra of Native and Unfolded Proteins 38
2.4.2	Measurement of Circular Dichroism 41
2.4.3	Evaluation of CD Data 42
	References 43

3	**Denaturation of Proteins by Urea and Guanidine Hydrochloride** 45
	C. Nick Pace, Gerald R. Grimsley, and J. Martin Scholtz
3.1	Historical Perspective 45
3.2	How Urea Denatures Proteins 45
3.3	Linear Extrapolation Method 48
3.4	$\Delta G(H_2O)$ 50
3.5	m-Values 55
3.6	Concluding Remarks 58
3.7	Experimental Protocols 59
3.7.1	How to Choose the Best Denaturant for your Study 59
3.7.2	How to Prepare Denaturant Solutions 59
3.7.3	How to Determine Solvent Denaturation Curves 60
3.7.3.1	Determining a Urea or GdmCl Denaturation Curve 62
3.7.3.2	How to Analyze Urea or GdmCl Denaturant Curves 63
3.7.4	Determining Differences in Stability 64
	Acknowledgments 65
	References 65

4	**Thermal Unfolding of Proteins Studied by Calorimetry** 70
	George I. Makhatadze
4.1	Introduction 70
4.2	Two-state Unfolding 71
4.3	Cold Denaturation 76
4.4	Mechanisms of Thermostabilization 77
4.5	Thermodynamic Dissection of Forces Contributing to Protein Stability 79
4.5.1	Heat Capacity Changes, ΔC_p 81
4.5.2	Enthalpy of Unfolding, ΔH 81
4.5.3	Entropy of Unfolding, ΔS 83
4.6	Multistate Transitions 84
4.6.1	Two-state Dimeric Model 85

4.6.2	Two-state Multimeric Model	86
4.6.3	Three-state Dimeric Model	86
4.6.4	Two-state Model with Ligand Binding	88
4.6.5	Four-state (Two-domain Protein) Model	90
4.7	Experimental Protocols	92
4.7.1	How to Prepare for DSC Experiments	92
4.7.2	How to Choose Appropriate Conditions	94
4.7.3	Critical Factors in Running DSC Experiments	94
	References	95
5	**Pressure–Temperature Phase Diagrams of Proteins**	99
	Wolfgang Doster and Josef Friedrich	
5.1	Introduction	99
5.2	Basic Aspects of Phase Diagrams of Proteins and Early Experiments	100
5.3	Thermodynamics of Pressure–Temperature Phase Diagrams	103
5.4	Measuring Phase Stability Boundaries with Optical Techniques	110
5.4.1	Fluorescence Experiments with Cytochrome c	110
5.4.2	Results	112
5.5	What Do We Learn from the Stability Diagram?	116
5.5.1	Thermodynamics	116
5.5.2	Determination of the Equilibrium Constant of Denaturation	117
5.5.3	Microscopic Aspects	120
5.5.4	Structural Features of the Pressure-denatured State	122
5.6	Conclusions and Outlook	123
	Acknowledgment	124
	References	124
6	**Weak Interactions in Protein Folding: Hydrophobic Free Energy, van der Waals Interactions, Peptide Hydrogen Bonds, and Peptide Solvation**	127
	Robert L. Baldwin	
6.1	Introduction	127
6.2	Hydrophobic Free Energy, Burial of Nonpolar Surface and van der Waals Interactions	128
6.2.1	History	128
6.2.2	Liquid–Liquid Transfer Model	128
6.2.3	Relation between Hydrophobic Free Energy and Molecular Surface Area	130
6.2.4	Quasi-experimental Estimates of the Work of Making a Cavity in Water or in Liquid Alkane	131
6.2.5	Molecular Dynamics Simulations of the Work of Making Cavities in Water	133
6.2.6	Dependence of Transfer Free Energy on the Volume of the Solute	134
6.2.7	Molecular Nature of Hydrophobic Free Energy	136

6.2.8	Simulation of Hydrophobic Clusters 137
6.2.9	ΔC_p and the Temperature-dependent Thermodynamics of Hydrophobic Free Energy 137
6.2.10	Modeling Formation of the Hydrophobic Core from Solvation Free Energy and van der Waals Interactions between Nonpolar Residues 142
6.2.11	Evidence Supporting a Role for van der Waals Interactions in Forming the Hydrophobic Core 144
6.3	Peptide Solvation and the Peptide Hydrogen Bond 145
6.3.1	History 145
6.3.2	Solvation Free Energies of Amides 147
6.3.3	Test of the Hydrogen-Bond Inventory 149
6.3.4	The Born Equation 150
6.3.5	Prediction of Solvation Free Energies of Polar Molecules by an Electrostatic Algorithm 150
6.3.6	Prediction of the Solvation Free Energies of Peptide Groups in Different Backbone Conformations 151
6.3.7	Predicted Desolvation Penalty for Burial of a Peptide H-bond 153
6.3.8	Gas–Liquid Transfer Model 154
	Acknowledgments 156
	References 156
7	**Electrostatics of Proteins: Principles, Models and Applications** 163
	Sonja Braun-Sand and Arieh Warshel
7.1	Introduction 163
7.2	Historical Perspectives 163
7.3	Electrostatic Models: From Microscopic to Macroscopic Models 166
7.3.1	All-Atom Models 166
7.3.2	Dipolar Lattice Models and the PDLD Approach 168
7.3.3	The PDLD/S-LRA Model 170
7.3.4	Continuum (Poisson-Boltzmann) and Related Approaches 171
7.3.5	Effective Dielectric Constant for Charge–Charge Interactions and the GB Model 172
7.4	The Meaning and Use of the Protein Dielectric Constant 173
7.5	Validation Studies 176
7.6	Systems Studied 178
7.6.1	Solvation Energies of Small Molecules 178
7.6.2	Calculation of pK_a Values of Ionizable Residues 179
7.6.3	Redox and Electron Transport Processes 180
7.6.4	Ligand Binding 181
7.6.5	Enzyme Catalysis 182
7.6.6	Ion Pairs 183
7.6.7	Protein–Protein Interactions 184
7.6.8	Ion Channels 185
7.6.9	Helix Macrodipoles versus Localized Molecular Dipoles 185
7.6.10	Folding and Stability 186
7.7	Concluding Remarks 189

Acknowledgments *190*
References *190*

8 Protein Conformational Transitions as Seen from the Solvent: Magnetic Relaxation Dispersion Studies of Water, Co-solvent, and Denaturant Interactions with Nonnative Proteins *201*
Bertil Halle, Vladimir P. Denisov, Kristofer Modig, and Monika Davidovic
8.1 The Role of the Solvent in Protein Folding and Stability *201*
8.2 Information Content of Magnetic Relaxation Dispersion *202*
8.3 Thermal Perturbations *205*
8.3.1 Heat Denaturation *205*
8.3.2 Cold Denaturation *209*
8.4 Electrostatic Perturbations *213*
8.5 Solvent Perturbations *218*
8.5.1 Denaturation Induced by Urea *219*
8.5.2 Denaturation Induced by Guanidinium Chloride *225*
8.5.3 Conformational Transitions Induced by Co-solvents *228*
8.6 Outlook *233*
8.7 Experimental Protocols and Data Analysis *233*
8.7.1 Experimental Methodology *233*
8.7.1.1 Multiple-field MRD *234*
8.7.1.2 Field-cycling MRD *234*
8.7.1.3 Choice of Nuclear Isotope *235*
8.7.2 Data Analysis *236*
8.7.2.1 Exchange Averaging *236*
8.7.2.2 Spectral Density Function *237*
8.7.2.3 Residence Time *239*
8.7.2.4 ^{19}F Relaxation *240*
8.7.2.5 Coexisting Protein Species *241*
8.7.2.6 Preferential Solvation *241*
References *242*

9 Stability and Design of α-Helices *247*
Andrew J. Doig, Neil Errington, and Teuku M. Iqbalsyah
9.1 Introduction *247*
9.2 Structure of the α-Helix *247*
9.2.1 Capping Motifs *248*
9.2.2 Metal Binding *250*
9.2.3 The 3_{10}-Helix *251*
9.2.4 The π-Helix *251*
9.3 Design of Peptide Helices *252*
9.3.1 Host–Guest Studies *253*
9.3.2 Helix Lengths *253*
9.3.3 The Helix Dipole *253*
9.3.4 Acetylation and Amidation *254*
9.3.5 Side Chain Spacings *255*

9.3.6	Solubility	256
9.3.7	Concentration Determination	257
9.3.8	Design of Peptides to Measure Helix Parameters	257
9.3.9	Helix Templates	259
9.3.10	Design of 3_{10}-Helices	259
9.3.11	Design of π-helices	261
9.4	Helix Coil Theory	261
9.4.1	Zimm-Bragg Model	261
9.4.2	Lifson-Roig Model	262
9.4.3	The Unfolded State and Polyproline II Helix	265
9.4.4	Single Sequence Approximation	265
9.4.5	N- and C-Caps	266
9.4.6	Capping Boxes	266
9.4.7	Side-chain Interactions	266
9.4.8	N1, N2, and N3 Preferences	267
9.4.9	Helix Dipole	267
9.4.10	3_{10}- and π-Helices	268
9.4.11	AGADIR	268
9.4.12	Lomize-Mosberg Model	269
9.4.13	Extension of the Zimm-Bragg Model	270
9.4.14	Availability of Helix/Coil Programs	270
9.5	Forces Affecting α-Helix Stability	270
9.5.1	Helix Interior	270
9.5.2	Caps	273
9.5.3	Phosphorylation	276
9.5.4	Noncovalent Side-chain Interactions	276
9.5.5	Covalent Side-chain interactions	277
9.5.6	Capping Motifs	277
9.5.7	Ionic Strength	279
9.5.8	Temperature	279
9.5.9	Trifluoroethanol	279
9.5.10	pK_a Values	280
9.5.11	Relevance to Proteins	281
9.6	Experimental Protocols and Strategies	281
9.6.1	Solid Phase Peptide Synthesis (SPPS) Based on the Fmoc Strategy	281
9.6.1.1	Equipment and Reagents	281
9.6.1.2	Fmoc Deprotection and Coupling	283
9.6.1.3	Kaiser Test	284
9.6.1.4	Acetylation and Cleavage	285
9.6.1.5	Peptide Precipitation	286
9.6.2	Peptide Purification	286
9.6.2.1	Equipment and Reagents	286
9.6.2.2	Method	286
9.6.3	Circular Dichroism	287
9.6.4	Acquisition of Spectra	288

9.6.4.1	Instrumental Considerations	288
9.6.5	Data Manipulation and Analysis	289
9.6.5.1	Protocol for CD Measurement of Helix Content	291
9.6.6	Aggregation Test for Helical Peptides	291
9.6.6.1	Equipment and Reagents	291
9.6.6.2	Method	292
9.6.7	Vibrational Circular Dichroism	292
9.6.8	NMR Spectroscopy	292
9.6.8.1	Nuclear Overhauser Effect	293
9.6.8.2	Amide Proton Exchange Rates	294
9.6.8.3	^{13}C NMR	294
9.6.9	Fourier Transform Infrared Spectroscopy	295
9.6.9.1	Secondary Structure	295
9.6.10	Raman Spectroscopy and Raman Optical Activity	296
9.6.11	pH Titrations	298
9.6.11.1	Equipment and Reagents	298
9.6.11.2	Method	298
	Acknowledgments	299
	References	299

10	**Design and Stability of Peptide β-Sheets**	**314**
	Mark S. Searle	
10.1	Introduction	314
10.2	β-Hairpins Derived from Native Protein Sequences	315
10.3	Role of β-Turns in Nucleating β-Hairpin Folding	316
10.4	Intrinsic ϕ, ψ Propensities of Amino Acids	319
10.5	Side-chain Interactions and β-Hairpin Stability	321
10.5.1	Aromatic Clusters Stabilize β-Hairpins	322
10.5.2	Salt Bridges Enhance Hairpin Stability	325
10.6	Cooperative Interactions in β-Sheet Peptides: Kinetic Barriers to Folding	330
10.7	Quantitative Analysis of Peptide Folding	331
10.8	Thermodynamics of β-Hairpin Folding	332
10.9	Multistranded Antiparallel β-Sheet Peptides	334
10.10	Concluding Remarks: Weak Interactions and Stabilization of Peptide β-Sheets	339
	References	340

11	**Predicting Free Energy Changes of Mutations in Proteins**	**343**
	Raphael Guerois, Joaquim Mendes, and Luis Serrano	
11.1	Physical Forces that Determine Protein Conformational Stability	343
11.1.1	Protein Conformational Stability [1]	343
11.1.2	Structures of the N and D States [2–6]	344
11.1.3	Studies Aimed at Understanding the Physical Forces that Determine Protein Conformational Stability [1, 2, 8, 19–26]	346
11.1.4	Forces Determining Conformational Stability [1, 2, 8, 19–27]	346

11.1.5	Intramolecular Interactions 347
11.1.5.1	van der Waals Interactions 347
11.1.5.2	Electrostatic Interactions 347
11.1.5.3	Conformational Strain 349
11.1.6	Solvation 350
11.1.7	Intramolecular Interactions and Solvation Taken Together 350
11.1.8	Entropy 351
11.1.9	Cavity Formation 352
11.1.10	Summary 353
11.2	Methods for the Prediction of the Effect of Point Mutations on in vitro Protein Stability 353
11.2.1	General Considerations on Protein Plasticity upon Mutation 353
11.2.2	Predictive Strategies 355
11.2.3	Methods 356
11.2.3.1	From Sequence and Multiple Sequence Alignment Analysis 356
11.2.3.2	Statistical Analysis of the Structure Databases 356
11.2.3.3	Helix/Coil Transition Model 357
11.2.3.4	Physicochemical Method Based on Protein Engineering Experiments 359
11.2.3.5	Methods Based only on the Basic Principles of Physics and Thermodynamics 364
11.3	Mutation Effects on in vivo Stability 366
11.3.1	The N-terminal Rule 366
11.3.2	The C-terminal Rule 367
11.3.3	PEST Signals 368
11.4	Mutation Effects on Aggregation 368
	References 369

I/2 Dynamics and Mechanisms of Protein Folding Reactions 377

12.1 Kinetic Mechanisms in Protein Folding 379
Annett Bachmann and Thomas Kiefhaber

12.1.1	Introduction 379
12.1.2	Analysis of Protein Folding Reactions using Simple Kinetic Models 379
12.1.2.1	General Treatment of Kinetic Data 380
12.1.2.2	Two-state Protein Folding 380
12.1.2.3	Complex Folding Kinetics 384
12.1.2.3.1	Heterogeneity in the Unfolded State 384
12.1.2.3.2	Folding through Intermediates 388
12.1.2.3.3	Rapid Pre-equilibria 391
12.1.2.3.4	Folding through an On-pathway High-energy Intermediate 393
12.1.3	A Case Study: the Mechanism of Lysozyme Folding 394
12.1.3.1	Lysozyme Folding at pH 5.2 and Low Salt Concentrations 394
12.1.3.2	Lysozyme Folding at pH 9.2 or at High Salt Concentrations 398
12.1.4	Non-exponential Kinetics 401

12.1.5	Conclusions and Outlook *401*
12.1.6	Protocols – Analytical Solutions of Three-state Protein Folding Models *402*
12.1.6.1	Triangular Mechanism *402*
12.1.6.2	On-pathway Intermediate *403*
12.1.6.3	Off-pathway Mechanism *404*
12.1.6.4	Folding Through an On-pathway High-Energy Intermediate *404*
	Acknowledgments *406*
	References *406*
12.2	**Characterization of Protein Folding Barriers with Rate Equilibrium Free Energy Relationships** *411*
	Thomas Kiefhaber, Ignacio E. Sánchez, and Annett Bachmann
12.2.1	Introduction *411*
12.2.2	Rate Equilibrium Free Energy Relationships *411*
12.2.2.1	Linear Rate Equilibrium Free Energy Relationships in Protein Folding *414*
12.2.2.2	Properties of Protein Folding Transition States Derived from Linear REFERs *418*
12.2.3	Nonlinear Rate Equilibrium Free Energy Relationships in Protein Folding *420*
12.2.3.1	Self-Interaction and Cross-Interaction Parameters *420*
12.2.3.2	Hammond and Anti-Hammond Behavior *424*
12.2.3.3	Sequential and Parallel Transition States *425*
12.2.3.4	Ground State Effects *428*
12.2.4	Experimental Results on the Shape of Free Energy Barriers in Protein Folding *432*
12.2.4.1	Broadness of Free Energy Barriers *432*
12.2.4.2	Parallel Pathways *437*
12.2.5	Folding in the Absence of Enthalpy Barriers *438*
12.2.6	Conclusions and Outlook *438*
	Acknowledgments *439*
	References *439*
13	**A Guide to Measuring and Interpreting ϕ-values** *445*
	Nicholas R. Guydosh and Alan R. Fersht
13.1	Introduction *445*
13.2	Basic Concept of ϕ-Value Analysis *445*
13.3	Further Interpretation of ϕ *448*
13.4	Techniques *450*
13.5	Conclusions *452*
	References *452*
14	**Fast Relaxation Methods** *454*
	Martin Gruebele
14.1	Introduction *454*

14.2	Techniques 455
14.2.1	Fast Pressure-Jump Experiments 455
14.2.2	Fast Resistive Heating Experiments 456
14.2.3	Fast Laser-induced Relaxation Experiments 457
14.2.3.1	Laser Photolysis 457
14.2.3.2	Electrochemical Jumps 458
14.2.3.3	Laser-induced pH Jumps 458
14.2.3.4	Covalent Bond Dissociation 459
14.2.3.5	Chromophore Excitation 460
14.2.3.6	Laser Temperature Jumps 460
14.2.4	Multichannel Detection Techniques for Relaxation Studies 461
14.2.4.1	Small Angle X-ray Scattering or Light Scattering 462
14.2.4.2	Direct Absorption Techniques 463
14.2.4.3	Circular Dichroism and Optical Rotatory Dispersion 464
14.2.4.4	Raman and Resonance Raman Scattering 464
14.2.4.5	Intrinsic Fluorescence 465
14.2.4.6	Extrinsic Fluorescence 465
14.3	Protein Folding by Relaxation 466
14.3.1	Transition State Theory, Energy Landscapes, and Fast Folding 466
14.3.2	Viscosity Dependence of Folding Motions 470
14.3.3	Resolving Burst Phases 471
14.3.4	Fast Folding and Unfolded Proteins 472
14.3.5	Experiment and Simulation 472
14.4	Summary 474
14.5	Experimental Protocols 475
14.5.1	Design Criteria for Laser Temperature Jumps 475
14.5.2	Design Criteria for Fast Single-Shot Detection Systems 476
14.5.3	Designing Proteins for Fast Relaxation Experiments 477
14.5.4	Linear Kinetic, Nonlinear Kinetic, and Generalized Kinetic Analysis of Fast Relaxation 477
14.5.4.1	The Reaction $D \rightleftharpoons F$ in the Presence of a Barrier 477
14.5.4.2	The Reaction $2A \rightleftharpoons A_2$ in the Presence of a Barrier 478
14.5.4.3	The Reaction $D \rightleftharpoons F$ at Short Times or over Low Barriers 479
14.5.5	Relaxation Data Analysis by Linear Decomposition 480
14.5.5.1	Singular Value Decomposition (SVD) 480
14.5.5.2	χ-Analysis 481
	Acknowledgments 481
	References 482

15	**Early Events in Protein Folding Explored by Rapid Mixing Methods** 491
	Heinrich Roder, Kosuke Maki, Ramil F. Latypov, Hong Cheng, and M. C. Ramachandra Shastry
15.1	Importance of Kinetics for Understanding Protein Folding 491
15.2	Burst-phase Signals in Stopped-flow Experiments 492
15.3	Turbulent Mixing 494

15.4	Detection Methods *495*	
15.4.1	Tryptophan Fluorescence *495*	
15.4.2	ANS Fluorescence *498*	
15.4.3	FRET *499*	
15.4.4	Continuous-flow Absorbance *501*	
15.4.5	Other Detection Methods used in Ultrafast Folding Studies *502*	
15.5	A Quenched-Flow Method for H-D Exchange Labeling Studies on the Microsecond Time Scale *502*	
15.6	Evidence for Accumulation of Early Folding Intermediates in Small Proteins *505*	
15.6.1	B1 Domain of Protein G *505*	
15.6.2	Ubiquitin *508*	
15.6.3	Cytochrome *c* *512*	
15.7	Significance of Early Folding Events *515*	
15.7.1	Barrier-limited Folding vs. Chain Diffusion *515*	
15.7.2	Chain Compaction: Random Collapse vs. Specific Folding *516*	
15.7.3	Kinetic Role of Early Folding Intermediates *517*	
15.7.4	Broader Implications *520*	
	Appendix *521*	
A1	Design and Calibration of Rapid Mixing Instruments *521*	
A1.1	Stopped-flow Equipment *521*	
A1.2	Continuous-flow Instrumentation *524*	
	Acknowledgments *528*	
	References *528*	
16	**Kinetic Protein Folding Studies using NMR Spectroscopy** *536*	
	Markus Zeeb and Jochen Balbach	
16.1	Introduction *536*	
16.2	Following Slow Protein Folding Reactions in Real Time *538*	
16.3	Two-dimensional Real-time NMR Spectroscopy *545*	
16.4	Dynamic and Spin Relaxation NMR for Quantifying Microsecond-to-Millisecond Folding Rates *550*	
16.5	Conclusions and Future Directions *555*	
16.6	Experimental Protocols *556*	
16.6.1	How to Record and Analyze 1D Real-time NMR Spectra *556*	
16.6.1.1	Acquisition *556*	
16.6.1.2	Processing *557*	
16.6.1.3	Analysis *557*	
16.6.1.4	Analysis of 1D Real-time Diffusion Experiments *558*	
16.6.2	How to Extract Folding Rates from 1D Spectra by Line Shape Analysis *559*	
16.6.2.1	Acquisition *560*	
16.6.2.2	Processing *560*	
16.6.2.3	Analysis *561*	
16.6.3	How to Extract Folding Rates from 2D Real-time NMR Spectra *562*	

16.6.3.1	Acquisition	563
16.6.3.2	Processing	563
16.6.3.3	Analysis	563
16.6.4	How to Analyze Heteronuclear NMR Relaxation and Exchange Data	565
16.6.4.1	Acquisition	566
16.6.4.2	Processing	567
16.6.4.3	Analysis	567
	Acknowledgments	569
	References	569

Part I, Volume 2

17	**Fluorescence Resonance Energy Transfer (FRET) and Single Molecule Fluorescence Detection Studies of the Mechanism of Protein Folding and Unfolding** 573	
	Elisha Haas	
	Abbreviations 573	
17.1	Introduction 573	
17.2	What are the Main Aspects of the Protein Folding Problem that can be Addressed by Methods Based on FRET Measurements? 574	
17.2.1	The Three Protein Folding Problems 574	
17.2.1.1	The Chain Entropy Problem 574	
17.2.1.2	The Function Problem: Conformational Fluctuations 575	
17.3	Theoretical Background 576	
17.3.1	Nonradiative Excitation Energy Transfer 576	
17.3.2	What is FRET? The Singlet–Singlet Excitation Transfer 577	
17.3.3	Rate of Nonradiative Excitation Energy Transfer within a Donor–Acceptor Pair 578	
17.3.4	The Orientation Factor 583	
17.3.5	How to Determine and Control the Value of R_o? 584	
17.3.6	Index of Refraction n 584	
17.3.7	The Donor Quantum Yield Φ_D^o 586	
17.3.8	The Spectral Overlap Integral J 586	
17.4	Determination of Intramolecular Distances in Protein Molecules using FRET Measurements 586	
17.4.1	Single Distance between Donor and Acceptor 587	
17.4.1.1	Method 1: Steady State Determination of Decrease of Donor Emission 587	
17.4.1.2	Method 2: Acceptor Excitation Spectroscopy 588	
17.4.2	Time-resolved Methods 588	
17.4.3	Determination of E from Donor Fluorescence Decay Rates 589	
17.4.4	Determination of Acceptor Fluorescence Lifetime 589	
17.4.5	Determination of Intramolecular Distance Distributions 590	

17.4.6	Evaluation of the Effect of Fast Conformational Fluctuations and Determination of Intramolecular Diffusion Coefficients 592	
17.5	Experimental Challenges in the Implementation of FRET Folding Experiments 594	
17.5.1	Optimized Design and Preparation of Labeled Protein Samples for FRET Folding Experiments 594	
17.5.2	Strategies for Site-specific Double Labeling of Proteins 595	
17.5.3	Preparation of Double-labeled Mutants Using Engineered Cysteine Residues (strategy 4) 596	
17.5.4	Possible Pitfalls Associated with the Preparation of Labeled Protein Samples for FRET Folding Experiments 599	
17.6	Experimental Aspects of Folding Studies by Distance Determination Based on FRET Measurements 600	
17.6.1	Steady State Determination of Transfer Efficiency 600	
17.6.1.1	Donor Emission 600	
17.6.1.2	Acceptor Excitation Spectroscopy 601	
17.6.2	Time-resolved Measurements 601	
17.7	Data Analysis 603	
17.7.1	Rigorous Error Analysis 606	
17.7.2	Elimination of Systematic Errors 606	
17.8	Applications of trFRET for Characterization of Unfolded and Partially Folded Conformations of Globular Proteins under Equilibrium Conditions 607	
17.8.1	Bovine Pancreatic Trypsin Inhibitor 607	
17.8.2	The Loop Hypothesis 608	
17.8.3	RNase A 609	
17.8.4	Staphylococcal Nuclease 611	
17.9	Unfolding Transition via Continuum of Native-like Forms 611	
17.10	The Third Folding Problem: Domain Motions and Conformational Fluctuations of Enzyme Molecules 611	
17.11	Single Molecule FRET-detected Folding Experiments 613	
17.12	Principles of Applications of Single Molecule FRET Spectroscopy in Folding Studies 615	
17.12.1	Design and Analysis of Single Molecule FRET Experiments 615	
17.12.1.1	How is Single Molecule FRET Efficiency Determined? 615	
17.12.1.2	The Challenge of Extending the Length of the Time Trajectories 617	
17.12.2	Distance and Time Resolution of the Single Molecule FRET Folding Experiments 618	
17.13	Folding Kinetics 619	
17.13.1	Steady State and trFRET-detected Folding Kinetics Experiments 619	
17.13.2	Steady State Detection 619	
17.13.3	Time-resolved FRET Detection of Rapid Folding Kinetics: the "Double Kinetics" Experiment 621	
17.13.4	Multiple Probes Analysis of the Folding Transition 622	
17.14	Concluding Remarks 625	

Acknowledgments 626
References 627

18 Application of Hydrogen Exchange Kinetics to Studies of Protein Folding 634
Kaare Teilum, Birthe B. Kragelund, and Flemming M. Poulsen
18.1 Introduction 634
18.2 The Hydrogen Exchange Reaction 638
18.2.1 Calculating the Intrinsic Hydrogen Exchange Rate Constant, k_{int} 638
18.3 Protein Dynamics by Hydrogen Exchange in Native and Denaturing Conditions 641
18.3.1 Mechanisms of Exchange 642
18.3.2 Local Opening and Closing Rates from Hydrogen Exchange Kinetics 642
18.3.2.1 The General Amide Exchange Rate Expression – the Linderstrøm-Lang Equation 643
18.3.2.2 Limits to the General Rate Expression – EX1 and EX2 644
18.3.2.3 The Range between the EX1 and EX2 Limits 646
18.3.2.4 Identification of Exchange Limit 646
18.3.2.5 Global Opening and Closing Rates and Protein Folding 647
18.3.3 The "Native State Hydrogen Exchange" Strategy 648
18.3.3.1 Localization of Partially Unfolded States, PUFs 650
18.4 Hydrogen Exchange as a Structural Probe in Kinetic Folding Experiments 651
18.4.1 Protein Folding/Hydrogen Exchange Competition 652
18.4.2 Hydrogen Exchange Pulse Labeling 656
18.4.3 Protection Factors in Folding Intermediates 657
18.4.4 Kinetic Intermediate Structures Characterized by Hydrogen Exchange 659
18.5 Experimental Protocols 661
18.5.1 How to Determine Hydrogen Exchange Kinetics at Equilibrium 661
18.5.1.1 Equilibrium Hydrogen Exchange Experiments 661
18.5.1.2 Determination of Segmental Opening and Closing Rates, k_{op} and k_{cl} 662
18.5.1.3 Determination of ΔG_{fluc}, m, and $\Delta G°_{unf}$ 662
18.5.2 Planning a Hydrogen Exchange Folding Experiment 662
18.5.2.1 Determine a Combination of t_{pulse} and pH_{pulse} 662
18.5.2.2 Setup Quench Flow Apparatus 662
18.5.2.3 Prepare Deuterated Protein and Chemicals 663
18.5.2.4 Prepare Buffers and Unfolded Protein 663
18.5.2.5 Check pH in the Mixing Steps 664
18.5.2.6 Sample Mixing and Preparation 664
18.5.3 Data Analysis 664
Acknowledgments 665
References 665

19	**Studying Protein Folding and Aggregation by Laser Light Scattering** 673
	Klaus Gast and Andreas J. Modler
19.1	Introduction 673
19.2	Basic Principles of Laser Light Scattering 674
19.2.1	Light Scattering by Macromolecular Solutions 674
19.2.2	Molecular Parameters Obtained from Static Light Scattering (SLS) 676
19.2.3	Molecular Parameters Obtained from Dynamic Light Scattering (DLS) 678
19.2.4	Advantages of Combined SLS and DLS Experiments 680
19.3	Laser Light Scattering of Proteins in Different Conformational States – Equilibrium Folding/Unfolding Transitions 680
19.3.1	General Considerations, Hydrodynamic Dimensions in the Natively Folded State 680
19.3.2	Changes in the Hydrodynamic Dimensions during Heat-induced Unfolding 682
19.3.3	Changes in the Hydrodynamic Dimensions upon Cold Denaturation 683
19.3.4	Denaturant-induced Changes of the Hydrodynamic Dimensions 684
19.3.5	Acid-induced Changes of the Hydrodynamic Dimensions 685
19.3.6	Dimensions in Partially Folded States – Molten Globules and Fluoroalcohol-induced States 686
19.3.7	Comparison of the Dimensions of Proteins in Different Conformational States 687
19.3.8	Scaling Laws for the Native and Highly Unfolded States, Hydrodynamic Modeling 687
19.4	Studying Folding Kinetics by Laser Light Scattering 689
19.4.1	General Considerations, Attainable Time Regions 689
19.4.2	Hydrodynamic Dimensions of the Kinetic Molten Globule of Bovine α-Lactalbumin 690
19.4.3	RNase A is Only Weakly Collapsed During the Burst Phase of Folding 691
19.5	Misfolding and Aggregation Studied by Laser Light Scattering 692
19.5.1	Overview: Some Typical Light Scattering Studies of Protein Aggregation 692
19.5.2	Studying Misfolding and Amyloid Formation by Laser Light Scattering 693
19.5.2.1	Overview: Initial States, Critical Oligomers, Protofibrils, Fibrils 693
19.5.2.2	Aggregation Kinetics of $A\beta$ Peptides 694
19.5.2.3	Kinetics of Oligomer and Fibril Formation of PGK and Recombinant Hamster Prion Protein 695
19.5.2.4	Mechanisms of Misfolding and Misassembly, Some General Remarks 698
19.6	Experimental Protocols 698
19.6.1	Laser Light Scattering Instrumentation 698

19.6.1.1	Basic Experimental Set-up, General Requirements	698
19.6.1.2	Supplementary Measurements and Useful Options	700
19.6.1.3	Commercially Available Light Scattering Instrumentation	701
19.6.2	Experimental Protocols for the Determination of Molecular Mass and Stokes Radius of a Protein in a Particular Conformational State	701
	Protocol 1 702	
	Protocol 2 704	
	Acknowledgments 704	
	References 704	
20	**Conformational Properties of Unfolded Proteins**	**710**
	Patrick J. Fleming and George D. Rose	
20.1	Introduction 710	
20.1.1	Unfolded vs. Denatured Proteins 710	
20.2	Early History 711	
20.3	The Random Coil 712	
20.3.1	The Random Coil – Theory 713	
20.3.1.1	The Random Coil Model Prompts Three Questions	716
20.3.1.2	The Folding Funnel 716	
20.3.1.3	Transition State Theory 717	
20.3.1.4	Other Examples 717	
20.3.1.5	Implicit Assumptions from the Random Coil Model	718
20.3.2	The Random Coil – Experiment 718	
20.3.2.1	Intrinsic Viscosity 719	
20.3.2.2	SAXS and SANS 720	
20.4	Questions about the Random Coil Model 721	
20.4.1	Questions from Theory 722	
20.4.1.1	The Flory Isolated-pair Hypothesis 722	
20.4.1.2	Structure vs. Energy Duality 724	
20.4.1.3	The "Rediscovery" of Polyproline II Conformation	724
20.4.1.4	P_{II} in Unfolded Peptides and Proteins 726	
20.4.2	Questions from Experiment 727	
20.4.2.1	Residual Structure in Denatured Proteins and Peptides	727
20.4.3	The Reconciliation Problem 728	
20.4.4	Organization in the Unfolded State – the Entropic Conjecture	728
20.4.4.1	Steric Restrictions beyond the Dipeptide 729	
20.5	Future Directions 730	
	Acknowledgments 731	
	References 731	
21	**Conformation and Dynamics of Nonnative States of Proteins studied by NMR Spectroscopy**	**737**
	Julia Wirmer, Christian Schlörb, and Harald Schwalbe	
21.1	Introduction 737	
21.1.1	Structural Diversity of Polypeptide Chains 737	

21.1.2	Intrinsically Unstructured and Natively Unfolded Proteins 739	
21.2	Prerequisites: NMR Resonance Assignment 740	
21.3	NMR Parameters 744	
21.3.1	Chemical shifts δ 745	
21.3.1.1	Conformational Dependence of Chemical Shifts 745	
21.3.1.2	Interpretation of Chemical Shifts in the Presence of Conformational Averaging 746	
21.3.2	J Coupling Constants 748	
21.3.2.1	Conformational Dependence of J Coupling Constants 748	
21.3.2.2	Interpretation of J Coupling Constants in the Presence of Conformational Averaging 750	
21.3.3	Relaxation: Homonuclear NOEs 750	
21.3.3.1	Distance Dependence of Homonuclear NOEs 750	
21.3.3.2	Interpretation of Homonuclear NOEs in the Presence of Conformational Averaging 754	
21.3.4	Heteronuclear Relaxation (^{15}N R_1, R_2, hetNOE) 757	
21.3.4.1	Correlation Time Dependence of Heteronuclear Relaxation Parameters 757	
21.3.4.2	Dependence on Internal Motions of Heteronuclear Relaxation Parameters 759	
21.3.5	Residual Dipolar Couplings 760	
21.3.5.1	Conformational Dependence of Residual Dipolar Couplings 760	
21.3.5.2	Interpretation of Residual Dipolar Couplings in the Presence of Conformational Averaging 763	
21.3.6	Diffusion 765	
21.3.7	Paramagnetic Spin Labels 766	
21.3.8	H/D Exchange 767	
21.3.9	Photo-CIDNP 767	
21.4	Model for the Random Coil State of a Protein 768	
21.5	Nonnative States of Proteins: Examples from Lysozyme, α-Lactalbumin, and Ubiquitin 771	
21.5.1	Backbone Conformation 772	
21.5.1.1	Interpretation of Chemical Shifts 772	
21.5.1.2	Interpretation of NOEs 774	
21.5.1.3	Interpretation of J Coupling Constants 780	
21.5.2	Side-chain Conformation 784	
21.5.2.1	Interpretation of J Coupling Constants 784	
21.5.3	Backbone Dynamics 786	
21.5.3.1	Interpretation of ^{15}N Relaxation Rates 786	
21.6	Summary and Outlook 793	
	Acknowledgments 794	
	References 794	
22	**Dynamics of Unfolded Polypeptide Chains** 809	
	Beat Fierz and Thomas Kiefhaber	
22.1	Introduction 809	

22.2	Equilibrium Properties of Chain Molecules	809
22.2.1	The Freely Jointed Chain	810
22.2.2	Chain Stiffness	810
22.2.3	Polypeptide Chains	811
22.2.4	Excluded Volume Effects	812
22.3	Theory of Polymer Dynamics	813
22.3.1	The Langevin Equation	813
22.3.2	Rouse Model and Zimm Model	814
22.3.3	Dynamics of Loop Closure and the Szabo-Schulten-Schulten Theory	815
22.4	Experimental Studies on the Dynamics in Unfolded Polypeptide Chains	816
22.4.1	Experimental Systems for the Study of Intrachain Diffusion	816
22.4.1.1	Early Experimental Studies	816
22.4.1.2	Triplet Transfer and Triplet Quenching Studies	821
22.4.1.3	Fluorescence Quenching	825
22.4.2	Experimental Results on Dynamic Properties of Unfolded Polypeptide Chains	825
22.4.2.1	Kinetics of Intrachain Diffusion	826
22.4.2.2	Effect of Loop Size on the Dynamics in Flexible Polypeptide Chains	826
22.4.2.3	Effect of Amino Acid Sequence on Chain Dynamics	829
22.4.2.4	Effect of the Solvent on Intrachain Diffusion	831
22.4.2.5	Effect of Solvent Viscosity on Intrachain Diffusion	833
22.4.2.6	End-to-end Diffusion vs. Intrachain Diffusion	834
22.4.2.7	Chain Diffusion in Natural Protein Sequences	834
22.5	Implications for Protein Folding Kinetics	837
22.5.1	Rate of Contact Formation during the Earliest Steps in Protein Folding	837
22.5.2	The Speed Limit of Protein Folding vs. the Pre-exponential Factor	839
22.5.3	Contributions of Chain Dynamics to Rate- and Equilibrium Constants for Protein Folding Reactions	840
22.6	Conclusions and Outlook	844
22.7	Experimental Protocols and Instrumentation	844
22.7.1	Properties of the Electron Transfer Probes and Treatment of the Transfer Kinetics	845
22.7.2	Test for Diffusion-controlled Reactions	847
22.7.2.1	Determination of Bimolecular Quenching or Transfer Rate Constants	847
22.7.2.2	Testing the Viscosity Dependence	848
22.7.2.3	Determination of Activation Energy	848
22.7.3	Instrumentation	849
	Acknowledgments	849
	References	849

23	**Equilibrium and Kinetically Observed Molten Globule States** 856	
	Kosuke Maki, Kiyoto Kamagata, and Kunihiro Kuwajima	
23.1	Introduction 856	
23.2	Equilibrium Molten Globule State 858	
23.2.1	Structural Characteristics of the Molten Globule State 858	
23.2.2	Typical Examples of the Equilibrium Molten Globule State 859	
23.2.3	Thermodynamic Properties of the Molten Globule State 860	
23.3	The Kinetically Observed Molten Globule State 862	
23.3.1	Observation and Identification of the Molten Globule State in Kinetic Refolding 862	
23.3.2	Kinetics of Formation of the Early Folding Intermediates 863	
23.3.3	Late Folding Intermediates and Structural Diversity 864	
23.3.4	Evidence for the On-pathway Folding Intermediate 865	
23.4	Two-stage Hierarchical Folding Funnel 866	
23.5	Unification of the Folding Mechanism between Non-two-state and Two-state Proteins 867	
23.5.1	Statistical Analysis of the Folding Data of Non-two-state and Two-state Proteins 868	
23.5.2	A Unified Mechanism of Protein Folding: Hierarchy 870	
23.5.3	Hidden Folding Intermediates in Two-state Proteins 871	
23.6	Practical Aspects of the Experimental Study of Molten Globules 872	
23.6.1	Observation of the Equilibrium Molten Globule State 872	
23.6.1.1	Two-state Unfolding Transition 872	
23.6.1.2	Multi-state (Three-state) Unfolding Transition 874	
23.6.2	Burst-phase Intermediate Accumulated during the Dead Time of Refolding Kinetics 876	
23.6.3	Testing the Identity of the Molten Globule State with the Burst-Phase Intermediate 877	
	References 879	
24	**Alcohol- and Salt-induced Partially Folded Intermediates** 884	
	Daizo Hamada and Yuji Goto	
24.1	Introduction 884	
24.2	Alcohol-induced Intermediates of Proteins and Peptides 886	
24.2.1	Formation of Secondary Structures by Alcohols 888	
24.2.2	Alcohol-induced Denaturation of Proteins 888	
24.2.3	Formation of Compact Molten Globule States 889	
24.2.4	Example: β-Lactoglobulin 890	
24.3	Mechanism of Alcohol-induced Conformational Change 893	
24.4	Effects of Alcohols on Folding Kinetics 896	
24.5	Salt-induced Formation of the Intermediate States 899	
24.5.1	Acid-denatured Proteins 899	
24.5.2	Acid-induced Unfolding and Refolding Transitions 900	
24.6	Mechanism of Salt-induced Conformational Change 904	
24.7	Generality of the Salt Effects 906	

24.8	Conclusion 907
	References 908

25	**Prolyl Isomerization in Protein Folding** 916
	Franz Schmid
25.1	Introduction 916
25.2	Prolyl Peptide Bonds 917
25.3	Prolyl Isomerizations as Rate-determining Steps of Protein Folding 918
25.3.1	The Discovery of Fast and Slow Refolding Species 918
25.3.2	Detection of Proline-limited Folding Processes 919
25.3.3	Proline-limited Folding Reactions 921
25.3.4	Interrelation between Prolyl Isomerization and Conformational Folding 923
25.4	Examples of Proline-limited Folding Reactions 924
25.4.1	Ribonuclease A 924
25.4.2	Ribonuclease T1 926
25.4.3	The Structure of a Folding Intermediate with an Incorrect Prolyl Isomer 928
25.5	Native-state Prolyl Isomerizations 929
25.6	Nonprolyl Isomerizations in Protein Folding 930
25.7	Catalysis of Protein Folding by Prolyl Isomerases 932
25.7.1	Prolyl Isomerases as Tools for Identifying Proline-limited Folding Steps 932
25.7.2	Specificity of Prolyl Isomerases 933
25.7.3	The Trigger Factor 934
25.7.4	Catalysis of Prolyl Isomerization During de novo Protein Folding 935
25.8	Concluding Remarks 936
25.9	Experimental Protocols 936
25.9.1	Slow Refolding Assays ("Double Jumps") to Measure Prolyl Isomerizations in an Unfolded Protein 936
25.9.1.1	Guidelines for the Design of Double Jump Experiments 937
25.9.1.2	Formation of U_S Species after Unfolding of RNase A 938
25.9.2	Slow Unfolding Assays for Detecting and Measuring Prolyl Isomerizations in Refolding 938
25.9.2.1	Practical Considerations 939
25.9.2.2	Kinetics of the Formation of Fully Folded IIHY-G3P* Molecules 939
	References 939

26	**Folding and Disulfide Formation** 946
	Margherita Ruoppolo, Piero Pucci, and Gennaro Marino
26.1	Chemistry of the Disulfide Bond 946
26.2	Trapping Protein Disulfides 947
26.3	Mass Spectrometric Analysis of Folding Intermediates 948
26.4	Mechanism(s) of Oxidative Folding so Far – Early and Late Folding Steps 949

26.5	Emerging Concepts from Mass Spectrometric Studies	950
26.5.1	Three-fingered Toxins	951
26.5.2	RNase A	953
26.5.3	Antibody Fragments	955
26.5.4	Human Nerve Growth Factor	956
26.6	Unanswered Questions	956
26.7	Concluding Remarks	957
26.8	Experimental Protocols	957
26.8.1	How to Prepare Folding Solutions	957
26.8.2	How to Carry Out Folding Reactions	958
26.8.3	How to Choose the Best Mass Spectrometric Equipment for Your Study	959
26.8.4	How to Perform Electrospray (ES)MS Analysis	959
26.8.5	How to Perform Matrix-assisted Laser Desorption Ionization (MALDI) MS Analysis	960
	References	961
27	**Concurrent Association and Folding of Small Oligomeric Proteins**	**965**
	Hans Rudolf Bosshard	
27.1	Introduction	965
27.2	Experimental Methods Used to Follow the Folding of Oligomeric Proteins	966
27.2.1	Equilibrium Methods	966
27.2.2	Kinetic Methods	968
27.3	Dimeric Proteins	969
27.3.1	Two-state Folding of Dimeric Proteins	970
27.3.1.1	Examples of Dimeric Proteins Obeying Two-state Folding	971
27.3.2	Folding of Dimeric Proteins through Intermediate States	978
27.4	Trimeric and Tetrameric Proteins	983
27.5	Concluding Remarks	986
	Appendix – Concurrent Association and Folding of Small Oligomeric Proteins	987
A1	Equilibrium Constants for Two-state Folding	988
A1.1	Homooligomeric Protein	988
A1.2	Heterooligomeric Protein	989
A2	Calculation of Thermodynamic Parameters from Equilibrium Constants	990
A2.1	Basic Thermodynamic Relationships	990
A2.2	Linear Extrapolation of Denaturant Unfolding Curves of Two-state Reaction	990
A2.3	Calculation of the van't Hoff Enthalpy Change from Thermal Unfolding Data	990
A2.4	Calculation of the van't Hoff Enthalpy Change from the Concentration-dependence of T_m	991
A2.5	Extrapolation of Thermodynamic Parameters to Different Temperatures: Gibbs-Helmholtz Equation	991

A3	Kinetics of Reversible Two-state Folding and Unfolding: Integrated Rate Equations *992*	
A3.1	Two-state Folding of Dimeric Protein *992*	
A3.2	Two-state Unfolding of Dimeric Protein *992*	
A3.3	Reversible Two-state Folding and Unfolding *993*	
A3.3.1	Homodimeric protein *993*	
A3.3.2	Heterodimeric protein *993*	
A4	Kinetics of Reversible Two-state Folding: Relaxation after Disturbance of a Pre-existing Equilibrium (Method of Bernasconi) *994*	
	Acknowledgments *995*	
	References *995*	
28	**Folding of Membrane Proteins** *998*	
	Lukas K. Tamm and Heedeok Hong	
28.1	Introduction *998*	
28.2	Thermodyamics of Residue Partitioning into Lipid Bilayers *1000*	
28.3	Stability of β-Barrel Proteins *1001*	
28.4	Stability of Helical Membrane Proteins *1009*	
28.5	Helix and Other Lateral Interactions in Membrane Proteins *1010*	
28.6	The Membrane Interface as an Important Contributor to Membrane Protein Folding *1012*	
28.7	Membrane Toxins as Models for Helical Membrane Protein Insertion *1013*	
28.8	Mechanisms of β-Barrel Membrane Protein Folding *1015*	
28.9	Experimental Protocols *1016*	
28.9.1	SDS Gel Shift Assay for Heat-modifiable Membrane Proteins *1016*	
28.9.1.1	Reversible Folding and Unfolding Protocol Using OmpA as an Example *1016*	
28.9.2	Tryptophan Fluorescence and Time-resolved Distance Determination by Tryptophan Fluorescence Quenching *1018*	
28.9.2.1	TDFQ Protocol for Monitoring the Translocation of Tryptophans across Membranes *1019*	
28.9.3	Circular Dichroism Spectroscopy *1020*	
28.9.4	Fourier Transform Infrared Spectroscopy *1022*	
28.9.4.1	Protocol for Obtaining Conformation and Orientation of Membrane Proteins and Peptides by Polarized ATR-FTIR Spectroscopy *1023*	
	Acknowledgments *1025*	
	References *1025*	
29	**Protein Folding Catalysis by Pro-domains** *1032*	
	Philip N. Bryan	
29.1	Introduction *1032*	
29.2	Bimolecular Folding Mechanisms *1033*	
29.3	Structures of Reactants and Products *1033*	
29.3.1	Structure of Free SBT *1033*	

29.3.2	Structure of SBT/Pro-domain Complex	*1036*
29.3.3	Structure of Free ALP	*1037*
29.3.4	Structure of the ALP/Pro-domain Complex	*1037*
29.4	Stability of the Mature Protease	*1039*
29.4.1	Stability of ALP	*1039*
29.4.2	Stability of Subtilisin	*1040*
29.5	Analysis of Pro-domain Binding to the Folded Protease	*1042*
29.6	Analysis of Folding Steps	*1043*
29.7	Why are Pro-domains Required for Folding?	*1046*
29.8	What is the Origin of High Cooperativity?	*1047*
29.9	How Does the Pro-domain Accelerate Folding?	*1048*
29.10	Are High Kinetic Stability and Facile Folding Mutually Exclusive?	*1049*
29.11	Experimental Protocols for Studying SBT Folding	*1049*
29.11.1	Fermentation and Purification of Active Subtilisin	*1049*
29.11.2	Fermentation and Purification of Facile-folding Ala221 Subtilisin from *E. coli*	*1050*
29.11.3	Mutagenesis and Protein Expression of Pro-domain Mutants	*1051*
29.11.4	Purification of Pro-domain	*1052*
29.11.5	Kinetics of Pro-domain Binding to Native SBT	*1052*
29.11.6	Kinetic Analysis of Pro-domain Facilitated Subtilisin Folding	*1052*
29.11.6.1	Single Mixing	*1052*
29.11.6.2	Double Jump: Renaturation–Denaturation	*1053*
29.11.6.3	Double Jump: Denaturation–Renaturation	*1053*
29.11.6.4	Triple Jump: Denaturation–Renaturation–Denaturation	*1054*
	References	*1054*
30	**The Thermodynamics and Kinetics of Collagen Folding**	*1059*
	Hans Peter Bächinger and Jürgen Engel	
30.1	Introduction	*1059*
30.1.1	The Collagen Family	*1059*
30.1.2	Biosynthesis of Collagens	*1060*
30.1.3	The Triple Helical Domain in Collagens and Other Proteins	*1061*
30.1.4	N- and C-Propeptide, Telopeptides, Flanking Coiled-Coil Domains	*1061*
30.1.5	Why is the Folding of the Triple Helix of Interest?	*1061*
30.2	Thermodynamics of Collagen Folding	*1062*
30.2.1	Stability of the Triple Helix	*1062*
30.2.2	The Role of Posttranslational Modifications	*1063*
30.2.3	Energies Involved in the Stability of the Triple Helix	*1063*
30.2.4	Model Peptides Forming the Collagen Triple Helix	*1066*
30.2.4.1	Type of Peptides	*1066*
30.2.4.2	The All-or-none Transition of Short Model Peptides	*1066*
30.2.4.3	Thermodynamic Parameters for Different Model Systems	*1069*
30.2.4.4	Contribution of Different Tripeptide Units to Stability	*1075*

30.2.4.5	Crystal and NMR Structures of Triple Helices	1076
30.2.4.6	Conformation of the Randomly Coiled Chains	1077
30.2.4.7	Model Studies with Isomers of Hydroxyproline and Fluoroproline	1078
30.2.4.8	Cis ⇌ trans Equilibria of Peptide Bonds	1079
30.2.4.9	Interpretations of Stabilities on a Molecular Level	1080
30.3	Kinetics of Triple Helix Formation	1081
30.3.1	Properties of Collagen Triple Helices that Influence Kinetics	1081
30.3.2	Folding of Triple Helices from Single Chains	1082
30.3.2.1	Early Work	1082
30.3.2.2	Concentration Dependence of the Folding of $(PPG)_{10}$ and $(POG)_{10}$	1082
30.3.2.3	Model Mechanism of the Folding Kinetics	1085
30.3.2.4	Rate Constants of Nucleation and Propagation	1087
30.3.2.5	Host–guest Peptides and an Alternative Kinetics Model	1088
30.3.3	Triple Helix Formation from Linked Chains	1089
30.3.3.1	The Short N-terminal Triple Helix of Collagen III in Fragment Col1–3	1089
30.3.3.2	Folding of the Central Long Triple Helix of Collagen III	1090
30.3.3.3	The Zipper Model	1092
30.3.4	Designed Collagen Models with Chains Connected by a Disulfide Knot or by Trimerizing Domains	1097
30.3.4.1	Disulfide-linked Model Peptides	1097
30.3.4.2	Model Peptides Linked by a Foldon Domain	1098
30.3.4.3	Collagen Triple Helix Formation can be Nucleated at either End	1098
30.3.4.4	Hysteresis of Triple Helix Formation	1099
30.3.5	Influence of cis–trans Isomerase and Chaperones	1100
30.3.6	Mutations in Collagen Triple Helices Affect Proper Folding	1101
	References	1101
31	**Unfolding Induced by Mechanical Force**	**1111**
	Jane Clarke and Phil M. Williams	
31.1	Introduction	1111
31.2	Experimental Basics	1112
31.2.1	Instrumentation	1112
31.2.2	Sample Preparation	1113
31.2.3	Collecting Data	1114
31.2.4	Anatomy of a Force Trace	1115
31.2.5	Detecting Intermediates in a Force Trace	1115
31.2.6	Analyzing the Force Trace	1116
31.3	Analysis of Force Data	1117
31.3.1	Basic Theory behind Dynamic Force Spectroscopy	1117
31.3.2	The Ramp of Force Experiment	1119
31.3.3	The Golden Equation of DFS	1121
31.3.4	Nonlinear Loading	1122

31.3.4.1	The Worm-line Chain (WLC)	*1123*
31.3.5	Experiments under Constant Force	*1124*
31.3.6	Effect of Tandem Repeats on Kinetics	*1125*
31.3.7	Determining the Modal Force	*1126*
31.3.8	Comparing Behavior	*1127*
31.3.9	Fitting the Data	*1127*
31.4	Use of Complementary Techniques	*1129*
31.4.1	Protein Engineering	*1130*
31.4.1.1	Choosing Mutants	*1130*
31.4.1.2	Determining $\Delta\Delta G_{D-N}$	*1131*
31.4.1.3	Determining $\Delta\Delta G_{TS-N}$	*1131*
31.4.1.4	Interpreting the Φ-values	*1132*
31.4.2	Computer Simulation	*1133*
31.5	Titin I27: A Case Study	*1134*
31.5.1	The Protein System	*1134*
31.5.2	The Unfolding Intermediate	*1135*
31.5.3	The Transition State	*1136*
31.5.4	The Relationship Between the Native and Transition States	*1137*
31.5.5	The Energy Landscape under Force	*1139*
31.6	Conclusions – the Future	*1139*
	References	*1139*

32 Molecular Dynamics Simulations to Study Protein Folding and Unfolding *1143*

Amedeo Caflisch and Emanuele Paci

32.1	Introduction	*1143*
32.2	Molecular Dynamics Simulations of Peptides and Proteins	*1144*
32.2.1	Folding of Structured Peptides	*1144*
32.2.1.1	Reversible Folding and Free Energy Surfaces	*1144*
32.2.1.2	Non-Arrhenius Temperature Dependence of the Folding Rate	*1147*
32.2.1.3	Denatured State and Levinthal Paradox	*1148*
32.2.1.4	Folding Events of Trp-cage	*1149*
32.2.2	Unfolding Simulations of Proteins	*1150*
32.2.2.1	High-temperature Simulations	*1150*
32.2.2.2	Biased Unfolding	*1150*
32.2.2.3	Forced Unfolding	*1151*
32.2.3	Determination of the Transition State Ensemble	*1153*
32.3	MD Techniques and Protocols	*1155*
32.3.1	Techniques to Improve Sampling	*1155*
32.3.1.1	Replica Exchange Molecular Dynamics	*1155*
32.3.1.2	Methods Based on Path Sampling	*1157*
32.3.2	MD with Restraints	*1157*
32.3.3	Distributed Computing Approach	*1158*
32.3.4	Implicit Solvent Models versus Explicit Water	*1160*
32.4	Conclusion	*1162*
	References	*1162*

33	**Molecular Dynamics Simulations of Proteins and Peptides: Problems, Achievements, and Perspectives** *1170*
	Paul Tavan, Heiko Carstens, and Gerald Mathias
33.1	Introduction *1170*
33.2	Basic Physics of Protein Structure and Dynamics *1171*
33.2.1	Protein Electrostatics *1172*
33.2.2	Relaxation Times and Spatial Scales *1172*
33.2.3	Solvent Environment *1173*
33.2.4	Water *1174*
33.2.5	Polarizability of the Peptide Groups and of Other Protein Components *1175*
33.3	State of the Art *1177*
33.3.1	Control of Thermodynamic Conditions *1177*
33.3.2	Long-range Electrostatics *1177*
33.3.3	Polarizability *1179*
33.3.4	Higher Multipole Moments of the Molecular Components *1180*
33.3.5	MM Models of Water *1181*
33.3.6	Complexity of Protein–Solvent Systems and Consequences for MM-MD *1182*
33.3.7	What about Successes of MD Methods? *1182*
33.3.8	Accessible Time Scales and Accuracy Issues *1184*
33.3.9	Continuum Solvent Models *1185*
33.3.10	Are there Further Problems beyond Electrostatics and Structure Prediction? *1187*
33.4	Conformational Dynamics of a Light-switchable Model Peptide *1187*
33.4.1	Computational Methods *1188*
33.4.2	Results and Discussion *1190*
	Summary *1194*
	Acknowledgments *1194*
	References *1194*

Part II, Volume 1

Contributors of Part II *LVIII*

1	**Paradigm Changes from "Unboiling an Egg" to "Synthesizing a Rabbit"** *3*
	Rainer Jaenicke
1.1	Protein Structure, Stability, and Self-organization *3*
1.2	Autonomous and Assisted Folding and Association *6*
1.3	Native, Intermediate, and Denatured States *11*
1.4	Folding and Merging of Domains – Association of Subunits *13*
1.5	Limits of Reconstitution *19*
1.6	In Vitro Denaturation-Renaturation vs. Folding in Vivo *21*

1.7	Perspectives	24
	Acknowledgements	26
	References	26

2	**Folding and Association of Multi-domain and Oligomeric Proteins**	**32**
	Hauke Lilie and Robert Seckler	
2.1	Introduction	32
2.2	Folding of Multi-domain Proteins	33
2.2.1	Domain Architecture	33
2.2.2	γ-Crystallin as a Model for a Two-domain Protein	35
2.2.3	The Giant Protein Titin	39
2.3	Folding and Association of Oligomeric Proteins	41
2.3.1	Why Oligomers?	41
2.3.2	Inter-subunit Interfaces	42
2.3.3	Domain Swapping	44
2.3.4	Stability of Oligomeric Proteins	45
2.3.5	Methods Probing Folding/Association	47
2.3.5.1	Chemical Cross-linking	47
2.3.5.2	Analytical Gel Filtration Chromatography	47
2.3.5.3	Scattering Methods	48
2.3.5.4	Fluorescence Resonance Energy Transfer	48
2.3.5.5	Hybrid Formation	48
2.3.6	Kinetics of Folding and Association	49
2.3.6.1	General Considerations	49
2.3.6.2	Reconstitution Intermediates	50
2.3.6.3	Rates of Association	52
2.3.6.4	Homo- Versus Heterodimerization	52
2.4	Renaturation versus Aggregation	54
2.5	Case Studies on Protein Folding and Association	54
2.5.1	Antibody Fragments	54
2.5.2	Trimeric Tail Spike Protein of Bacteriophage P22	59
2.6	Experimental Protocols	62
	References	65

3	**Studying Protein Folding in Vivo**	**73**
	I. Marije Liscaljet, Bertrand Kleizen, and Ineke Braakman	
3.1	Introduction	73
3.2	General Features in Folding Proteins Amenable to in Vivo Study	73
3.2.1	Increasing Compactness	76
3.2.2	Decreasing Accessibility to Different Reagents	76
3.2.3	Changes in Conformation	77
3.2.4	Assistance During Folding	78
3.3	Location-specific Features in Protein Folding	79
3.3.1	Translocation and Signal Peptide Cleavage	79
3.3.2	Glycosylation	80

3.3.3	Disulfide Bond Formation in the ER	81
3.3.4	Degradation	82
3.3.5	Transport from ER to Golgi and Plasma Membrane	83
3.4	How to Manipulate Protein Folding	84
3.4.1	Pharmacological Intervention (Low-molecular-weight Reagents)	84
3.4.1.1	Reducing and Oxidizing Agents	84
3.4.1.2	Calcium Depletion	84
3.4.1.3	ATP Depletion	85
3.4.1.4	Cross-linking	85
3.4.1.5	Glycosylation Inhibitors	85
3.4.2	Genetic Modifications (High-molecular-weight Manipulations)	86
3.4.2.1	Substrate Protein Mutants	86
3.4.2.2	Changing the Concentration or Activity of Folding Enzymes and Chaperones	87
3.5	Experimental Protocols	88
3.5.1	Protein-labeling Protocols	88
3.5.1.1	Basic Protocol Pulse Chase: Adherent Cells	88
3.5.1.2	Pulse Chase in Suspension Cells	91
3.5.2	(Co)-immunoprecipitation and Accessory Protocols	93
3.5.2.1	Immunoprecipitation	93
3.5.2.2	Co-precipitation with Calnexin ([84]; adapted from Ou et al. [85])	94
3.5.2.3	Co-immunoprecipitation with Other Chaperones	95
3.5.2.4	Protease Resistance	95
3.5.2.5	Endo H Resistance	96
3.5.2.6	Cell Surface Expression Tested by Protease	96
3.5.3	SDS-PAGE [13]	97
	Acknowledgements	98
	References	98
4	**Characterization of ATPase Cycles of Molecular Chaperones by Fluorescence and Transient Kinetic Methods**	**105**
	Sandra Schlee and Jochen Reinstein	
4.1	Introduction	105
4.1.1	Characterization of ATPase Cycles of Energy-transducing Systems	105
4.1.2	The Use of Fluorescent Nucleotide Analogues	106
4.1.2.1	Fluorescent Modifications of Nucleotides	106
4.1.2.2	How to Find a Suitable Analogue for a Specific Protein	108
4.2	Characterization of ATPase Cycles of Molecular Chaperones	109
4.2.1	Biased View	109
4.2.2	The ATPase Cycle of DnaK	109
4.2.3	The ATPase Cycle of the Chaperone Hsp90	109
4.2.4	The ATPase Cycle of the Chaperone ClpB	111
4.2.4.1	ClpB, an Oligomeric ATPase With Two AAA Modules Per Protomer	111

4.2.4.2	Nucleotide-binding Properties of NBD1 and NBD2	*111*
4.2.4.3	Cooperativity of ATP Hydrolysis and Interdomain Communication	*114*
4.3	Experimental Protocols	*116*
4.3.1	Synthesis of Fluorescent Nucleotide Analogues	*116*
4.3.1.1	Synthesis and Characterization of (P_β)MABA-ADP and (P_γ)MABA-ATP	*116*
4.3.1.2	Synthesis and Characterization of N8-MABA Nucleotides	*119*
4.3.1.3	Synthesis of MANT Nucleotides	*120*
4.3.2	Preparation of Nucleotides and Proteins	*121*
4.3.2.1	Assessment of Quality of Nucleotide Stock Solution	*121*
4.3.2.2	Determination of the Nucleotide Content of Proteins	*122*
4.3.2.3	Nucleotide Depletion Methods	*123*
4.3.3	Steady-state ATPase Assays	*124*
4.3.3.1	Coupled Enzymatic Assay	*124*
4.3.3.2	Assays Based on $[\alpha\text{-}^{32}P]$-ATP and TLC	*125*
4.3.3.3	Assays Based on Released P_i	*125*
4.3.4	Single-turnover ATPase Assays	*126*
4.3.4.1	Manual Mixing Procedures	*126*
4.3.4.2	Quenched Flow	*127*
4.3.5	Nucleotide-binding Measurements	*127*
4.3.5.1	Isothermal Titration Calorimetry	*127*
4.3.5.2	Equilibrium Dialysis	*129*
4.3.5.3	Filter Binding	*129*
4.3.5.4	Equilibrium Fluorescence Titration	*130*
4.3.5.5	Competition Experiments	*132*
4.3.6	Analytical Solutions of Equilibrium Systems	*133*
4.3.6.1	Quadratic Equation	*133*
4.3.6.2	Cubic Equation	*134*
4.3.6.3	Iterative Solutions	*138*
4.3.7	Time-resolved Binding Measurements	*141*
4.3.7.1	Introduction	*141*
4.3.7.2	One-step Irreversible Process	*142*
4.3.7.3	One-step Reversible Process	*143*
4.3.7.4	Reversible Second Order Reduced to Pseudo-first Order	*144*
4.3.7.5	Two Simultaneous Irreversible Pathways – Partitioning	*146*
4.3.7.6	Two-step Consecutive (Sequential) Reaction	*148*
4.3.7.7	Two-step Binding Reactions	*150*
	References	*152*
5	**Analysis of Chaperone Function in Vitro**	*162*
	Johannes Buchner and Stefan Walter	
5.1	Introduction	*162*
5.2	Basic Functional Principles of Molecular Chaperones	*164*
5.2.1	Recognition of Nonnative Proteins	*166*

5.2.2	Induction of Conformational Changes in the Substrate	167
5.2.3	Energy Consumption and Regulation of Chaperone Function	169
5.3	Limits and Extensions of the Chaperone Concept	170
5.3.1	Co-chaperones	171
5.3.2	Specific Chaperones	171
5.4	Working with Molecular Chaperones	172
5.4.1	Natural versus Artificial Substrate Proteins	172
5.4.2	Stability of Chaperones	172
5.5	Assays to Assess and Characterize Chaperone Function	174
5.5.1	Generating Nonnative Conformations of Proteins	174
5.5.2	Aggregation Assays	174
5.5.3	Detection of Complexes Between Chaperone and Substrate	175
5.5.4	Refolding of Denatured Substrates	175
5.5.5	ATPase Activity and Effect of Substrate and Cofactors	176
5.6	Experimental Protocols	176
5.6.1	General Considerations	176
5.6.1.1	Analysis of Chaperone Stability	176
5.6.1.2	Generation of Nonnative Proteins	177
5.6.1.3	Model Substrates for Chaperone Assays	177
5.6.2	Suppression of Aggregation	179
5.6.3	Complex Formation between Chaperones and Polypeptide Substrates	183
5.6.4	Identification of Chaperone-binding Sites	184
5.6.5	Chaperone-mediated Refolding of Test Proteins	186
5.6.6	ATPase Activity	188
	Acknowledgments	188
	References	189
6	**Physical Methods for Studies of Fiber Formation and Structure**	**197**
	Thomas Scheibel and Louise Serpell	
6.1	Introduction	197
6.2	Overview: Protein Fibers Formed in Vivo	198
6.2.1	Amyloid Fibers	198
6.2.2	Silks	199
6.2.3	Collagens	199
6.2.4	Actin, Myosin, and Tropomyosin Filaments	200
6.2.5	Intermediate Filaments/Nuclear Lamina	202
6.2.6	Fibrinogen/Fibrin	203
6.2.7	Microtubules	203
6.2.8	Elastic Fibers	204
6.2.9	Flagella and Pili	204
6.2.10	Filamentary Structures in Rod-like Viruses	205
6.2.11	Protein Fibers Used by Viruses and Bacteriophages to Bind to Their Hosts	206
6.3	Overview: Fiber Structures	206

6.3.1	Study of the Structure of β-sheet-containing Proteins	207
6.3.1.1	Amyloid	207
6.3.1.2	Paired Helical Filaments	207
6.3.1.3	β-Silks	207
6.3.1.4	β-Sheet-containing Viral Fibers	208
6.3.2	α-Helix-containing Protein Fibers	209
6.3.2.1	Collagen	209
6.3.2.2	Tropomyosin	210
6.3.2.3	Intermediate Filaments	210
6.3.3	Protein Polymers Consisting of a Mixture of Secondary Structure	211
6.3.3.1	Tubulin	211
6.3.3.2	Actin and Myosin Filaments	212
6.4	Methods to Study Fiber Assembly	213
6.4.1	Circular Dichroism Measurements for Monitoring Structural Changes Upon Fiber Assembly	213
6.4.1.1	Theory of CD	213
6.4.1.2	Experimental Guide to Measure CD Spectra and Structural Transition Kinetics	214
6.4.2	Intrinsic Fluorescence Measurements to Analyze Structural Changes	215
6.4.2.1	Theory of Protein Fluorescence	215
6.4.2.2	Experimental Guide to Measure Trp Fluorescence	216
6.4.3	Covalent Fluorescent Labeling to Determine Structural Changes of Proteins with Environmentally Sensitive Fluorophores	217
6.4.3.1	Theory on Environmental Sensitivity of Fluorophores	217
6.4.3.2	Experimental Guide to Labeling Proteins With Fluorophores	218
6.4.4	1-Anilino-8-Naphthalensulfonate (ANS) Binding to Investigate Fiber Assembly	219
6.4.4.1	Theory on Using ANS Fluorescence for Detecting Conformational Changes in Proteins	219
6.4.4.2	Experimental Guide to Using ANS for Monitoring Protein Fiber Assembly	220
6.4.5	Light Scattering to Monitor Particle Growth	220
6.4.5.1	Theory of Classical Light Scattering	221
6.4.5.2	Theory of Dynamic Light Scattering	221
6.4.5.3	Experimental Guide to Analyzing Fiber Assembly Using DLS	222
6.4.6	Field-flow Fractionation to Monitor Particle Growth	222
6.4.6.1	Theory of FFF	222
6.4.6.2	Experimental Guide to Using FFF for Monitoring Fiber Assembly	223
6.4.7	Fiber Growth-rate Analysis Using Surface Plasmon Resonance	223
6.4.7.1	Theory of SPR	223
6.4.7.2	Experimental Guide to Using SPR for Fiber-growth Analysis	224
6.4.8	Single-fiber Growth Imaging Using Atomic Force Microscopy	225

6.4.8.1	Theory of Atomic Force Microscopy	225
6.4.8.2	Experimental Guide for Using AFM to Investigate Fiber Growth	225
6.4.9	Dyes Specific for Detecting Amyloid Fibers	226
6.4.9.1	Theory on Congo Red and Thioflavin T Binding to Amyloid	226
6.4.9.2	Experimental Guide to Detecting Amyloid Fibers with CR and Thioflavin Binding	227
6.5	Methods to Study Fiber Morphology and Structure	228
6.5.1	Scanning Electron Microscopy for Examining the Low-resolution Morphology of a Fiber Specimen	228
6.5.1.1	Theory of SEM	228
6.5.1.2	Experimental Guide to Examining Fibers by SEM	229
6.5.2	Transmission Electron Microscopy for Examining Fiber Morphology and Structure	230
6.5.2.1	Theory of TEM	230
6.5.2.2	Experimental Guide to Examining Fiber Samples by TEM	231
6.5.3	Cryo-electron Microscopy for Examination of the Structure of Fibrous Proteins	232
6.5.3.1	Theory of Cryo-electron Microscopy	232
6.5.3.2	Experimental Guide to Preparing Proteins for Cryo-electron Microscopy	233
6.5.3.3	Structural Analysis from Electron Micrographs	233
6.5.4	Atomic Force Microscopy for Examining the Structure and Morphology of Fibrous Proteins	234
6.5.4.1	Experimental Guide for Using AFM to Monitor Fiber Morphology	234
6.5.5	Use of X-ray Diffraction for Examining the Structure of Fibrous Proteins	236
6.5.5.1	Theory of X-Ray Fiber Diffraction	236
6.5.5.2	Experimental Guide to X-Ray Fiber Diffraction	237
6.5.6	Fourier Transformed Infrared Spectroscopy	239
6.5.6.1	Theory of FTIR	239
6.5.6.2	Experimental Guide to Determining Protein Conformation by FTIR	240
6.6	Concluding Remarks	241
	Acknowledgements	242
	References	242
7	**Protein Unfolding in the Cell**	254
	Prakash Koodathingal, Neil E. Jaffe, and Andreas Matouschek	
7.1	Introduction	254
7.2	Protein Translocation Across Membranes	254
7.2.1	Compartmentalization and Unfolding	254
7.2.2	Mitochondria Actively Unfold Precursor Proteins	256
7.2.3	The Protein Import Machinery of Mitochondria	257
7.2.4	Specificity of Unfolding	259

7.2.5	Protein Import into Other Cellular Compartments 259	
7.3	Protein Unfolding and Degradation by ATP-dependent Proteases 260	
7.3.1	Structural Considerations of Unfoldases Associated With Degradation 260	
7.3.2	Unfolding Is Required for Degradation by ATP-dependent Proteases 261	
7.3.3	The Role of ATP and Models of Protein Unfolding 262	
7.3.4	Proteins Are Unfolded Sequentially and Processively 263	
7.3.5	The Influence of Substrate Structure on the Degradation Process 264	
7.3.6	Unfolding by Pulling 264	
7.3.7	Specificity of Degradation 265	
7.4	Conclusions 266	
7.5	Experimental Protocols 266	
7.5.1	Size of Import Channels in the Outer and Inner Membranes of Mitochondria 266	
7.5.2	Structure of Precursor Proteins During Import into Mitochondria 266	
7.5.3	Import of Barnase Mutants 267	
7.5.4	Protein Degradation by ATP-dependent Proteases 267	
7.5.5	Use of Multi-domain Substrates 268	
7.5.6	Studies Using Circular Permutants 268	
	References 269	
8	**Natively Disordered Proteins** 275	
	Gary W. Daughdrill, Gary J. Pielak, Vladimir N. Uversky, Marc S. Cortese, and A. Keith Dunker	
8.1	Introduction 275	
8.1.1	The Protein Structure-Function Paradigm 275	
8.1.2	Natively Disordered Proteins 277	
8.1.3	A New Protein Structure-Function Paradigm 280	
8.2	Methods Used to Characterize Natively Disordered Proteins 281	
8.2.1	NMR Spectroscopy 281	
8.2.1.1	Chemical Shifts Measure the Presence of Transient Secondary Structure 282	
8.2.1.2	Pulsed Field Gradient Methods to Measure Translational Diffusion 284	
8.2.1.3	NMR Relaxation and Protein Flexibility 284	
8.2.1.4	Using the Model-free Analysis of Relaxation Data to Estimate Internal Mobility and Rotational Correlation Time 285	
8.2.1.5	Using Reduced Spectral Density Mapping to Assess the Amplitude and Frequencies of Intramolecular Motion 286	
8.2.1.6	Characterization of the Dynamic Structures of Natively Disordered Proteins Using NMR 287	
8.2.2	X-ray Crystallography 288	
8.2.3	Small Angle X-ray Diffraction and Hydrodynamic Measurements 293	

8.2.4	Circular Dichroism Spectropolarimetry	297
8.2.5	Infrared and Raman Spectroscopy	299
8.2.6	Fluorescence Methods	301
8.2.6.1	Intrinsic Fluorescence of Proteins	301
8.2.6.2	Dynamic Quenching of Fluorescence	302
8.2.6.3	Fluorescence Polarization and Anisotropy	303
8.2.6.4	Fluorescence Resonance Energy Transfer	303
8.2.6.5	ANS Fluorescence	305
8.2.7	Conformational Stability	308
8.2.7.1	Effect of Temperature on Proteins with Extended Disorder	309
8.2.7.2	Effect of pH on Proteins with Extended Disorder	309
8.2.8	Mass Spectrometry-based High-resolution Hydrogen-Deuterium Exchange	309
8.2.9	Protease Sensitivity	311
8.2.10	Prediction from Sequence	313
8.2.11	Advantage of Multiple Methods	314
8.3	Do Natively Disordered Proteins Exist Inside Cells?	315
8.3.1	Evolution of Ordered and Disordered Proteins Is Fundamentally Different	315
8.3.1.1	The Evolution of Natively Disordered Proteins	315
8.3.1.2	Adaptive Evolution and Protein Flexibility	317
8.3.1.3	Phylogeny Reconstruction and Protein Structure	318
8.3.2	Direct Measurement by NMR	320
8.4	Functional Repertoire	322
8.4.1	Molecular Recognition	322
8.4.1.1	The Coupling of Folding and Binding	322
8.4.1.2	Structural Plasticity for the Purpose of Functional Plasticity	323
8.4.1.3	Systems Where Disorder Increases Upon Binding	323
8.4.2	Assembly/Disassembly	325
8.4.3	Highly Entropic Chains	325
8.4.4	Protein Modification	327
8.5	Importance of Disorder for Protein Folding	328
8.6	Experimental Protocols	331
8.6.1	NMR Spectroscopy	331
8.6.1.1	General Requirements	331
8.6.1.2	Measuring Transient Secondary Structure in Secondary Chemical Shifts	332
8.6.1.3	Measuring the Translational Diffusion Coefficient Using Pulsed Field Gradient Diffusion Experiments	332
8.6.1.4	Relaxation Experiments	332
8.6.1.5	Relaxation Data Analysis Using Reduced Spectral Density Mapping	333
8.6.1.6	In-cell NMR	334
8.6.2	X-ray Crystallography	334
8.6.3	Circular Dichroism Spectropolarimetry	336

Acknowledgements 337
References 337

9 The Catalysis of Disulfide Bond Formation in Prokaryotes 358
Jean-Francois Collet and James C. Bardwell

9.1 Introduction 358
9.2 Disulfide Bond Formation in the E. coli Periplasm 358
9.2.1 A Small Bond, a Big Effect 358
9.2.2 Disulfide Bond Formation Is a Catalyzed Process 359
9.2.3 DsbA, a Protein-folding Catalyst 359
9.2.4 How is DsbA Re-oxidized? 361
9.2.5 From Where Does the Oxidative Power of DsbB Originate? 361
9.2.6 How Are Disulfide Bonds Transferred From DsbB to DsbA? 362
9.2.7 How Can DsbB Generate Disulfide by Quinone Reduction? 364
9.3 Disulfide Bond Isomerization 365
9.3.1 The Protein Disulfide Isomerases DsbC and DsbG 365
9.3.2 Dimerization of DsbC and DsbG Is Important for Isomerase and Chaperone Activity 366
9.3.3 Dimerization Protects from DsbB Oxidation 367
9.3.4 Import of Electrons from the Cytoplasm: DsbD 367
9.3.5 Conclusions 369
9.4 Experimental Protocols 369
9.4.1 Oxidation-reduction of a Protein Sample 369
9.4.2 Determination of the Free Thiol Content of a Protein 370
9.4.3 Separation by HPLC 371
9.4.4 Tryptophan Fluorescence 372
9.4.5 Assay of Disulfide Oxidase Activity 372
References 373

10 Catalysis of Peptidyl-prolyl cis/trans Isomerization by Enzymes 377
Gunter Fischer

10.1 Introduction 377
10.2 Peptidyl-prolyl cis/trans Isomerization 379
10.3 Monitoring Peptidyl-prolyl cis/trans Isomerase Activity 383
10.4 Prototypical Peptidyl-prolyl cis/trans Isomerases 388
10.4.1 General Considerations 388
10.4.2 Prototypic Cyclophilins 390
10.4.3 Prototypic FK506-binding Proteins 394
10.4.4 Prototypic Parvulins 397
10.5 Concluding Remarks 399
10.6 Experimental Protocols 399
10.6.1 PPIase Assays: Materials 399
10.6.2 PPIase Assays: Equipment 400
10.6.3 Assaying Procedure: Protease-coupled Spectrophotometric Assay 400

10.6.4	Assaying Procedure: Protease-free Spectrophotometric Assay 401
	References 401

11	**Secondary Amide Peptide Bond *cis/trans* Isomerization in Polypeptide Backbone Restructuring: Implications for Catalysis** 415
	Cordelia Schiene-Fischer and Christian Lücke
11.1	Introduction 415
11.2	Monitoring Secondary Amide Peptide Bond *cis/trans* Isomerization 416
11.3	Kinetics and Thermodynamics of Secondary Amide Peptide Bond *cis/trans* Isomerization 418
11.4	Principles of DnaK Catalysis 420
11.5	Concluding Remarks 423
11.6	Experimental Protocols 424
11.6.1	Stopped-flow Measurements of Peptide Bond *cis/trans* Isomerization 424
11.6.2	Two-dimensional ^{1}H-NMR Exchange Experiments 425
	References 426

12	**Ribosome-associated Proteins Acting on Newly Synthesized Polypeptide Chains** 429
	Sabine Rospert, Matthias Gautschi, Magdalena Rakwalska, and Uta Raue
12.1	Introduction 429
12.2	Signal Recognition Particle, Nascent Polypeptide–associated Complex, and Trigger Factor 432
12.2.1	Signal Recognition Particle 432
12.2.2	An Interplay between Eukaryotic SRP and Nascent Polypeptide–associated Complex? 435
12.2.3	Interplay between Bacterial SRP and Trigger Factor? 435
12.2.4	Functional Redundancy: TF and the Bacterial Hsp70 Homologue DnaK 436
12.3	Chaperones Bound to the Eukaryotic Ribosome: Hsp70 and Hsp40 Systems 436
12.3.1	Sis1p and Ssa1p: an Hsp70/Hsp40 System Involved in Translation Initiation? 437
12.3.2	Ssb1/2p, an Hsp70 Homologue Distributed Between Ribosomes and Cytosol 438
12.3.3	Function of Ssb1/2p in Degradation and Protein Folding 439
12.3.4	Zuotin and Ssz1p: a Stable Chaperone Complex Bound to the Yeast Ribosome 440
12.3.5	A Functional Chaperone Triad Consisting of Ssb1/2p, Ssz1p, and Zuotin 440
12.3.6	Effects of Ribosome-bound Chaperones on the Yeast Prion [PSI^+] 442
12.4	Enzymes Acting on Nascent Polypeptide Chains 443

12.4.1	Methionine Aminopeptidases *443*	
12.4.2	N^α-acetyltransferases *444*	
12.5	A Complex Arrangement at the Yeast Ribosomal Tunnel Exit *445*	
12.6	Experimental Protocols *446*	
12.6.1	Purification of Ribosome-associated Protein Complexes from Yeast *446*	
12.6.2	Growth of Yeast and Preparation of Ribosome-associated Proteins by High-salt Treatment of Ribosomes *447*	
12.6.3	Purification of NAC and RAC *448*	
	References *449*	

Part II, Volume 2

13	**The Role of Trigger Factor in Folding of Newly Synthesized Proteins** *459*	
	Elke Deuerling, Thomas Rauch, Holger Patzelt, and Bernd Bukau	
13.1	Introduction *459*	
13.2	In Vivo Function of Trigger Factor *459*	
13.2.1	Discovery *459*	
13.2.2	Trigger Factor Cooperates With the DnaK Chaperone in the Folding of Newly Synthesized Cytosolic Proteins *460*	
13.2.3	In Vivo Substrates of Trigger Factor and DnaK *461*	
13.2.4	Substrate Specificity of Trigger Factor *463*	
13.3	Structure–Function Analysis of Trigger Factor *465*	
13.3.1	Domain Structure and Conservation *465*	
13.3.2	Quaternary Structure *468*	
13.3.3	PPIase and Chaperone Activity of Trigger Factor *469*	
13.3.4	Importance of Ribosome Association *470*	
13.4	Models of the Trigger Factor Mechanism *471*	
13.5	Experimental Protocols *473*	
13.5.1	Trigger Factor Purification *473*	
13.5.2	GAPDH Trigger Factor Activity Assay *475*	
13.5.3	Modular Cell-free *E. coli* Transcription/Translation System *475*	
13.5.4	Isolation of Ribosomes and Add-back Experiments *483*	
13.5.5	Cross-linking Techniques *485*	
	References *485*	

14	**Cellular Functions of Hsp70 Chaperones** *490*	
	Elizabeth A. Craig and Peggy Huang	
14.1	Introduction *490*	
14.2	"Soluble" Hsp70s/J-proteins Function in General Protein Folding *492*	
14.2.1	The Soluble Hsp70 of *E. coli*, DnaK *492*	
14.2.2	Soluble Hsp70s of Major Eukaryotic Cellular Compartments *493*	
14.2.2.1	Eukaryotic Cytosol *493*	
14.2.2.2	Matrix of Mitochondria *494*	
14.2.2.3	Lumen of the Endoplasmic Reticulum *494*	

14.3	"Tethered" Hsp70s/J-proteins: Roles in Protein Folding on the Ribosome and in Protein Translocation	495
14.3.1	Membrane-tethered Hsp70/J-protein	495
14.3.2	Ribosome-associated Hsp70/J-proteins	496
14.4	Modulating of Protein Conformation by Hsp70s/J-proteins	498
14.4.1	Assembly of Fe/S Centers	499
14.4.2	Uncoating of Clathrin-coated Vesicles	500
14.4.3	Regulation of the Heat Shock Response	501
14.4.4	Regulation of Activity of DNA Replication-initiator Proteins	502
14.5	Cases of a Single Hsp70 Functioning With Multiple J-Proteins	504
14.6	Hsp70s/J-proteins – When an Hsp70 Maybe Isn't Really a Chaperone	504
14.6.1	The Ribosome-associated "Hsp70" Ssz1	505
14.6.2	Mitochondrial Hsp70 as the Regulatory Subunit of an Endonuclease	506
14.7	Emerging Concepts and Unanswered Questions	507
	References	507
15	**Regulation of Hsp70 Chaperones by Co-chaperones**	**516**
	Matthias P. Mayer and Bernd Bukau	
15.1	Introduction	516
15.2	Hsp70 Proteins	517
15.2.1	Structure and Conservation	517
15.2.2	ATPase Cycle	519
15.2.3	Structural Investigations	521
15.2.4	Interactions With Substrates	522
15.3	J-domain Protein Family	526
15.3.1	Structure and Conservation	526
15.3.2	Interaction With Hsp70s	530
15.3.3	Interactions with Substrates	532
15.4	Nucleotide Exchange Factors	534
15.4.1	GrpE: Structure and Interaction with DnaK	534
15.4.2	Nucleotide Exchange Reaction	535
15.4.3	Bag Family: Structure and Interaction With Hsp70	536
15.4.4	Relevance of Regulated Nucleotide Exchange for Hsp70s	538
15.5	TPR Motifs Containing Co-chaperones of Hsp70	540
15.5.1	Hip	541
15.5.2	Hop	542
15.5.3	Chip	543
15.6	Concluding Remarks	544
15.7	Experimental Protocols	544
15.7.1	Hsp70s	544
15.7.2	J-Domain Proteins	545
15.7.3	GrpE	546
15.7.4	Bag-1	547

15.7.5	Hip 548
15.7.6	Hop 549
15.7.7	Chip 549
	References 550

16   **Protein Folding in the Endoplasmic Reticulum Via the Hsp70 Family** 563
Ying Shen, Kyung Tae Chung, and Linda M. Hendershot

16.1	Introduction 563
16.2	BiP Interactions with Unfolded Proteins 564
16.3	ER-localized DnaJ Homologues 567
16.4	ER-localized Nucleotide-exchange/releasing Factors 571
16.5	Organization and Relative Levels of Chaperones in the ER 572
16.6	Regulation of ER Chaperone Levels 573
16.7	Disposal of BiP-associated Proteins That Fail to Fold or Assemble 575
16.8	Other Roles of BiP in the ER 576
16.9	Concluding Comments 576
16.10	Experimental Protocols 577
16.10.1	Production of Recombinant ER Proteins 577
16.10.1.1	General Concerns 577
16.10.1.2	Bacterial Expression 578
16.10.1.3	Yeast Expression 580
16.10.1.4	Baculovirus 581
16.10.1.5	Mammalian Cells 583
16.10.2	Yeast Two-hybrid Screen for Identifying Interacting Partners of ER Proteins 586
16.10.3	Methods for Determining Subcellular Localization, Topology, and Orientation of Proteins 588
16.10.3.1	Sequence Predictions 588
16.10.3.2	Immunofluorescence Staining 589
16.10.3.3	Subcellular Fractionation 589
16.10.3.4	Determination of Topology 590
16.10.3.5	*N*-linked Glycosylation 592
16.10.4	Nucleotide Binding, Hydrolysis, and Exchange Assays 594
16.10.4.1	Nucleotide-binding Assays 594
16.10.4.2	ATP Hydrolysis Assays 596
16.10.4.3	Nucleotide Exchange Assays 597
16.10.5	Assays for Protein–Protein Interactions in Vitro/in Vivo 599
16.10.5.1	In Vitro GST Pull-down Assay 599
16.10.5.2	Co-immunoprecipitation 600
16.10.5.3	Chemical Cross-linking 600
16.10.5.4	Yeast Two-hybrid System 601
16.10.6	In Vivo Folding, Assembly, and Chaperone-binding Assays 601
16.10.6.1	Monitoring Oxidation of Intrachain Disulfide Bonds 601
16.10.6.2	Detection of Chaperone Binding 602

	Acknowledgements 603
	References 603
17	**Quality Control In Glycoprotein Folding** 617
	E. Sergio Trombetta and Armando J. Parodi
17.1	Introduction 617
17.2	ER N-glycan Processing Reactions 617
17.3	The UDP-Glc:Glycoprotein Glucosyltransferase 619
17.4	Protein Folding in the ER 621
17.5	Unconventional Chaperones (Lectins) Are Present in the ER Lumen 621
17.6	In Vivo Glycoprotein-CNX/CRT Interaction 623
17.7	Effect of CNX/CRT Binding on Glycoprotein Folding and ER Retention 624
17.8	Glycoprotein-CNX/CRT Interaction Is Not Essential for Unicellular Organisms and Cells in Culture 627
17.9	Diversion of Misfolded Glycoproteins to Proteasomal Degradation 629
17.10	Unfolding Irreparably Misfolded Glycoproteins to Facilitate Proteasomal Degradation 632
17.11	Summary and Future Directions 633
17.12	Characterization of N-glycans from Glycoproteins 634
17.12.1	Characterization of N-glycans Present in Immunoprecipitated Samples 634
17.12.2	Analysis of Radio-labeled N-glycans 636
17.12.3	Extraction and Analysis of Protein-bound N-glycans 636
17.12.4	GII and GT Assays 637
17.12.4.1	Assay for GII 637
17.12.4.2	Assay for GT 638
17.12.5	Purification of GII and GT from Rat Liver 639
	References 641
18	**Procollagen Biosynthesis in Mammalian Cells** 649
	Mohammed Tasab and Neil J. Bulleid
18.1	Introduction 649
18.1.1	Variety and Complexity of Collagen Proteins 649
18.1.2	Fibrillar Procollagen 650
18.1.3	Expression of Fibrillar Collagens 650
18.2	The Procollagen Biosynthetic Process: An Overview 651
18.3	Disulfide Bonding in Procollagen Assembly 653
18.4	The Influence of Primary Amino Acid Sequence on Intracellular Procollagen Folding 654
18.4.1	Chain Recognition and Type-specific Assembly 654
18.4.2	Assembly of Multi-subunit Proteins 654
18.4.3	Coordination of Type-specific Procollagen Assembly and Chain Selection 655

18.4.4	Hypervariable Motifs: Components of a Recognition Mechanism That Distinguishes Between Procollagen Chains?	*656*
18.4.5	Modeling the C-propeptide	*657*
18.4.6	Chain Association	*657*
18.5	Posttranslational Modifications That Affect Procollagen Folding	*658*
18.5.1	Hydroxylation and Triple-helix Stability	*658*
18.6	Procollagen Chaperones	*658*
18.6.1	Prolyl 4-Hydroxylase	*658*
18.6.2	Protein Disulfide Isomerase	*659*
18.6.3	Hsp47	*660*
18.6.4	PPI and BiP	*661*
18.7	Analysis of Procollagen Folding	*662*
18.8	Experimental Part	*663*
18.8.1	Materials Required	*663*
18.8.2	Experimental Protocols	*664*
	References	*668*

19	**Redox Regulation of Chaperones**	***677***
	Jörg H. Hoffmann and Ursula Jakob	
19.1	Introduction	*677*
19.2	Disulfide Bonds as Redox-Switches	*677*
19.2.1	Functionality of Disulfide Bonds	*677*
19.2.2	Regulatory Disulfide Bonds as Functional Switches	*679*
19.2.3	Redox Regulation of Chaperone Activity	*680*
19.3	Prokaryotic Hsp33: A Chaperone Activated by Oxidation	*680*
19.3.1	Identification of a Redox-regulated Chaperone	*680*
19.3.2	Activation Mechanism of Hsp33	*681*
19.3.3	The Crystal Structure of Active Hsp33	*682*
19.3.4	The Active Hsp33-Dimer: An Efficient Chaperone Holdase	*683*
19.3.5	Hsp33 is Part of a Sophisticated Multi-chaperone Network	*684*
19.4	Eukaryotic Protein Disulfide Isomerase (PDI): Redox Shuffling in the ER	*685*
19.4.1	PDI, A Multifunctional Enzyme in Eukaryotes	*685*
19.4.2	PDI and Redox Regulation	*687*
19.5	Concluding Remarks and Outlook	*688*
19.6	Appendix – Experimental Protocols	*688*
19.6.1	How to Work With Redox-regulated Chaperones in Vitro	*689*
19.6.1.1	Preparation of the Reduced Protein Species	*689*
19.6.1.2	Preparation of the Oxidized Protein Species	*690*
19.6.1.3	In Vitro Thiol Trapping to Monitor the Redox State of Proteins	*691*
19.6.2	Thiol Coordinating Zinc Centers as Redox Switches	*691*
19.6.2.1	PAR-PMPS Assay to Quantify Zinc	*691*
19.6.2.2	Determination of Zinc-binding Constants	*692*
19.6.3	Functional Analysis of Redox-regulated Chaperones in Vitro/in Vivo	*693*
19.6.3.1	Chaperone Activity Assays	*693*

19.6.3.2	Manipulating and Analyzing Redox Conditions in Vivo	694
	Acknowledgements 694	
	References 694	

20	**The *E. coli* GroE Chaperone**	**699**
	Steven G. Burston and Stefan Walter	
20.1	Introduction 699	
20.2	The Structure of GroEL 699	
20.3	The Structure of GroEL-ATP 700	
20.4	The Structure of GroES and its Interaction with GroEL	701
20.5	The Interaction Between GroEL and Substrate Polypeptides	702
20.6	GroEL is a Complex Allosteric Macromolecule	703
20.7	The Reaction Cycle of the GroE Chaperone	705
20.8	The Effect of GroE on Protein-folding Pathways	708
20.9	Future Perspectives 710	
20.10	Experimental Protocols 710	
	Acknowledgments 719	
	References 719	

21	**Structure and Function of the Cytosolic Chaperonin CCT**	**725**
	José M. Valpuesta, José L. Carrascosa, and Keith R. Willison	
21.1	Introduction 725	
21.2	Structure and Composition of CCT 726	
21.3	Regulation of CCT Expression 729	
21.4	Functional Cycle of CCT 730	
21.5	Folding Mechanism of CCT 731	
21.6	Substrates of CCT 735	
21.7	Co-chaperones of CCT 739	
21.8	Evolution of CCT 741	
21.9	Concluding Remarks 743	
21.10	Experimental Protocols 743	
21.10.1	Purification 743	
21.10.2	ATP Hydrolysis Measurements 744	
21.10.3	CCT Substrate-binding and Folding Assays 744	
21.10.4	Electron Microscopy and Image Processing 744	
	References 747	

22	**Structure and Function of GimC/Prefoldin**	**756**
	Katja Siegers, Andreas Bracher, and Ulrich Hartl	
22.1	Introduction 756	
22.2	Evolutionary Distribution of GimC/Prefoldin	757
22.3	Structure of the Archaeal GimC/Prefoldin	757
22.4	Complexity of the Eukaryotic/Archaeal GimC/Prefoldin	759
22.5	Functional Cooperation of GimC/Prefoldin With the Eukaryotic Chaperonin TRiC/CCT 761	

22.6	Experimental Protocols 764	
22.6.1	Actin-folding Kinetics 764	
22.6.2	Prevention of Aggregation (Light-scattering) Assay 765	
22.6.3	Actin-binding Assay 765	
	Acknowledgements 766	
	References 766	
23	**Hsp90: From Dispensable Heat Shock Protein to Global Player** 768	
	Klaus Richter, Birgit Meinlschmidt, and Johannes Buchner	
23.1	Introduction 768	
23.2	The Hsp90 Family in Vivo 768	
23.2.1	Evolutionary Relationships within the Hsp90 Gene Family 768	
23.2.2	In Vivo Functions of Hsp90 769	
23.2.3	Regulation of Hsp90 Expression and Posttranscriptional Activation 772	
23.2.4	Chemical Inhibition of Hsp90 773	
23.2.5	Identification of Natural Hsp90 Substrates 774	
23.3	In Vitro Investigation of the Chaperone Hsp90 775	
23.3.1	Hsp90: A Special Kind of ATPase 775	
23.3.2	The ATPase Cycle of Hsp90 780	
23.3.3	Interaction of Hsp90 with Model Substrate Proteins 781	
23.3.4	Investigating Hsp90 Substrate Interactions Using Native Substrates 783	
23.4	Partner Proteins: Does Complexity Lead to Specificity? 784	
23.4.1	Hop, p23, and PPIases: The Chaperone Cycle of Hsp90 784	
23.4.2	Hop/Sti1: Interactions Mediated by TPR Domains 787	
23.4.3	p23/Sba1: Nucleotide-specific Interaction with Hsp90 789	
23.4.4	Large PPIases: Conferring Specificity to Substrate Localization? 790	
23.4.5	Pp5: Facilitating Dephosphorylation 791	
23.4.6	Cdc37: Building Complexes with Kinases 792	
23.4.7	Tom70: Chaperoning Mitochondrial Import 793	
23.4.8	CHIP and Sgt1: Multiple Connections to Protein Degradation 793	
23.4.9	Aha1 and Hch1: Just Stimulating the ATPase? 794	
23.4.10	Cns1, Sgt2, and Xap2: Is a TPR Enough to Become an Hsp90 Partner? 796	
23.5	Outlook 796	
23.6	Appendix – Experimental Protocols 797	
23.6.1	Calculation of Phylogenetic Trees Based on Protein Sequences 797	
23.6.2	Investigating the in Vivo Effect of Hsp90 Mutations in *S. cerevisiae* 797	
23.6.3	Well-characterized Hsp90 Mutants 798	
23.6.4	Investigating Activation of Heterologously Expressed Src Kinase in *S. cerevisiae* 800	
23.6.5	Investigation of Heterologously Expressed Glucocorticoid Receptor in *S. cerevisiae* 800	

23.6.6	Investigation of Chaperone Activity	801
23.6.7	Analysis of the ATPase Activity of Hsp90	802
23.6.8	Detecting Specific Influences on Hsp90 ATPase Activity	803
23.6.9	Investigation of the Quaternary Structure by SEC-HPLC	804
23.6.10	Investigation of Binding Events Using Changes of the Intrinsic Fluorescence	806
23.6.11	Investigation of Binding Events Using Isothermal Titration Calorimetry	807
23.6.12	Investigation of Protein-Protein Interactions Using Cross-linking	807
23.6.13	Investigation of Protein-Protein Interactions Using Surface Plasmon Resonance Spectroscopy	808
	Acknowledgements	810
	References	810

24 Small Heat Shock Proteins: Dynamic Players in the Folding Game 830
Franz Narberhaus and Martin Haslbeck

24.1	Introduction	830
24.2	α-Crystallins and the Small Heat Shock Protein Family: Diverse Yet Similar	830
24.3	Cellular Functions of α-Hsps	831
24.3.1	Chaperone Activity in Vitro	831
24.3.2	Chaperone Function in Vivo	835
24.3.3	Other Functions	836
24.4	The Oligomeric Structure of α-Hsps	837
24.5	Dynamic Structures as Key to Chaperone Activity	839
24.6	Experimental Protocols	840
24.6.1	Purification of sHsps	840
24.6.2	Chaperone Assays	843
24.6.3	Monitoring Dynamics of sHsps	846
	Acknowledgements	847
	References	848

25 Alpha-crystallin: Its Involvement in Suppression of Protein Aggregation and Protein Folding 858
Joseph Horwitz

25.1	Introduction	858
25.2	Distribution of Alpha-crystallin in the Various Tissues	858
25.3	Structure	859
25.4	Phosphorylation and Other Posttranslation Modification	860
25.5	Binding of Target Proteins to Alpha-crystallin	861
25.6	The Function of Alpha-crystallin	863
25.7	Experimental Protocols	863
25.7.1	Preparation of Alpha-crystallin	863
	Acknowledgements	870
	References	870

26	**Transmembrane Domains in Membrane Protein Folding, Oligomerization, and Function** *876*	
	Anja Ridder and Dieter Langosch	
26.1	Introduction *876*	
26.1.1	Structure of Transmembrane Domains *876*	
26.1.2	The Biosynthetic Route towards Folded and Oligomeric Integral Membrane Proteins *877*	
26.1.3	Structure and Stability of TMSs *878*	
26.1.3.1	Amino Acid Composition of TMSs and Flanking Regions *878*	
26.1.3.2	Stability of Transmembrane Helices *879*	
26.2	The Nature of Transmembrane Helix-Helix Interactions *880*	
26.2.1	General Considerations *880*	
26.2.1.1	Attractive Forces within Lipid Bilayers *880*	
26.2.1.2	Forces between Transmembrane Helices *881*	
26.2.1.3	Entropic Factors Influencing Transmembrane Helix–Helix Interactions *882*	
26.2.2	Lessons from Sequence Analyses and High-resolution Structures *883*	
26.2.3	Lessons from Bitopic Membrane Proteins *886*	
26.2.3.1	Transmembrane Segments Forming Right-handed Pairs *886*	
26.2.3.2	Transmembrane Segments Forming Left-handed Assemblies *889*	
26.2.4	Selection of Self-interacting TMSs from Combinatorial Libraries *892*	
26.2.5	Role of Lipids in Packing/Assembly of Membrane Proteins *893*	
26.3	Conformational Flexibility of Transmembrane Segments *895*	
26.4	Experimental Protocols *897*	
26.4.1	Biochemical and Biophysical Techniques *897*	
26.4.1.1	Visualization of Oligomeric States by Electrophoretic Techniques *898*	
26.4.1.2	Hydrodynamic Methods *899*	
26.4.1.3	Fluorescence Resonance Transfer *900*	
26.4.2	Genetic Assays *901*	
26.4.2.1	The ToxR System *901*	
26.4.2.2	Other Genetic Assays *902*	
26.4.3	Identification of TMS-TMS Interfaces by Mutational Analysis *903*	
	References *904*	

Part II, Volume 3

27	**SecB** *919*	
	Arnold J. M. Driessen, Janny de Wit, and Nico Nouwen	
27.1	Introduction *919*	
27.2	Selective Binding of Preproteins by SecB *920*	
27.3	SecA-SecB Interaction *925*	
27.4	Preprotein Transfer from SecB to SecA *928*	
27.5	Concluding Remarks *929*	
27.6	Experimental Protocols *930*	
27.6.1	How to Analyze SecB-Preprotein Interactions *930*	

27.6.2	How to Analyze SecB-SecA Interaction *931*
	Acknowledgements *932*
	References *933*

28 **Protein Folding in the Periplasm and Outer Membrane of *E. coli* *938***
Michael Ehrmann
28.1 Introduction *938*
28.2 Individual Cellular Factors *940*
28.2.1 The Proline Isomerases FkpA, PpiA, SurA, and PpiD *941*
28.2.1.1 FkpA *942*
28.2.1.2 PpiA *942*
28.2.1.3 SurA *943*
28.2.1.4 PpiD *943*
28.2.2 Skp *944*
28.2.3 Proteases and Protease/Chaperone Machines *945*
28.2.3.1 The HtrA Family of Serine Proteases *946*
28.2.3.2 *E. coli* HtrAs *946*
28.2.3.3 DegP and DegQ *946*
28.2.3.4 DegS *947*
28.2.3.5 The Structure of HtrA *947*
28.2.3.6 Other Proteases *948*
28.3 Organization of Folding Factors into Pathways and Networks *950*
28.3.1 Synthetic Lethality and Extragenic High-copy Suppressors *950*
28.3.2 Reconstituted in Vitro Systems *951*
28.4 Regulation *951*
28.4.1 The Sigma E Pathway *951*
28.4.2 The Cpx Pathway *952*
28.4.3 The Bae Pathway *953*
28.5 Future Perspectives *953*
28.6 Experimental Protocols *954*
28.6.1 Pulse Chase Immunoprecipitation *954*
Acknowledgements *957*
References *957*

29 **Formation of Adhesive Pili by the Chaperone-Usher Pathway *965***
Michael Vetsch and Rudi Glockshuber
29.1 Basic Properties of Bacterial, Adhesive Surface Organelles *965*
29.2 Structure and Function of Pilus Chaperones *970*
29.3 Structure and Folding of Pilus Subunits *971*
29.4 Structure and Function of Pilus Ushers *973*
29.5 Conclusions and Outlook *976*
29.6 Experimental Protocols *977*
29.6.1 Test for the Presence of Type 1 Piliated *E. coli* Cells *977*
29.6.2 Functional Expression of Pilus Subunits in the *E. coli* Periplasm *977*
29.6.3 Purification of Pilus Subunits from the *E. coli* Periplasm *978*

29.6.4	Preparation of Ushers 979	
	Acknowledgements 979	
	References 980	

30 **Unfolding of Proteins During Import into Mitochondria** 987
Walter Neupert, Michael Brunner, and Kai Hell
30.1 Introduction 987
30.2 Translocation Machineries and Pathways of the Mitochondrial Protein Import System 988
30.2.1 Import of Proteins Destined for the Mitochondrial Matrix 990
30.3 Import into Mitochondria Requires Protein Unfolding 993
30.4 Mechanisms of Unfolding by the Mitochondrial Import Motor 995
30.4.1 Targeted Brownian Ratchet 995
30.4.2 Power-stroke Model 995
30.5 Studies to Discriminate between the Models 996
30.5.1 Studies on the Unfolding of Preproteins 996
30.5.1.1 Comparison of the Import of Folded and Unfolded Proteins 996
30.5.1.2 Import of Preproteins With Different Presequence Lengths 999
30.5.1.3 Import of Titin Domains 1000
30.5.1.4 Unfolding by the Mitochondrial Membrane Potential $\Delta\Psi$ 1000
30.5.2 Mechanistic Studies of the Import Motor 1000
30.5.2.1 Brownian Movement of the Polypeptide Within the Import Channel 1000
30.5.2.2 Recruitment of mtHsp70 by Tim44 1001
30.5.2.3 Import Without Recruitment of mtHsp70 by Tim44 1002
30.5.2.4 MtHsp70 Function in the Import Motor 1003
30.6 Discussion and Perspectives 1004
30.7 Experimental Protocols 1006
30.7.1 Protein Import Into Mitochondria in Vitro 1006
30.7.2 Stabilization of the DHFR Domain by Methotrexate 1008
30.7.3 Import of Precursor Proteins Unfolded With Urea 1009
30.7.4 Kinetic Analysis of the Unfolding Reaction by Trapping of Intermediates 1009
References 1011

31 **The Chaperone System of Mitochondria** 1020
Wolfgang Voos and Nikolaus Pfanner
31.1 Introduction 1020
31.2 Membrane Translocation and the Hsp70 Import Motor 1020
31.3 Folding of Newly Imported Proteins Catalyzed by the Hsp70 and Hsp60 Systems 1026
31.4 Mitochondrial Protein Synthesis and the Assembly Problem 1030
31.5 Aggregation versus Degradation: Chaperone Functions Under Stress Conditions 1033
31.6 Experimental Protocols 1034

31.6.1	Chaperone Functions Characterized With Yeast Mutants 1034
31.6.2	Interaction of Imported Proteins With Matrix Chaperones 1036
31.6.3	Folding of Imported Model Proteins 1037
31.6.4	Assaying Mitochondrial Degradation of Imported Proteins 1038
31.6.5	Aggregation of Proteins in the Mitochondrial Matrix 1038
	References 1039

32   Chaperone Systems in Chloroplasts   1047
Thomas Becker, Jürgen Soll, and Enrico Schleiff

32.1	Introduction 1047
32.2	Chaperone Systems within Chloroplasts 1048
32.2.1	The Hsp70 System of Chloroplasts 1048
32.2.1.1	The Chloroplast Hsp70s 1049
32.2.1.2	The Co-chaperones of Chloroplastic Hsp70s 1051
32.2.2	The Chaperonins 1052
32.2.3	The HSP100/Clp Protein Family in Chloroplasts 1056
32.2.4	The Small Heat Shock Proteins 1058
32.2.5	Hsp90 Proteins of Chloroplasts 1061
32.2.6	Chaperone-like Proteins 1062
32.2.6.1	The Protein Disulfide Isomerase (PDI) 1062
32.2.6.2	The Peptidyl-prolyl *cis* Isomerase (PPIase) 1063
32.3	The Functional Chaperone Pathways in Chloroplasts 1065
32.3.1	Chaperones Involved in Protein Translocation 1065
32.3.2	Protein Transport Inside of Plastids 1070
32.3.3	Protein Folding and Complex Assembly Within Chloroplasts 1071
32.3.4	Chloroplast Chaperones Involved in Proteolysis 1072
32.3.5	Protein Storage Within Plastids 1073
32.3.6	Protein Protection and Repair 1074
32.4	Experimental Protocols 1075
32.4.1	Characterization of Cpn60 Binding to the Large Subunit of Rubisco via Native PAGE (adopted from Ref. [6]) 1075
32.4.2	Purification of Chloroplast Cpn60 From Young Pea Plants (adopted from Ref. [203]) 1076
32.4.3	Purification of Chloroplast Hsp21 From Pea (*Pisum sativum*) (adopted from [90]) 1077
32.4.4	Light-scattering Assays for Determination of the Chaperone Activity Using Citrate Synthase as Substrate (adopted from [196]) 1078
32.4.5	The Use Of *Bis*-ANS to Assess Surface Exposure of Hydrophobic Domains of Hsp17 of *Synechocystis* (adopted from [202]) 1079
32.4.6	Determination of Hsp17 Binding to Lipids (adopted from Refs. [204, 205]) 1079
	References 1081

33   An Overview of Protein Misfolding Diseases   1093
Christopher M. Dobson

| 33.1 | Introduction 1093 |

33.2	Protein Misfolding and Its Consequences for Disease	1094
33.3	The Structure and Mechanism of Amyloid Formation	1097
33.4	A Generic Description of Amyloid Formation	1101
33.5	The Fundamental Origins of Amyloid Disease	1104
33.6	Approaches to Therapeutic Intervention in Amyloid Disease	1106
33.7	Concluding Remarks 1108	
	Acknowledgements 1108	
	References 1109	
34	**Biochemistry and Structural Biology of Mammalian Prion Disease**	**1114**
	Rudi Glockshuber	
34.1	Introduction 1114	
34.1.1	Prions and the "Protein-Only" Hypothesis 1114	
34.1.2	Models of PrP^{Sc} Propagation 1115	
34.2	Properties of PrP^C and PrP^{Sc} 1117	
34.3	Three-dimensional Structure and Folding of Recombinant PrP 1120	
34.3.1	Expression of the Recombinant Prion Protein for Structural and Biophysical Studies 1120	
34.3.2	Three-dimensional Structures of Recombinant Prion Proteins from Different Species and Their Implications for the Species Barrier of Prion Transmission 1120	
34.3.2.1	Solution Structure of Murine PrP 1120	
34.3.2.2	Comparison of Mammalian Prion Protein Structures and the Species Barrier of Prion Transmission 1124	
34.3.3	Biophysical Characterization of the Recombinant Prion Protein 1125	
34.3.3.1	Folding and Stability of Recombinant PrP 1125	
34.3.3.2	Role of the Disulfide Bond in PrP 1127	
34.3.3.3	Influence of Point Mutations Linked With Inherited TSEs on the Stability of Recombinant PrP 1129	
34.4	Generation of Infectious Prions in Vitro: Principal Difficulties in Proving the Protein-Only Hypothesis 1131	
34.5	Understanding the Strain Phenomenon in the Context of the Protein-Only Hypothesis: Are Prions Crystals? 1132	
34.6	Conclusions and Outlook 1135	
34.7	Experimental Protocols 1136	
34.7.1	Protocol 1 [53, 55] 1136	
34.7.2	Protocol 2 [54] 1137	
	References 1138	
35	**Insights into the Nature of Yeast Prions**	**1144**
	Lev Z. Osherovich and Jonathan S. Weissman	
35.1	Introduction 1144	
35.2	Prions as Heritable Amyloidoses 1145	
35.3	Prion Strains and Species Barriers: Universal Features of Amyloid-based Prion Elements 1149	

35.4	Prediction and Identification of Novel Prion Elements *1151*
35.5	Requirements for Prion Inheritance beyond Amyloid-mediated Growth *1154*
35.6	Chaperones and Prion Replication *1157*
35.7	The Structure of Prion Particles *1158*
35.8	Prion-like Structures as Protein Interaction Modules *1159*
35.9	Experimental Protocols *1160*
35.9.1	Generation of Sup35 Amyloid Fibers in Vitro *1160*
35.9.2	Thioflavin T–based Amyloid Seeding Efficacy Assay (Adapted from Chien et al. 2003) *1161*
35.9.3	AFM-based Single-fiber Growth Assay *1162*
35.9.4	Prion Infection Protocol (Adapted from Tanaka et al. 2004) *1164*
35.9.5	Preparation of Lyticase *1165*
35.9.6	Protocol for Counting Heritable Prion Units (Adapted from Cox et al. 2003) *1166*
	Acknowledgements *1167*
	References *1168*
36	**Polyglutamine Aggregates as a Model for Protein-misfolding Diseases** *1175*
	Soojin Kim, James F. Morley, Anat Ben-Zvi, and Richard I. Morimoto
36.1	Introduction *1175*
36.2	Polyglutamine Diseases *1175*
36.2.1	Genetics *1175*
36.2.2	Polyglutamine Diseases Involve a Toxic Gain of Function *1176*
36.3	Polyglutamine Aggregates *1176*
36.3.1	Presence of the Expanded Polyglutamine Is Sufficient to Induce Aggregation in Vivo *1176*
36.3.2	Length of the Polyglutamine Dictates the Rate of Aggregate Formation *1177*
36.3.3	Polyglutamine Aggregates Exhibit Features Characteristic of Amyloids *1179*
36.3.4	Characterization of Protein Aggregates in Vivo Using Dynamic Imaging Methods *1180*
36.4	A Role for Oligomeric Intermediates in Toxicity *1181*
36.5	Consequences of Misfolded Proteins and Aggregates on Protein Homeostasis *1181*
36.6	Modulators of Polyglutamine Aggregation and Toxicity *1184*
36.6.1	Protein Context *1184*
36.6.2	Molecular Chaperones *1185*
36.6.3	Proteasomes *1188*
36.6.4	The Protein-folding "Buffer" and Aging *1188*
36.6.5	Summary *1189*
36.7	Experimental Protocols *1190*
36.7.1	FRAP Analysis *1190*
	References *1192*

37	**Protein Folding and Aggregation in the Expanded Polyglutamine Repeat Diseases** *1200*	
	Ronald Wetzel	
37.1	Introduction *1200*	
37.2	Key Features of the Polyglutamine Diseases *1201*	
37.2.1	The Variety of Expanded PolyGln Diseases *1201*	
37.2.2	Clinical Features *1201*	
37.2.2.1	Repeat Expansions and Repeat Length *1202*	
37.2.3	The Role of PolyGln and PolyGln Aggregates *1203*	
37.3	PolyGln Peptides in Studies of the Molecular Basis of Expanded Polyglutamine Diseases *1205*	
37.3.1	Conformational Studies *1205*	
37.3.2	Preliminary in Vitro Aggregation Studies *1206*	
37.3.3	In Vivo Aggregation Studies *1206*	
37.4	Analyzing Polyglutamine Behavior With Synthetic Peptides: Practical Aspects *1207*	
37.4.1	Disaggregation of Synthetic Polyglutamine Peptides *1209*	
37.4.2	Growing and Manipulating Aggregates *1210*	
37.4.2.1	Polyglutamine Aggregation by Freeze Concentration *1210*	
37.4.2.2	Preparing Small Aggregates *1211*	
37.5	In vitro Studies of PolyGln Aggregation *1212*	
37.5.1	The Universe of Protein Aggregation Mechanisms *1212*	
37.5.2	Basic Studies on Spontaneous Aggregation *1213*	
37.5.3	Nucleation Kinetics of PolyGln *1215*	
37.5.4	Elongation Kinetics *1218*	
37.5.4.1	Microtiter Plate Assay for Elongation Kinetics *1219*	
37.5.4.2	Repeat-length and Aggregate-size Dependence of Elongation Rates *1220*	
37.6	The Structure of PolyGln Aggregates *1221*	
37.6.1	Electron Microscopy Analysis *1222*	
37.6.2	Analysis with Amyloid Dyes Thioflavin T and Congo Red *1222*	
37.6.3	Circular Dichroism Analysis *1224*	
37.6.4	Presence of a Generic Amyloid Epitope in PolyGln Aggregates *1225*	
37.6.5	Proline Mutagenesis to Dissect the Polyglutamine Fold Within the Aggregate *1225*	
37.7	Polyglutamine Aggregates and Cytotoxicity *1227*	
37.7.1	Direct Cytotoxicity of PolyGln Aggregates *1228*	
37.7.1.1	Delivery of Aggregates into Cells and Cellular Compartments *1229*	
37.7.1.2	Cell Killing by Nuclear-targeted PolyGln Aggregates *1229*	
37.7.2	Visualization of Functional, Recruitment-positive Aggregation Foci *1230*	
37.8	Inhibitors of polyGln Aggregation *1231*	
37.8.1	Designed Peptide Inhibitors *1231*	
37.8.2	Screening for Inhibitors of PolyGln Elongation *1231*	
37.9	Concluding Remarks *1232*	
37.10	Experimental Protocols *1233*	

37.10.1	Disaggregation of Synthetic PolyGln Peptides	*1233*
37.10.2	Determining the Concentration of Low-molecular-weight PolyGln Peptides by HPLC *1235*	
	Acknowledgements *1237*	
	References *1238*	

38 **Production of Recombinant Proteins for Therapy, Diagnostics, and Industrial Research by in Vitro Folding** *1245*
Christian Lange and Rainer Rudolph

38.1	Introduction *1245*	
38.1.1	The Inclusion Body Problem *1245*	
38.1.2	Cost and Scale Limitations in Industrial Protein Folding	*1248*
38.2	Treatment of Inclusion Bodies *1250*	
38.2.1	Isolation of Inclusion Bodies *1250*	
38.2.2	Solubilization of Inclusion Bodies *1250*	
38.3	Refolding in Solution *1252*	
38.3.1	Protein Design Considerations *1252*	
38.3.2	Oxidative Refolding With Disulfide Bond Formation *1253*	
38.3.3	Transfer of the Unfolded Proteins Into Refolding Buffer *1255*	
38.3.4	Refolding Additives *1257*	
38.3.5	Cofactors in Protein Folding *1260*	
38.3.6	Chaperones and Folding-helper Proteins *1261*	
38.3.7	An Artificial Chaperone System *1261*	
38.3.8	Pressure-induced Folding *1262*	
38.3.9	Temperature-leap Techniques *1263*	
38.3.10	Recycling of Aggregates *1264*	
38.4	Alternative Refolding Techniques *1264*	
38.4.1	Matrix-assisted Refolding *1264*	
38.4.2	Folding by Gel Filtration *1266*	
38.4.3	Direct Refolding of Inclusion Body Material *1267*	
38.5	Conclusions *1268*	
38.6	Experimental Protocols *1268*	
38.6.1	Protocol 1: Isolation of Inclusion Bodies *1268*	
38.6.2	Protocol 2: Solubilization of Inclusion Bodies *1269*	
38.6.3	Protocol 3: Refolding of Proteins *1270*	
	Acknowledgements *1271*	
	References *1271*	

39 **Engineering Proteins for Stability and Efficient Folding** *1281*
Bernhard Schimmele and Andreas Plückthun

39.1	Introduction *1281*	
39.2	Kinetic and Thermodynamic Aspects of Natural Proteins *1281*	
39.2.1	The Stability of Natural Proteins *1281*	
39.2.2	Different Kinds of "Stability" *1282*	
39.2.2.1	Thermodynamic Stability *1283*	

39.2.2.2	Kinetic Stability	1285
39.2.2.3	Folding Efficiency	1287
39.3	The Engineering Approach	1288
39.3.1	Consensus Strategies	1288
39.3.1.1	Principles	1288
39.3.1.2	Examples	1291
39.3.2	Structure-based Engineering	1292
39.3.2.1	Entropic Stabilization	1294
39.3.2.2	Hydrophobic Core Packing	1296
39.3.2.3	Charge Interactions	1297
39.3.2.4	Hydrogen Bonding	1298
39.3.2.5	Disallowed Phi-Psi Angles	1298
39.3.2.6	Local Secondary Structure Propensities	1299
39.3.2.7	Exposed Hydrophobic Side Chains	1299
39.3.2.8	Inter-domain Interactions	1300
39.3.3	Case Study: Combining Consensus Design and Rational Engineering to Yield Antibodies with Favorable Biophysical Properties	1300
39.4	The Selection and Evolution Approach	1305
39.4.1	Principles	1305
39.4.2	Screening and Selection Technologies Available for Improving Biophysical Properties	1311
39.4.2.1	In Vitro Display Technologies	1313
39.4.2.2	Partial in Vitro Display Technologies	1314
39.4.2.3	In Vivo Selection Technologies	1315
39.4.3	Selection for Enhanced Biophysical Properties	1316
39.4.3.1	Selection for Solubility	1316
39.4.3.2	Selection for Protein Display Rates	1317
39.4.3.3	Selection on the Basis of Cellular Quality Control	1318
39.4.4	Selection for Increased Stability	1319
39.4.4.1	General Strategies	1319
39.4.4.2	Protein Destabilization	1319
39.4.4.3	Selections Based on Elevated Temperature	1321
39.4.4.4	Selections Based on Destabilizing Agents	1322
39.4.4.5	Selection for Proteolytic Stability	1323
39.5	Conclusions and Perspectives	1324
	Acknowledgements	1326
	References	1326

Index 1334

Contributors of Part II

James C. Bardwell
Department of Molecular
Cellular and Developmental Biology
University of Michigan
830 North University Avenue
Ann Arbor, MI 48109
USA

Thomas Becker
Botanisches Institut
Ludwig-Maximilians-Universität München
Menzinger Str. 67
80638 München
Germany

Ineke Braakman
Department of Bio-Organic Chemistry I
Utrecht University
Padualaan 8
3584 CH Utrecht
The Netherlands

Andreas Bracher
Department of Cellular Biochemistry
Max Planck Institute of Biochemistry
Am Klopferspitz 18a
82152 Martinsried
Germany

Michael Brunner
Biochemistry Center (BZH)
Heidelberg University
Im Neuenheimer Feld 328
69120 Heidelberg
Germany

Johannes Buchner
Institut für Organische Chemie und
Biochemie
Technische Universität München
Lichtenbergstrasse 4
85747 Garching
Germany

Bernd Bukau
ZMBH–Zentrum für Molekulare Biologie
Heidelberg
Ruprecht-Karls-Universität Heidelberg
Im Neuenheimer Feld 282
69120 Heidelberg
Germany

Neil J. Bulleid
Division of Biochemistry
School of Biological Sciences
University of Manchester
Oxford Road
Manchester M13 9PT
United Kingdom

Steven G. Burston
Department of Biochemistry
University of Bristol
University Walk
Bristol BS8 1TD
United Kingdom

José L. Carrascosa
Centro Nacional de Biotecnología
Consejo Superior de Investigaciones
Científicas
Campus de la Universidad Autónoma de
Madrid
Spain

Kyung Tae Chung
Department of Applied Life Sciences
DongEui University
Pusan 614-714
South Korea

Jean-Francois Collet
Department of Molecular, Cellular and
Developmental Biology
University of Michigan
830 North University Avenue
Ann Arbor, MI 48109
USA

Mark S. Cortese
Center for Computational Biology and
Bioinformatics
University of Indiana School of Medicine
Indianapolis, IN 46202
USA

Elizabeth A. Craig
Department of Biochemistry
University of Wisconsin – Madison
433 Babcock Drive
Madison, WI 53706
USA

Gary W. Daughdrill
Department of Microbiology, Molecular
Biology and Biochemistry
University of Idaho
Moscow, ID 83844
USA

Janny de Wit
Department of Molecular Microbiology
Groningen Biomolecular Sciences and
Biotechnology Institute
University of Groningen
Kerklaan 30
9751 NN Haren
The Netherlands

Elke Deuerling
ZMBH – Zentrum für Molekulare Biologie
Heidelberg
Universität Heidelberg
Im Neuenheimer Feld 282
69120 Heidelberg
Germany

Christopher M. Dobson
Department of Chemistry
University of Cambridge
Lensfield Road
Cambridge CB2 1EW
United Kingdom

Arnold J. M. Driessen
Department of Molecular Microbiology
Groningen Biomolecular Sciences and
Biotechnology Institute
University of Groningen
Kerklaan 30
9751 NN Haren
The Netherlands

A. Keith Dunker
Center for Computational Biology and
Bioinformatics
University of Indiana School of Medicine
714 N. Senate Avenue
Indianapolis, IN 46202
USA

Michael Ehrmann
School of Biosciences
Cardiff University
Museum Ave.
Cardiff CF10 3US
United Kingdom

Gunter Fischer
Max Planck Research Unit for Enzymology
of Protein Folding
Weinbergweg 22
06120 Halle/Saale
Germany

Matthias Gautschi
ETHZ–Swiss Federal Institute of Technology
ETH Hönggerberg
8093 Zürich
Switzerland

Rudi Glockshuber
Institut für Molekularbiologie und Biophysik
ETH Hönggerberg, HPK E17
8093 Zürich
Switzerland

F. Ulrich Hartl
Department of Cellular Biochemistry
Max Planck Institute of Biochemistry
Am Klopferspitz 18a
82152 Martinsried
Germany

Martin Haslbeck
Institut für Organische Chemie und
Biochemie
Technische Universität München
Lichtenbergstr. 4
85747 Garching
Germany

Kai Hell
Institut für Physiologische Chemie
Ludwig-Maximilians-Universität München
Butenandtstr. 5B
81377 München
Germany

Linda M. Hendershot
Department of Tumor Cell Biology
St. Jude Children's Research Hospital
332 N. Lauderdale
Memphis, TN 38105
USA

Jörg H. Hoffmann
Department of Molecular, Cellular and
Developmental Biology
University of Michigan
830 North University Avenue
Ann Arbor, MI 48109
USA

Joseph Horwitz
Jules Stein Eye Institute
UCLA School of Medicine
100 Stein Plaza
Los Angeles, CA 90095-7008
USA

Peggy Huang
Department of Biochemistry
University of Wisconsin – Madison
433 Babcock Drive
Madison, WI 53706
USA

Rainer Jaenicke
School of Crystallography
Birkbeck College
University of London
Malet Street
London, WC 1E 7HX
United Kingdom

Neil E. Jaffe
Department of Biochemistry, Molecular
Biology and Cell Biology
Northwestern University
2145 Sheridan Road
Evanston, IL 60208
USA

Ursula Jakob
Department of Molecular, Cellular and
Developmental Biology
University of Michigan
830 North University Avenue
Ann Arbor, MI 48109
USA

Soojin Kim
Department of Biochemistry, Molecular
Biology and Cell Biology
Northwestern University
2205 Tech Drive
Evanston, IL 60208
USA

Bertrand Kleizen
Department of Bio-Organic Chemistry I
Utrecht University
Padualaan 8
3584 CH Utrecht
The Netherlands

Prakash Koodathingal
Department of Biochemistry, Molecular
Biology and Cell Biology
Northwestern University
2145 Sheridan Road
Evanston, IL 60208
USA

Christian Lange
Institut für Biotechnologie
Martin-Luther-Universität Halle/Wittenberg
Kurt-Mothes-Straße 3
06120 Halle/Saale
Germany

Hauke Lilie
Institut für Biotechnologie
Martin-Luther-Universität Halle/Wittenberg
Kurt-Mothes-Straße 3
06120 Halle
Germany

Dieter Langosch
Lehrstuhl Chemie der Biopolymere
Technische Universität München
Weihenstephaner Berg 3
85354 Freising
Germany

Marije Liscaljet
Department of Bio-Organic Chemistry I
Utrecht University
Padualaan 8
3584 CH Utrecht
The Netherlands

Christian Lücke
Max Planck Research Unit for Enzymology of
Protein Folding
Weinbergweg 22
06120 Halle/Saale
Germany

Andreas Matouschek
Department of Biochemistry, Molecular
Biology and Cell Biology
Northwestern University
2145 Sheridan Road
Evanston, IL 60208
USA

Matthias P. Mayer
ZMBH–Zentrum für Molekulare Biologie
Ruprecht-Karls-Universität Heidelberg
Im Neuenheimer Feld 282
69120 Heidelberg
Germany

Birgit Meinlschmidt
Institut für Organische Chemie und Biochemie
Technische Universität München
Lichtenbergstrasse 4
85747 Garching
Germany

Richard I. Morimoto
Department of Biochemistry, Molecular Biology and Cell Biology
Northwestern University
2205 Tech Drive
Evanston, IL 60208
USA

James F. Morley
Department of Biochemistry, Molecular Biology and Cell Biology
Northwestern University
2205 Tech Drive
Evanston, IL 60208
USA

Franz Narberhaus
Ruhr-Universität Bochum
Lehrstuhl für Biologie der Mikroorganismen
Gebäude NDEF 06/783
44780 Bochum
Germany

Walter Neupert
Institut für Physiologische Chemie
Ludwig-Maximilians-Universität München
Butenandtstr. 5B
81377 München
Germany

Nico Nouwen
Department of Molecular Microbiology
Groningen Biomolecular Sciences and Biotechnology Institute
University of Groningen
Kerklaan 30
9751 NN Haren
The Netherlands

Lev Z. Osherovich
Department of Biochemistry
University of California, San Francisco
600 16th Street
San Franciscso, CA 94143
USA

Armando J. Parodi
Fundacion Instituto Leloir
Avda Patricias Argentinas 435
Buenos Aires C1405BWE
Argentina

Holger Patzelt
ZMBH – Zentrum für Molekulare Biologie Heidelberg
Universität Heidelberg
Im Neuenheimer Feld 282
69120 Heidelberg
Germany

Anat Peres Ben-Zvi
Department of Biochemistry, Molecular Biology and Cell Biology
Northwestern University
2205 Tech Drive
Evanston, IL 60208
USA

Nikolaus Pfanner
Institut für Biochemie und Molekularbiologie
Universität Freiburg
Hermann-Herder-Str. 7
79104 Freiburg
Germany

Gary J. Pielak
Department of Chemistry and
Department of Biochemistry and Biophysics
University of North Carolina
Chapel Hill, NC 27599
USA

Andreas Plückthun
Biochemisches Institut
Universität Zürich
Winterthurerstr. 190
8057 Zürich
Switzerland

Magdalena Rakwalska
Institut für Biochemie und Molekularbiologie
Universität Freiburg
Hermann-Herder-Str. 7
79104 Freiburg
Germany

Thomas Rauch
ZMBH – Zentrum für Molekulare Biologie Heidelberg
Universität Heidelberg
Im Neuenheimer Feld 282
69120 Heidelberg
Germany

Uta Raue
Institut für Biochemie und Molekularbiologie
Universität Freiburg
Hermann-Herder-Str. 7
79104 Freiburg
Germany

Jochen Reinstein
Max Planck Institute for Medical Resarch
Jahnstr. 29
69120 Heidelberg
Germany

Klaus Richter
Institut für Organische Chemie und Biochemie
Technische Universität München
Lichtenbergstrasse 4
85747 Garching
Germany

Anja Ridder
Lehrstuhl Chemie der Biopolymere
Technische Universität München
Weihenstephaner Berg 3
85354 Freising
Germany

Sabine Rospert
Institut für Biochemie und Molekularbiologie
Universität Freiburg
Hermann-Herder-Str. 7
79104 Freiburg
Germany

Rainer Rudolph
Institut für Biotechnologie
Martin-Luther-Universität Halle/Wittenberg
Kurt Mothes Strasse 3
06120 Halle/Saale
Germany

Thomas Scheibel
Institut für Organische Chemie und Biochemie
Technische Universität München
Lichtenbergstr. 4
85747 Garching
Germany

Cordelia Schiene-Fischer
Max Planck Research Unit for Enzymology of Protein Folding
Weinbergweg 22
06120 Halle/Saale
Germany

Bernhard Schimmele
Biochemisches Institut
Universität Zürich
Winterthurerstr. 190
8057 Zürich
Switzerland

Sandra Schlee
Max-Planck-Institute for Medical Resarch
Jahnstr. 29
69120 Heidelberg
Germany

Enrico Schleiff
Botanisches Institut
Ludwig-Maximilians-Universität München
Menzinger Str. 67
80638 München
Germany

Robert Seckler
Institut für Physikalische Biochemie
Universität Potsdam
Karl Liebknecht Strasse 24
14476 Golm
Germany

Louise C. Serpell
School of Life Sciences
University of Sussex
Falmer, BN1 9QG
United Kingdom

Ying Shen
Department of Tumor Cell Biology
St. Jude Children's Research Hospital
332 N. Lauderdale
Memphis, TN 38105
USA

Katja Siegers
Department of Cellular Biochemistry
Max Planck Institute of Biochemistry
Am Klopferspitz 18a
82152 Martinsried
Germany

Jürgen Soll
Botanisches Institut
Ludwig-Maximilians-Universität München
Menzinger Str. 67
80638 München
Germany

Mohammed Tasab
Division of Biochemistry
School of Biological Sciences
University of Manchester
Oxford Road
Manchester M13 9PT
United Kingdom

E. Sergio Trombetta
Department of Cell Biology
Yale University School of Medicine
PO Box 208002
New Haven, CT 06520-8002
USA

Vladimir N. Uversky
Institute for Biological Instrumentation
Russian Academy of Sciences
142292 Puschchino, Moscow Region
Russia

José M. Valpuesta
Centro Nacional de Biotecnología
Consejo Superior de Investigaciones
Científicas
Campus de la Universidad Autónoma de
Madrid
Spain

Michael Vetsch
Institut für Molekularbiologie und Biophysik
ETH Hönggerberg, HPK E17
8093 Zürich
Switzerland

Wolfgang Voos
Institut für Biochemie und Molekularbiologie
Universität Freiburg
Hermann-Herder-Str. 7
79104 Freiburg
Germany

Stefan Walter
Institut für Organische Chemie und
Biochemie
Technische Universität München
Lichtenbergstr. 4
85747 Garching
Germany

Jonathan S. Weissman
Department of Cellular and Molecular
Pharmacology
University of California, San Francisco
600 16th Street
San Francisco, CA 94143
USA

Ronald Wetzel
Graduate School of Medicine
University of Tennessee
1924 Alcoa Highway
Knoxville, TN 37920
USA

Keith R. Willison
Centre for Cell and Molecular Biology
Institute of Cancer Research
Chester Beatty Laboratories
237 Fulham Road
Chelsea, London SW3 6JB
United Kingdom

Part II

1
Paradigm Changes from "Unboiling an Egg" to "Synthesizing a Rabbit"

Rainer Jaenicke

1.1
Protein Structure, Stability, and Self-organization

Max Perutz, in reviewing Anson and Mirsky's discovery [1–3] that hemoglobin fully recovers its functional properties after reversible acid denaturation, felt that "making a protein alive again is as impossible as inventing the *perpetuum mobile*"; this was in 1940 [4]. As we know today from 75 years practice since his first experiments, Anson (Figure 1.1) was right, despite Kailin's protest that "boiling an egg does not denature its protein, it kills it!" [5]: proteins do have the capacity to spontaneously and autonomously organize their three-dimensional structure. At Anson's time this was a model, because the antivitalist creed that the composition of proteins is revealed by analytical chemistry and their behavior, by physical chemistry (which in the last analysis depends on their structure [6]) was no more than a guiding principle for chemists interested in biological material [7]. Svedberg had just published the determination of the molecular weight of the first protein, equine hemoglobin, making use of his newly developed analytical ultracentrifuge [8]. With this data, it became clear that proteins are monodisperse entities with well-defined molecular masses. To determine the detailed 3D structure of hemoglobin at atomic resolution still took more than a generation [9]. Thus, the first folding experiments were based on the assumption that an isolated homogeneous component of mammalian blood, obtained by fractional crystallization, not only had the intrinsic property of self-organization but also allowed one to quantify its physiological activity in vivo in terms of cooperative ligand binding, allostery, Bohr effect etc., following the reductionist's creed and methodology. With this paradigm in mind, the alchemy of protein folding addressed a wide range of fundamental biological problems, tacitly assuming that the results of in vitro studies are biologically relevant. As sometimes happens, a few results of this bold hypothesis turned out to be practicable in various ways, so that, in the long run, the scope switched from pure physical to biological chemistry. After a generation or two, research and development in biotechnology shifted the dimensions of in vitro folding experiments from microliter pipettes and Eppendorf vials to pipelines and $5–10\text{-m}^3$ tanks, nowadays used in the production of pharmaceutically or technologically im-

Protein Folding Handbook. Part II. Edited by J. Buchner and T. Kiefhaber
Copyright © 2005 WILEY-VCH Verlag GmbH & Co. KGaA, Weinheim
ISBN: 3-527-30784-2

Fig. 1.1. Pioneers in the early days of protein folding: left, M. L. ("Tim") Anson (1901–1968), right, Chris B. Anfinsen (1916–1995).

portant proteins. Frequently the final products are modified in one way or the other to alter their long-term stability, their specificity, or other properties.

In this context, Anfinsen's (Figure 1.1) classical experiments influenced the philosophy of his followers right from the beginning by the fact that he started from *reduced* ribonuclease A (RNaseA) rather than the natural enzyme with its cystine cross-bridges intact. In spite of the chemical modification, the reshuffling process led to the correctly cross-linked native state, thus proving that all the information required for folding and stabilizing the native three-dimensional conformation of proteins is encoded in their amino acid sequence [10]. Extending this conclusion to the entire structural hierarchy of proteins, it was postulated that the primary structure directs not only the formation of multiple noncovalent ("weak") and covalent ("strong") bonds within a single polypeptide chain but also the interactions between the subunits of oligomeric proteins, driving the sequential folding-association reaction and stabilizing the final tertiary and quaternary structure.

In distinguishing between weak and strong interactions, it is important to note that globular proteins in their natural aqueous environment show only marginal Gibbs free energies of stabilization. With numerical data on the order of 50 kJ mol^{-1}, they are the equivalent of a small number of hydrogen bonds or hydrophobic interactions, or perhaps just one or two ion pairs. This holds despite the fact that thousands of atoms may be involved in large numbers of attractive and repulsive interactions, in total adding up to molecular energies a million times larger than the above average value [11]. The reason for this apparent discrepancy is that proteins are *multifunctional*: they need to fold, they serve specific modes of action, and they provide amino acid pools in both catabolic and anabolic processes.

With this in mind, the point is that, given one single device, it is impossible to simultaneously *maximize* the efficiency of all these "functions"; nature has to compromise, and since stability, biological activity, and turnover all require a certain degree of flexibility, stability (equivalent to rigidity) cannot win.

Recent advances in our understanding of the different types of weak interactions involved in the self-organization and stabilization of proteins were derived from theoretical studies on the one hand ([12–15], cf. R. L. Baldwin, Vol. I/1) and from the analysis of complete genomes of phylogenetically related mesophiles and (hyper-)thermophiles, as well as systematic mutant studies, on the other [16–18]. Apart from the latter "quasi-Linnaean approach," protein engineering provides us with a wealth of alternatives to crosscheck the genomix results [19–23]; in combining and applying these approaches to technologically relevant systems, increases in thermal stability of up to 50 °C have been accomplished [24]. In contrast to such success stories, Kauzmann's catalog of weak intermolecular interactions [25] has not changed since 1959, and a 2004 "balance sheet of ΔG_{stab} contributions to folding" would be as tentative or postdictive as J. L. Finney and N. C. Pace's classical attempts to predict ΔG_{stab} for RNase and lysozyme in 1982 and 1996 [26, 27]. What has become clear in recent years is that increases in overall packing density, helicity, and proline content (i.e., destabilization of the unfolded state) and improved formation of networks of H-bonds and ion-pairs are important increments of protein stabilization. In the long-standing controversy regarding the prevalence of the various contributions, evidently all components are significant. In the case of hydrophobic interactions, entropic (water-release) *and* enthalpic (London-van der Waals) effects seem to be equally important [28, 29]. That hydrogen bonding and ion pairs are essential is now generally accepted [27, 30, 31]; surprisingly, recent evidence seems to indicate that polar-group burial also makes a positive contribution to the free energy of stabilization [32].

As has been mentioned, the balance of the attractive and repulsive forces yields only marginal free energies of stabilization. They may be affected by mutations or changes in the solvent, the latter often highly significant due to their effect on the unfolding *kinetics*. The reason is that by adding specific ions or other ligands, the free energy of activation (ΔG^{+}_{stab}) of the N → U transition, i.e., the free-energy difference between the native and the transition states, may be drastically increased. The resulting "kinetic stabilization" has been shown to be essential for the extrinsic stabilization or longevity of the native state as well as for the folding competence of conjugated and ultrastable proteins [16, 33, 34].

From the thermodynamic point of view, protein folding is a hierarchical process, driven by the accumulation of increments of free energy from local interactions among neighboring residues, secondary structural elements, domains, and subunits (Scheme 1.1). Domains represent independent folding units. Correspondingly, the folding kinetics divide into the collapse of subdomains and domains and their merging to form the compact tertiary fold. Establishing the particular steps allows the folding pathway for a given protein to be elucidated. In proceeding to oligomeric proteins, docking of structured monomers is the final step. In agreement with this mechanism, in vitro experiments have shown that the overall fold-

Structural levels: primary → secondary/supersecondary → tertiary → quaternary structure
Interactions: short-range and long-range (= short-range through-chain and through-space) [11, 12, 42]
Folding pathway: next-neighbor interactions → collapsed unfolded state and/or molten globule state [43, 44] → docking of domains [45, 46] → assembly [47, 48]
Intermediates: kernels/molten globule → folded/unfolded subdomains and domains → native-like structured monomers → stepwise assembly, e.g., nM → n/2 M_2 → n/4 M_4 for a tetramer
Off-pathway reactions: misfolding, domain swapping, misassembly, aggregation [34, 36–40]

Scheme 1.1. Hierarchy of protein folding and association

ing and association reaction can be quantified by a consecutive uni-bimolecular kinetic scheme. In a first folding step, the subunits need to expose specific interfaces to recognize the correct partners; then, after complementary docking sites have been formed, association to well-defined oligomers can occur in a concentration-dependent fashion. Their affinity depends on the same types of weak interactions that participate in the stabilization of the monomeric entities. No wonder that intra- and intermolecular interactions can replace each other, causing misassembly and subsequent aggregation rather than proper quaternary structure formation. At high subunit concentrations, when folding becomes rate-limiting and association becomes diffusion-controlled, aggregation is expected to become predominant. In fact, early refolding studies (focusing on the structure-function relationship of oligomeric proteins) proved that reactivation yielded optimum curves, confirming the consecutive mechanism with aggregation as a competing side reaction [35–37] (Figure 1.2). Later, when "inclusion bodies" were identified as typical byproducts of protein overexpression in vivo, it became clear that the sequential folding/association mechanism (including the kinetic partitioning between the two processes) also holds for the nascent protein when it leaves the ribosome. It took 10 more years to fully appreciate that evolution took care of the problem by providing folding catalysts and heat shock proteins as chaperones [41]. Evidently, they shift the kinetic partitioning between assembly and aggregation toward the native state. So far, the detailed co-translational and post-translational processes *in the cell* have withstood a sound kinetic and structural analysis.

1.2
Autonomous and Assisted Folding and Association

The contributions to the present volume deal with the above side reactions in the self-organization of proteins as well as the helper proteins involved in either avoidance of misfolding or the catalysis of rate-determining steps along the folding pathway. Reading the seminal papers, and having in mind the procedures that were applied in early protein denaturation-renaturation experiments, it becomes clear that the pioneers in the field were aware of the possible role of accessory components long before they were discovered: *templates* assisting the nascent polypeptide chain

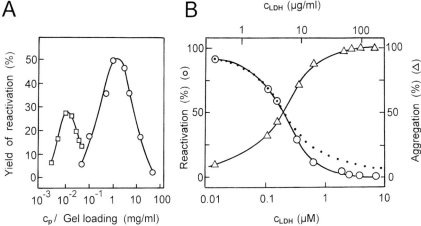

Fig. 1.2. Kinetic competition between folding and aggregation. (A) Effect of protein concentration on the renaturation of α-glucosidase after urea denaturation, illustrating the partitioning between "renaturation" and aggregation. Denaturation in 8 M urea, 10 mM K-phosphate pH 7.7, 1 h incubation at 20 °C; reactivation in 10 mM Na-phosphate pH 7.6, 10 °C, 30 mM NaCl, and 8% ethylene glycol, measured after ≥24 h renaturation. (□) refers to wild-type α-glucosidase and (○) to the fusion protein containing a C-terminal hexa-arginine tail, which allows the stabilization of the enzyme on heparin-Sepharose. The yield of reactivation is expressed as a percentage of the activity of the corresponding native enzyme (data from Ref. [38]). (B) With increasing protein concentration, the yield of reactivation of lactate dehydrogenase after acid denaturation in 0.1 M H_3PO_4, pH 2, and subsequent renaturation in phosphate buffer pH 7.0 decreases (○) and aggregation increases (△) in a complementary way [37]. As indicated by the dotted line, to a first approximation, the kinetic competition can be modeled by Eq. (5), i.e., by sequential unimolecular folding ($k_1 = 0.01$ s^{-1}) and (diffusion-controlled) bimolecular aggregation ($k_2 = 10^5$ M^{-1} s^{-1}), with molar concentrations based on a subunit molecular mass of 35 kDa [39].

on its pathway toward the native conformation, or *shuffling enzymes* and *foldases* as catalysts were postulated or even isolated early in the game [49, 50]. However, they were ignored for the simple reason that there seemed to be no real need for helpers, because the spontaneous reaction was autonomous and fast enough to proceed without accessory components. It took more than two decades to change the paradigm from Cohn and Edsall's purely physicochemical approach [7] to considerations comparing optimal in vitro conditions with the situation in the cell [51, 52]. That catalysts or chaperones might be biologically significant was suggested by three observations (Figure 1.3). First, experiments with mouse microsomes depleted of Anfinsen's "shuffling enzyme" proved *protein disulfide isomerase* (PDI) to be essential for protein folding in the endoplasmic reticulum (ER) [55]. Second, the unexpected bimodal N ↔ I ↔ U kinetics of the two-state N ↔ U equilibrium transition of ribonuclease (RNaseA) and its physicochemical characteristics suggested proline isomerization to be involved in protein folding [56, 57]; in fact, the enzyme *peptidyl-prolyl cis-trans isomerase* (PPI) was found to catalyze rate-limiting steps on the folding pathway of proteins [58–60]. Third, increasing numbers of

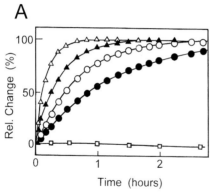 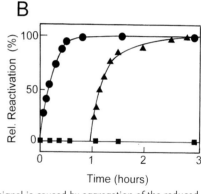

Fig. 1.3. Folding catalysts and chaperones. (A) Acceleration of the oxidative refolding of ribonuclease T1 by PPI and PDI, monitored by tryptophan fluorescence at 320 nm: 2.5 μM RNaseT1 in 0.1 M Tris-HCl pH 7.8, 0.2 M guanidinium chloride, 4 mM reduced/0.4 mM oxidized glutathione, 25 °C. Re-oxidation in the absence of PPI and PDI (●), in the presence of 1.4 μM PPI (○), in the presence of 1.6 μM PDI (▲), in the presence of 1.4 μM PPI plus 1.6 μM PDI (△), and in the presence of 10 mM dithioerythritol (to block disulfide bond formation) (□); the slight decrease in the latter signal is caused by aggregation of the reduced and unfolded protein. Curves calculated for single first-order reactions with time constants $\tau = 4300$ s (●), 2270 s (○), 1500 s (▲), and 650 s (△) [53]. (B) Reactivation of citrate synthase (CS) after preceding denaturation in 6 M guanidinium chloride in the presence of GroEL without GroES and ATP (■), with GroES and ATP (●), and with GroES and ATP added after 55 min (▲). Parallel light-scattering measurements show that in the absence of the components of the GroE system, CS forms inactive aggregates [54].

stress-response proteins (Hsps) were found to be ubiquitous and by no means restricted to heat stress; evidently, they played a role as *molecular chaperones*, i.e., they mediated or promoted correct protein folding by transiently interacting with the nascent polypeptide or with the protein at an early stage of the folding process, without becoming components of its final functional state [61] (see the respective chapters 4, 5, 14–16, 20–24, 31, 32). Evidently, this definition also holds for helper or scaffold proteins that are known to aid the genetically determined *morphopoiesis* of complex assembly systems such as phage [34].

The given examples stand for the end of an era that might be called the heroic age of protein chemistry, with names such as M. Calvin, E. J. Cohn, J. T. Edsall, G. Embden, H. A. Krebs, K. U. Linderstrøm-Lang, F. Lipmann, F. Lynen, O. Meyerhoff, A. Szent-Györgyi, H. Theorell, and O. Warburg, to mention just a few. They developed the methodology to purify proteins by fractionation and to systematically search for their respective biological activities. In proceeding from the inventory of the relevant molecules and the elucidation of their structures, they identified the molecular partners, measuring the rate and equilibrium constants for each reaction. As the next steps, the molecules were localized in the various compartments of live cells, physiological tests for the participation in specific cellular processes were developed, and mathematical models for understanding the system's behavior were formulated. Crowning the reductionist approach, the data were finally put together, ending up with the sequences and cycles of reactions that constitute the awesome networks of metabolic pathways and information transfer, including their regulation and energetics.

Faced with these Herculean efforts on the one hand, and the pressure in present-day paper production on the other, we reach the latest paradigm change: from the reductionist detail to the simplistic concept that by replacing the atomic complexity of molecules with the information pool encoded in genomes, proteomes and their corollaries at the metabolic, regulatory, and developmental levels might help to solve the riddle of life just by simplifying proteins and their interactions to spheres, squares, triangles, and arrows (as symbols for interactions).[1]

In the minds of some of the above "heroes," this change started long before the -*omix* age: in 1948, at the end of his lectures on *The Nature of Life*, the alternative to mere speculation or getting your hands dirty with experiments led Albert Szent-Györgyi to ironically promise his audience that next time he would pull a synthetic rabbit out of his pocket [64].

In the present context of protein folding vs. protein misfolding, it seems appropriate to start with a critical discussion of the physicochemical principles of self-organization, with special emphasis on the kinetic competition of folding and aggregation that led nature to evolve folding catalysts and chaperones. Combining the structural hierarchy of proteins with the processes involved in folding and association (cf. Scheme 1.1), it has been shown that commonly small (single-domain) proteins can fold without helpers, in many cases undergoing fully reversible equilibrium transitions without significant amounts of intermediates. In contrast, large (multi-domain) proteins and subunit assemblies, owing to the above-mentioned kinetic competition of intra- and intermolecular interactions in the processes of folding and association, require careful optimization to reach high yields; at high protein levels and in the absence of accessory proteins, off-pathway reactions take over (cf. Figure 1.2B) [46–48, 61].

To the uninitiated, the above hierarchical scheme might suggest that the self-organization of proteins necessarily occurs along a compulsory pathway with well-separated consecutive steps, in contrast to the alternative random-search hypothesis, which favors the idea that a jigsaw puzzle might be a more appropriate way to model protein folding [65]. When inspecting available kinetic data, it becomes clear that there is a wide variance of mechanisms that suggest "energy landscapes" rather than simple two-dimensional energy profiles as adequate pictographs to illustrate folding paths [13, 66–68] (Figure 1.4). Relevant experimental facts include

---

1 For a critical discussion of the dominating role of the "pictorial molecular paradigm" in the biosciences and its philosophical implications, including the question as to whether chemical formulae and graphical representations of biopolymers such as proteins are objective representations of reality, see Refs. [62, 63].

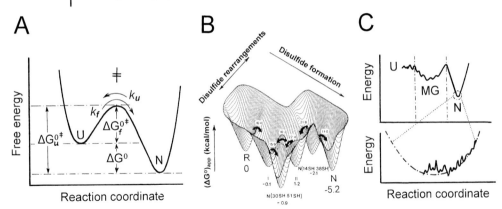

Fig. 1.4. Free-energy profiles and energy landscapes for protein folding and unfolding. (A) Relationships between the free energies of activation for folding ($\Delta G^{0\ddagger}_f$), and unfolding ($\Delta G^{0\ddagger}_u$) and of the equilibrium free energy (ΔG^0), using conventional transition-state theory. (B) "Energy-landscape" presentation describing the oxidative in vitro folding pathway of basic pancreatic trypsin inhibitor (BPTI) according to Creighton [68, 69]. The apparent free energies (in kcal mol^{-1}) are given for the fully reduced (R), intermediate (I and II), and folded (N) forms of the protein and for the transition states of the $SH \rightarrow SS$ reshuffling reaction. The reaction coordinate is represented by the three-dimensional downhill sequence of arrows. The species *I* and *II* include the one- and two-disulfide intermediates; $N_{(14SH,38SH)}$ and $N_{(30SH-51SH)}$ are native-like intermediates [34]. (C) Hypothetical conformational energy plot illustrating local energy minima and barriers along a possible protein-folding pathway. The lower "rugged potential" depicts the vicinity of the native state [13].

the following. (1) Folding as a cooperative process rarely allows true intermediates to be accumulated to a level suitable for a detailed characterization; this finding does not exclude ordered pathways but just requires a proper choice of conditions to optimize the population of intermediates. (2) For methodological reasons, the analysis of next-neighbor interactions in the folding process is prone to "in vitro artifacts," because anomalous intermediates may be populated (see the classical studies devoted to the folding pathway of basic pancreatic trypsin inhibitor [69] and to the early events in the flash-induced folding of cytochrome *c* [70, 71]). (3) For lysozyme, it has been shown that folding in fact follows multiple pathways [72]. (4) Finally, in the case of bacterial luciferase, it appears that the kinetic competition of folding and association of the $\alpha\beta$-heterodimer constitute a trap on the folding path that guides the individual subunits to the energy minimum of the active dimer [73]. Taken together, these and other examples seem to indicate that individual proteins show individual folding characteristics, an observation that also holds for the stability and stabilization of proteins. Evidently, any generalization in the world of proteins has its limitations, so that firm statements regarding "the in vitro/in vivo issue" or the wealth of expressions for "different types of molten globules" or the distinction between "short-range and long-range intermolecular forces" should either be ignored or at least taken with a grain of salt.

1.3
Native, Intermediate, and Denatured States

Before entering today's jungle of protein folding, with its roots deep in the solid ground of topology, thermodynamics, molecular dynamics, and kinetics from the microsecond to the minutes, hours, or days time range [74] (cf. Vol. I) and its youngest branches in present-day mainstream biotechnology and molecular medicine, it seems appropriate to briefly remind the reader what the starting points look like and what kind of problems motivated the aborigines to settle in the woods before the age of genomics, protein design, and folding diseases. The challenging questions were as follows. Is there a correlation between the 1D information at the level of the gene and the 3D structure of the correspondding protein? What does denaturation of proteins mean? What do the native and denatured states imply with respect to the folding mechanism of proteins? What are the best denaturation-renaturation conditions with minimal side reactions, and which criteria are best suited to characterize the authentic (native) state? What are the limits of in vitro protein folding and reconstitution, with respect to size, compartmentation, and structural complexity? Here are just a few short answers that might help in setting the frame for the crucial questions concerning the "folding pathology" in vitro and in vivo.

Owing to the large size of protein molecules and the unresolved problem of how to summarize the weak attractive and repulsive interatomic forces in a unique potential function, theorists have been unsuccessful in the a priori solution of the protein-folding code that would allow one to translate amino acid sequences into their corresponding three-dimensional structures. Considering the present-day output of solved X-ray and NMR structures *per anno* and the promise of developing high-throughput techniques, "cracking the protein folding code" has evidently lost its former challenging priority. The reason for this is simply that in order to understand the physical and functional properties of a protein at the atomic level or to acquire insight into the mechanistic details of enzyme catalysis, a resolution of <2 Å is required. This level of accuracy cannot be accomplished by any predictive approach [75].

The question of whether the physical state of a protein obtained either from the crystalline state or at high protein concentration in dilute buffer is related to or indistinguishable from its functional state in vivo has been extensively discussed since Kendrew, Perutz, and Phillips came up with their first 3D structures. Hardly any scientific result has ever been challenged to this extent; the final answer has been fundamentally positive [76], with one important limitation. Owing to their marginal stability in solution, any protein undergoes steady $N \leftrightarrow U$ transitions between the native and unfolded states. Under physiological conditions, the less-structured state is unstable and the protein reassumes its native configuration. Thus, proteins in their native functional state are dynamic systems exhibiting a high degree of local flexibility; ligand binding, allosteric transitions, and catalytic mechanisms depend on the interconversion of multiple states. The detailed quantitative understanding must necessarily take into account the structures of these

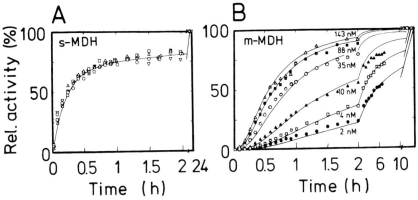

Fig. 1.8. Alternative mechanisms of the sequential folding and association of a dimeric protein: a case study on cytoplasmic (s-MDH) and mitochondrial malate dehydrogenase (m-MDH). (A) Reconstitution kinetics of s-MDH in 0.2 M phosphate buffer pH 7.6 after denaturation in 6 M guanidinium chloride in the concentration range between 1 and 13 μg mL^{-1}. The process is governed by rate-determining folding, yielding structured monomers that (in a subsequent diffusion-controlled step) form the active dimer [34]. (B) Reconstitution kinetics of m-MDH after denaturation in 1 M glycine/H$_3$PO$_4$, pH 2.3; renaturation as in (A), at 0.07 (●), 0.14 (□), 0.35 (▲), 1.2 (○), 3.1 (■), and 5.0 (△) μg mL^{-1}. Curves calculated according to Eq. (3), with $k_1 = 6.5 \times 10^{-4}$ s^{-1} and $k_2 = 3 \times 10^4$ M^{-1} s^{-1} (20 °C) [97]. Varying the mode of denaturation (low pH, high guanidinium chloride, urea) or adding the coenzyme (NAD$^+$) has no effect on the renaturation mechanism [100].

folding and misassembly (cf. "aggregation" in Figure 1.2), association generally contributes as a rate-determining step only below the concentration limit at which the production of structured monomers becomes rate-limiting. If a slow folding reaction at the level of the monomer determines the overall reaction, and association is diffusion-controlled, no concentration dependence can be detected in spite of the bimolecular step; the same holds if an assembly undergoes slow intramolecular rearrangements to finally reach the native state after slow reshuffling [51].

Coming back to the "kinetic partitioning," three questions are of interest: (1) what is the committed step in aggregate formation, (2) when is the "structured monomer" committed to end up as the native protein, and (3) what is known about the structure of aggregates and their constituent polypeptide chains? Regarding the first two points, commitment to aggregation was shown to be a fast reaction, whereas the kinetics of the commitment to renaturation follows precisely the slow kinetics of overall reactivation. This means that early collapsed conformers are much more sensitive to aggregation than later species; after a certain intermediate state has been formed, slow shuffling leads "one way" to the native state. Only the native state is fully protected from misassembly [47, 102]. Concerning the structure of aggregates, electron microscopy and circular dichroism showed that wrong subunit interactions give rise to irregular networks with a broad distribution of highly structured particles at least 10 times the size of the native proteins. At high concentrations, (e.g., at the cellular levels of recombinant proteins giving rise to inclusion bodies), they may form gel-like phases, which in the early days of protein downstream processing were appropriately named "refractile bodies." They resemble the native protein in certain spectral properties, as far as turbidity allows such conclusions [37, 103].

1.5
Limits of Reconstitution

The assumption that "unboiling the egg," after preceding deactivation, randomization, and coagulation (or whatever Max Perutz might have associated with "boiling"), leads to the unrestrained recovery of proteins to their "initial native state" has been accepted as a working hypothesis since Anson and Mirsky's days [1]. In 1944, when the first review on protein denaturation was published, the authors came to the conclusion that "... certain proteins are capable of reverting from the denatured insoluble state to a soluble form which resembles the parent native protein in one or more properties [while] other proteins appear to be incapable of this conversion" [104]. Thirty years later, when the author used this argument, Harold Scheraga's reply was "you didn't try long enough."[3] In the meanwhile, the situation had shifted from the playground of two lonely researchers at the Princeton Institute for Animal and Plant Pathology to a topologically and functionally important hot issue, called the protein-folding problem [105]. To give an example, the question at which state of association on the folding-association pathway an oligomeric enzyme gains enzymatic or regulatory properties required optimal reversibility of both dissociation-reassociation and deactivation-reactivation. Optimization was pure alchemy, blended with some insight into standard denaturation mechanisms and the physicochemical basis of protein stability. There were at least five variables, and the initial and final products had to be carefully compared using all available physical and biochemical characteristics [106]. Proving the native state the proper way, i.e., by X-ray analysis, was the only thing we were unable to do; later we found out that the commercial "crystallized enzyme" we had started out with had been purified using a routine denaturation-renaturation cycle.

As mentioned earlier (cf. section 1.3), different denaturants yield different denatured states. Commonly, maximum yields of renaturation are obtained after "complete randomization" (e.g., at high guanidinium chloride concentration), under essentially irreversible conditions. Exceedingly low protein levels are necessary to minimize aggregation. To escape this highly uneconomical requirement, i.e., to increase the steady-state concentration of the refolding protein at low levels of aggregation-competent intermediates, recycling, pulse dilution, and immobilization techniques were devised [84]. False intermediates, trapped under strongly native

---

[3] Twenty years later, his suggestion might have been "try molecular chaperones or any other accessory protein," because there *are* proteins that have resisted renaturation to this day.

conditions, may be destabilized at optimally chosen residual denaturant conditions or in the presence of 0.5–1.0 M arginine; the rationale behind these unexpectedly efficient recipes is that under strongly native conditions, weakly destabilizing agents help the kinetically trapped aggregation-competent intermediates and aggregates to overcome the activation barrier between the aggregated state and the global energy minimum of the renatured protein [47, 84].

There is no doubt that in general proteins do undergo reversible unfolding and that under optimal conditions the product of renaturation is indistinguishable from the native starting material [34]. As has been shown, side reactions may compete with full recovery. They can be minimized by carefully optimizing the denaturation-renaturation conditions: oxidation by properly adjusting the pH and the redox state of the solvent, proteolysis by adding inhibitors, misassembly by reducing the protein level below the critical concentration of aggregate formation, etc. In asking what the limits of in vitro protein folding and reconstitution are regarding size, structural complexity, and compartmentation in the cell, evolution may teach us how the observed high efficiency of protein self-organization is correlated with the basic architecture of proteins and their genes. In this context, the following questions have been addressed. (1) How is the domain structure of large polypeptide chains correlated with the codon usage and the organization of their corresponding genes? (2) Is there a correlation between the occurrence of domains and the state of association of polypeptides that would clearly support the domain-swapping hypothesis? (3) Does intrinsic form determination of the subunits govern the self-assembly of multimeric proteins? (4) At which level of complexity are assembly structures such as ribosomes or phage and virus particles determined by extrinsic morphopoietic factors or vectorial transcriptional programs? (5) Subunit exchange in oligomeric proteins requires highly homologous subunit interfaces; here the question arises, how can the reciprocity be analyzed to understand the occurrence of homo- and hetero-oligomeric quaternary structures? (6) Finally, is the specific compartmentation of isoenzymes within the cell correlated with specific folding-association mechanisms? The results have been summarized in a number of reviews [33, 34, 46, 51, 90, 106].

In the present context, a few comments may suffice. Domains have been shown to fold as independent entities that, in certain cases, may give rise to large assemblies [34, 47]. Intrinsic form determination holds not only for high-molecular-weight assemblies of a single macromolecular species (e.g., apoferritin, TMV protein, and coats of other viruses) but also for heteromultimeric systems (such as multi-enzyme complexes) [34, 107, 108]. In the case of highly complex multicomponent systems, nature either provides *scaffold proteins* or "aids assembly" by the sequential co-transcriptional supply of protein components. Mimicking this *genetically determined morphopoiesis*, successful in vitro reconstitution was accomplished by sequentially adding "primary," "secondary," "tertiary," etc., components on the one hand and by optimizing the solvent conditions on the other [34, 108, 109]. That nature makes use of the limited specificity of subunit contacts became clear when the five isoenzymes of lactate dehydrogenase were discovered in the 1960s. The general question of how specific the recognition of complementary protein surfaces is in oligomeric proteins is highly significant because, in the cell, nascent polypeptide chains fold in the presence of high levels of other proteins, including chains coded by the same gene but not synchronized in their translation. Thus, the problem arises whether and to what extent chimeric assemblies of structurally related proteins might occur, either as intermediates or final products. In using two dimeric NAD-dependent dehydrogenases (lactate dehydrogenase from *Limulus* and malate dehydrogenase from pig mitochondria), the result of careful denaturation-renaturation experiments was negative: both the time course and concentration dependence, as well as the yield of reactivation in the absence and in the presence of the prospective hybridization partner, remained unaltered. Similarly, equilibrium experiments did not indicate any exchange of subunits [110]. Similar experiments, using mitochondrial and cytoplasmic malate dehydrogenase, showed that the two isoenzymes could be easily synchronized with respect to their folding and reactivation kinetics; however, their dimerization mechanism differs widely, so that hybrid formation again does not occur [100, 111, R. Jaenicke, unpublished results]. The conclusion that no hybrids are formed is by no means trivial from the perspective of the present-day results of protein engineering, with circular permutants of domains, fragmentation of polypeptides into their (hypothetical) functional modules, and similar acrobatics in our daily routine.

1.6
In Vitro Denaturation-Renaturation vs. Folding in Vivo

In the last chapter I tried to summarize certain limits where the physicochemist's optimism that biological morphogenesis is just a variant of abiotic self-organization might be wrong. At this point it is amusing to meet respectable biologists who, faced with the "in vitro/in vivo issue," happily welcome Aristotle again, assuming that in the crowded cell the vitalistic doctrine is better than Descartes' radicalism. Having in mind that in vivo folding is part of the cellular process of protein biosynthesis and subsequent compartmentation, undoubtedly, there are significant differences between the co- and post-translational structure formation on the one hand, and optimally controlled reconstitution processes involving mature proteins in vitro on the other. There are two questions: are these differences relevant, and can we combine available fragmentary evidence from in vitro translation data with in vivo data on other cell-biological events related to self-organization, growth, and differentiation? In his excellent review on protein folding in the cell, Freedman [112] related the paradigm change from Anfinsen's dogma "one sequence → one 3D structure" to the "Anfinsen-cage" concept in chaperon research to two experimental observations: (1) the formation of misfolded recombinant proteins in microbial host cells and (2) the fate of translation products and their ultimate functional intra- or extracellular location. The wish to understand protein folding in the context of the wide variety of co- and posttranslational processes clearly emphasized that in vitro refolding studies on isolated unfolded proteins provide an incomplete model for describing protein folding in the cell. In the context of the above observations, two examples may be taken to prove that the model studies do provide essential background against which the cellular events can be understood. One deals with the attempt to gain direct insight into the detailed kinetic mechanism of the folding and assembly of the tailspike endorhamnosidase from *Salmonella* bacteriophage P22 in vitro and in the cell. Making use of temperature-sensitive mutants, spectral and pulse-labeling experiments showed that in both sets of experiments the rate-limiting folding intermediates are formed at identical rates [47] (Figure 1.9). In the other example, it was shown that misfolding and subsequent inclusion-body formation at high levels of recombinant

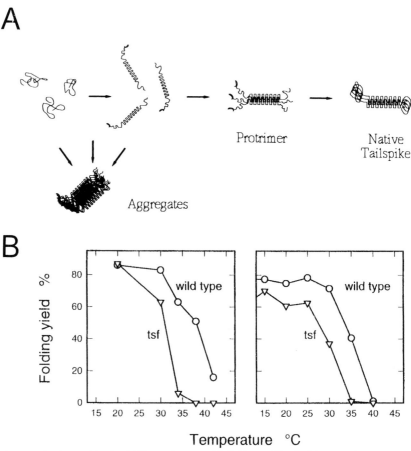

Fig. 1.9. In vitro and in vivo folding and assembly of the trimeric tailspike endorhamnosidase (Tsp) of *Salmonella* bacteriophage P22, making use of the temperature-sensitive folding and aggregation of folding intermediates. (A) Schematic representation of the folding/assembly/aggregation pathways. (B) Folding yields of wild-type and mutant Tsp upon biosynthesis in vivo on the one hand (left), and dilution from denaturant in vitro (right) on the other. The similarity of the data shows that the in vitro reconstitution applies to the situation in the living cell [113, 114].

proteins can be quantitatively described by kinetic partitioning between one rate-determining folding step (k_1) and parallel diffusion-controlled aggregation (k_2), in accordance with Eq. (4) in the simplified form

$$N \xleftarrow{k_1} U/M_u \xrightarrow{k_2} A \qquad (5)$$

(cf. Figure 1.2B) [39].

The observation that overexpression often leads to the deposition of recombinant proteins in the host cell may favor the idea that under normal physiological conditions in vivo, structure formation yields exclusively native protein. This assumption is incorrect because, in the eukaryotic cell, trafficking through the ER and the Golgi and subsequent quality control may reduce the yield considerably [115]. Thus, in mechanistic terms, refolding does not differ significantly from in vivo folding. The same holds for the directionality of translation in the cell, which, on first sight, seems to be fundamentally different from the integral folding of the complete polypeptide chain in renaturation experiments. Here, due to the involvement of chaperone proteins in the initial folding events, structure formation in vivo does not necessarily occur co-translationally in a vectorial fashion. Furthermore, circular permuted proteins and protein fragments have been shown to fold into stable, functional conformers, indicating that multiple folding pathways exist (altogether leading to the same 3D structure) and that the original N-terminus is not required for the folding process. In this context, inverting the direction of either protein biosynthesis (by Merrifield synthesis) or protein folding (by immobilization) has been shown to end up with the native functional protein: with Merrifield synthesis, one may argue that the synthesis from the C- to the N-terminal end includes coupling-uncoupling and protection-deprotection reactions under nonnative conditions, but the final product is still in its native state [116]. In order to promote the direction of folding in either the natural (N → C) or the inverse (C → N) direction, immobilization of standard proteins at their N- or C-termini (by histidine or arginine tags, combined with ion-exchange chromatography) was applied. In both cases, denaturation-renaturation cycles have shown that the products were the same, independent of the free end of the polypeptide chain [38].[4]

In the present context, the above *one sequence → one 3D structure* dogma is incomplete without stressing the importance of the solvent environment. In contrast to the alchemy of solvent variations that have been worked out in order to optimize the downstream processing of recombinant proteins, evolutionary adaptation has fixed the protein inventory of mesophiles and extremophiles to specific cellular solvent conditions. Making use of the expression of recombinant extremophile proteins in mesophilic hosts, limits can be defined about how far physical variables may be varied without interfering with the "foldability" of guest proteins. It is evident that early intermediates on the folding path are most accessible toward delete-

---

4 An additional spin-off of this approach was that immobilization allowed the kinetic partitioning to be shifted from A to N (cf. Eq. (5)): the tagged proteins showed not only significantly increased yields of reactivation but also enhanced intrinsic stabilities.

rious solvent effects. Thus, if the native structure can be reached either after recombinant expression or after refolding in vitro, intermediates and subunits must be stable under the respective conditions. Considering temperature, it turns out that alterations in the folding conditions often have surprisingly little effect. A drastic example is the expression of active enzymes from hyperthermophilic bacteria in *Escherichia coli*, in which the temperature difference ΔT_{opt} between the host and the guest may amount to >60 °C. Here, it has been shown that denaturation-renaturation experiments at up to 100 °C yield the fully active protein, indistinguishable from its initial native state, proving that all folding-association intermediates must be stable over the whole temperature range [16]. When comparing homologous proteins from mesophiles and thermophiles, it is interesting to note that thermophilic proteins are often kinetically stabilized, showing drastically decreased unfolding rates but unchanged folding kinetics [18], no doubt highly advantageous considering short-term exposure to lethal temperatures in hydrothermal vents on the one hand, and kinetic competition of protein folding and proteolytic degradation on the other.

Little is known about the effects of the other extreme physical parameters on protein folding. Because the cytosolic pH is close to neutrality, even in the case of extreme acidophiles and alkalophiles, standard in vitro conditions are close to those in the cell. Halophilic proteins seem to require salt, not only as stabilizing agent but also for folding and assembly [117]. For non-halophilic proteins, the ionic strength is not as critical, apart from general Hofmeister effects that may be used to optimize the in vitro folding conditions. Regarding possible effects of high hydrostatic pressure on cellular processes, it is obvious that the adaptive effort to cope with the biologically relevant extremes of pressure (<120 MPa) can be ignored in comparison to the adaptation to low temperature prevailing in the deep sea [118]. Finally, viscosity should be considered because of macromolecular crowding in the cytosol. As has been mentioned, glycerol, sucrose, and other carbohydrates have been used to simulate frictional effects on domain pairing. In connection with protein folding, information is scarce, especially because polyols not only increase the viscosity of the solvent but also enhance the stability of the solute owing to excluded-volume effects and preferential solvation; there is no way to alter viscosity alone.

1.7
Perspectives

As we have witnessed, interest in the protein-folding problem started from the sensational resurrection of denatured proteins in Tim Anson's hands, long before the first 3D structure of a globular protein was elucidated. After self-organization had become an accepted fact, experimental studies were centered mainly on the investigation of the mechanisms of folding and association of model proteins in terms of the complete description of the unfolded (nascent) and final (native) states, including all accessible intermediates along the U → N transition. The main obstacle in reaching this goal was the elusive nature of the folding protein, especially the question of whether folding is in fact a sequential reaction with well-populated intermediates or a multiple-pathway process, as in the case of a jigsaw puzzle [119]. What was important at this point was the observation that model peptides as well as separate protein fragments and domains were able to form ordered structures, supporting the idea that local structures and modules may serve as "seeds" in the folding process. That such nuclei did not necessarily adopt the same conformation in unrelated protein structures was no argument against their significance as intermediates. Specific ligands such as metal ions were found to be essential as guides along the folding path; interestingly, in cases like these, estimated cellular levels were found to be close to the optimum concentration in in vitro refolding experiments [34].

Regarding folding in the cell, in the early days it was tacitly assumed that the in vitro mechanism could be directly transferred to describe the folding and association of the nascent polypeptide as the final step of translation. Obvious discrepancies were clear from the beginning; for example, "unscrambling" of ribonuclease took unreasonably long, in vitro reactivation at standard cellular protein levels ended up with aggregates rather than active material, certain proteins were found to be inaccessible to "renaturation," etc. The discovery of a whole spectrum of folding catalysts and molecular chaperones resolved most of these challenges because they were clearly correlated with the three established rate-limiting steps in the overall folding reaction: PDI with disulfide shuffling, PPI with proline isomerization, and molecular chaperones with the kinetic partitioning between folding and aggregation. Based on mutant studies and on the fact that under normal physiological conditions several percent of the cellular proteins are molecular chaperones, there could hardly be any doubt that all three classes of proteins were physiologically essential. As a matter of fact, complementing the in vitro experiments with the accessory components allowed the in vitro results to mimic the reality in the cell.

No new biophysical principles emerged from all these studies; on the contrary, the better the mechanisms of folding catalysts and molecular chaperones were understood, the more it became clear that the basic principles of catalysis, allosteric regulation, and compartmentation hold also in the cellular environment. In spite of that, attempts at quantifying protein folding in vivo in physicochemical terms failed because the cytosol represents a *complex macromolecular multicomponent system under crowding conditions*, the full description of which still goes beyond the presently available repertoire of experimental methods. As in other "real systems" such as the real gas or concentrated simple electrolytes, the quantitative treatment of the non-ideality of highly charged polyelectrolytes in aqueous multicomponent systems is still beyond the reach of a sound theoretical treatment.

Twenty-five years ago, at the end of the first International Conference on Protein Folding in Regensburg, Cyrus Chothia quoted Johannes Kepler's treatise *On the Six-cornered Snowflake* as the first "theory of biological structures." In his essay Kepler argued that a material's structure arises from its intrinsic properties and that in biological structures it is important to consider the structure-function relationship [120]. In this context, focusing on protein folding, at the 1979 conference attempts were made to combine the intrinsic physical and chemical properties of proteins and the laws of kinetics and thermodynamics in order to predict how they determine the structure and activity. At that time, "simple protein molecules" such as basic pancreatic trypsin inhibitor, ribonuclease, and lactate dehydrogenase were discussed as adequate models, the argument being that the central problem at all levels of biological structure is to understand how the intrinsic entropy of its various constituents is overcome to form particular stable structures in a finite time and how the physical and chemical properties of the materials determine the way in which they function. Cyrus, the theoretician, ended by vaguely announcing that "the next time we meet in Regensburg we shall talk about more biological systems such as the red cell membrane or coated vesicles" [121]. I do not have to explain why this second conference never materialized. Now, after a quarter of a century, two former Regensburg graduate students are keeping Cyrus' promise, figuring as editors of the present volume. In fact, this volume covers the state of the art, from the principles, dynamics, and mechanisms of protein stability, protein design, and structure prediction to the function and regulation of chaperones and research and development in the fields of engineering protein folding and stability and refolding technology, an amazing horizon having in mind the first steps in protein folding in the dark age of biocolloidology, when the chemical establishment still considered proteins to be loose aggregates of smaller entities, unflinchingly opposing the term *macromolecule*.

Acknowledgements

This perspectives article is dedicated to my teachers, Camille Montfort, Hans Kautsky, Joachim Stauff, and Max A. Lauffer, and to my students, Rainer Rudolph and Franz Schmid, who taught me more than I was able to teach them.

References

1 M. L. ANSON, A. E. MIRSKY, On some general properties of proteins. *J. gen. Physiol.* **1925**, *9*, 169–179.
2 M. L. ANSON, A. E. MIRSKY, The equilibria between native and denatured hemoglobin in salicylate solutions and the theoretical consequences of the equilibrium between native and denatured protein. *J. gen Physiol.* **1934**, *7*, 399–408.
3 M. L. ANSON, Protein denaturation and the properties of protein groups. *Adv. Protein Chem.* **1945**, *2*, 361–386.
4 M. F. PERUTZ, Unboiling an egg. *Discovery* **1940**, March, reprinted in [5] p. 14.
5 M. F. PERUTZ, Chairman's Introduction. In: Protein Folding (R. JAENICKE, Ed.) Elsevier/North-Holland Biomed. Press, Amsterdam, New York **1980**, 13.
6 E. J. COHN, The physical chemistry of the proteins. *Physiol. Rev.* **1925**, *5*, 349–437.
7 C. TANFORD, Cohn and Edsall: Physical chemistry conclusively supports a protein model. *Biophys. Chem.* **2003**, *100*, 81–90.

8 T. Svedberg, R. Fåhraeus, A new method for the determination of the molecular weight of the proteins. *J. Amer. Chem. Soc.* **1926**, *48*, 430–438.

9 M. F. Perutz, M. G. Rossmann, A. F. Cullis, H. Muirhead, G. Will, A. C. T. North, Structure of haemoglobin: A three-dimensional Fourier synthesis at 5.5 Å resolution obtained by X-ray analysis. *Nature* **1960**, *185*, 416–422.

10 C. B. Anfinsen, The limited digestion of ribonuclease with pepsin. *J. Biol. Chem.* **1956**, *221*, 405–412.

11 R. L. Baldwin, D. Eisenberg, Protein stability. In Protein Engineering (D. L. Oxender, C. F. Fox, Eds.) A. R. Liss, New York **1987**, 127–148.

12 K. A. Dill, Dominant forces in protein folding. *Biochemistry* **1990**, *29*, 7133–7155.

13 M. Karplus, E. Shakhnovich, Protein folding: Theoretical studies of thermodynamics and dynamics. In: Protein Folding (T. E. Creighton, Ed.) W. H. Freeman, New York, **1992**, 127–195.

14 A. Fersht, Forces between molecules, and binding energies. In: Structure and Mechanism in Protein Science, W. H. Freeman, New York, **1999**, 324–348.

15 T. Lazaridis, M. Karplus, Thermodynamics of protein folding. *Biophys. Chem.* **2003**, *100*, 367–395.

16 R. Jaenicke, R. Sterner, Life at high temperature. In: The Prokaryotes, 3rd Ed. (M. Dworkin et al., Eds.) An Evolving Electronic Resource, Latest Update Release 3.9, Springer-New York **2002**, 1–56.

17 R. Jaenicke, G. Böhm, Stabilization of proteins: What extremophiles teach us about protein stability. *Curr. Opin. Struct. Biol.* **1998**, *8*, 738–748.

18 D. Perl, C. Welker, T. Schindler, M. A. Marahiel, R. Jaenicke, F. X. Schmid, Rapid two-state folding is conserved in cold-shock proteins from mesophiles, thermophiles and hyperthermophiles. *Nature Struct. Biol.* **1998**, *5*, 229–235.

19 D. Perl, U. Müller, U. Heinemann, F. X. Schmid, Two exposed amino acid residues confer thermostability on a cold shock protein. *Nature Struct. Biol.* **2000**, *7*, 380–383.

20 F. H. Arnold (Ed.) Evolutionary Protein Design, *Adv. Protein Chem.* **2001**, Vol. 55.

21 P. L. Wintrode, F. H. Arnold, Temperature adaptation of enzymes: Lessons from laboratory evolution. *Adv. Protein Chem.* **2001**, *55*, 161–226.

22 W. P. C. Stemmer, Rapid evolution of a protein in vitro by DNA shuffling. *Nature* **1994**, *370*, 389–391.

23 M. Lehmann, C. Loch, A. Middendorf, D. Studer, S. F. Lassen, L. Pasamontes, A. P. van Loon, M. Wyss, The consensus concept for thermostability of proteins: Further proof of concept. *Protein Eng.* **2002**, 403–411.

24 N. Declerck, M. Machius, P. Joyet, G. Wiegand, R. Huber, C. Gaillardin, Hyperthermostabilization of *Bacillus licheniformis* α-amylase and modulation of its stability over a 50 degrees C temperature range. *Protein Eng.* **2003**, *16*, 287–293.

25 W. Kauzmann, Some factors in the interpretation of protein denaturation. *Adv. Protein Chem.* **1959**, *14*, 1–63.

26 J. L. Finney, Solvent effects in biomolecular processes. In: Biophysics of Water (F. Franks, S. Mathias, Eds.), J. Wiley & Sons Chichester **1982**, 55–58.

27 N. C. Pace, B. A. Shirley, M. McNutt, K. Gajiwala, Forces contributing to the conformational stability of proteins. *FASEB J.* **1996**, *10*, 75–83.

28 G. I. Makhatadze, P. L. Privalov, Hydration effects in protein unfolding. *Biophys. Chem.* **1994**, *51*, 291–309.

29 G. I. Makhatadze, P. L. Privalov, Energetics of protein structure. *Adv. Protein Chem.* **1995**, *47*, 307–425.

30 R. L. Baldwin, John Schellman and his scientific work. *Biophys. Chem.* **2002**, *101–102*, 9–13.

31 M. F. Perutz, H. Raidt, Stereochemical basis of heat stability in bacterial ferredoxins and in hemoglobin A2. *Nature* **1975**, *255*, 256–259.

32 C. N. Pace, Polar group burial contributes more to protein stability

than non-polar group burial. *Biochemistry* **2001**, *40*, 310–313.

33 R. JAENICKE, C. SLINGSBY, Eye lens crystallins and their bacterial homologs. *Crit. Rev. Biochem. Mol. Biol.* **2001**, *36*, 435–499.

34 R. JAENICKE, Folding and association of proteins. *Progr. Biophys. Mol. Biol.* **1987**, *49*, 117–237.

35 R. JAENICKE, Reassociation and reactivation of LDH from the unfolded subunits. *Eur. J. Biochem.* **1974**, *46*, 149–155.

36 G. ORSINI, M. E. GOLDBERG, The renaturation of reduced chymotrypsinogen A in GdmCl: Refolding vs. aggregation. *J. Biol. Chem.* **1978**, *253*, 3453–3458.

37 G. ZETTLMEISSL, R. RUDOLPH, R. JAENICKE, Reconstitution of LDH: Non-covalent aggregation vs. reactivation. I. Physical properties and kinetics of aggregation. *Biochemistry* **1979**, *18*, 5567–5571.

38a G. STEMPFER, B. HÖLL-NEUGEBAUER, E. KOPETZKI, R. RUDOLPH, Improved refolding of an immobilized fusion protein. *Nature Biotechnol.* **1996**, *14*, 329–334.

38b G. STEMPFER, B. HÖLL-NEUGEBAUER, R. RUDOLPH, A fusion protein designed for non-covalent immobilization: Stability, enzymatic activity and use in an enzyme reactor. *Nature Biotechnol.* **1996**, *14*, 481–484.

39 T. KIEFHABER, R. RUDOLPH, H.-H. KOHLER, J. BUCHNER, Protein aggregation in vitro and in vivo: A quantitative model of the kinetic competition between folding and aggregation. *Bio/Technology* **1991**, *9*, 825–829.

40 E. DE BERNARDEZ CLARK, E. SCHWARZ, R. RUDOLPH, Inhibition of aggregation side reactions during in vitro protein folding. *Meth. Enzymol.* **1999**, *309*, 217–236.

41 R. I. MORIMOTO, A. TISSIÈRES, C. GEORGOPOULOS, The stress response: Function of the proteins, and perspectives. *Cold Spring Harbor Monograph Ser.* **1990**, *19*, 1–36.

42 K. P. MURPHY, Non-covalent forces important to the conformational stability of protein structures. In: Protein Stability and Folding (B. A. SHIRLEY, Ed.), *Meth. Mol. Biol.* Vol. 40, Humana Press, Totowa **1995**, 1–34.

43 O. B. PTITSYN, Molten globule and protein folding. *Adv. Protein Chem.* **1995**, *49*, 83–229.

44 M. ARAI, K. KUWAJIMA, Role of the molten globule state in protein folding. *Adv. Protein Chem.* **2000**, *53*, 209–282.

45 J. JANIN, S. J. WODAK, Structural domains in proteins and the role in the dynamics of protein function. *Progr. Biophys. Mol. Biol.* **1983**, *42*, 21–78.

46 M. P. SCHLUNEGGER, M. J. BENNETT, D. EISENBERG, Oligomer formation by three-dimensional domain swapping: A model for protein assembly and misassembly. *Adv. Protein Chem.* **1997**, *50*, 61–122.

47 R. JAENICKE, R. SECKLER, Protein misassembly in vitro. *Adv. Protein Chem.* **1997**, *50*, 1–59.

48 R. JAENICKE, H. LILIE, Folding and association of oligomeric and multimeric proteins. *Adv. Protein Chem.* **2000**, *53*, 329–401.

49 R. F. GOLDBERGER, C. J. EPSTEIN, C. B. ANFINSEN, Acceleration of reactivation of reduced bovine pancreatic ribonuclease by a microsomal system from rat liver. *J. Biol. Chem.* **1963**, *238*, 628–635.

50 P. VENETIANER, F. B. STRAUB, The enzymic reactivation of reduced ribonuclease. *Biochem. Biophys. Acta* **1963**, *67*, 166–168.

51 R. JAENICKE, Protein folding and association: Significance of in vitro studies for self-organization and targeting in the cell. *Curr. Topics Cellular Regulation* **1996**, *34*, 209–314.

52 R. JAENICKE, Protein self-organization in vitro and in vivo: Partitioning between physical biochemistry and cell biology. *Biol. Chem.* **1998**, *379*, 237–243.

53 E. R. SCHÖNBRUNNER, F. X. SCHMID, Peptidyl-prolyl *cis-trans* isomerase improves the efficiency of protein disulfide isomerase as a catalyst of protein folding. *Proc. Natl. Acad. Sci. USA* **1992**, *89*, 4510–4513.

54 J. Buchner, M. Schmidt, M. Fuchs, R. Jaenicke, R. Rudolph, F. X. Schmid, T. Kiefhaber, GroE facilitates refolding of citrate synthase by suppressing aggregation. *Biochemistry* **1991**, *30*, 1586–1591.

55 R. B. Freedman, Protein disulfide isomerase: multiple roles in the modification of nascent secretory proteins. *Cell* **1989**, *57*, 1069–1072.

56 J. F. Brandts, H. R. Halvorson, M. Brennan, Consideration of the possibility that the slow step in protein denaturation reactions is due to *cis-trans* isomerism of protein residues. *Biochemistry* **1975**, *14*, 4953–4963.

57 F. X. Schmid, R. L. Baldwin, Acid catalysis of the formation of the slow-folding species of RNaseA: Evidence that the reaction is proline isomerization. *Proc. Natl. Acad. Sci. USA* **1978**, *75*, 4764–4768.

58 G. Fischer, H. Bang, The refolding of urea denatured RNaseA is catalyzed by peptidyl-prolyl *cis-trans* isomerase. *Biochim. Biophys. Acta* **1985**, *828*, 39–42.

59 K. Lang, F. X. Schmid, G. Fischer, Catalysis of protein folding by prolyl isomerase. *Nature* **1987**, *329*, 268–270.

60 T. Kiefhaber, H. P. Grunert, U. Hahn, F. X. Schmid, Replacement of a *cis* proline simplifies the mechanism of RNase T1 folding. *Biochemistry* **1990**, *29*, 6475–6480.

61 R. J. Ellis, S. M. van der Vies, Molecular chaperones. *Annu. Rev. Biochem.* **1991**, *60*, 321–347.

62 R. Rorty, Philosophy and the Mirror of Nature. Princeton Univ. Press, Princeton, **1979**.

63 P.-L. Luisi, R. M. Thomas, The pictorial molecular paradigm: Pictorial communication in the chemical and biological sciences. *Naturwissenschaften* **1990**, *77*, 67–74.

64 A. Szent-Györgyi, The Nature of Life, **1948**, Academic Press, New York.

65 S. C. Harrison, R. Durbin, Is there a single pathway for the folding of a polypeptide chain? *Proc. Natl. Acad. Sci. USA* **1985**, *82*, 4028–4030.

66 K. A. Dill, S. Bromberg, K. Yue, K. M. Fiebig, D. P. Yee, P. D. Thomas, H. S. Chan, Principles of protein folding: A perspective from simple exact models. *Protein Sci.* **1995**, *4*, 561–602.

67 L. A. Mirny, V. Abkevich, E. I. Shakhnovich, Universality and diversity of the protein folding scenario: A comprehensive analysis with the aid of a lattic model. *Folding and Design* **1996**, *1*, 103–116.

68 J. N. Onuchic, Z. Luthey-Schulten, P. G. Wolynes, Theory of protein folding: The energy landscape perspective. *Annu. Rev. Phys. Chem.* **1997**, *48*, 545–600.

69 T. E. Creighton, Protein folding coupled to disulphide-bond formation. In: Mechanisms of Protein Folding, 2nd Ed. (R. H. Pain, Ed.) Oxford University Press, Oxford **2000**, 250–278.

70 S. J. Hagen, J. Hofrichter, A. Szabo, W. A. Eaton, Diffusion-limited contact formation in unfolded cytochrome c: Estimating the maximum rate of protein folding. *Proc. Natl. Acad. Sci. USA* **1996**, *93*, 11615–11617.

71 W. A. Eaton, V. Munoz, S. J. Hagen, G. S. Jas, L. J. Lapidus, E. R. Henry, J. Hofrichter, Fast kinetics and mechanism in protein folding. *Annu. Rev. Biophys. Biomol. Str.* **2000**, *29*, 327–359.

72 S. E. Radford, C. M. Dobson, P. A. Evans, The folding of hen lysozyme involves partially structured intermediates and multiple pathways. *Nature* **1992**, *358*, 302–307.

73 J. F. Sinclair, M. M. Ziegler, T. O. Baldwin, Kinetic partitioning during protein folding yields multiple native states. *Nature Struct. Biol.* **1994**, *1*, 320–326.

74 O. Bieri, T. Kiefhaber, Elementary processes in protein folding. *Biol. Chem.* **1999**, *380*, 923–929.

75 E. E. Eaton (Ed.), 4th Meeting on the CASP, Proteins: Structure, Function and Genetics, **2001**, *45*, S5, 1–199.

76 J. A. Rupley, The comparison of protein structure in crystal and in solutions. In: Structure and Stability of Biological Macromolecules (S. N.

Timasheff, G. D. Fasman, Eds.) Marcel Dekker, New York **1965**, 291–352.

77 H. Frauenfelder, H. Hartmann, M. Karplus, I. D. Kuntz, J. Kuriyan, F. Parak, G. A., Petsko, D. Ringe, R. F. Tilton, Jr., M. L. Connolly, N. Max, Thermal expansion of a protein. *Biochemistry* **1987**, *26*, 254–261.

78 P. A. Rejto, S. T. Freer, Protein conformational substates from X-ray crystallography. *Progr. Biophys. Mol. Biol.* **1996**, *66*, 167–196.

79 Q. Xue, E. S. Yeung, Difference in the chemical reactivity of individual molecules of an enzyme. *Nature* **1995**, *373*, 681–683.

80 B. Schuler, E. A. Lipman, W. A. Eaton, Probing the free energy surface for protein folding with single-molecule fluorescence spectroscopy. *Nature* **2002**, *419*, 743–747.

81 G. D. Rose (Ed.), Unfolded Proteins. *Adv. Protein Chem.* **2002**, Vol. 62.

82 C. Tanford, Protein denaturation: Characterization of the denatured state and the transition from the native to denatured state. *Adv. Protein Chem.* **1968**, *23*, 121–282; Theoretical models for the mechanism of denaturation, *ibid.* **1970**, *24*, 1–95.

83 C. Branden, J. Tooze, Building blocks of protein structures. In: Introduction to Protein Structure, 2nd Ed. Garland Publ., New York, **1998**, 8–12.

84 R. Rudolph, G. Böhm, H. Lilie, R. Jaenicke, in: Protein function: A practical approach (T. E. Creighton, Ed.) 2nd Ed., IRL Press, Oxford, **1997**, 57–99.

85 T. Dams, R. Jaenicke, Stability and folding of DHFR from the hyperthermophilic bacterium *Thermotoga maritima*. *Biochemistry* **1999**, *38*, 9169–9178.

86 M. Kretschmar, E.-M. Mayr, R. Jaenicke, Kinetic and thermodynamic stabilization of the $\beta\gamma$-crystallin homolog spherulin 3a from *Physarum polycephalum* by Ca^{2+}. *J. Mol. Biol.* **1999**, *289*, 701–705.

87 S. D. Hoeltzli, C. Frieden, Refolding of 6^{19}F tryptophan labelled *Escherichia coli* DHFR in the presence of ligand: A stopped-flow NMR spectroscopy study. *Biochemistry* **1998**, *37*, 387–398.

88 F. X. Schmid, Kinetics of unfolding and refolding of single-domain proteins. In: Protein Folding (T. E. Creighton, Ed.) Freeman, New York, **1992**, 197–241.

89 O. Bieri, T. Kiefhaber, Kinetic models in protein folding. In: Mechanisms of Protein Folding, 2nd Ed. (R. H. Pain, Ed.) Oxford University Press, Oxford **2000**, 34–64.

90 R. Jaenicke, Folding and stability of domain proteins. *Progr. Biophys. Mol. Biol.* **1999**, *71*, 155–241.

91 R. Rudolph, R. Siebendritt, G. Nesslauer, A. K. Sharma, R. Jaenicke, Folding of an all-β-protein: Independent domain folding in γII crystallin from calf eye lens. *Proc. Natl. Acad. Sci. USA* **1990**, *87*, 4625–4629.

92 R. Jaenicke, Protein folding: Local structures, domains, subunits and assemblies. *Biochemistry* **1991**, *30*, 3147–3161.

93 M. G. Rossmann, P. Argos, Protein folding. *Annu. Rev. Biochem.* **1981**, *50*, 497–532.

94 J.-R. Garel, Folding of large proteins: Multidomain and multisubunit proteins. In: Protein Folding (T. E. Creighton, Ed.) W. H. Freeman, New York, **1992**, 405–454.

95 G. Kern, D. Kern, R. Jaenicke, R. Seckler, Kinetics of folding and association of differently glycosylated variants of invertase from *Saccharomyces cerevisiae*. *Protein Sci.* **1993**, *2*, 1862–1868.

96 L. C. Wu, P. L. Kim, A specific hydrophobic core in the α-lactalbumin molten globule. *J. Mol. Biol.* **1998**, *280*, 175–182.

97 R. Jaenicke, Intermolecular forces in the process of heat aggregation of globular proteins and the correlation between aggregation and denaturation. *J. Polymer Sci.*, **1967**, *16*, 2143–2160.

98 L. Pauling, Aggregation of proteins. *Disc. Faraday Soc.*, **1953**, *13*, 170–176.

99 J. D. BERNAL, Structural arrangements of macromolecules. *Disc. Faraday Soc.*, **1958**, *25*, 7–18.

100 R. JAENICKE, R. RUDOLPH, I. HEIDER, Quaternary structure, subunit activity and in vitro association of porcine mitochondrial malate dehydrogenase. *Biochemistry* **1979**, *18*, 1217–1223.

101 R. RUDOLPH, I. HEIDER, E. WESTHOF, R. JAENICKE, Mechanism of refolding and reactivation of LDH after dissociation in various solvent media. *Biochemistry* **1977**, *16*, 3384–3390.

102 M. E. GOLDBERG, R. RUDOLPH, R. JAENICKE, A kinetic study of the competition between renaturation and aggregation during the refolding of denatured-reduced egg-white lysozyme. *Biochemistry* **1991**, *30*, 2790–2797.

103 G. A. BOWDEN, A. M. PAREDES, G. GEORGIOU, Structure and morphology of protein inclusion bodies in *Escherichia coli. Biotechnology* **1991**, *9*, 725–730.

104 H. NEURATH, J. P. GREENSTEIN, F. W. PUTNAM, J. O. ERICKSON, The chemistry of protein denaturation. *Chem. Reviews* **1944**, *34*, 157–265.

105 R. JAENICKE (Ed.), Protein Folding, Elsevier/North-Holland Biomed. Press, Amsterdam, New York, **1980**.

106 R. JAENICKE, R. RUDOLPH, Refolding and association of oligomeric proteins. *Meth. Enzymol.* **1986**, *131*, 218–250.

107 M. GERL, R. JAENICKE, J. M. A. SMITH, P. M. HARRISON, Selfassembly of apoferritin from horse spleen after reversible chemical modification with 2.3-dimethylmaleic anhydride. *Biochemistry* **1988**, *27*, 4089–4096.

108 G. E. W. WOLSTENHOLME, M. O'CONNOR (Eds.), Principles of Biomolecular Organization, Ciba Foundation Symp. 1965, Churchill Ltd., London **1966**.

109 K. H. NIERHAUS, Assembly of the prokaryotic ribosome. In: Protein Synthesis and Ribosome Structure (K. H. NIERHAUS, D. WILSON, Eds.), Wiley-VCH, Weinheim, **2004**.

110 M. GERL, R. RUDOLPH, R. JAENICKE, Mechanism and specificity of reconstitution of dimeric lactate dehydrogenase from *Limulus polyphemus.* *Biol. Chem. Hoppe-Seyler* **1985**, *366*, 447–454.

111 R. RUDOLPH, I. FUCHS, R. JAENICKE, Reassociation of dimeric c-MDH is determined by slow and very fast folding reactions. *Biochemistry* **1986**, *25*, 1662–1667.

112 R. B. FREEDMAN, Protein folding in the cell. In: Protein Folding (T. E. CREIGHTON, Ed.) Freeman, New York, **1992**, 455–539.

113 M. DANNER, R. SECKLER, Mechanism of phage P22 tailspike protein folding mutations. *Protein Sci.* **1993**, *2*, 1869–1881.

114 M. DANNER, A. FUCHS, S. MILLER, R. SECKLER, Folding and assembly of phage p22 tailspike endorhamnosidase lacking the N-terminal, head-binding domain. *Eur. J. Biochem.* **1993**, *215*, 653–661.

115 A. HELENIUS, T. MARQUART, I. BRAAKMAN, The endoplasmic reticulum as a protein-folding compartment. *Trends Cell Biol.* **1992**, *2*, 227–231.

116 B. M. MERRIFIELD, Life During a Golden Age of Peptide Chemistry: The Concept and Development of Solid-phase Peptide Synthesis. ACS, Washington, **1993**.

117 K. HECHT, R. JAENICKE, Malate dehydrogenase from the extreme halophilic archaebacterium *Halobacterium marismortui*: Reconstitution of the enzyme after denaturation and dissociation in various denaturants. *Biochemistry* **1989**, *28*, 4979–4985.

118 M. GROSS, R. JAENICKE, Proteins under pressure: The influence of high hydrostatic pressure on structure, function and assembly of proteins and protein complexes. *Eur. J. Biochem.* **1994**, *221*, 617–630.

119 R. L. BALDWIN, The nature of protein folding pathways. *J. Biomol. NMR* **1995**, *5*, 103–109.

120 J. KEPLER, STRENA, Seu De Nive Sexangula, G. Tampach, Francofurti a.M., **1611**.

121 C. CHOTHIA, Closing remarks. In: R. JAENICKE (Ed.) Protein Folding Elsevier/North-Holland Biomed. Pess, Amsterdam, New York, **1980**, 583–585.

2
Folding and Association of Multi-domain and Oligomeric Proteins

Hauke Lilie and Robert Seckler

2.1
Introduction

Folding and association of nascent or refolding polypeptide chains are spontaneous and autonomous processes. All information required for the formation of the native three-dimensional structure of a given protein or protein complex is encoded in the amino acid sequence. The structure of a protein is the result of an optimum partitioning of the nonpolar and polar parts of the polypeptide chain between regions of high and low dielectric constant in the solvent environment [1]. In aqueous solution this leads to a minimization of hydrophobic surface area. This basic principle governs both structure formation of monomeric proteins and association of subunits of oligomeric proteins.

In many cases the biological activity of proteins depends on their association state. The association can result in homo- or hetero-oligomers that may be either stable or only transiently formed during their specific functional cycle. The Protein Data Bank (pdb) of protein structures contains more than 2000 oligomeric structures, the largest of which comprise virus shells, and homomeric or heteromeric protein complexes containing more than 20 subunits, such as ferritin (24 identical subunits, [2]), the chaperonin complex GroEL/ES (21 subunits $\alpha14\beta7$, [3]), and dihydrolipoyl acetyltransferase (60 subunits, [4]).

For quite a number of multi-domain and even oligomeric proteins, folding and assembly pathways as well as their thermodynamic stabilities and dissociation constants have been determined. This chapter deals with the general principles of structure formation and stabilization of multi-domain and oligomeric proteins and the experimental approaches to analyze these reactions. Some closely related topics, which are not part of this chapter, are summarized in other chapters of this book. These comprise virus assembly (Chapter 5), collagen folding in vitro (Chapter 21 in Part I) and in vivo (Chapter 26), formation of protein fibrils (Chapters 31–36), and the catalysis of rate-limiting steps of folding such as peptide bond isomerization (Chapter 27) or disulfide bond formation (Chapter 29).

Protein Folding Handbook. Part II. Edited by J. Buchner and T. Kiefhaber
Copyright © 2005 WILEY-VCH Verlag GmbH & Co. KGaA, Weinheim
ISBN: 3-527-30784-2

2.2 Folding of Multi-domain Proteins

2.2.1 Domain Architecture

Domains are compact substructures within protein molecules. There are different definitions of "domain" based on analysis of the three-dimensional structure of the respective protein, the autonomous folding units of a protein, and functions such as cofactor binding or resistance to proteolysis. Interestingly, in a number of cases the boundaries of protein domains correlate with intron-exon boundaries on the genetic level [5, 6]. This suggests that exon shuffling (and thus domain shuffling) is an important mechanism in generating functional diversity during evolution of multi-domain proteins [1].

A hierarchical classification of domain structures (CATH) has been proposed, using as parameters the *secondary structure composition* (C for class), the *gross arrangement of secondary structure* (A for architecture), the *sequential connectivities* (T for topology), and the *structural and functional similarity* (H for homology) [7–9]. By this procedure 27 000 known domain structures can be grouped into 1800 sequence families, 700 fold groups, 245 superfamilies, and about 30 different architectures (Figure 2.1, see p. 36/37) of a total of three major classes (all-α, α/β, all-β).

Based on the three-dimensional structure and structure acquisition of proteins, domains are considered as cooperative units in the folding process. In this concept, the individual domains of a multi-domain protein can be considered simply as small, one-domain proteins that fold to a native-like structure independently of the remaining part of the protein. This holds for both in vitro folding and co-translational folding. The latter has been shown for the folding and assembly of nascent antibody chains [10]. Based on this kind of vectorial mechanism with well-defined assembly modules, large polypeptide chains speed up folding by many orders of magnitude. At the same time, possible wrong long-range interactions that would lead to an incorrectly folded protein are minimized.

Methods used to investigate the structure formation process of multi-domain proteins are the same as for simple model proteins and will not be discussed here. The most important techniques are listed in Table 2.1. A detailed description of some of these methods can be found in several chapters of Volume I of this handbook.

Detailed analyses on the structure, thermodynamic stability, and folding of multi-domain proteins have been performed, e.g., for γB-crystallin, lysozyme, α-lactalbumin, the light chain of antibodies, and many other proteins [11, 12]. From these studies it can be concluded that in most cases the domains fold independently kinetically and that they are thermodynamically stable entities. The complete refolding, however, often requires an additional kinetic phase involving intramolecular domain association. If the interaction sites of the domains are very

Tab. 2.1. Experimental approaches to study folding of domain proteins.

Assay	Methods	Information
Spectroscopy	Circular dichroism (CD)	
	In the far UV region	secondary structure information
	In the near UV region	environment of aromatic side chains
	Fluorescence	tertiary and quaternary structure
	NMR (combined with H/D exchange) and 2D-NMR	changes of the environment of single amino acids
Function	Enzymatic activity and/or ligand binding	monitoring the native state (sometimes can be active as well)
Stability	Limited proteolysis or denaturant	only folded domains or the native state might be resistant to proteolytic degradation or increased denaturant concentration

large, then domain-domain interactions may contribute significantly to the stability of the respective domains, and thus the isolated domains are thermodynamically unstable on their own. In such cases the folding and structure of the partner domain represent an essential template for structure formation of the other domain [13, 14].

One might suggest that proteins with very similar structure should show similar folding characteristics. This, however, is not the case. An intriguing example is the comparison of hen egg white lysozyme (HEWL) and α-lactalbumin. Both proteins with a molecular mass of approximately $Mr = 14\,400$ are structurally almost identical, containing an α-helical and a β-sheet domain. However, both their denaturant-induced equilibrium transitions and their folding characteristics are quite different. HEWL shows an equilibrium transition according to a classical two-state model (cf. Chapter 3 in Part I), i.e., only the native and the completely denatured state of the protein are populated in the transition region. In contrast, α-lactalbumin shows a three-state unfolding transition (in the absence of Ca^{2+} ions) according to the model $N \leftrightarrow I \leftrightarrow U$, where N is the native state, I is a partially structured folding intermediate, and U is the completely denatured protein [15]. The equilibrium intermediate I retains the α-helical domain, whereas the β-sheet domain is denatured. This same kind of intermediate can also be observed during the refolding kinetics of α-lactalbumin. It has been used as a model structure for the definition of the molten globule, a type of intermediate often observed in the folding of proteins (cf. Chapter 13 in Part I). Under strong refolding conditions (low residual concentration of denaturant), HEWL also folds via intermediates, which, however, do not resemble a classical molten globule [16]. Furthermore, in the case of HEWL, the intermediate is not obligatory; the protein can fold directly from the denatured to the native state. Due to kinetic competition, a fraction of the denatured molecules folds to an intermediate state, which subsequently undergoes a structural conversion to the native state [17].

2.2.2
γ-Crystallin as a Model for a Two-domain Protein

γ-Crystallin as an eye-lens protein of vertebrates belongs to a superfamily of β-sheeted proteins with a Greek key topology. It is a two-domain protein of a molecular mass of 21 kDa. The two domains, connected by a short peptide linker, are highly homologous (cf. Figure 2.5). γ-Crystallin is extremely stable, retaining its native structure at pH 1–10 or in 7 M urea at neutral pH [18]. The denaturant-induced equilibrium transition at neutral pH yields a broad transition and a partially irreversible denaturation. At acidic pH, however, the overall denaturation process splits into two transitions, both of which are fully reversible [18–20]. Making use of sedimentation analysis, intrinsic protein fluorescence, and circular dichroism, the urea-dependent equilibrium unfolding at pH 2 can be quantitatively described by a three-state transition $N \leftrightarrow I \leftrightarrow U$, with the intermediate populated at 2–4 M urea (Figure 2.2).

The kinetic data on folding and unfolding of γ-crystallin fit well to the biphasic equilibrium transition. Below 3 M urea, the denaturation kinetic is monophasic, representing the reaction $N \to I$. At higher urea concentrations, the unfolding reaction becomes biphasic, the faster reaction corresponding to the structural conversion $N \to I$ and the slower one to $I \to U$. Similar effects are observed for the refolding: at 3 M urea, the folding reaction is monophasic ($U \to I$), while under strongly native conditions both reactions $U \to I$ and $I \to N$ can be observed. On the other hand, when starting either denaturation or refolding from the intermediate state, the respective kinetics are monophasic. Summarizing the dependence of the rate constants (or relaxation times) of folding and unfolding on the urea concentration, two separate V-shaped profiles (chevron plot) are obtained (Figure 2.2). The respective reactions correspond to structural transition of $N \to I$, $I \to N$, $U \to I$, and $I \to U$.

Structural interpretation of these data, however, is difficult. It is not possible to assign any of the equilibrium transitions and kinetic reactions to a certain domain of this two-domain protein. In order to correlate the mechanism with the specific structural properties of γ-crystallin, the recombinantly produced, isolated domains were investigated [19]. Both the C-terminal and the N-terminal domains have been shown to be monomeric in their isolated states as well as if simultaneously incubated even at very high protein concentrations. The thermodynamic stability and folding and unfolding reactions of the isolated N-terminal domain of γ-crystallin resemble those of the intermediate I of the full-length protein [18, 19], indicating that the intermediate I consists of a native-like structured N-terminal and a completely unfolded C-terminal domain. The isolated C-terminal domain, however, is much less stable than it is in the context of the whole protein (Figure 2.3). In fact, even in denaturant-free buffer, a significant fraction of the isolated C-terminal domain is unfolded.

The difference in the stability of the C-terminal domain in its isolated form and in the context of the full-length protein implies a significant contribution of the do-

Fig. 2.1. Architectures of domains for the mainly α, mainly β, and $\alpha\beta$ classes of the CATH database [7–9].

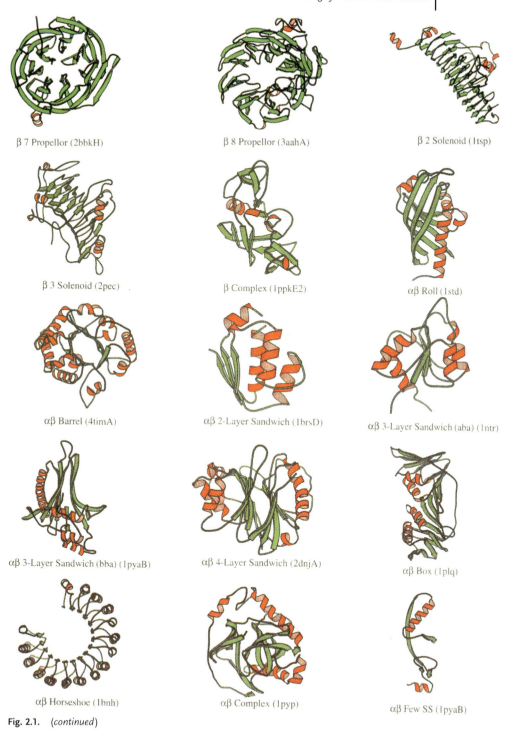

Fig. 2.1. (continued)

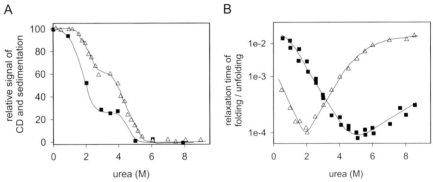

Fig. 2.2. Stability and folding of γ-crystallin (adapted from Ref. [18]). (A) Urea-induced equilibrium transition of γ-crystallin. The protein at a concentration of 0.2 mg mL^{-1} in 10 mM HCl (pH 2), 0.1 M NaCl was incubated at different urea concentrations. Subsequently, the samples were analyzed using CD (△) and sedimentation (s20,w: ■). (B) Kinetics of folding/unfolding of γ-crystallin (△,■) measured by fluorescence and HPLC gel filtration. The relaxation times of the kinetics are plotted against the urea concentration used for folding/unfolding.

main interactions (ΔG_{int}) to the overall free energy of stabilization (ΔG_γ), in addition to the sum of the intrinsic stabilities of the isolated domains ($\Delta G_N, \Delta G_C$):

$$\Delta G_\gamma = \Delta G_N + \Delta G_C + \Delta G_{int}$$

Under the given conditions (pH 2) the stability ΔG_C amounts to only -4.7 kJ mol^{-1}, whereas the ΔG_{int} is -16 kJ mol^{-1}. Thus, at low pH the C-terminal do-

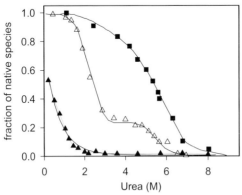

Fig. 2.3. Stability of a single domain of γ-crystallin. Urea-induced equilibrium transition of γ-crystallin and its single domains measured in 10 mM HCl pH 2, 0.1 M NaCl. Complete γ-crystallin (△), the isolated N-terminal domain (■), and the isolated C-terminal domain (▲).

main of γ-crystallin is mainly stabilized by an intramolecular association with the N-terminal domain. This intramolecular interface is characterized by a well-defined hydrophobic core of Phe 56, Val 132, Leu 145, and Val 170. If Phe 56 of the N-terminal domain is replaced by either Ala, thus creating a cavity in the hydrophobic core, or Asp, placing a non-saturated charge in the contact site, the C-terminal domain is considerably destabilized, again emphasizing the importance of the interaction free energy of stabilization (ΔG_{int}) in stabilizing the structure of the C-terminal domain of γ-crystallin.

2.2.3
The Giant Protein Titin

Titin is a protein located in the sarcomere of vertebrate striated muscle cells. A single molecule of titin spans half the sarcomere from the Z-disc to the M-line, extending about 1 µm in length (Figure 2.4). The protein consists mostly of several hundred copies of β-sheet domains of the immunoglobulin type and of fibronectin type III (Fn3); the molecular mass ranges from 3 MDa to 4 MDa, depending on the source of titin. The function of titin varies along the sarcomere. In the M-line it forms an integral part of the protein meshwork, while in the A-band it probably regulates the assembly of the thick filaments and might be involved in the interfilament interaction of myosin and the actin/nebulin thin filament [21, 22]. In the I-band, titin serves as an elastic connection between the Z-disc and the thick filaments. It is hypothesized that these elastic properties depend on partial unfolding/refolding of titin during muscle tension.

In order to obtain insight into the molecular properties of this giant protein, single immunoglobulin and Fn3 domains as well as tandem repeats of these two domains have been cloned and biophysically analyzed. The individual domains vary considerably in sequence. Using fluorescence and circular dichroism as probes in equilibrium unfolding experiments, the thermodynamic stability of a variety of these domains was determined to be in the range of 8.6–42 kJ mol^{-1} [23]. Similarly, the rates of chemical unfolding of individual domains vary by almost six orders of magnitude [23]. It has been speculated that this dramatic variation in folding behavior is important in ensuring independent folding of domains in the multi-domain protein titin.

Site-directed mutagenesis of each position in a protein together with a detailed analysis of the folding and unfolding kinetics of these mutants permit the structural mapping of the transition state of folding (θ-value analysis, cf. Chapter 2 in Part I). Such an analysis for the titin domain I27 reveals that the transition state of folding is highly structured and remarkably native-like regarding its solvent accessibility. Only a small region close to the N-terminal region of the domain remains completely unfolded.

How do the molecular properties of the single domains of titin correlate with the function of the whole molecule? Titin is involved in muscle tension and elasticity and thus is subjected to mechanical forces. The elasticity of a protein can be measured by atomic force microscopy (AFM) [22]. To this end, the protein is fixed at

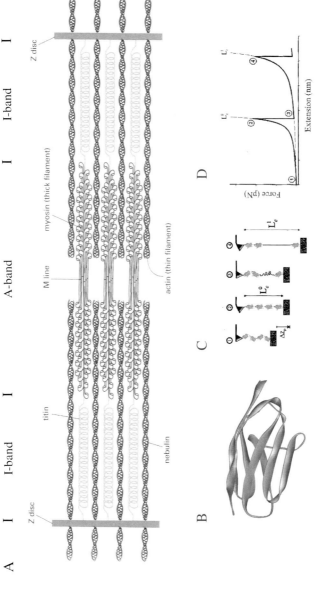

Fig. 2.4. (A) Schematic representation of a sarcomere of striated muscle. (B–D) Single-molecule AFM measurements of an engineered polyprotein of the titin domain I27. (B) Structure of the I27 Ig-like domain. (C, D) Schematic representation of the sequence of events during the stretching of I27 multi-domain protein. Stretching the ends of the polyprotein with an atomic force microscope sequentially unfolds the protein modules, generating a saw-tooth pattern in the force-extension relationship that reveals the mechanical characteristics of the protein. (1) An anchored multi-domain protein composed of four I27 domains. The protein is relaxed. (2) Stretching this protein requires a force that is measured as a deflection of the cantilever. (3) The applied force triggers unfolding of a domain, increasing the contour length of the protein and relaxing the cantilever back to its resting position. (4) Further stretching removes the slack and brings the protein to its new contour length Lc 1.

one end on a surface and at the other end on the tip of a so-called cantilever (Figure 2.4). Moving the cantilever slowly and on an Angstrom scale from the surface and thus continuously increasing the mechanical force leads to a stretching of the fixed protein. The resistance of the protein to this stretching results in a bending of the cantilever that is optically detected. At some point a protein domain unfolds, the resistance is partially released, and the cantilever returns to its starting position. This procedure is repeated as often as domains are located between the static surface and the cantilever. The experimental result is a sawtooth-shaped curve, in which the width of a single sawtooth reflects the difference in length of the folded and unfolded domains (28 nm for the titin domain I27) and the height of the peak corresponds to the mechanical force necessary for domain unfolding (Figure 2.4). The immunoglobulin domain I27 of titin unfolds at 210 pN, and the Fn3 domain I28 unfolds at ca. 260 pN. However, the hetero-oligomer $(I27-I28)_4$ is slightly stabilized, hence the mechanical force to denature the domain I28 in this construct increases to 306 pN [24]. Such a stabilization of the Fn3 domain I28 in the context of the neighboring I27 domain was also observed using chemical denaturation methods [25]. Comparison of the chemically induced unfolding and the mechanically induced unfolding reveals similar mechanisms of denaturation. The forced unfolding starts with the detachment of the N-terminal β-strand from the otherwise stable β-sheet. This N-terminal β-strand is not structured in the transition state of chemical denaturation. Site-directed mutagenesis studies show that the region important for kinetic stability is very similar in both cases, although there are also significant differences in the forced and chemically induced unfolding transition states [26, 27].

Presently, it seems that chemical and mechanical denaturation processes are related, at least for proteins evolved to respond to mechanical stress. However, in the case of barnase, a globular two-domain protein, mechanically and chemically induced unfolding seem to proceed quite differently on the structural level [28].

2.3
Folding and Association of Oligomeric Proteins

2.3.1
Why Oligomers?

Oligomers consist of noncovalently associated subunits of identical origin (homo-oligomers) or different origin (hetero-oligomers). Two or more monomers form dimers, trimers, tetramers, etc. Higher oligomers may possess topological units (e.g., the dodecameric aspartate transcarbamylase (ATCase) from *E. coli* consists of two catalytic trimers and three regulatory dimers). Tetramers of glyceraldehyde-3-phosphate dehydrogenase and yeast pyruvate decarboxylase are formed as dimers of dimers. Compared to other tetramerization pathways such as the addition of monomers to dimers or trimers, the specific topology of a tetramer as a dimer of dimers leads to kinetics of association that seem to be less prone to competition with off-pathway reactions of nonnative assemblies or accumulation of association intermediates [29].

The advantage of oligomeric proteins compared to their non-associated monomeric counterparts may be found on the functional, structural, and genetic levels. Oligomerization permits an allosteric regulation of enzyme activity. In multi-enzyme complexes, the catalytic efficiency of the overall process may be significantly increased. This would especially be the case if labile intermediates were channeled from one active site to the other. These functional advantages, however, can be conferred not only by oligomeric structures but also by large multi-domain proteins. Indeed, a few such very large multi-domain proteins consisting of more than 10 000 amino acids are known [30]. There is no fundamental structural or functional distinction between domains and subunits. The eukaryotic fatty acid synthase contains the whole assembly line of fatty acid synthesis [31, 32]; similarly, many non-ribosomal peptides are synthesized by very large proteins (e.g., the cyclosporin synthetase, in which all the different enzyme functions are located in one polypeptide chain) [33]. However, these very large multi-domain proteins are rare: most large protein structures are formed by association of subunits. Oligomerization of small gene products is by far more economical; this is quite obvious in the case of viruses, where a small protein, and thus a small gene, is sufficient for building up a large virus capsid of the size of several megadaltons. Furthermore, errors occurring during protein synthesis can be eliminated more easily in the case of oligomeric structures. Another possible advantage of oligomeric protein complexes versus large multi-domain proteins may result from differences in folding. It has been shown that domains in the context of a larger polypeptide chain fold more slowly than the same domain in its isolated state [34–36]. This difference may result from nonnative long-range interactions with other parts of the polypeptide chain and/or diffusional limitations of a long polypeptide during folding. In the case of the central domain of bacteriophage P22 tail spike protein, the folding of this β-helix domain is accelerated about twofold upon removal of the N-terminal, structurally unrelated domain. This slightly accelerated folding increases the yield of folding significantly both in vitro and in vivo [36, 37].

2.3.2
Inter-subunit Interfaces

The increasing number of high-resolution crystal structures of oligomeric proteins and protein complexes available in the Protein Data Bank allows a statistical analysis of the properties of interfaces of subunits and domains [38–41]. The majority of the subunit interfaces have a surface of 1000–3000 Å^2. The major part of this surface is usually hydrophobic, conveying the stability of interaction; polar interactions, including hydrogen bonding and ionic interactions, as well as surface complementarity serve as specificity factors of interaction [43, 44]. In fact, the molecular properties of the interface of stably associated subunits regarding the distribution of hydrophobic versus polar residues are very similar to that of the hydrophobic core of proteins and, conversely, very different from that of the solvent-accessible surface. In contrast to the interfaces responsible for stable subunit association, the contact sites of proteins that interact only transiently with other proteins are less hydrophobic [41].

A qualitative analysis of 136 subunit interfaces of homodimeric proteins [45] showed that about 30% are characterized by a well-defined hydrophobic core, surrounded by polar interactions [45, 46]. However, the majority of interfaces consist of multiple small, hydrophobic patches interspersed by hydrogen bonds and polar interactions [45].

Such large differences in interface structure might correspond to differences in interface stability and in folding and assembly mechanisms, as proposed by Nussinov and coworkers [40, 47]. This is illustrated by the tetrameric yeast pyruvate decarboxylase (PDC). The topology of PDC is reminiscent of a dimer of dimers. The interface between the dimers that form the tetramer is largely polar, and the tetramer spontaneously dissociates to dimers upon dilution. The dissociation constant of this dimer-tetramer equilibrium is 8.1 µM [48]. The interface between the monomers, however, is strongly hydrophobic; the dimer does not dissociate into monomers just by dilution of the protein. Only in the presence of 2–3 M urea can folded monomers be populated [48]. Unfortunately, there are only very few cases where thermodynamic, kinetic, and crystallographic data of a certain protein can be combined in order to assess these questions. Recent studies suggest that, in a given interface, a few amino acids may play a dominant role in determining the free energy of stabilization of the interaction. These residues are called hot spots of the binding surface [46].

Whereas in most cases the association of proteins is mediated by globular, native-like structured domains, a growing number of protein-protein interactions have been discovered in which one of the partners is unstructured in its isolated state. These natively unstructured proteins can be grouped into two classes, one with random coil structures and the other with a pre-molten globule structure [13]. Only upon association with their targets do these proteins develop an ordered and biologically active structure. Examples are several ribosomal proteins that transform to a rigid well-folded conformation only in the presence of the ribosomal RNA [49, 50]. In the case of cyclin-dependent kinase inhibitor p21, the N-terminal fragment lacks stable secondary and tertiary structure in the free solution state. However, when bound to its target, Cdk 2, it adopts a well-ordered conformation [51].

Besides the usual subunit interaction via globular domains, there is also another major type of protein-protein interaction mediated by special helical structures: the so-called coiled coils [52]. A detailed description of this interaction that may lead to dimers, trimers, or tetramers, depending on the sequence of the coiled coils, can be found in Chapter 18 in part I.

Furthermore, completely artificially designed peptides have been described, conferring heterodimerization of proteins fused to them. These peptides are based on only a stretch of eight charged amino acids and an additional cysteine. Two of these peptides containing complementary charged side chains associate with each other due to their polyionic interaction; subsequently, the cysteines can form a disulfide bond, thus stabilizing the heterodimerization [53–55].

2.3.3
Domain Swapping

As pointed out by Bennett et al. [56–58], the gradual accumulation of random mutations on the surface of a monomeric protein that are required to create a suitable subunit interface and stabilize a dimer cannot be accomplished within a biologically feasible time. At least in some cases the evolution of domains offers a key to understanding the mechanism to proceed from monomers to dimers. Starting from a two-domain monomer, domain swapping permits the transition from monomers to dimers (or any other association state) simply by switching from an inter-domain to an inter-subunit interaction. In this scheme, the interface that is almost impossible to create by random mutagenesis in evolution is available a priori. Based on the already-existing interaction site, additional mutations might then stabilize the domain-swapped dimer compared to the monomer (Figure 2.5).

The concept of domain swapping may be illustrated by $\beta\gamma$-crystallins. $\beta\gamma$-crystallins are members of the superfamily of β-sheeted proteins containing the Greek key motif. γ-crystallin is monomeric even at extremely high protein concentrations, whereas β-crystallin forms dimers or higher oligomers. Both show highly homologous two-domain structures. Comparison of the structures of γ- and β-crystallins suggests that the β dimers are the product of domain swapping and association of a γ ancestor (Figure 2.5). This hypothesis is supported by the fact that 60% of the interface residues of the N- and C-terminal domains of γ-crystallin are identical or highly conserved compared to those of the N- and C-terminal domains of the two subunits of β-crystallin [59].

Domain swapping does not correlate necessarily with structural domains. Instead, the swapped portion can vary from a large globular domain, as described for $\beta\gamma$-crystallins, to a structural element as short as a single helix or β-strand. In fact, of the ca. 40 structurally characterized cases of domain-swapped proteins, most swapped domains are either at the N- or C-terminus and are diverse in their primary and secondary structure [60]. Hsp33, for example, forms a dimer in which a C-terminal arm of one monomer consisting of two α-helices wraps around a second monomer [61, 62]. This C-terminal arm is connected to the remaining part of the protein by a putatively flexible hinge region. The chaperone-active dimer is in equilibrium with a less active or even inactive monomer ($K_D = 0.6$ µM, [63]). A detailed structural and functional analysis of Hsp33 is presented in Chapter 21. Unfortunately, the structure of the Hsp33 monomer is not known. Thus, at present it is only speculation that in the monomeric structure the C-terminal arm folds back on the monomer.

Higher association states of domain-swapped proteins are found in several virus capsids, in which capsomers swap β-strands, or pathologic aggregates in the case of α1-antitrypsin [64].

Although domain swapping is an intriguing concept in understanding evolution

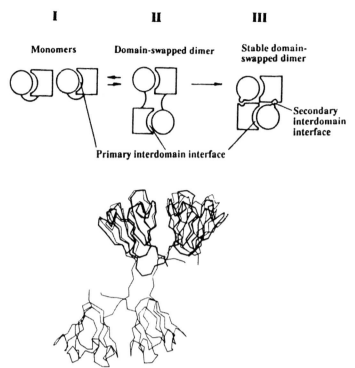

Fig. 2.5. Domain swapping. (A) The monomer (I) with its primary inter-domain interface may participate in interchain interactions, generating domain-swapped dimmers (II). Stabilizing mutations in the linker region allow stable dimers (III) to be formed [56]. (B) β-crystallin dimer (thin line) as potential product of swapped domains of γ-crystallin (thick line) (adapted from Ref. [11]).

of protein dimerization and oligomerization, it can be applied only in a limited number of cases. It cannot explain hetero-oligomerization of subunits encoded by widely differing genes as these subunits do not share a common inter-domain interface or different parallel routes of dimerization, as shown for bovine seminal ribonuclease, which exists in both a domain-swapped and a non-swapped dimeric form [65, 66].

2.3.4
Stability of Oligomeric Proteins

The laws of thermodynamics require that association equilibria be shifted towards the dissociated state at low protein concentrations. Indeed, for several oligomers it has been shown that the dissociation constants of dimerization are in the nanomolar to micromolar range, and thus the dissociation can be achieved directly by dilution experiments. Examples are phosphofructokinase [67], dimeric Arc repressor [68], Hsp33 [63], and yeast hexokinase [69, 70]. In some cases this association equilibrium is functionally and physiologically relevant. Dimeric Hsp33 is active as a molecular chaperone, whereas the monomeric form is less active or non-active [63, 71]. The K_D describing the monomer-dimer equilibrium is in the range of the physiological concentration of Hsp33 in *E. coli* [63]. In the case of yeast hexokinase, the monomeric state of the protein is less active than the dimeric form. Interestingly, the dimerization is favored in the presence of the substrate glucose [69].

Most oligomeric proteins, however, dissociate only at concentrations too low to observe dissociation by mere dilution [1]. This finding suggests that subunit interactions account for a large fraction of the conformational stability of oligomeric proteins; sub-nanomolar dissociation equilibrium constants for dimerization correspond to standard free energy changes of >50 kJ mol^{-1}. In such cases, dissociation is achievable only by denaturation. Whereas the denaturant-induced equilibrium transition of denaturation of monomeric proteins is independent of the protein concentration, the midpoint of transition of oligomeric proteins varies with protein concentration. In the simplest case of a reversibly denaturing dimeric protein, this transition can be ascribed to the reaction $N_2 \leftrightarrow 2U$. The linear extrapolation method, used for analysis of the transition (cf. Chapter 3 in Part I), yields a straight line with a slope depending on the protein concentration used (Figure 2.6). These lines describing the denaturation transition at different protein concentrations will

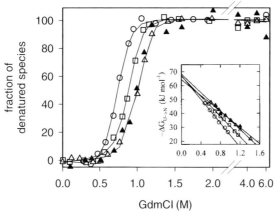

Fig. 2.6. Guanidine-dependent equilibrium transitions of the dimeric antibody domain CH3. Samples were incubated for six days at 20 °C in 0.1 M Tris-HCl (pH 8.0) and varying GdmCl concentrations. For the fluorescence measurements the excitation wavelength was set to 280 nm and emission was analyzed at 355 nm. The CH3 dimer concentrations were 0.2 µM (open circles), 1.6 µM (open squares), and 4.1 µM (open triangles). For the CD experiment, the ellipticity was recorded at 213 nm (filled triangles) at a protein concentration of 4.1 µM. All measurements were carried out at 20 °C. Inset: From the transitions, the stability ΔG could be calculated at the respective GdmCl concentrations. These ΔG values were plotted against the respective GdmCl concentration and linear extrapolated to 0 M GdmCl [42].

intersect the y-axis at the same value that corresponds to the stability of the dimeric native state in the absence of denaturant [72].

Oligomeric states often are not only thermodynamically stable but also kinetically stabilized, i.e., the activation energy of dissociation under native conditions is very high. Dissociation rates in denaturant-free buffer have been estimated by extrapolation to be as low as 10^{-14} s^{-1} [73].

2.3.5
Methods Probing Folding/Association

The reconstitution of oligomeric proteins comprises folding of subunits and their association. Therefore, all methods suited to analyze folding of proteins and domains can be applied to oligomeric proteins as well (cf. Table 2.1). In the following section, only methods that detect association reactions are discussed. In using these methods, one should be very careful, because these techniques do not always distinguish between specific association and unspecific aggregation. The latter, however, will often be found to compete with proper refolding and association.

2.3.5.1 Chemical Cross-linking

Highly reactive bifunctional chemicals can be used to covalently cross-link subunits of an oligomeric structure with each other. In the case that cross-linking proceeds much faster than oligomerization, it is possible to get time-resolved information of the association reaction by cross-linking aliquots of the reaction at different time points. The amount of cross-linked oligomers can subsequently be analyzed on SDS-PAGE. Glutaraldehyde has been shown to be ideally suited as a cross-linker for several reasons. It forms a Schiff base with lysine side chains, which are usually located on the surface of a folded subunit. The reaction is very fast and efficient. It is not dependent on a specific distance between two lysine residues because glutaraldehyde exists in differently polymerized forms in aqueous solutions; therefore, the distance between the two reactive aldehyde groups of glutaraldehyde varies with its polymerization state. For a given protein, the glutaraldehyde concentration, reaction time, temperature, and protein concentration need to be optimized in order to guarantee that non-associated subunits are not cross-linked artificially. A protocol for glutaraldehyde cross-linking is given in the Appendix.

2.3.5.2 Analytical Gel Filtration Chromatography

Gel filtration is a standard technique to determine the molecular mass of proteins. Using it for kinetic analyses of association, however, poses problems. Gel filtration is a comparatively slow method; the time resolution of kinetic measurement will therefore be low. The protein analyzed will be diluted on the column, thus shifting a possible association equilibrium towards the dissociated species. Only if the dissociation of the oligomer is much slower than the time it takes to carry out the gel-filtration chromatography will significant results be obtained (the rate of dissociation might be slowed down by cooling the column). Gel-filtration chromatography has been used successfully to characterize, among others, the reconstitution of β-galactosidase [74], P22 tail spike protein [75], and an antibody Fab fragment [76].

2.3.5.3 Scattering Methods

Light scattering, small-angle X-ray scattering, and neutron scattering provide information on particle size and shape. They can be used online without the necessity of taking aliquots during reconstitution for analysis. However, scattering methods need high protein concentrations (small-angle X-ray and neutron scattering, several milligrams per milliliter; light scattering, depending on the size of the oligomer, not less than 0.1 mg mL^{-1}). At these high protein concentrations, unspecific aggregation is often favored over correct reconstitution. Light scattering has been used predominantly to monitor formation of large assemblies such as viruses or protein filaments. More often it is used as a method to follow aggregation [77, 78].

2.3.5.4 Fluorescence Resonance Energy Transfer

Fluorescence resonance energy transfer (FRET) can be measured if the emission spectrum of a fluorophore overlaps with the absorption spectrum of another fluorophore. Then the energy of the excited donor fluorophore will not be emitted as fluorescence but will be transferred to the acceptor fluorophore. In this setup, the donor fluorophore will be excited and the fluorescence of the acceptor is measured. The transfer efficiency is inversely proportional to the sixth power of the distance between the two fluorophores, thus permitting measurement of distances within proteins [79] and monitoring of protein complex formation [80]. FRET measurements have to be carefully controlled. The fluorescence dyes that are coupled to the respective subunits are bulky hydrophobic chromophores and might induce association or aggregation of the coupled proteins artificially.

The FRET technique has become increasingly important to study protein-protein interaction in the cellular context. To this end, the proteins in question are fused to different variants of the green fluorescent protein, e.g., BFP and GFP. If these fusion proteins associate in vivo, FRET can be observed and quantified by fluorescence microscopy.

2.3.5.5 Hybrid Formation

The concentration of unassembled subunits can be measured by their ability to form hybrid oligomers. Originally, the technique used a large excess of chemically modified or isoenzyme subunits added after varied reconstitution time to trap unassembled subunits [81–84]. Problems such as slow folding of added subunits and preferential formation of homo-oligomers are associated with this approach. More-reliable data may be obtained with naturally occurring or site-directed mutant proteins with altered net charge but unaffected folding and association kinetics [37, 75, 85]. The technique may be illustrated using the trimeric tail spike protein from phage P22. Wild-type and mutant tail spikes with altered charge were separately denatured by urea at pH 3. Subsequently, both proteins were diluted separately into neutral buffer to remove the denaturants and start refolding. After different times, aliquots from both solutions were mixed, reconstitution of native trimers was allowed to reach completion, and the differently charged hybrids and homo-oligomers were separated by electrophoresis [75]. In samples mixed early during renaturation, wild-type homotrimers, hybrids, and mutant homotrimers formed at the statistically expected ratio of 1:2:2:1, whereas in samples mixed late, only homo-oligomers were detected. The half-time of assembly can be estimated from the amount of hybrid present in samples mixed at intermediate renaturation times.

It should be noted that hybrid formation measures only stably associated protomers; those in rapid exchange between loosely associated oligomers will be equivalent to free subunits.

2.3.6
Kinetics of Folding and Association

2.3.6.1 General Considerations

The reconstitution of oligomeric proteins from the denatured state generally constitutes a series of uni- and bimolecular steps. For a dimeric protein, reconstitution involves the following steps (neglecting possible off-pathway side reactions):

$$2U \rightarrow 2M \rightarrow M_2 \rightarrow N_2$$

where U is the denatured state. M represents a folded monomer, M_2 a nonnative dimer, and N_2 the native dimer. The folding reactions preceding subunit assembly are identical to those of single-domain or multi-domain proteins. They consist of a hydrophobic collapse, formation of secondary structure, and the merging of domains, leading to association-competent monomers. As in the case of monomeric proteins, these reactions may occur as sequential or parallel reactions on multiple pathways [86, 87]. All these reactions follow first-order kinetics and do not depend on protein concentration. In contrast, the association of subunits ($2M \rightarrow M_2$) is a bimolecular and therefore probably a second-order reaction. Subsequently, the dimer may need to undergo further unimolecular rearrangements to reach the final native state ($M_2 \rightarrow N_2$).

The assembly of larger oligomers can be described accordingly as a series of multiple uni- and bimolecular steps. The assembly of a homotrimer is expected to proceed via a dimeric intermediate:

$$3M \leftrightarrow M_2 + M \rightarrow M_3$$

In this reaction scheme, the first association resulting in a dimer, in which only one of two interfaces is buried, is considered a fast equilibrium. This first association step is unfavorable and the dimeric intermediate is not populated in equilibrium. The second association step will pull this equilibrium towards the native trimeric state. In this concept, the overall process follows apparent third-order kinetics [84].

A tetramer may be envisioned as a dimer of dimers. Thus, the scheme for this reaction is just an extension of that shown for a dimeric protein [1, 88]:

$$4U \rightarrow 4M \rightarrow 2M'_2 \rightarrow 2M_2 \rightarrow M_4 \rightarrow N_4$$

In the case of multimers such as fibrillary proteins or virus capsids, the simplest model comprises a stepwise addition of monomers to the growing chain:

$$2A \leftrightarrow A_2, A_2 + A \leftrightarrow A_3, A_3 + A \leftrightarrow A_4, \ldots$$

In this scheme the equilibrium constants of association may be all identical. However, the first steps of multimerization are often thermodynamically disfavored, i.e., the equilibrium is on the side of the non-associated species. Only after reaching a certain size of the oligomer structure does further addition of monomers to this nucleus become thermodynamically favorable. This would lead to an overall assembly process characterized by a nucleation/polymerization reaction.

2.3.6.2 Reconstitution Intermediates

Because the half-time of association is concentration-dependent, the nature of the rate-limiting step of reconstitution of oligomeric proteins may change with protein concentration [1]. At high protein concentrations, the formation of association-competent subunits is rate-limiting; at low protein concentrations, dimerization is rate-limiting, and the overall reaction follows second-order kinetics. Under these conditions, highly structured, association-competent subunits often are populated kinetically during refolding/assembly.

Such intermediates can be observed spectroscopically. In general, secondary structure as monitored by far-UV circular dichroism is formed quickly upon folding/association. Furthermore, the secondary structural content of the folded subunits does not change significantly upon subsequent association. Thus, the change in CD signal follows first-order kinetics, reaching the signal of the native protein as subunit folding is completed. In contrast, if reconstitution is analyzed by reactivation, i.e., regain of enzyme activity, no reactivation is observed during subunit folding provided that the non-associated subunits do not possess enzyme activity by themselves. The folding of the subunits results in a lag-phase in regain of activity; reactivation occurs upon association. Thus, the overall reactivation is described by a uni-bimolecular process that follows a sigmoidal time course, with the lag-phase determined by first-order subunit folding (Figure 2.7) [89].

Another method often used to analyze protein folding is fluorescence. However, whereas circular dichroism monitors secondary structure formation almost exclusively on the subunit level and, in most cases, enzymatic activity corresponds to the native state after assembly, fluorescence may detect both subunit folding and association, depending on the localization of the fluorophores in the quaternary structure. Therefore, the interpretation of fluorescence data might be more difficult. On the other hand it allows the detection of the association process even if the protein is enzymatically inactive.

In some cases the folded but non-associated subunit already shows ligand bind-

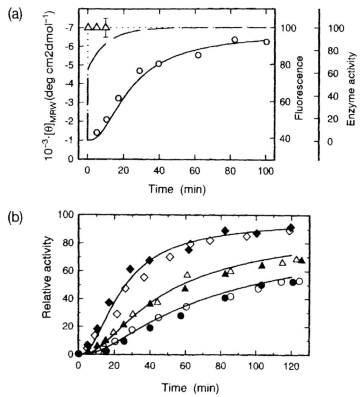

Fig. 2.7. Folding and association pathway of yeast invertase. (a) Changes in circular dichroism (open triangles), fluorescence emission (broken line), and enzymatic activity (open circles) occur with very different kinetics during the reconstitution of homodimeric yeast invertase. (b) The first-order folding rate obtained from the fluorescence change can be used to fit a sequential uni-bimolecular mode to the reactivation data obtained at different protein concentrations (full lines). Subunit concentrations were 8.5 nM (circles), 17 nM (triangles), and 68 nM (diamonds) [89].

ing or partial enzyme activity. Examples are mammalian aldolases [90], lactate dehydrogenase [91], the P22 tail spike protein [92], and yeast hexokinases [69, 70].

Monomeric folding intermediates of oligomeric proteins often possess large interfaces for subunit association. This subunit interaction is quite similar to the domain-domain interactions in multi-domain proteins. As shown before for γ-crystallin, the association can contribute substantially to the overall stability of the subunits. Thus, it is not surprising that folded subunits usually are not detectable as equilibrium unfolding intermediates if the native protein is subjected to high temperature or chemical denaturants [1, 88]. Structured monomers of lactate dehydrogenase have been obtained at low pH in the presence of the stabilizing salt sodium sulfate [93–95]. Another possibility to obtain stably folded monomeric subunits is based on protein engineering, as has been demonstrated for triose phosphate isomerase [96, 97], tetrameric human aldolase [98], and P22 tail spike protein [92]. In all these cases the engineered monomeric form retained some enzyme activity, but, as expected, the variants were strongly destabilized compared to the oligomeric wild-type form.

2.3.6.3 Rates of Association

The maximum rate of association of two molecules is determined by their random diffusional collision. For a common protein domain, this limit is about 10^9 M^{-1} s^{-1} [99]. Indeed, this limit is reached in the binding of chaperones to their substrate proteins. Chaperones recognize a wide variety of substrate proteins mainly by hydrophobic interactions. Since these interactions do not need a pre-orientation of chaperone and substrate, most of the diffusional collisions result in complex formation. However, this is quite different for the specific association of folded subunits. Here, the pre-formed interfaces require an exact rotational alignment for association, and this reduces the diffusion-controlled rate by several orders of magnitude. Brownian dynamics simulations estimate the rate of association of model proteins to be 10^6 M^{-1} s^{-1} [100]. Most second-order rate constants determined experimentally are even smaller (Table 2.2). There are, however, two examples published—the core domain of trp repressor and the arc repressor—that show faster dimerization than expected (Table 2.2) [101, 102]. In both cases folding is tightly coupled to association; the association-competent monomers are only partially structured.

2.3.6.4 Homo- Versus Heterodimerization

The molecular interaction of subunits in oligomers can be considered as specific as the formation of the hydrophobic core of a protein. The structure of a protein or subunit determines the kinetic and thermodynamic characteristics of its interaction with other molecules. Therefore, a competition between homo- and heterodimerization is expected only if the different subunits of the oligomer are highly homologous.

Tab. 2.2. Rates of association of some oligomeric proteins.

Protein	Subunit mass (kDa)	Reaction	k (M^{-1} s^{-1})	Reference
Triosephosphate isomerase	27	2M → D	3×10^5	1
Malate dehydrogenase (mitochondrial)	33	2M → D	3×10^4	1
Invertase	76	2M → D	1×10^4	89
β-Galactosidase	116	2M → D	4×10^3	74
Alcohol dehydrogenase (liver)	40	2M → D	2×10^3	1
Phosphoglycerate mutase	27	2M → D	6×10^3	103
		2D → T	3×0^4	103
Lactate dehydrogenase	36	2D → T	2×10^4	1
Arc repressor	62	2M → D	1×10^7	102
Trp aporepressor	12	2M → D	3×10^8	101

The most prominent example of such a competition is the reconstitution of heterodimeric bacterial luciferase from *Vibrio harveyi*. The two subunits consisting of a $(\beta\alpha)_8$ barrel are structurally homologous and probably result from gene duplication [104]. The association is mediated by two helices from each subunit forming a four-helix bundle at the contact site [105, 106]. The reconstitution of luciferase from the urea-denatured state is fully reversible. It comprises as rate-limiting steps the folding of the subunits on the monomeric level, the association of the subunits to an inactive heterodimer (at low protein concentrations), and a conformational rearrangement on the dimeric level leading to the native state [107–110]. As shown by far-UV circular dichroism, the inactive dimer seems to be incompletely folded.

Whereas the reconstitution of luciferase is quantitative if the subunits are refolded together, refolding of the subunits separately with a subsequent assembly does not lead to active heterodimer [107, 111]. In a detailed study, it could be shown that active heterodimeric luciferase can be assembled upon folding of the β-subunit in the presence of the pre-folded α-subunit, but not vice versa [73]. In the absence of the α-subunit, the β-subunit forms a homodimer that is almost as stable thermodynamically as the heterodimer. Furthermore, the β_2 homodimer is kinetically very stable, and under native conditions the dissociation does not occur on a biological or experimental timescale [73].

Upon reconstitution, the association of the β_2 homodimer and the $\alpha\beta$ heterodimer compete with each other. The almost exclusive formation of the heterodimer is due to the kinetics of association: the rate of heterodimerization is more than tenfold higher than that of the β_2 homodimerization. The structures of the homo- and heterodimer are almost identical regarding the contacts within the interfaces; most of the inter-subunit hydrogen bonds and water-mediated contacts are conserved between the two dimers [105, 106]. Thus, from the structural point of view, there is no obvious explanation for the different association kinetics of homo- and heterodimeric luciferase.

Mammalian lactate dehydrogenase (LDH) may serve as another example of the competition of homo- and hetero-oligomerization. LDH is a tetrameric protein that exists in two homomeric variants: in skeletal muscle (LDH-M_4) and in the heart (LDH-H_4). If these two isoforms are reconstituted together from the denatured monomers, the product is not only the two homomeric tetramers M_4 and H_4 but also all types of hybrids. Even more surprising, the distribution of all the variants fits to a statistical process, e.g., the ratio of M_4, M_3H_1, M_2H_2, M_1H_3, and H_4 is 1:4:6:4:1 [1, 94]. The beauty of this system is that it can be used to analyze the two different association steps upon reconstitution; at different time points of the reconstitution of LDH-M_4, an excess of LDH-H subunits can be added to the reaction. From the ratios of the hybrids reconstituted after addition of LDH-H subunits to the refolding LDH-M, the amount of monomeric, dimeric, and tetrameric M-subunits can be calculated [94].

Distinguishing homo- from hetero-oligomers is not always a simple matter, because in order to obtain competition of homo- and hetero-oligomerization, the subunits have to be highly homologous. The different hybrids of LDH could be separated by native electrophoresis, and in the case of luciferase the two subunits show different pI values, allowing a separation of the different subunits by ion-exchange chromatography [109]. Furthermore, enzymatic activity or varying stability of the different species may be used to distinguish between homo- and hetero-oligomers.

2.4
Renaturation versus Aggregation

Folding and association of multi-domain and oligomeric proteins are complex processes involving partially structured folding intermediates. These processes can be accompanied by nonspecific side reactions leading to aggregation. The molecular basis for aggregation is that intermediates in refolding/reassembly may engage in wrong intermolecular rather than in correct intramolecular or assembly interactions. Thus, aggregation is at least a second-order or even higher-order reaction, depending strongly on the concentration of folding intermediates [112, 113]. Therefore, the kinetic partitioning between folding and aggregation favors aggregation at high protein concentrations. In contrast, at low protein concentrations, folding dominates over aggregation (Figure 2.8).

Using reduced lysozyme as a model system, it could be shown that commitment to aggregation is a fast reaction preceding the commitment to renaturation; early folding intermediates were especially prone to aggregation [114]. After a certain intermediate state is reached, slow conformational changes lead exclusively to the native state (Scheme 2.1 [114]).

As shown by circular dichroism and Fourier transform infrared spectroscopy (FTIR), aggregated proteins possess a high content of secondary structure and probably of native-like structured protein domains [115, 116].

Aggregation is not an artifact of in vitro refolding; the same phenomenon occurs in vivo. It is especially obvious in the case of inclusion body formation upon high-level expression of recombinant proteins in microorganisms and the accumulation of fibrillary aggregates in human diseases [115–118]. Cells have developed a set of chaperones and folding catalysts to deal with aggregation. These two protein classes are discussed in detail in several chapters of this book.

2.5
Case Studies on Protein Folding and Association

2.5.1
Antibody Fragments

Antibodies are hetero-oligomeric proteins consisting of two light chains and two heavy chains. These four polypeptides are covalently linked by several disulfide bonds. The light chain consists of two domains and the heavy chain of four domains, with each domain adopting a typical immunoglobulin-like topology of a two-sheeted β structure. Folding of these domains has been shown to be a highly

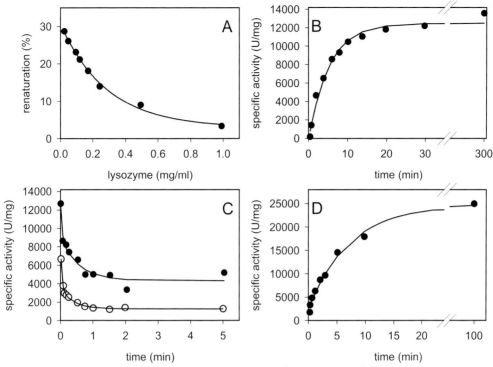

Fig. 2.8. Folding and aggregation of reduced lysozyme. (A) Yield of enzymatic activity as a function of turkey lysozyme concentration during oxidative refolding. (B) Kinetics of reactivation of turkey lysozyme. (C) Kinetics of commitment of aggregation of denatured and reduced lysozyme. Starting protein concentration for renaturation was 0.185 mg mL^{-1} (●) and 0.74 mg mL^{-1} (○). At different time points of renaturation, the samples were further diluted 10-fold and incubated overnight. (D) Kinetics of commitment of renaturation. Renaturation was performed with denatured and reduced turkey lysozyme. At different time points, a high concentration of denatured and reduced hen lysozyme was added. After completion of renaturation, the activity of turkey lysozyme was determined (adapted from Ref. [114]).

cooperative and reversible process involving an intermediate with already native-like secondary structure [119, 120].

The two antibody domains of the light chain (κ-type) and the C$_H$1 and C$_H$3 domain of the heavy chain (γ1-type) contain conserved *cis*-prolines. In the denatured

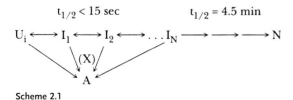

Scheme 2.1

state (depending on the denaturation time) equilibration between the *cis* and *trans* configuration occurs. Since the two configurations are separated by a high-energy barrier, re-isomerization is a slow process. This is reflected by an additional slow folding phase observed during refolding [119] that can be accelerated by prolyl isomerases [121, 122]. As soon as immunoglobulin domains are covalently linked, as in the case of the immunoglobulin light chain, which is composed of two domains, an additional slow folding phase is detected that may be due to interactions of the individual domains [120]. An additional level of complexity is observed in the case of the antibody Fab fragment, which consists of the light chain and the two N-terminal domains of the heavy chains, the Fd fragment (Figure 2.9). The two polypeptide chains associate noncovalently via contact sites between both the variable domains and the constant domains of each chain. Precise assembly of the two

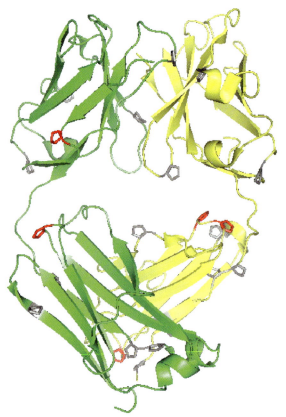

Fig. 2.9. Structure of the Fab fragment of the monoclonal antibody MAK33 (κ/IgG1) (pdb: 1FHE5, [123]). The light chain and the Fd chain are colored in green and yellow, respectively. Proline residues in *trans* configuration are highlighted in gray, those in *cis* configuration in red. The intermolecular disulfide bond connecting the C-terminus of the two chains is not resolved in the structure.

chains is required for antigen-binding activity [124, 125]. Furthermore, a disulfide bond connects the C-terminal domains of both chains covalently.

Despite this molecular complexity, the regain of antigen-binding activity of a Fab fragment, measured by ELISA, follows simple first-order kinetics [76, 122]. Spectroscopic analyses, however, reveal a complex folding behavior, which is expected for a multi-domain protein. The Fab fragment contains 22 prolyl residues, five of them in the *cis* configuration in the native state. Hence, upon refolding all these prolines need to isomerize to their native peptide bond configuration; this reaction dominates the overall folding process. Consequently, as described for the single-antibody domains, the folding of the Fab fragment can be catalyzed by peptidyl-prolyl-*cis/trans* isomerases [122]. A detailed description of peptidyl-prolyl-*cis/trans* isomerases and their effect on protein folding is given in Chapter 27 of this book.

If the disulfide between the two chains of the Fab fragment is deleted (with the remaining intra-domain disulfides still intact), then the reconstitution comprises both folding of the four domains and association of the two subunits. In this case, the regain of antigen-binding activity follows a sigmoidal time course (Figure 2.10). The kinetics of reconstitution does not depend on protein concentration, indicating that the association of the subunits is not rate-limiting [76]. Instead, the reconstitution process can be described by a serial first-order reaction with one of the phases identical to the proline-determined refolding of the disulfide-bonded Fab described before ($k = 0.04$ min^{-1}, Scheme 2.2). The other phase describes the slow formation of an association-competent state of the light chain.

It is known that structure formation of the two domains of the light chain is a fast process compared to the overall folding/reconstitution of the whole Fab frag-

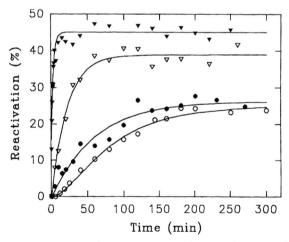

Fig. 2.10. Folding of Fab, not containing the C-terminal disulfide bond between the two chains. Refolding was achieved by diluting the denatured protein into 0.1 M Tris, pH 8, either starting from a long-term denatured protein in the absence (○) or presence (▽) of a fivefold molar excess of native light chain or starting from short-term denatured protein in the absence (●) or presence (▼) of a fivefold molar excess of native light chain. The data were fitted to a single first-order or a serial first-order reaction [76].

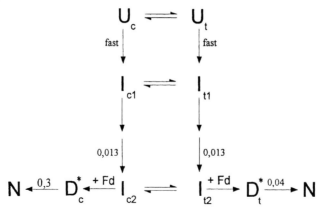

Scheme 2.2. Folding/reconstitution of the non-disulfide-bonded Fab fragment. U and N donate the denatured and the native state of the Fab fragment. Fd marks the position at which the association of light chain and Fd along the reconstitution pathway takes place. D describes an intermediate of associated light chain and Fd fragment, and I represents non-associated folding intermediates. The subscribed c and t reflect the *cis* and *trans* configuration of the rate-limiting proline peptide bond, respectively. The values given at two of the arrows indicate the rate constants (min^{-1}) of the respective reactions (adapted from Ref. [76]).

ment, even though folding of the light chain itself is limited by prolyl isomerization [121]. If the prolyl isomerization is catalyzed (or excluded from the folding/assembly reaction by short-term denaturation), the time course of reconstitution of the Fab fragment switches from a sigmoidal to a monophasic first-order kinetic; prolyl isomerization is not rate-limiting anymore. The yield of reconstitution does not change. The yield is determined by aggregation of the Fd fragment. Because peptidyl-prolyl-*cis/trans* isomerases catalyze the rate-limiting step occurring after subunit association of the Fab fragment, they do not effect the kinetic competition of off-pathway aggregation and productive reconstitution.

The situation changes if the Fab reconstitution is performed in the presence of an excess of native light chain. Under these conditions, the rate-limiting step before subunit association vanishes, and the regain of Fab activity becomes identical to the prolyl isomerization-determined folding of the Fab fragment, in which the two chains are disulfide-linked. This shows that (1) the rate-limiting step of reconstitution on the monomeric level is the folding of the light chain, presumably corresponding to a specific intramolecular pairing of the two folded domains [120], and (2) the rate-limiting prolyl isomerization on the level of the already associated subunits occurs in the Fd part of the Fab fragment (Scheme 2.2).

The folding/reconstitution of the disulfide-bonded and the non-disulfide-bonded Fab fragment comprise domain folding and domain interactions that are identical on the molecular level; however, in one case the domain pairing occurs intramolecularly, in the other case intermolecularly. The rate-limiting folding of the light chain can be observed only for the non-disulfide-bonded Fab fragment, clearly indicating that in the disulfide-bonded form the folding of the light chain is intramolecularly chaperoned by the nearby Fd part of the Fab fragment. On the other hand, the reconstitution of the non-disulfide-bonded Fab shows that the isomerization of at least one prolyl residue of the Fd part obligatorily occurs only after subunit association. Thus, only the association with the folded light chain provides the structural information for the isomerization state of the respective proline in the Fd part. This is an intriguing example of how folding and association of subunits are interrelated.

Whereas in the case of the Fab fragment the association of the two polypeptides precedes prolyl isomerization, it is different for the antibody domain C_H3. The isolated C_H3 domain forms a dimer. Under conditions where prolyl isomerization does not contribute to the folding kinetics, formation of the β-sandwich structure is a slow process, even compared to other antibody domains, while the subsequent association of the folded monomers is fast. After long-term denaturation, the majority of the unfolded C_H3 molecules reach the native state in two serial reactions involving the re-isomerization of one proline bond to the *cis*-configuration. The folded species with the wrong isomer accumulates as a monomeric intermediate; the proline isomerization is a prerequisite for dimerization of the C_H3 domain [42].

2.5.2
Trimeric Tail spike Protein of Bacteriophage P22

The tail spike protein (TSP) of bacteriophage P22 from *Salmonella typhimurium* has been a paradigm in studying protein folding and association in vivo since Jonathan King's early attempts to solve the problem of phage morphogenesis by genetic methods [126–130]. The beauty of both the system and the approach is that it has provided insight not only into the principles of protein self-organization in the cytoplasm but also into the mechanisms of misfolding, aggregation, and inclusion body formation [85, 115, 131–134]. It has been shown that in vitro reconstitution is clearly related to the situation in the living cell [115, 132].

The tail spike is a multifunctional trimeric protein composed of identical 72-kDa polypeptide chains, attached to the viral capsid by their 108-residue N-terminal domain. Six of these trimers assemble onto the virus head to form the tail, completing the infectious phase of the phage. In the absence of heads, tail spikes accumulate as soluble protein in the bacterial cytosol. Their assembly and the competing formation of inclusion bodies at high temperature have been studied in detail using temperature-sensitive (*ts*) mutants. In this connection, the assembly of viable phage was used as an in vitro assay for proper folding. Other means to follow the folding and assembly pathway of TSP, in addition to spectroscopic and hydrodynamic techniques, are its endorhamnosidase activity (directed toward the O-antigen of the host), its reactivity with monoclonal antibodies, and the thermal stability and detergent-resistance of the mature protein.

Crystal structures of wild-type TSP and mutant variants of the protein have been determined at high resolution [135, 136]. The major part of the trimer represents a tein concentrations, kinetic competition of folding and association occurs. In the case of TSP, decreasing the stability of an on-pathway intermediate by mutation or high temperature leads to accumulation of less-structured or misfolded conformations that aggregate at a faster rate.

As TSP has been the paradigm for folding studies in vitro and in vivo, this might be the right place to consider briefly the significance of chaperones on the above reconstitution reactions. Incompletely folded TSP chains can be trapped on the *E. coli* chaperonin GroEL in the absence of ATP. However, even an excess of GroEL/GroES increases the yield of TSP reconstitution only marginally at temperatures higher than 30 °C [140]. Overexpression of the chaperone system in *Salmonella* does not suppress *tsf* mutations [141, 142]. On recombinant expression in *E. coli*, TSP mutants impaired in folding or assembly are proteolyzed more rapidly when GroE is simultaneously overexpressed, thereby preventing aggregation of TSP in inclusion bodies [134]. The irreversible aggregation reaction itself has been studied by initiating refolding at intermediate denaturant concentrations at which a large fraction of refolding proteins can be induced to aggregate at low temperature [93, 115, 132, 143, 144]. The misassembly process can be described as a nucleated linear polymerization reaction involving partially folded or misfolded rather than fully unfolded polypeptide chains. It is specific so that TSP does not form mixed aggregates with other proteins (e.g., P22 coat protein), in agreement with the observation that overexpression or heterologous expression of proteins in bacteria commonly leads to inclusion bodies consisting of a single or very few polypeptide species [145].

2.6
Experimental Protocols

Protocol 1: Monitoring Aggregation by Light Scattering

1. Place a stirrable cuvette with the appropriate renaturation buffer in a fluorometer or photometer and adjust the temperature.
2. Set the excitation and emission wavelength to 500 nm. At this wavelength neither protein nor buffer shows any spectroscopic signal. Therefore, any signal observed results from scattering of the light by particles such as aggregates. Start a time-dependent measurement of the light-scattering signal until a stable base line is reached.
3. Dilute the denatured protein to an appropriate protein concentration (e.g., 10–100 µg mL^{-1}) into the renaturation buffer under vigorous stirring and follow the kinetics of aggregation by light scattering. Depending on the folding kinetics of the respective protein, the aggregation can occur on a minute or hour time range.
4. Aggregation can be minimized by variation of the protein concentration, temperature, buffer components and pH, and addition of solubilizing chemicals such as molar concentrations of arginine [113].

Protocol 2: Cross-linking by Glutaraldehyde

1. Dialyze the native protein against a suitable buffer, which should not contain reactive amino groups (e.g., Tris is not suitable).
2. Incubate the protein at concentrations between 5 and 100 µg mL^{-1} with 100 mM glutaraldehyde at room temperature for 2 min. Depending on the amount of cross-linking, the concentration of glutaraldehyde, the temperature, and the time can be varied.
3. Further cross-linking is blocked by addition of 200 mM NaBH$_4$ in 0.1 M NaOH.
4. The cross-linking product is analyzed by SDS-PAGE. The samples at low protein concentration have to be precipitated before. Add 0.01 volume 10% Na-desoxycholate (dissolved in slightly alkaline buffer). Subsequently, add 0.01 volume of 85% phosphoric acid. Incubate for 15 min at room temperature. Harvest the precipitate by centrifugation and resolubilize the pellet in Laemmli buffer.
5. For reconstitution experiments, choose the protein concentration at which the cross-linking is quantitative without formation of any higher molecular mass products (unspecific cross-linking).
6. Initiate reconstitution by diluting the denatured protein into the refolding buffer at the respective protein concentration.
7. Take aliquots at different time points of reconstitution and cross-link the protein as described before.
8. The bands of monomeric and oligomeric species can be quantified by densitometry of Coomassie-stained gels. (Band intensities of silver-stained gels are proportional to the amount of loaded protein only over a very narrow range and are only rarely suited for quantitative analysis.)

Protocol 3: Determination of the Dissociation Constant of a Homodimer by Analytical Ultracentrifugation

1. Dialyze the protein extensively against an appropriate buffer. If the analytical ultracentrifuge is equipped with an absorbance unit, the buffer should not absorb strongly at the detection wavelengths (230 nm and 280 nm). The buffer should contain at least 50 mM salt.
2. Dilute the protein with dialysis buffer to concentrations between 10 and 1000 µg mL^{-1}. Using an eight-hole rotor and six-channel center pieces, 21 samples (150 µL each) can be measured simultaneously.
3. Choose an appropriate rotor speed. Sedimentation equilibrium is reached when the radial distribution of the protein no longer changes over time (identical scans over a time span of 5 h). Depending on the sample volume and the size of the protein, this may take 24–72 h. Ideally, at equilibrium the protein should be completely depleted from the meniscus of the solution. The sedimentation equilibrium can be run at temperatures between 4 °C and 40 °C.
4. Measure the radial distribution of the protein at different wavelength (usually 280 nm and 230 nm).

5. Calculate the apparent molecular mass Mr from the equilibrium profile. This can be done using the standard software provided with the ultracentrifuge. A manual method would be the analysis of the slope of a plot of ln A versus r^2 according to Eq. (1)

$$Mr = (d \ln A/dr^2) \times 2RT/(1 - v\rho)\omega^2 \qquad (1)$$

with A being the absorbance, r the distance to the center of the rotor, v the partial specific volume of the protein, ρ the density of the solvent, and ω the angular velocity. The partial specific volume of the protein can be calculated from its amino acid composition [146].

6. If a monomer-dimer equilibrium exits in the respective concentration range, the apparent molecular mass Mr will vary with the protein concentration. A semi-logarithmic plot of log cp (in molar units) against Mr results in a sigmoidal curve. The dissociation constant K_D can be calculated by fitting the data to the following equations:

$$Mr(app) = ((c_M{}^*M) + c_D{}^*D))/(c_M + c_D) \qquad (2)$$

where c_M and c_D are the concentrations of the monomeric and dimeric forms, respectively. M and D represent the molecular mass of the monomer and dimer. c_D can be expressed as a function of c_M and the dissociation constant, K_D (Eq. (3)):

$$c_D = c_M{}^2/K_D \qquad (3)$$

c_M is a function of the total protein concentration c_{total} according to Eq. (4):

$$c_M = -0.5{}^*K_D/2 + [(0.5{}^*K_D/2)^2 + 0.5{}^*K_D{}^*c_{total}]^{0.5} \qquad (4)$$

Protocol 4: Thermodynamic Stability of Dimers

The interaction of subunits often contributes significantly to the overall stability of a protein. In some cases, dissociation of the subunits leads to complete denaturation of the protein. Under these conditions, the stability of the oligomeric protein can be inferred from a denaturant-induced transition curve (cf. Chapter 3 in Part I). The analysis of such data, however, is slightly different from that of monomeric proteins.

1. Prepare stock solutions of the native protein, the denatured protein (e.g., in 6 M GdmCl or 10 M urea), and the denaturation buffer (either GdmCl or urea; for preparation of denaturation buffer, compare Chapter 3 in Part I).
2. Dilute the protein into different concentrations of denaturant and allow the solutions to equilibrate. Depending on the protein, this may take only a few hours or several days. Usually a transition curve should consist of 20–30 different denaturant concentrations. To assay the reversibility of the process, the tran-

sition has to be set up starting from both the native and the denatured protein, respectively.

3. Measure the transition at different protein concentrations (e.g., 10–100 µg mL^{-1}) and using different methods (e.g., enzyme activity, fluorescence, CD). Only if the transition is reversible and the different methods yield the same transition curve (at a given protein concentration) can the data be analyzed according to a two-state model. (A reversible transition with different transition curves measured by different methods can be analyzed by a three-state model [15].) The higher the protein concentration, the more the transition midpoint is shifted to higher denaturant concentrations.

4. Calculate the equilibrium constant K_U for each data point within the transitions. K_U is defined as:

$$K_U = [D^2]/[N] = 2P_t[f^2_d/(1 - f_d)] \quad (5)$$

In this equation, P_t corresponds to the total monomer concentration of the protein and f_d to the fraction of denatured protein.

5. Calculate the free energy of stabilization ΔG at the different denaturant concentrations from the equilibrium constants using the equation

$$\Delta G = -RT \ln K_U \quad (6)$$

6. Obtain $\Delta G_{U \to N}$ in the absence of denaturant by linear extrapolation of the data to 0 M denaturant. If the two-state analysis is valid, the linear extrapolation of the ΔG values of transitions measured at different protein concentrations should intersect the y-axis at the same point (compare Figure 2.6).

References

1 JAENICKE, R. (1987). Folding and association of proteins. *Prog. Biophys. Mol. Biol.*, 49, 117–237.

2 LAWSON, D. M., ARTYMIUK, P. J., YEWDALL, S. J., SMITH, I. M., LIVINGSTONE, I. C., TREFFRY, A., LUZZAGO, A., LEVI, S., AROSIO, P., CESARENI et al. (1991). Solving the structure of human H ferritin by genetically engineering intermolecular crystal contacts. *Nature*, 349, 541–544.

3 XU, Z., HORWICH, A. L. & SIGLER, P. B. (1997). The crystal structure of the asymmetric GroEL-GroES-(ADP)7 chaperonin complex. *Nature*, 388, 741–750.

4 IZARD, T., AEVARSSON A., ALLEN, M. D., WESTPHAL, A. H., PERHAM, R. N., DE KOK, A. & HOL, W. G. (1999). Principles of quasi-equivalence and euclidean geometry govern the assembly of cubic and dodecahedral cores of pyruvate dehydrogenase complexes. *Proc. Natl Acad. Sci. USA*, 96, 1240–1245.

5 GILBERT, W. (1978). Why genes in pieces? *Nature*, 271, 501.

6 GILBERT, W. (1985). Genes-in-pieces revisited. *Science*, 228, 823–824.

7 ORENGO, C. A., MICHIE, A. D., JONES, S., JONES, D. T., SWINDELLS, M. B. & THORNTON, J. M. (1997). CATH – a hierarchic classification of protein domain structures. *Structure*, 5, 1093–1108.

8 THORNTON, J. M., ORENGO, C. A., TODD, A. E. & PEARL, F. M. (1999). Protein folds, functions and evolution. *J. Mol. Biol.*, 293, 333–342.

9 ORENGO, C. A., BRAY, J. E., BUCHAN, D. W., HARRISON, A., LEE, D., PEARL, F. M., SILLITOE, I., TODD, A. E. & THORNTON, J. M. (2002). The CATH protein family database: a resource for structural and functional annotation of genomes. *Proteomics*, 2, 11–21.

10 BERGMAN, L. W. & KUEHL, M. W. (1979). Formation of intermolecular disulfide bonds on nascent immunoglobulin polypeptides. *J. Biol. Chem.*, 254, 5690–5694.

11 JAENICKE, R. (1999) Stability and folding of domain proteins. *Prog. Biophys. Mol. Biol.*, 71, 155–241.

12 KUWAJIMA, Y. (1989). The molten globule state as a clue for understanding the folding and cooperativity of globular-protein structure. *Proteins*, 6, 87–103.

13 UVERSKY, V. N. (2002). Natively unfolded proteins: A point where biology waits for physics. *Prot. Science*, 11, 739–756.

14 NAMBA, K. (2001). Roles of partly unfolded conformations in macromolecular self-assembly. *Genes to Cells*, 6, 1–12.

15 IKEGUCHI, M., KUWAJIMA, K. & SUGAI, S. (1986). Ca2+-induced alteration in the unfolding behavior of alpha-lactalbumin. *J. Biochem.*, 99, 1191–1201.

16 SEGEL, D. J., BACHMANN, A., HOFRICHTER, J., HODGSON, K. O., DONIACH, S. & KIEFHABER, T. (1999). Characterization of transient intermediates in lysozyme folding with time-resolved small-angle X-ray scattering. *J. Mol. Biol.*, 288, 489–499.

17 WILDEGGER, G. & KIEFHABER, T. (1997). Three-state model for lysozyme folding: triangular folding mechanism with an energetically trapped intermediate. *J. Mol. Biol.*, 270, 294–304.

18 RUDOLPH, R., SIEBENDRITT, R., NESSLAUER, G., SHARMA, A. K. & JAENICKE, R. (1990). Folding of an all-beta protein: independent domain folding in gamma II-crystallin from calf eye lens. *Proc. Natl Acad. Sci. USA*, 88, 2854–2858.

19 MAYR, E. M., JAENICKE, R. & GLOCKSHUBER, R. (1997). The domains in gamma B-crystallin: identical fold-different stabilities. *J. Mol. Biol.*, 269, 260–269.

20 PALME, S., SLINGSBY, C. & JAENICKE, R. (1997). Mutational analysis of hydrophobic domain interactions in gamma B-crystallin from bovine eye lens. *Protein Sci.*, 6, 1529–1536.

21 GUITERREZ-CRUZ, G., VAN HEERDEN, A. H. & WANG, K. (2001). Modular motif, structural folds and affinity profiles of the PEVK segment of the human fetal skeletal muscle titin. *J. Biol. Chem.*, 276, 7442–7449.

22 SMITH, D. A. & REDFORD, S. (2000). Protein folding: Pulling back the frontiers. *Current Biol.*, 10, 662–664.

23 HEAD, J. G., HOUMEIDA, A., KNIGHT, P. J., CLARKE, A. R., TRINICK, J. & BRADY, R. L. (2001). Stability and folding rates of domains spanning the large A-band super-repeat of titin. *Biophys. J.*, 81, 1570–1579.

24 LI, H., OBERHAUSER, A. F., FOWLER, S. B., CLARKE, J. & FERNANDEZ, J. M. (2000). Atomic force microscopy reveals the mechanical design of a modular protein. *Proc. Natl Acad. Sci. USA*, 97, 6527–6531.

25 POLITOU, A. S., GAUTEL, M., IMPROTA, M., VANGELISTA, L. & PASTORE, A. (1996). The elastic I band region of titin is assembled in a "modular" fashion by weakly interacting Ig-like domains. *J. Mol. Biol.*, 255, 604–616.

26 FOWLER, S. B. & CLARKE, J. (2001). Mapping the folding pathway of an immunoglobulin domain: structural detail from Phi value analysis and movement of the transition state. *Structure*, 9, 355–366.

27 BEST, R. B., FOWLER, S. B., HERRERA, J. L., STEWARD, A., PACI, E. & CLARKE, J. (2003). Mechanical unfolding of a titin Ig domain: structure of the transition state revealed by combining atomic force microscopy, protein engineering and molecular dynamics simulations. *J. Mol. Biol.*, 330, 867–877.

28 Best, B., Li, B., Steward, A., Daggett, V. & Clarke, J. (2001). Can non-mechanical proteins withstand force? Stretching barnase by atomic force microscopy and molrcular dynamics simlation. *Biophys. J.*, 81, 2344–2356.

29 Powers, E. T. & Powers, D. L. (2004). A perspective of mechanisms of protein tetramer formation. *Biophys. J.*, 85, 3587–3599.

30 Seckler, R. (2000). Assembly of multi-subunit strucutres. In: Mechanisms of Protein Folding, Ed.: R. H. Pain, 2nd ed., Oxford Press, pp. 279–308.

31 Branden, C.-I. and Tooze, J. (1999). Introduction to protein structure, (2nd edn). Garland, New York, p. 410.

32 Smith, S. (1994). The animal fatty add synthase: one gene, one polypeptide, seven enzymes. *FASEB J.*, 8, 1248–1259.

33 Weber, G., Schorgendorfer, K., Schneider-Scherzer, E. & Leitner, E. (1994). The peptide synthetase catalyzing cyclosporin production in Tolypocladium niveum is encoded by a giant 45.8-kilobase open reading frame. *Curr. Genet.*, 26, 120–125.

34 Dautry-Varsat, A. & Garel, J. R. (1978). Refolding of a bifunctional enzyme and its monofunctional fragment. *Proc. Natl Acad. Sci. USA*, 75, 5979–5982.

35 Creighton, T. E. (1992). Proteins: structures and molecular properties, (2nd edn). W. H. Freeman, New York, p. 309–325.

36 Miller, S., Schuler, B. & Seckler, R. (1998). Phage P22 tailspike protein: removal of head binding domain unmasks effects of folding mutations on native-state thermal stability. *Prot. Sci.*, 7, 2223–2232.

37 Danner, M., Fuchs, A., Miller, S. and Seckler, R. (1993). Folding and assembly of phage P22 tailspike endo-rhamnosidase lacking the N-terminal, head-binding domain. *Eur. J. Biochem.*, 215, 653–661.

38 Jones, S. & Thornton, J. M. (1997). Analysis of protein-protein interaction sites using surface patches. *J. Mol. Biol.*, 272, 121–132.

39 Tsai, C. J., Lin, S. L., Wolfson, H. J. & Nussinov, R. (1997). Studies of protein-protein interfaces: a statistical analysis of the hydrophobic effect. *Protein Sci.*, 6, 53–64.

40 Tsai, C. J., Xu, D. & Nussinov, R. (1997). Structural motifs at protein-protein interfaces: protein cores versus two-state and three-state model complexes. *Prot. Sci.*, 6, 1793–1805.

41 Le Conte, L., Chothia, C. & Ianin, J. (1999). The atomic structure of protein-protein recognition sites. *J. Mol. Biol.*, 285, 2177–2198.

42 Thies, M., Mayer, S., Augustine, J. G., Frederick, C. A., Lilie, H. & Buchner, J. (1999). Folding and association of the antibody domain $C_H 3$: Prolyl isomerization preceeds dimerization. *J. Mol. Biol.*, 293, 67–79.

43 Chothia, C. and Janin, J. (1975). Principles of protein-protein recognition. *Nature*, 256, 705–708.

44 Dill, K. A. (1990). Dominant forces in protein folding. *Biochemistry*, 29, 7133–7155.

45 Larsen, T. A., Olson, A. J. & Goodsell, D. S. (1998). Morphology of protein-protein interfaces. *Structure*, 6, 421–427.

46 DeLano, W. L. (2002). Unravelling hot spots in the binding interface: progress and challenges. *Curr. Opin. Struct. Biol.*, 12, 14–20.

47 Xu, D., Tsai, C. J. & Nussinov, R. (1998). Mechanism and evolution of protein dimerization. *Protein Sci.*, 7, 533–544.

48 Killenberg-Jabs, M., Jabs, A., Lilie, H., Golbik, R. & Hubner, G. (2001). Active oligomeric states of pyruvate decarboxylase and their functional characterization. *Eur. J. Biochem.*, 268, 1698–1704.

49 Venyaminov, S. Yu., Gudkov, A. T., Gogia, Z. V. & Tumanova, L .G. (1981). Absorption and cicular dichroism spectra of individual proteins from Escherichia coli ribosomes. Pushkino, Russia.

50 Yusupov, M. M., Yusupova, G. Z., Baucom, A., Lieberman, K., Earnest, T. N., Cate, J. H. & Noller, H. F. (2001). *Science*, 292, 883–896.

51 KRIWACKI, R. W., HENGST, L., TENNANT, L., REED, S. I. & WRIGHT, P. E. (1996). Structural studies of p21 in the free and Cdk2-bound state: Conformational disorder mediates binding diversity. *Prog. Natl Acad. Sci. USA*, 93, 11504–11509.

52 YU, Y. B. (2002). Coiled coils: stability, specificity, and drug delivery potential. *Adv. Drug Deliv. Rev.*, 54, 1113–1129.

53 RICHTER, S., STUBENRAUCH, K., LILIE, H. & RUDOLPH, R. (2001). Polyionic fusion peptides function as specific dimerization motifs. *Prot. Engineering*, 14, 775–783.

54 STUBENRAUCH, K., GLEITER, S., BRINKMANN, U., RUDOLPH, R. & LILIE, H. (2001). Tumor cell specific targeting and gene transfer by recombinant polyoma virus like particle/antibody conjugates. *Biochem. J.*, 356, 867–873.

55 KLEINSCHMIDT, M., RUDOLPH, R. & LILIE, H. (2003). Design of a modular immunotoxin connected by polyionic adapter peptides. *J. Mol. Biol.*, 327, 445–452.

56 BENNETT, M. J., CHOE, S. & EISENBERG, D. (1994). Domain swapping: entangling alliances between proteins. *Proc. Natl Acad. Sci. USA*, 91, 3127–3131.

57 BENNETT, M. J., SCHLUNEGGER, M. P. & EISENBERG, D. (1995). 3D domain swapping: a mechanism for oligomer assembly. *Protein Sci.*, 4, 2455–2468.

58 SCHLUNEGGER, M. P., BENNETT, M. J. & EISENBERG, D. (1997). Oligomer formation by 3D domain swapping: a model for protein assembly and misassembly. *Adv. Protein Chem.*, 50, 61–122.

59 BAX, B., LAPATTO, R., NALINI, V., DRIESSEN, H., LINDLEY, P. V., MAHADEVAN, D., BLUNDELL, T. L. & SLINGSBY, C. (1990). X-ray analysis of beta B2-crystallin and evolution of oligomeric lens proteins. *Nature*, 347, 776–779.

60 LIU, Y. & EISENBERG, D. (2202). 3D domain swapping: As domains continue to swap. *Prot. Science*, 11, 1285–1299.

61 VIJAYALAKSHMI, J., MUKHERGEE, M. K., GRAUMANN, J., JAKOB, U. & SAPER, M. A. (2001). The 2.2 A crystal structure of Hsp33: A heat shock protein with a redox-regulated chaperone activity. *Structure*, 9, 367–375.

62 KIM, S. J., JEONG, D. G., CHI, S. W., LEE, J. S. & RYA, S. E. (2001). Crystal structure of proteolytic fragments of the redox-sensitive Hsp33 with constitutive chaperone activity. *Nat. Struct. Biol.*, 8, 459–466.

63 GRAUMANN, J., LILIE, H., TANG, X., TUCKER, K. A., HOFFMANN, J., JANAKIRAMAN, V., SAPER, M., BARDWELL, J. C. A. & JAKOB, U. (2001). Activation of the redox regulated chaperone Hsp33 – a two step mechanism. *Structure*, 9, 377–387.

64 LOMAS, D. A., EVANS, D. L., FINCH, J. T. & CARRELL, R. W. (1992). The mechanism of Z alpha 1-antitrypsin accumulation in the liver. *Nature*, 357, 541–542.

65 D'ALESSIO, G. (1995). Oligomer evolution in action? *Nat. Struct. Biol.*, 2, 11–13.

66 PICCOLI, R., TAMBURRINI, M., PICCIALLI, G., DI DONATO, A., PARENTE, A. & D'ALESSIO, G. (1992). The dual-mode quaternary structure of seminal RNase. *Proc. Natl Acad. Sci. USA*, 89, 1870–1874.

67 KONO, N., UYEDA, K. & OLIVER, R. M. (1973). Chicken liver phosphofructokinase. I. Crystallization and physicochemical properties. *J. Biol. Chem.*, 248, 8592–8602.

68 BOWIE, J. U. & SAUER, R. T. (1989). Equilibrium dissociation and unfolding of the Arc repressor dimer. *Biochemistry*, 28, 7139–7143.

69 GOLBIK, R., NAUMANN, M., OTTO, A., MÜLLER, E.-C., BEHLKE, J., REUTER, R., HÜBNER, G. & KRIEGEL, T. M. (2001). Regulation of phosphotransferase activity of hexokinase 2 from Saccharomyces cerevisiae by modification at serine-14. *Biochemistry*, 40, 1083–1090.

70 BAER, D., GOLBIK, R., HUEBNER, G., LILIE, H., MUELLER, E. C., NAUMANN, M., OTTO, A., REUTER, R., BREUNIG, K. D. & KRIEGEL, T. M. (2003). The unique hexokinase of Kluyveromyces lactis: Isolation, molecular characteri-

zation and evaluation of a role in glucose signaling. *J. Biol. Chem.*, 278, 39280–39286.

71 HOFFMANN, J. H., LINKE, K., GRAF, P. C., LILIE, H. & JAKOB, U. (2004). Identification of a redox regulated chaperone network. *EMBO J.*, 23, 160–168.

72 NEET, K. E. & TIMM, D. E. (1994). Conformational stability of dimeric proteins: quantitative studies by equilibrium denaturation. *Protein Sci.*, 3, 2167–2174.

73 SINCLAIR, J. F., ZIEGLER, M. M. & BALDWIN, T. O. (1994). Kinetic partitioning during protein folding yields multiple native states. *Nature Struct. Biol.*, 1, 320–326.

74 NICHTL, A., BUCHNER, J., JAENICKE, R., RUDOLPH, R. & SCHEIBEL, T. (1998). Folding and association of beta-galactosidase. *J. Mol. Biol.*, 282, 1083–1091.

75 FUCHS, A., SEIDERER, C. & SECKLER, R. (1991). In vitro folding pathway of phage P22 tailspike protein. *Biochemistry*, 30, 6598–6604.

76 LILIE, H., RUDOLPH, R. & BUCHNER, J. (1995). Association of antibody chains at different stages of folding; Prolyl isomerization occurs after formation of quaternary structure. *J. Mol. Biol.*, 248, 190–201.

77 CLELAND, J. L. & WANG, D. I. (1990). Refolding and aggregation of bovine carbonic anhydrase B: quasi-elastic light scattering analysis. *Biochemistry*, 29, 11072–11078.

78 BUCHNER, J., GRALLERT, H. & JAKOB, U. (1998). Analysis of chaperone function using citrate synthase as nonnative substrate protein. *Methods Enzymol.*, 290, 323–338.

79 STRYER, L. (1978) Fluorescence energy transfer as a spectroscopic ruler. *Annu. Rev. Biochem.*, 47, 819–846.

80 HASSIEPEN, U., FEDERWISCH, M., MULDERS, T., LENZ, V. J., GATTNER, H. G., KRUGER, P. & WOLLMER, A. (1998). Analysis of protein self-association at constant concentration by fluorescence-energy transfer. *Eur. J. Biochem.*, 255, 580–587.

81 TENENBAUM-BAYER, H. & LEVITZKI, A. (1976). The refolding of lactate dehydrogenase subunits and their assembly to the functional tetrarner. *Biochim. Biophys. Acta*, 445, 261–279.

82 CARDENAS, J. M., HUBBARD, D. R. and ANDERSON, S. (1977). Subunit structure and hybrid formation of bovine pyruvate kinases. *Biochemistry*, 16, 191.

83 GROSSMAN, S. H., PYLE, J. & STEINER, R. J. (1981). Kinetic evidence for active monomers during the reassembly of denatured creatine kinase. *Biochemistry*, 20, 6122–6128.

84 BURNS, D. L. & SCHACHMAN, H. K. (1982). Assembly of the catalytic trimers of aspartate transcarbamoylase from folded monomers. *J. Biol. Chem.*, 257, 8638–8647.

85 DANNER, M. & SECKLER, R. (1993). Mechanism of phage P22 tailspike protein folding mutations. *Protein Sci.*, 2, 1869–1881.

86 WEISSMAN, J. S. (1995). All roads lead to Rome? The multiple pathways of protein folding. *Chem. Biol.*, 2, 255–260.

87 DILL, K. A. & CHAN, H. S. (1997). From Levinthal to pathways to funnels. *Nat. Struct. Biol.*, 4, 10–19.

88 JAENICKE, R. & LILIE, H. (2000). Folding and association of oligomeric and multimeric proteins. *Adv. Prot. Chem.*, 53, 329–401.

89 KERN, G., KERN, D., JAENICKE, R. & SECKLER, R. (1993). Kinetics of folding and association of differently glycosylated variants of invertase from Saccharomyces cerevisiae. *Prot. Science*, 2, 1862–1867.

90 RUDOLPH, R., WESTHOF, E. & JAENICKE, R. (1977). Kinetic analysis of the reactivation of rabbit muscle aldolase after denaturation with guanidine-HCl. *FEBS Lett.*, 73, 204–206.

91 OPITZ, U., RUDOLPH, R., JAENICKE, R., ERICSSON, L. & NEURATH, H. (1987). Proteolytic dimers of porcine muscle lactate dehydrogenase: characterization, folding, and reconstitution of the truncated and nicked polypeptide chain. *Biochemistry*, 26, 1399–13406.

92 MILLER, S., SCHULER, B. & SECKLER,

R. (1998). A reversibly unfolding fragment of P22 tailspike protein with native structure: the isolated beta-helix domain. *Biochemistry*, 37, 9160–9168.

93 JAENICKE, R. & RUDOLPH, R. (1986). Refolding and association of oligomeric proteins. *Methods Enzymol.*, 131, 218–250.

94 HERMANN, R., RUDOLPH, R. & JAENICKE, R. (1982). The use of subunit hybridization to monitor the reassociation of porcine lactate dehydrogenase after acid dissociation. *Hoppe Seylers Z. Physiol. Chem.*, 363, 1259–1265.

95 GERL, M., RUDOLPH, R. & JAENICKE, R. (1985). Mechanism and specificity of reconstitution of dimeric lactate dehydrogenase from Limulus polyphemus. *Biol. Chem. Hoppe Seyler*, 366, 447–454.

96 BORCHERT, T. V., ABAGYAN, R., JAENICKE, R. & WIERENGA, R. K. (1994). Design, creation, and characterization of a stable, monomeric triosephosphate isomerase. *Proc. Natl Acad. Sci. USA*, 91, 1515–1518.

97 SCHLIEBS, W., THANKI, N., JAENICKE, R. & WIERENGA, R. K. (1997). A double mutation at the tip of the dimer interface loop of triosephosphate isomerase generates active monomers with reduced stability. *Biochemistry*, 36, 9655–9662.

98 BEERNINK, P. T. & TOLAN, D. R. (1996). Disruption of the aldolase A tetramer into catalytically active monomers. *Proc. Natl Acad. Sci. USA*, 93, 5374–5379.

99 KOREN, R. & HAMMES, G. G. (1976). A kinetic study of protein-protein interactions. *Biochemistry*, 15, 1165–1171.

100 NORTHRUP, S. H. & ERICKSON, H. P. (1992). Kinetics of protein-protein association explained by Brownian dynamics computer simulation. *Proc. Natl Acad. Sci. USA*, 89, 3338–3342.

101 GLOSS, L. M. & MATTHEWS, C. R. (1998). Mechanism of folding of the dimeric core domain of Escherichia coli trp repressor: a nearly diffusion-limited reaction leads to the formation of an on-pathway dimeric intermediate. *Biochemistry*, 37, 15990–15998.

102 MILLA, M. E. & SAUER, R. T. (1994). P22 Arc repressor: folding kinetics of a single-domain, dimeric protein. *Biochemistry*, 33, 1125–1131.

103 HERRMANN, R., RUDOLPH, R., JAENICKE, R., PRICE, N. C. & SCOBBIE, A. (1983). The reconstitution of denatured phosphoglycerate mutase. *J. Biol. Chem.*, 258, 11014–11019.

104 FISHER, A. J., RAUSHEL, F. M., BALDWIN, T. O. & RAYMENT, I. (1995). Three-dimensional structure of bacterial luciferase from Vibrio harveyi at 2.4 A resolution. *Biochemistry*, 34, 6581–6586.

105 THODEN, J. B., HOLDEN, H. M., FISHER, A. J., SINCLAIR, J. F., WESENBERG, G., BALDWIN, T. O. & RAYMENT, I. (1997). Structure of the beta 2 homodimer of bacterial luciferase from Vibrio harveyi: X-ray analysis of a kinetic protein folding trap. *Protein Sci.*, 6, 13–23.

106 TANNER, J. J., MILLER, M. D., WILSON, K. S., TU, S. C. & KRAUSE, K. L. (1997). Structure of bacterial luciferase beta 2 homodimer: implications for flavin binding. *Biochemistry*, 36, 665–672.

107 WADDLE, J. J., JOHNSTON, T. C. & BALDWIN, T. O. (1987). Polypeptide folding and dimerization in bacterial-luciferase occur by a concerted mechanism in vivo. *Biochemistry*, 26, 4917–4921.

108 ZIEGLER, M. M., GOLDBERG, M. E., CHAFFOTTE, A. F. & BALDWIN, T. O. (1993). Refolding of luciferase subunits from urea and assembly of the active heterodimer. Evidence for folding intermediates that precede and follow the dimerization step on the pathway to the active form of the enzyme. *J. Biol. Chem.*, 268, 10760–10765.

109 BALDWIN, T. O., ZIEGLER, M. M., CHAFFOTTE, A. F. & GOLDBERG, M. E. (1993). Contribution of folding steps involving the individual subunits of bacterial luciferase to the assembly of the active heterodimeric enzyme. *J. Biol. Chem.*, 268, 10766–10772.

110 CLARK, A. C., SINCLAIR, J. F. & BALDWIN, T. O. (1993). Folding of

bacterial luciferase involves a non-native heterodimeric intermediate in equilibrium with the native enzyme and the unfolded subunits. *J. Biol. Chem.*, 268, 10773–10779.
111 SINCLAIR, J. F., WADDLE, J. J., WADDILL, E. F. & BALDWIN, T. O. (1993). Purified native subunits of bacterial luciferase are active in the bioluminescence reaction but fail to assemble into the alpha beta structure. *Biochemistry*, 32, 5036–5044.
112 ZETTLMEISSL, G., RUDOLPH, R. & JAENICKE, R. (1979). Reconstitution of lactic dehydrogenase. Noncovalent aggregation vs. reactivation. 1. Physical properties and kinetics of aggregation. *Biochemistry*, 18, 5567–5571.
113 DE BERNADEZ CLARK, E., SCHWARZ, E. & RUDOLPH, R. (1999). Inhibition of Aggregation Side Reactions during in vitro Protein Folding. *Methods in Enzymology*, 309, 217–236.
114 GOLDBERG, M. E., RUDOLPH, R. & JAENICKE, R. (1991). A kinetic study of the competition between renaturation and aggregation during the refolding of denatured-reduced egg white lysozyme. *Biochemistry*, 30, 2790–2797.
115 JAENICKE, R. & SECKLER, R. (1997). Protein misassembly in vitro. *Adv. Protein Chem.*, 50, 1–59.
116 WETZEL, R. (1997). Domain stability in immunoglobulin light chain deposition disorders. *Adv. Protein Chem.*, 50, 183–242.
117 SUNDE, M. & BLAKE, C. (1997). The structure of amyloid fibrils by electron microscopy and X-ray diffraction. *Adv. Protein Chem.*, 50, 123–159.
118 KIEFHABER, T., RUDOLPH, R., KOHLER, H. H. & BUCHNER, J. (1991). Protein aggregation in vitro and in vivo: a quantitative model of the kinetic competition between folding and aggregation. *Biotechnology (N.Y.)*, 9, 825–829.
119 GOTO, Y. & HAMAGUCHI, K. (1982). Unfolding and refolding of the constant fragment of the immunoglobulin light chain. *J. Mol. Biol.*, 156, 891–910.
120 TSUNENAGA, M., GOTO, Y., KAWATA, Y. & HAMAGUCHI, K. (1987). Unfolding and refolding of a type kappa immunoglobulin light chain and its variable and constant fragments. *Biochemistry*, 26, 6044–6051.
121 LANG, K., SCHMID, F. X. & FISCHER, G. (1987). Catalysis of protein folding by prolyl isomerase. *Nature*, 329, 268–270.
122 LILIE, H., LANG, K., RUDOLPH, R. & BUCHNER, J. (1993). Prolyl isomerases catalyze antibody folding in vitro. *Prot. Sci.*, 2, 1490–1496.
123 AUGUSTINE, J. G., DE LA CALLE, A., KNARR, G., BUCHNER, J. & FREDERICK, C. A. (2001). The crystal structure of the fab fragment of the monoclonal antibody MAK33. Implications for folding and interaction with the chaperone bip. *J. Biol. Chem.*, 276, 3287–3294.
124 PADLAN, E. A., COHEN, G. H. & DAVIES, D. R. (1986). Antibody Fab assembly: the interface residues between C_H1 and C_L. *Mol. Immunol.*, 23, 951–960.
125 CHOTHIA, C., NOVOTNY, Y., BRUCCOLERI, R. & KARPLUS, M. (1985). Domain association in immunoglobulin molecules. The packing of variable domains. *J. Mol. Biol.*, 186, 651–663.
126 SMITH, D. H., BERGET, P. B. & KING, J. (1980). Temperature-sensitive mutants blocked in the folding or subunit assembly of the bacteriophage P22 tail-spike protein. I. Fine-structure mapping. *Genetics*, 96, 331–352.
127 GOLDENBERG, D. P. & KING, J. (1981). Temperature-sensitive mutants blocked in the folding or subunit of the bacteriophage P22 tail spike protein. II. Active mutant proteins matured at 30 degrees C. *J. Mol. Biol.*, 145, 633–651.
128 SMITH, D. H. & KING, J. (1981). Temperature-sensitive mutants blocked in the folding or subunit assembly of the bacteriophage P22 tail spike protein. III. Intensive polypeptide chains synthesized at 39 degrees C. *J. Mol. Biol.*, 145, 653–676.
129 KING, J. and YU, M. H. (1986). Mutational analysis of protein folding

pathways: the P22 tailspike endorhamnosidase. *Methods Enzymol.*, 131, 250–266.

130 BETTS, S. & KING, J. (1999). There's a right way and a wrong way: in vivo and in vitro folding, misfolding and subunit assembly of the P22 tailspike. *Structure Fold. Des.*, 7, 131–139.

131 MITRAKI, A., FANE, B., HAASE PETTINGELL, C., STURTEVANT, J. & KING, J. (1991). Global suppression of protein folding defects and inclusion body formation. *Science*, 253, 54–58.

132 MITRAKI, A., DANNER, M., KING, J. & SECKLER, R. (1993). Temperature-sensitive mutations and second-site suppressor substitutions affect folding of the P22 tailspike protein in vitro. *J. Biol. Chem.*, 268, 20071–20075.

133 SCHULER, B. & SECKLER, R. (1998). P22 tailspike folding mutants revisited: effects on the thermodynamic stability of the isolated beta-helix domain. *J. Mol. Biol.*, 281, 227–234.

134 SECKLER, R. (1998). Folding and function of repetitive structure in the homotrimeric phage P22 tailspike protein. *J. Struct. Biol.*, 122, 216–222.

135 STEINBACHER, S., SECKLER, R., MILLER, S., STEIPE, B., HUBER, R. & REINEMER, P. (1994). Crystal structure of P22 tailspike protein: interdigitated subunits in a thermostable trimer. *Science*, 265, 383–386.

136 STEINBACHER, S., MILLER, S., BAXA, U., BUDISA, N., WEINTRAUB, A., SECKLER, R. & HUBER, R. (1997). Phage P22 tailspike protein: crystal structure of the head-binding domain at 2.3 A, fully refined structure of the endorhamnosidase at 1.56 A resolution, and the molecular basis of O-antigen recognition and cleavage. *J. Mol. Biol.*, 267, 865–880.

137 BAXA, U., STEINBACHER, S., WEINTRAUB, A., HUBER, R. & SECKLER, R. (1999). Mutations improving the folding of phage P22 tailspike protein affect its receptor binding activity. *J. Mol. Biol.*, 293, 693–701.

138 BEISSINGER, M., LEE, S. C., STEINBACHER, S., REINEMER, P., HUBER, R., YU, M. H. & SECKLER, R. (1995). Mutations that stabilize folding intermediates of phage P22 tailspike protein: folding in vivo and in vitro, stability, and structural context. *J. Mol. Biol.*, 249, 185–194.

139 GOLDENBERG, D. & KING, J. (1982). Trimeric intermediate in the in vivo folding and subunit assembly of the tail spike endorhamnosidase of bacteriophage P22. *Proc. Natl Acad. Sci. USA*, 79, 3403–3407.

140 BRUNSCHIER, R., DANNER, M. & SECKLER, R. (1993). Interactions of phage P22 tailspike protein with GroE molecular chaperones during refolding in vitro. *J. Biol. Chem.*, 268, 2767–2772.

141 GORDON, C. L., SATHER, S. K., CASJENS, S. & KING, J. (1994). Selective in vivo rescue by GroEL/ES of thermolabile folding intermediates to phage P22 structural proteins. *J. Biol. Chem.*, 269, 27941–27951.

142 SATHER, S. K. & KING, J. (1994). Intracellular trapping of a cytoplasmic folding intermediate of the phage P22 tailspike using iodoacetamide. *J. Biol. Chem.*, 269, 25268–25276.

143 SPEED, M. A., WANG, D. I. & KING, J. (1995). Multimeric intermediates in the pathway to the aggregated inclusion body state for P22 tailspike polypeptide chains. *Protein Sci.*, 4, 900–908.

144 SPEED, M. A., MORSHEAD, T., WANG, D. I. & KING, J. (1997). Conformation of P22 tailspike folding and aggregation intermediates probed by monoclonal antibodies. *Protein Sci.*, 6, 99–108.

145 SPEED, M. A., WANG, D. I. & KING, J. (1996). Specific aggregation of partially folded polypeptide chains: the molecular basis of inclusion body composition. *Nat. Biotechnol.*, 14, 1283–1287.

146 DURCHSCHLAG, H. (1986). Specific volumes of biological macromolecules and some other molecules of biological interest. In: Thermodynamic data for biochemistry and biotechnology, ed.: H.-J. HINZ, Springer Verlag, Heidelberg, pp. 45–127.

3
Studying Protein Folding in Vivo

I. Marije Liscaljet, Bertrand Kleizen, and Ineke Braakman

3.1
Introduction

To be biologically active, proteins must fold into their three-dimensional, native structure. Christian Anfinsen and others showed decades ago that all information required for a protein to attain this three-dimensional structure resides in its primary sequence [1]. Irrespective of whether a protein folds in an intact cell, in vivo, or in a test tube, in vitro, essentially the same end product is obtained. Yet, many differences exist between the two. While in vivo folding starts co-translationally [2, 3], in vitro (re)folding starts from the complete protein molecule. The folding environment is different as well: in vivo proteins are folded in an extremely crowded environment [4–6], containing hundreds of milligrams of protein per milliliter, in contrast to in vitro folding where the protein is purified and folds in a diluted system. In vitro, folding often is inefficient under biological conditions such as physiological temperatures, with a large fraction of the protein misfolding and precipitating. In vivo, proteins are assisted during folding, and because of chaperones and folding enzymes [7–9], the folding process is more efficient and results in less aggregation than in vitro [10]. In vitro folding can take place in milliseconds, proteins can be studied directly at the molecular level, and there are no barriers such as membranes between experimenter and the folding proteins. Success rate is highest with relatively small, single-domain proteins, whereas especially mammalian proteins usually are large, consisting of more than one domain. Although in vitro folding experiments provide information on molecular details of the folding process, in vivo studies are essential for a complete view on protein folding. This chapter will discuss how protein folding in a cell can be studied and manipulated, with a slight focus on the endoplasmic reticulum.

3.2
General Features in Folding Proteins Amenable to in Vivo Study

In contrast to in vitro protein folding, protein folding in vivo can be studied only via indirect methods. Instead of direct measurements on the protein itself, the fold-

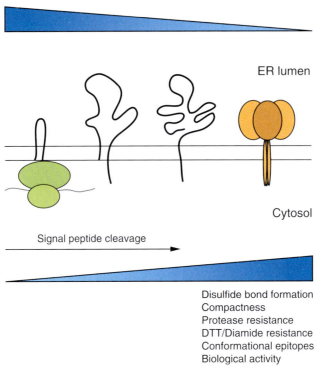

Fig. 3.1. Studying protein folding in the ER. Newly synthesized proteins that are folding in the ER lose and gain various features, which can be used to determine their folding status; examples are given in this figure. Folding starts on the nascent chain and continues after chain termination, during or after which the protein may oligomerize.

ing status of a protein can be followed through changes in shape, associations with other proteins, co- and posttranslational modifications, and its compartmental location (Figure 3.1). During in vivo folding, proteins are not purified but are part of the cell "soup." So far the best way to follow folding of a small population of newly synthesized proteins in a cell is by labeling them with ^{35}S-labeled cysteine and/or -methionine. In this way, changes in a small, synchronized protein population can be followed with time.

A pulse-chase experiment (Figure 3.2) in intact cells is an assay in which kinetics and characteristics of sequential steps are determined in vivo. Newly synthesized proteins are labeled with radioactive amino acids for a very short pulse time, because folding starts during synthesis and one would ideally follow folding from the very first folding intermediates. Labeled proteins are "chased" by incubation with unlabeled amino acids. After different times of chase, cells and supernatants can be collected and cooled on ice, and free cysteines are blocked with an alkylating

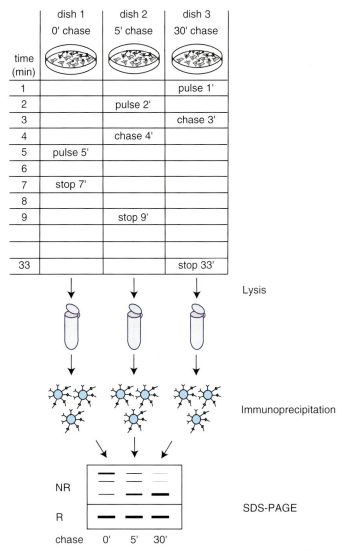

Fig. 3.2. Generic pulse-chase time schedule. To study folding and disulfide bond formation of a newly synthesized protein in the ER, adherent cells expressing the protein of interest are pulse labeled for a short period and chased with unlabeled amino acids; one dish is needed per time point. The experimental design is such that dishes can be pulse labeled and chased in parallel. After different times of chase, free cysteines are blocked with an alkylating agent and cells are lysed in a detergent-containing buffer. The protein is isolated using immunoprecipitation, and samples are analyzed using non-reducing (NR) and reducing (R) SDS-PAGE to visualize disulfide bond formation and changes in molecular weight.

agent. Cells are lysed in a detergent-containing buffer and the protein under study is isolated by immunoprecipitation. Samples can be analyzed by various types of gel electrophoresis, including native gels suitable for detection of complexes, and reducing and non-reducing SDS-PAGE to detect folding intermediates with a variable set of disulfide bonds (see the Appendix and [11]). Other changing features of folding proteins, such as their compactness or resistance to proteolytic digestion, can be determined with this pulse-chase assay as well.

Other labeling assays that can be used to study protein folding include the pulse-chase assay in suspension cells, microsomes, or semi-permeabilized cell systems [12] in which mRNA is translated in vitro in the presence of a source of ER membranes (see Chapter 18) and the recently developed in vitro chase assay in which a protein is translated in vivo and chased in a detergent lysate (Maggioni et al., manuscript submitted). These assays will be discussed in the Appendix.

3.2.1
Increasing Compactness

One of the general features of a protein during folding is its increasing compactness. During the folding process of a soluble protein, all hydrophobic side chains will strive to be buried inside the molecule and all hydrophilic ones to be exposed on the outside; hydrogen bonds are formed within the protein and between exposed residues and water. Electrostatic, van der Waals, and hydrophobic interactions play a major role during the folding process, and disulfide bridges will form within the protein when conformation and environmental conditions allow. The end product of the folding process will be the native conformation of the protein, which is assumed to be the most compact, energy-favorable form. To study the increase in compactness of proteins, non-denaturing gel electrophoresis can be used. The formation of oligomers, protein complexes, and transient protein-protein interactions can be detected on a native gel as well as by sucrose velocity gradient centrifugation. To monitor disulfide bond formation within or between polypeptide chains, non-reducing denaturing gel electrophoresis (SDS-PAGE) [13] is an effective technique. In the assays described in this chapter, the folding process of a (disulfide bond containing glyco)protein can be followed kinetically.

3.2.2
Decreasing Accessibility to Different Reagents

Due to the increasing compactness of a protein on its folding pathway, it will become less accessible to various reagents including proteases. Changes in protein conformation usually result in decreased sensitivity to protease digestion, since potential cleavage sites are buried inside the protein during the folding process. To limit the appearance of protease-specific bands, a range of proteases should be used optimally, some of which cleave frequently in a protein. Examples of the latter are proteinase K, TPCK-trypsin, and TLCK-chymotrypsin. Limited proteolysis was first used to identify domains in immunoglobulin [14] and is a very powerful tool.

The target protein is exposed to a protease, and in a time course the generation of stable products, i.e. protected inaccessible domains, can be monitored. To identify these domains, the digestion pattern of full-length protein can be compared to that of expressed isolated fragments of the protein, or domain-specific antibodies can be used to identify protease-resistant fragments.

Other reagents that can be used to obtain more information about the folding status of a protein include diazenedicarboxylic acid bis(N,N-dimethylamide) (diamide) and 4,4'-dithiodipyridine (4-DPS) [15, 16]. Diamide is an oxidant that, when added to the chase medium, will over-oxidize all compartments in the cell, resulting in oxidation of all cellular thiols, including those in proteins and glutathione. Since only proteins with free sulfhydryl groups are sensitive to diamide treatment, in the secretory pathway, only folding intermediates will oxidize, as most mature folded proteins have disulfide bonds. If treated and non-treated proteins are analyzed by non-reducing SDS-PAGE, folding intermediates and the native protein can be discriminated [17–19] (Braakman and Helenius, in preparation).

Like diamide, reducing agents such as dithiothreitol (DTT) or 2-mercaptoethanol (2ME) can be used in a pulse-chase assay to determine the folding status of a protein. The influenza virus hemagglutinin (HA) [20, 21] and other proteins with disulfide bonds [22, 23] undergo a conversion from DTT sensitivity to DTT resistance during their conformational maturation in the ER. In combination with conformation-sensitive and epitope-specific antibodies, DTT sensitivity is very useful in determining the folding hierarchy in a protein.

3.2.3
Changes in Conformation

During folding, domains are formed within a protein, coinciding with changes in conformation. Antigenic epitopes become exposed, are masked, or form because distant parts of the polypeptide chain start to interact. Conformational changes can be followed by probing the folding protein with conformation-sensitive antibodies. Contrary to general belief, completely conformation-sensitive antibodies are rare. Polyclonal as well as monoclonal antibodies generally recognize a limited conformation range of a protein, depending on how the antigen used to produce the antibody was obtained. Protein bands cut out of SDS-PA gels, when injected into rabbits, yield antibodies that work well in Western blotting (against SDS-denatured antigen) but that often fail to work in immunofluorescence, EM, or immunoprecipitation, which are conditions that allow proteins to remain more or less native. Antibodies generated against native, properly folded and assembled antigens often fail to interact with folding intermediates or SDS-denatured protein. Antibodies against peptides often fail to bind protein, because the peptide is not representative of the protein conformation or the peptide is not exposed in the complete protein. Therefore, an antibody needs to be characterized before it becomes a powerful tool in protein-folding studies.

Another assay that can be used concerns biological function of the folding protein: this includes binding to ligands and other proteins or an increase in enzymatic activity. To test, for instance, whether the protein can bind its receptor, the receptor may be coupled to Sepharose beads, which can be used in a precipitation assay to isolate the desired protein, or, alternatively, interaction can be tested in an ELISA-like assay.

3.2.4
Assistance During Folding

All information required for a protein to attain its final three-dimensional structure resides in its primary sequence [1]. However, complex proteins need the assistance of molecular chaperones and folding enzymes to reach their native structure efficiently without formation of (large) aggregates. Molecular chaperones interact with nonnative states of proteins; they are important during folding of newly synthesized polypeptides where they facilitate rate-limiting steps, stabilize unfolded proteins, and prevent unwanted intra- and interchain interactions, which could lead to aggregation [24–26]. Classical chaperones in the ER, such as BiP [27], recognize exposed hydrophobic patches [28, 29], while calnexin and calreticulin recognize monoglucosylated glycan chains on proteins in the ER [30]. BiP is a member of the Hsp70 family; ATP binding and hydrolysis are crucial for the cycle of binding and release of substrate [31, 32]. The preference for binding of either BiP or calnexin/calreticulin is thought to depend on the position of the first N-linked glycan in the polypeptide chain [33]. When N-glycans are present in the first ~50 amino acids, a protein in general will associate with calnexin and calreticulin first, whereas proteins with N-glycans downstream in the sequence may bind BiP or other chaperones first. Other observations, however, suggest that calnexin/calreticulin may interact with non-glycosylated proteins as well [34, 35]. As a rule, a protein's association with molecular chaperones will decrease its folding rate but increase folding efficiency and thereby yield.

The ER in addition possesses a wealth of enzymes that assist proteins during folding. Protein disulfide isomerase (PDI) (56 kDa) [36] is a member of the thioredoxin superfamily that acts catalytically in both the formation and reduction of disulfide bonds, and hence in disulfide bond rearrangements. Another PDI family member is ERp57, which facilitates disulfide bond formation in newly synthesized proteins [37] and works in a complex with either calnexin or calreticulin [38]. A different class of enzymes comprises the peptidyl-prolyl isomerases (PPIases) [39], which catalyze cis/trans isomerization of peptide bonds N-terminally to proline residues [40] and thereby increase folding rates [41]. The ribosome attaches amino acids in the trans conformation, which requires an isomerization step for all cis prolines in the final folded structure. PPIases belong to three structurally diverse families: the cyclophilins (inhibited by cyclosporin A [CsA]), the FK506-binding proteins (inhibited by FK506 and rapamycin), and the parvulins. They reside in the cytoplasm as well as in various eukaryotic organelles including the ER. In 1991 Steinmann et al. [42] and Lodish and Kong [43] showed that the folding of procollagen I and transferrin is slowed down by CsA due to inhibition of cyclophilin.

Folding proteins can be assisted by chaperones and folding enzymes from different compartments. If a protein is soluble, only the chaperones and folding enzymes that are located in the same compartment as the protein will assist folding. Membrane-bound or membrane-spanning proteins, however, will have domains in different compartments. For example, both P-glycoprotein and the cystic fibrosis transmembrane conductance regulator (CFTR) are multiple-membrane-spanning proteins that are members of the ABC transporter family. For both P-glycoprotein and CFTR, approximately 10–15% of the protein is located in the ER lumen and ER membrane, whereas the bulk is located in the cytosol. The only ER chaperone that interacts with these membrane proteins is calnexin [44, 45]. On the cytosolic side, the most abundant chaperone Hsp70 assists in folding these proteins [46, 47]. Interestingly, when folding attempts fail, Hsp70 also promotes proteasomal degradation of misfolded CFTR by targeting the E3 ubiquitin ligase CHIP to this protein [48]. Cytosolic Hsp90 also has been demonstrated to be involved in degrading CFTR [49]. Since both ER and cytosolic chaperones impose a quality control on this type of membrane protein, it will be of great interest to investigate how events at the two sides of a membrane are communicated and coordinated.

To study a possible interaction of a substrate protein with folding enzymes and chaperones, co-immunoprecipitations (see protocols) can be done using antibodies against folding factors; if interactions are too weak or too transient, chemical cross-linkers may stabilize the complex during analysis. To examine the role of a particular chaperone on the folding of a substrate protein, one may examine the effect of its increased or decreased expression in the cell or its presence or absence in a folding assay. To test whether an ATP-dependent chaperone is involved, ATP can be depleted from cells. Section 3.4 will address these and other conditions that may affect the folding process.

3.3
Location-specific Features in Protein Folding

3.3.1
Translocation and Signal Peptide Cleavage

One location-specific feature of protein folding is its potential coupling to translocation into an organelle and subsequent signal peptide cleavage. Proteins that are targeted to mitochondria, chloroplasts, and the ER normally carry a signal peptide at their N-terminus.

Mitochondria have two membranes: the inner membrane, which encloses the matrix space, and the outer membrane, which is in contact with the cytosol. The majority of mitochondrial proteins are encoded in the cell nucleus and need to be imported from the cytosol via the macromolecular Tim/Tom complex into the mitochondrial matrix, the mitochondrial inner membrane, or the intermembrane space [50].

Chloroplasts are chlorophyll-containing, double membrane-bounded organelles that are present in all higher plants. They possess the same compartments and membranes as mitochondria but have an extra subcompartment, the thylakoid space, which is surrounded by the thylakoid membrane. Like mitochondria, proteins are posttranslationally imported via the Tic/Toc complex [51].

Proteins that need to be imported into the ER lumen of eukaryotes are translocated from the cytosol through the translocon. Co-translational translocation is signal recognition particle (SRP)-dependent [52]. SRP binds to the signal sequence on the nascent chain, the ribosome/nascent polypeptide/SRP complex docks onto the SRP receptor at the ER membrane, and via multiple steps, the protein is translocated into the ER lumen. Oligosaccharyl transferase (OST) attaches glycans to the nascent chains, and in most proteins the signal peptide is cleaved off co-translationally by the signal peptidase present in the membrane with its active site in the lumen of the ER. Up to now, only few exceptions have been reported in which signal peptide removal is (at least partially) posttranslational, all of them viral glycoproteins [53, 54]. Posttranslational translocation is SRP-independent; this type of translocation needs additional factors to associate with the translocon complex.

Proteolytic cleavage events in mitochondria, chloroplasts, and the ER can be followed by mobility changes of the protein in reducing SDS-PAGE. If the signal peptide of a certain protein is cleaved off, the protein will have an increased electrophoretic mobility. Signal peptide cleavage in nascent chains is not detectable in a one-dimension gel, but it can be followed in two-dimensional SDS-PAGE, where the protein sample is digested with an SDS-resistant protease between the first and second dimension [55].

To test which population of full-length folding intermediates has lost its signal sequence in the ER (when cleavage is posttranslational), the process can be correlated to disulfide bond formation by using another two-dimensional SDS-PAGE system [56]. Samples are run non-reduced in the first dimension (horizontal) and subsequently are reduced in the second dimension (vertical). In this gel all proteins lacking disulfides will run on the diagonal. If a protein has intrachain disulfide bonds, it will run faster in the first (nonreducing) dimension than in the second (reducing) dimension and thus will end up above the diagonal. Proteins that form interchain disulfide-linked complexes will run below the diagonal. Antibodies against the signal peptide itself or antibodies that recognize the N-terminus of the protein after, but not before, cleavage can be used to monitor the process of signal peptide cleavage more directly [57]. Translocation of proteins in a cell-free translation organellar system can be demonstrated by protease-resistance. A newly synthesized protein that is not translocated yet will be degraded by (membrane-impermeable) proteases added to the membranes but will be protected from digestion once translocated.

3.3.2
Glycosylation

Glycan moieties on proteins are essential for folding, sorting, and targeting of glycoproteins through the secretory pathway to various cellular compartments. N-linked glycans affect local conformation of the polypeptide chain they are attached to; increase local solubility of proteins, thereby preventing aggregation; are important for the interaction with the two lectin chaperones in the ER, calnexin and calreticulin; and are important for targeting misfolded proteins for degradation. Hence, for most glycoproteins, their N-linked glycans are indispensable for proper folding.

The glycosylation process starts at the cytosolic side of the ER, where monosaccharides are added to the lipid intermediate dolichol phosphate up to $Man_5GlcNac_2$-PP-Dol. This precursor is translocated to the lumenal side, where it is elongated to $Glc_3Man_9GlcNAc_2$-PP-Dol. The oligosaccharyl transferase then transfers the 14-mer to an asparagine residue in the consensus sequence N-X-S/T of a nascent polypeptide chain, wherein X is any amino acid except proline [58]. Glucosidase I (GI, a membrane-bound enzyme) removes the first glucose and glucosidase II (GII, a soluble enzyme), the second and third. Monoglucosylated oligosaccharides associate with the ER-resident lectins calnexin (membrane-bound) and calreticulin (soluble), which are in a complex with ERp57 [38]. When GII cleaves the last glucose, the protein is released from calnexin/calreticulin and, if correctly folded, can leave the ER. UDP-Glc:glycoprotein glucosyltransferase (GT), a soluble ER protein of 160 kDa, is a sensor of glycoprotein conformation and reglucosylates glycans close to un- or misfolded amino acid stretches [59–61]. By reglucosylating the protein, it becomes a ligand again for calnexin/calreticulin and therefore a substrate for this quality-control cycle [9]. Two ER mannosidases can remove one or two mannoses in the ER before the protein is transported to the Golgi, where further oligosaccharide processing proceeds and proteins can be O-linked glycosylated. Alternatively, a permanently misfolded protein may be targeted for degradation from the ER through glycan recognition (see Section 3.3.4).

The enzyme endoglycosidase H (Endo H) often is used to monitor the movement of newly synthesized glycoproteins from the ER to the Golgi complex. Glycans on proteins remain sensitive to digestion by Endo H as long as they are in the ER and in early regions of the Golgi, but they become resistant beyond. Endo H digestion after the immunoprecipitation of a glycoprotein allows conclusions to be made on the protein's location. If a protein is highly heterogeneously glycosylated, e.g., HIV envelope, which has ∼30 N-linked glycans, Endo H can also be used after the immunoprecipitation to increase mobility of folding intermediates in the ER and to bring about a collapse of smeary bands into one sharp band in a gel, which enables identification and quantitation. The enzyme PNGase F in principle removes all glycans, irrespective of composition, and theoretically is the better deglycosylation enzyme. Its activity changes the glycosylated asparagine into an aspartic acid, however, which in some proteins decreases mobility, increases fuzziness of bands, and may confuse the pattern. The enzyme of choice needs to be determined for every protein one studies. To gather more information on how glycosylation affects protein folding, glycosylation site mutants and enzymes involved in glycosylation can be used (see Section 3.4).

3.3.3
Disulfide Bond Formation in the ER

The lumen of the ER supports a relatively oxidizing environment that facilitates the formation of disulfide bonds in folding proteins. Disulfide bonds are thought to stabilize proteins during and after folding, but this is an oversimplification of their role. In the cytosol the ratio between reduced glutathione (GSH) and oxidized glutathione (GSSG) is ∼60:1, while in the ER this ratio is 3:1 [62]. Glutathione is the major redox buffer in the ER. Therefore, glutathione was thought for a long time to be responsible for generating and maintaining the redox potential in the ER. This view changed with the identification of the FAD-binding [63] ER oxidoreductase 1 (Ero1p) in *Saccharomyces cerevisiae* [64, 65] and its homologues Ero1α and Ero1β in more complex eukaryotes [66, 67]. Disulfide-linked complexes between Ero1p and PDI have been captured in yeast as well as mammalian cells. In addition, disulfide-linked complexes were found between PDI and a newly synthesized secretory protein [68]. Ero1 transfers disulfides to (or rather, electrons from) substrate proteins via protein disulfide isomerase (PDI). Other oxidoreductases, such as Erv proteins [69] and PDI homologues, play a role as well [70].

Formation of disulfide bonds in a folding protein can be followed through a pulse-chase assay. In combination with an immunoprecipitation and reducing and non-reducing SDS-PAGE, the formation of disulfides can be monitored (see the Appendix and [11]). To study which enzymes are involved during disulfide bond formation and isomerization, co-immunoprecipitations can be performed with different antibodies against folding enzymes with or without prior cross-linking. Furthermore, the folding enzyme concentration can be changed, as is described below.

If the studied protein contains disulfide bonds, an alkylating agent must always be added to the stop solution and lysis buffer of the pulse-chase assay to block free sulfhydryl groups [71, 72]. This will prevent formation of (additional) disulfide bonds in the protein after cell lysis. Three commonly used blocking agents are iodoacetamide (IAM), iodoacetic acid (IAC), and *N*-ethyl maleimide (NEM). IAM and IAC bind relatively slowly, but irreversibly. IAC is negatively charged, which decreases mobility of proteins it is attached to. NEM is smaller than IAM and IAC; it binds fast but is under certain conditions partially reversible. It is therefore less suitable when prolonged incubation periods at 37 °C are required. Since all alkylating agents have different characteristics, the alkylation protocol may be optimized for each individual protein.

3.3.4
Degradation

In the ER, a stringent quality-control system operates that discriminates between correctly folded proteins, misfolded proteins, and unassembled protein subunits. This prevents misfolded proteins from leaving the ER, where they can cause harm to the cell or even to the complete organism. The ER-associated degradation (ERAD) pathway ensures ubiquitin-mediated degradation of ER-associated misfolded proteins [73]. Trimming of the α1,2-linked mannose of the middle branch by mannosidase I to $Man_8GlcNac_2$ uncovers a degradation signal [74, 75]. EDEM (ER degradation-enhancing α-mannosidase-like protein) functions in the ERAD pathway by accepting terminally misfolded glycoproteins from calnexin [76, 77]. The substrate is retro-translocated into the cytosol, where proteins are deglycosylated, ubiquitinated, and then degraded by the 26S proteasome [73]. Post-ER quality-control systems exist as well: proteins can be transported to the Golgi or the plasma membrane before they are degraded, most often by vacuolar/lysosomal proteases.

Proteins can be degraded after synthesis as well as during their translation, which is undetectable in a pulse-chase assay. The net result is a lower amount of labeled protein, which can be visualized only through the use of a proteasome inhibitor during the pulse labeling. If the total signal of labeled protein at the end of the pulse is higher in the presence than in the absence of the inhibitor, co-translational degradation apparently does happen in the control situation. Degradation *after* translation can be followed on gel, because the total protein signal will decrease during the chase. Proteasome inhibitors and lysosomal enzyme inhibitors can be added to prove degradation and identify the site of breakdown. A useful system to study protein folding without degradation is an in vitro translation system in the absence or presence of a membrane source, the latter ranging from enriched organelles to digitonin-permeabilized cells. The hemin present in the reticulocyte lysate inhibits proteasomal activity.

3.3.5
Transport from ER to Golgi and Plasma Membrane

Different methods can be used to monitor transport of a protein out of the ER to the Golgi complex or the plasma membrane. Endo H resistance is a good tool to monitor the movement of newly synthesized glycoproteins from ER (sensitive to digestion by Endo H) to Golgi (resistant to Endo H). Resistance of newly synthesized proteins to reduction by DTT (or oxidation by diamide) can be used to determine the folding status of a protein. If a protein is transported very rapidly from ER to Golgi, which then coincides with rapid and massive changes in electrophoretic mobility, such as O-glycosylation [78] or N-glycan modification [79], and one wants to trap the protein in the ER, in vitro translation in the presence of microsomes or semi-intact cells [12] can be used, since these systems lack ER-to-Golgi transport.

Cell surface arrival of proteins can be shown using protease digestion on intact cells, biotinylation of intact cells, or indirect immunofluorescence on fixed cells. Indirect immunofluorescence allows detection of possible locations of a protein in the cell, but this steady-state method does not allow any conclusion on precursor-product status, or on kinetics of processes. To follow a traveling protein with time throughout a living cell, it may be fused to a fluorescent marker protein such as GFP. For this purpose, the protein population needs to be synchronized by for example release from a protein synthesis block (cycloheximide incubation and removal) [80], a disulfide-bond formation block (DTT incubation and removal) [20], or use of a conditional mutant of either the protein under study or a transport protein such as the Sec mutants often used in *S. cerevisiae* [81]. Receptor or ligand binding can be tested biochemically, either in the cell lysate or on intact cells, after the protein's arrival on the plasma membrane, and when the studies concern a soluble protein, protein secretion can be measured by collecting the chase medium.

3.4
How to Manipulate Protein Folding

3.4.1
Pharmacological Intervention (Low-molecular-weight Reagents)

To study the role of folding factors in vivo, their activity needs to be manipulated. Drugs act fast and do not allow time for genetic adaptation of cells, but they may not inhibit to 100% and may exhibit pleiotropic effects. Genetic changes of cells circumvent this lack of specificity. One particular folding factor can be changed, but compensatory regulation processes may completely change the protein composition of the cell. With these limitations in mind, both high- and low-molecular-weight manipulations can increase our understanding of protein folding in cells.

3.4.1.1 Reducing and Oxidizing Agents

Reducing and oxidizing agents not only are useful to determine the conformation of a folding protein (as discussed in Section 3.2), but also they can be used as a tool to manipulate protein folding. If separation of protein translation and protein folding is desired, for example, to study protein folding in the absence of ATP or to determine whether co-translational disulfide bond formation is needed for a certain protein to fold, a reducing agent such as DTT can be added to the pulse medium to prevent disulfide bond formation during protein synthesis. Furthermore, reducing and oxidizing agents can be added to the chase medium to determine the effect of reduction/oxidation on protein folding and (posttranslational) signal peptide cleavage.

3.4.1.2 Calcium Depletion

In addition to its role in protein folding, the ER has an important role in calcium signaling and in sequestering calcium from the cytosol. The ER contains many calcium-binding proteins, which have a large capacity for calcium and help keep the cellular internal calcium concentration constant. The first identified calcium-binding protein was calsequestrin in the sarcoplasmic reticulum of striated muscle [82]. Its C-terminal domain binds calcium with low affinity but with high capacity. The calcium level in the ER is important not only for signaling but also for proteins traveling through the ER. During folding they are assisted by calcium-binding molecular chaperones such as BiP (Grp78), Grp94 (endoplasmin), and the lectins calnexin and calreticulin. When the level of calcium in the ER drops, some of the newly synthesized proteins misfold and aggregate [83]. To study the effect of calcium on protein folding, calcium in the cell can be depleted using various chemicals. A23187 is a divalent cation ionophore [84] that can be added to calcium-free starvation, pulse, and chase media. Thapsigargin is a potent cell permeable, IP3-independent, intracellular calcium releaser, as it efficiently inhibits all members of the Ca-ATPase family of calcium pumps, which pump calcium into the ER [85, 86]. Calcium chelators such as BAPTA, EDTA, and EGTA can be used as well to study the effect of calcium (and calcium-binding chaperones) on protein folding.

3.4.1.3 ATP Depletion

ATP is a common currency of energy in all cells, and its presence is needed for various reactions and processes, including protein folding [87]. The ER contains many chaperones such as BiP, the lumenal Hsp70 of the ER, which has an N-terminal ATP-binding domain and a C-terminal peptide-binding domain and whose activities are tightly coupled. In the Hsp70 family of chaperones, association and dissociation with the substrate protein are controlled by ATP binding and hydrolysis. To test the effect of ATP(-dependent chaperones) on protein folding, ATP generation can be blocked. 2-Deoxy-glucose inhibits glycolysis and, thereby, anaerobic ATP generation. It can be added to glucose-free cell culture medium and has been shown to be effective in pulse chase experiments [87]. Because 2-deoxy-glucose in some studies was shown to inhibit glycosylation [88], the incubation time and concentration need to be determined carefully. Sodium azide and the uncoupling protonophore carbonyl cyanide 3-chlorophenylhydrazone (CCCP) [89] poison mitochondria, and thus inhibit aerobic ATP generation. Inhibition of both glycolysis and oxidative phosphorylation by combination treatment usually depletes ATP effectively from living cells [87]. After depletion, ATP levels in the cell lysates can be measured, e.g., with a luciferase-luciferin assay in a scintillation counter or luminometer [21, 90].

3.4.1.4 Cross-linking

Chemical cross-linkers are useful for multiple purposes in protein science, such as the stabilization of protein-chaperone complexes. There are two prominent groups of cross-linkers: homobifunctional cross-linking reagents, which have two identical reactive groups, and heterobifunctional cross-linking reagents, in which the reactive groups are chemically distinct. Further variation is present in spacer arm length, cleavability, or membrane permeability, and cross-linkers may react chemically or photochemically upon UV illumination. Whereas membrane-permeable cross-linkers are suitable for cross-linking in the cell, membrane-impermeable reagents are useful for cross-linking at the cell surface and in cell lysates. Cross-linkers with shorter spacer arms (4–8 Å) often are used for intramolecular cross-linking, and reagents with longer spacers are favored for intermolecular cross-linking. The most frequently used cross-linkers at the moment are DSP (dithio-*bis*-succinimidylpropionate), a water-insoluble, homobifunctional *N*-hydroxysuccimide ester with a spacer arm length of 12 Å, which is thiol-cleavable and primary amine reactive and is used in many applications [91], and BMH (*bis*-maleimidohexane), a water insoluble, homobifunctional, non-cleavable cross-linker with a spacer arm length of 16 Å, which is reactive towards sulfhydryl groups and is also used in many applications [92, 93]. Since there is a wide variety in characteristics and potential applications of cross-linkers, it is desirable to compare different cross-linkers before protein-folding studies are performed.

3.4.1.5 Glycosylation Inhibitors

Inhibitors that prevent glycan synthesis or modify the carbohydrate portion of the glycoprotein are useful to determine function of the glycan or change fate of a protein. A wide variety in glycosylation inhibitors exists (see also chapter 17), each blocking a specific step in glycosylation. Tunicamycin, for instance, was isolated from *Streptomyces lysosuperificus* in the early 1970s [94]; it inhibits transfer of N-acetylglucosamine to dolichol phosphate, thereby completely blocking N-linked glycosylation, which most often causes misfolding of glycoproteins. The misfolding of many proteins in the ER can lead to upregulation of classical chaperones and folding-facilitating proteins through a so-called unfolded protein response (UPR) [95].

Other than complete inhibition by tunicamycin, which can trigger the UPR, more subtle inhibitors can be used, e.g., castanospermine. This is a plant alkaloid that was isolated from the seeds of the Australian tree *Castanospermum australe*, which inhibits glucosidase I and II. An example of a glucosidase I-specific inhibitor is australine, whereas deoxynojirimycin inhibits glucosidase II better than glucosidase I. Glucosidase I and II inhibitors can be useful in preventing association of the folding protein with the molecular chaperones calnexin and calreticulin, which bind only to monoglucosylated proteins. If a prolonged association with calnexin or calreticulin is desired, a glucosidase II inhibitor can be added during the chase period in a pulse-chase assay. This prevents cleavage of the last glucose, such that the glycoprotein will remain bound to the lectins calnexin and calreticulin. If a mannosidase I inhibitor such as kifunensine is added during the pulse chase, degradation of misfolded glycoproteins can be prevented.

Glycosylation inhibitors are useful reagents in determining the role of oligosaccharides during protein folding, but because they often do not inhibit 100% and because some inhibitors can have an effect on protein synthesis, it is very important to always include the proper controls when they are used in experiments.

3.4.2
Genetic Modifications (High-molecular-weight Manipulations)

3.4.2.1 Substrate Protein Mutants
To study the effect of certain amino acids or modifications on the folding pattern of a protein, various mutations can be made in the cDNA encoding the protein. To investigate the effect of glycosylation on a protein, N-glycosylation consensus sequences can be removed or created by site-directed mutagenesis. Mutations of choice are the exchange of Asn for Gln or the downstream Ser or Thr for Ala; to exclude the effect of an amino acid change rather than of a lacking glycan, both strategies can be used and compared. This allows study of the role of every individual glycosylation site for proper folding, alone or in combination. Additional glycosylation sites can be generated to study the effect of hyperglycosylation. The addition of glycosylation consensus sequences is also used to determine the topology of a protein [96]; interpretation of such data needs care, however, because a glycan may change the topology or fate of a protein.

The role of a specific disulfide bond for protein folding can be studied by changing the cysteines into another residue using site-directed mutagenesis. By creating single-cysteine and double (-cystine) mutations, the effect of removal of a complete disulfide bond on protein folding can be compared to the removal of a single cysteine, hence the presence of an odd number of cysteines. Cysteines are best changed into alanines, because they leave "only" a small gap in the protein structure, i.e., where the thiol atom resided. Although serines are thought to resemble cysteines more closely than alanines, their higher hydrophilicity and their hydrogen atoms may cause problems when incorporated into the tightly packed core of a protein. Methionines and valines in some instances are better replacements for cysteines, but they may be too large in other sites. To be certain that the result reflects the removal of a disulfide bond rather than the change of an amino acid in the sequence, it is wise to substitute cysteine for more than one other amino acid and compare the effects.

Many other mutants can be thought of. The signal peptide of a protein can be modified to determine which residues are responsible for signal peptide cleavage, and the cleavage site can be changed to determine the effect of timing and cleavage on folding. In addition, the original signal sequence can be replaced completely by the signal sequence of another protein, and folding kinetics and secretion can be studied, as was done for HIV envelope [53]. To determine sequence and hierarchy of domain folding within a protein, and to determine whether domains fold independently of each other, single domains can be expressed, and C-terminal and N-terminal truncation mutants can be studied.

When protein mutations result in clinical phenotypes, as in many inherited diseases, it is interesting to focus on those mutations that result in folding defects.

3.4.2.2 Changing the Concentration or Activity of Folding Enzymes and Chaperones

To determine the effect of molecular chaperones and folding enzymes on the folding of a protein, the protein can be overexpressed by infection/transfection. Cells can be transiently transfected using a plasmid encoding the protein of interest behind an appropriate promoter. High expression can be achieved using virus-based expression systems; one example is the recombinant vaccinia virus expressing T7 polymerase in combination with a plasmid containing the protein of interest behind the T7 promoter. Stable, overexpressing cell lines can be used as well; one example is the Dorner cell lines [97], which overexpress a particular ER chaperone, such as PDI, BiP, or Grp94.

Because of the redundancy of many chaperones and folding enzymes, overexpression of a folding factor may not lead to a desired effect. Lowering expression or activity may be more successful. In cases where molecular details of the activity of a folding factor are known, dominant-negative mutants have been generated: the "trap" mutants of chaperones such as GroEL [98] and BiP [99] and active-site mutants of enzymes such as Ero [66]. Many cell lines exist in which a particular folding factor is absent or incapacitated. Examples are the CHO-derived Lec cell lines [100], which include the glucosidase I-negative Lec23 cells [101]. Other examples are the BW 5147 mouse lymphoma-derived PhaR2.7 cell line, which is glucosidase II-deficient (because of which the lectins calnexin and calreticulin cannot bind to their substrates anymore) [102], and the human T lymphoblastoid cell line CEM.NKR [103], which is resistant to natural killer cells (NK cells) and has no calnexin.

The most dramatic effect can be expected when a folding factor is completely absent. In various organisms such as yeast, genes can be deleted. In mammalian cell lines, this is still a procedure with little success. Alternatives are cell lines derived from knockout animals, which are becoming increasingly available. In other organisms or cell lines, depletion of a gene is the maximum reachable. Gene expression can be suppressed by siRNA, which was first discovered in the nematode worm *C. elegans*; this animal is capable of sequence-specific gene silencing in response to double-stranded RNA (dsRNA) [104]. The technology has been adapted for use in mammalian cells, and an increasing number of siRNA constructs and libraries are becoming available, which makes siRNA a powerful tool to study how chaperones and folding enzymes influence protein folding.

3.5
Experimental Protocols

3.5.1
Protein-labeling Protocols

3.5.1.1 Basic Protocol Pulse Chase: Adherent Cells

In vivo, conformational changes during folding, disulfide bond formation, signal peptide cleavage, other proteolytic processing steps, glycosylation and glycan modifications, transport to other compartments, and some aspects of biological activity all can be studied in intact suspension cells or in adherent cells growing in monolayers in tissue culture dishes [11]. The adherent cells are easiest to use because they allow multiple wash steps in a short period of time, which are needed for short pulse and chase times or for studies requiring frequent changes of media with very different composition. When cells do not adhere or when low volumes of media are desirable (e.g., when expensive additives are needed), most of the analyses can be done in suspension, as described in the alternate protocol. Prepare the cells expressing the protein of choice in 35- or 60-mm dishes. For each time point, at least one dish is needed; the protocol is written for 60-mm dishes. On the day of the pulse chase, the cell monolayer needs to be almost confluent (80–90%).

Prepare all solutions needed. Also prepare a pulse-chase scheme on which all steps in the procedure (pulse, chase, stop) are noted. We use a 1-min scheme, meaning that every minute one action (pulse, chase, or stop) is performed. If aspiration is fast (which depends on the vacuum system used) and the experimenter is experienced, 30-s intervals become possible.

1. Prepare the pulse-chase setup, which includes a water bath at the desired temperature, with tube racks and 1–2 mm water above them, which can be maintained at this temperature. *Make sure the water level is just above the racks, such*

that the bottom of a tissue culture dish is completely in contact with the water but dishes are not floating when they are without cover.
2. Wash the cells with 2 mL wash buffer and add 2 mL starvation medium. Incubate for 15–40 min. at 37 °C in the presence of 5% CO_2.
3. Place the dishes on the rack in the water bath. *Make sure there are no air bubbles below the dishes.*
4. After 15 min pre-incubation with starvation medium, start the first pulse labeling. The staggering schedule (Figure 3.2) should allow labeling of the last dish within ∼40 min of the start of starvation. Pulse label the cells one dish at a time: aspirate depletion medium, add 400 µL labeling medium at time 0 s. to the center of the dish, swirl gently, and incubate for the desired labeling time on the water bath. For pulse times longer than 15 min, a larger volume and incubation (on a rocker) in a 37 °C incubator in the presence of 5% CO_2 is recommended. *The pulse time optimally is equal to or shorter than the synthesis time of the protein under study (an average of 4–5 amino acid residues per second). Labeling time should be long enough to detect the protein.*
5. Add 2 mL chase medium at precisely the end of the pulse (i.e., at time 2'0" if a 2-min pulse is aimed for). Swirl gently to mix; the labeling stops immediately upon addition of chase medium. Aspirate this mix of labeling and chase medium and add another 2 mL of chase medium. For short chases: incubate on the water bath. For chases longer than 30 min, incubate dishes in the incubator. For 0' chase: add 2 mL chase medium at precisely the end of the pulse. Swirl gently to mix. Aspirate chase medium, place the dish immediately on ice, and add 2 mL of ice-cold stop buffer. *This is best achieved by putting an aluminum plate on top of an ice pan full of ice, with a wet paper towel on top to allow optimal contact between the plastic of the cell culture dish and the ice-cold surface.*
6. Stop the chase: aspirate chase medium at precisely the endpoint of the chase time, place the dish on the aluminum plate on ice, and immediately add 2 mL of ice-cold stop buffer. *If the studied protein will be secreted, collect the chase medium and use it for immunoprecipitation.*
7. Just before lysis: wash the cells again with 2 mL ice-cold stop buffer.
8. Aspirate the dish as dry as possible without letting cells warm up and add 600 µL ice-cold lysis buffer.
9. Scrape the cell lysate and nuclei off the dish with a cell scraper and transfer the lysate to a microcentrifuge tube.
10. Centrifuge lysates at 16 000 g at 4 °C for 10 min to spin down the nuclei. Use the supernatant directly for immunoprecipitations or transfer the supernatant to a new microcentrifuge tube, snap freeze, and store at −80 °C.

Buffers for the Pulse Chase

Wash buffer: Hank's Balanced Salt Solution (Invitrogen) or PBS with calcium and magnesium. Keep at 37 °C.

Starvation medium: cysteine- and methionine-free tissue culture medium (ICN) supplemented with 10 mM Na-HEPES (pH 7.4). Keep at 37 °C.

Pulse medium: cysteine- and methionine-free medium containing 10 mM Na-HEPES (pH 7.4), 125 µCi ^{35}S-cysteine/methionine per milliliter. Keep at 37 °C. *To manipulate protein folding, chemicals such as glycosylation inhibitors, proteasome inhibitors, and DTT can be added to the pulse medium.*

Chase medium: complete tissue culture medium for the appropriate cell line (also a 1/1 mix with serum-free medium is possible) supplemented with 10 mM Na-HEPES (pH 7.4), 5 mM cysteine, and 5 mM methionine. Keep at 37 °C. *If protein-folding kinetics is to be studied, 1 mM cycloheximide needs to be added to the chase medium to stop elongation of unfinished nascent chains; without it, the amount of labeled full-length protein may increase during the chase, even though incorporation of ^{35}S label is stopped completely. To manipulate protein folding, chemicals such as glycosylation inhibitors, proteasome inhibitors, DTT, diamide, or cross-linkers can be added to the chase medium, or ATP can be depleted.*

Stop buffer: Hank's Balanced Salt Solution (Invitrogen) or PBS with calcium and magnesium, supplemented with 20 mM alkylating agent (IAC, IAM, or NEM). Keep at 0 °C. *See Section 3.3.3.*

Lysis buffer: PBS (pH 7.4) or similar salt-containing buffer, containing 0.5% Triton X-100, 1 mM EDTA, 20 mM alkylating agent, 1 mM PMSF, and 10 µg µL^{-1} each of chymostatin, leupeptin, antipain, and pepstatin. Keep at 0 °C. *Add PMSF to the lysis buffer just before lysis. PMSF is highly unstable in water and has a half-life of only a few hours on ice. Different proteins may need the use of different protease inhibitors. The combination in the protocol is an empirically determined one that works well for proteins used in our lab. Salt concentration should be isotonic to prevent nuclei from disrupting. When noncovalent interactions are studied, other detergents may work better. Examples include CHAPS, octyl glucoside, or digitonin. When the protein associates with detergent-resistant membranes, the use of Triton X-100 and CHAPS should be avoided. Octyl glucoside is one of the detergents that will solubilize all membranes.*

Postponed Posttranslational Folding To separate translation from the folding process, e.g., to study the effect of factors that would affect translation, such as ATP depletion, oxidative folding can be postponed until after synthesis by addition of a reducing agent to the pulse medium [20]. This will prevent disulfide bond formation during the pulse. When the reducing agent is removed after the pulse, disulfide bond formation and folding can proceed. For various proteins, oxidative folding can be postponed until after synthesis without it creating a problem for folding.

1. Add DTT to the pulse medium to a concentration of 5–10 mM and pre-incubate for 5 min with DTT before the pulse. *The DTT concentration needs to be titrated for each protein, but usually 5 mM should be sufficient. Since labeling efficiency might decrease in DTT, labeling time may need to be increased. DTT may affect cellular ATP levels during long incubations, depending on the cell line. Therefore, a comparison with co-translational folding always needs to be made and ATP levels need to be measured.*

DTT Resistance When proteins fold and become more compact, this is usually reflected by resistance to reduction by DTT in vivo. Influenza hemagglutinin, for example, acquires DTT resistance when it is a completely oxidized monomer that is trimerization-competent but not yet a trimer [21].

1. Label the protein (see pulse-chase protocol) and chase for different times.
2. After the chase: perform in a parallel dish with the same chase time an additional chase of 5 min with 5 mM DTT. Compare this sample with the sample in which the protein is chased without DTT.

Diamide Resistance

1. Label the protein (see pulse-chase protocol) and chase for different times.
2. After the chase: perform in a parallel dish with the same chase time an additional chase of 5 min with 5 mM diamide. Compare this sample with the sample in which the protein is chased without diamide.

3.5.1.2 Pulse Chase in Suspension Cells

When cells do not adhere properly or when small volumes are desired, cells can be pulse labeled in solution. In this assay the different chase samples can be collected from one tube. Wash steps after the pulse and chase are not included, because the centrifugation steps would take too much time. Prior to starting the experiment it is necessary to determine:

- the minimum volume for incubating the cells during the pulse (x μL), (*This is cell line dependent.*)
- the number of chase points (y), and
- the desired sample volume (z). We used, for example, $x = 300$ μL, $y = 3$, and $z = 500$ μL.

1. Transfer suspension cells to a sterile 50-mL tube with cap. We used, for example, $\sim 10^6$ cells per time point.
2. Pellet cells at 500 g for 4 min at RT. Resuspend cells in $2y$ mL depletion medium and pellet cells again. Resuspend cells in $2y$ mL depletion medium. Incubate cells for 15 min at 37 °C in a CO_2 incubator.
3. Pellet cells at 500 g for 4 min at RT. Resuspend cells in x μL depletion medium, change to appropriate tube if necessary, and place in a water bath at 37 °C or the desired temperature for the experiment.
4. Add 50–100 μCi ^{35}S-labeled methionine and cysteine per time point and mix gently to start the pulse. Incubate for the pulse period.
5. Add $\geq yz$ μL chase medium. The total volume should be slightly more than $y \times z$ μL to allow for fluid loss due to evaporation during the experiment. Mix by gently pipetting up and down.
6. Immediately take the first sample (z μL). Transfer to microcentrifuge tube with pre-prepared z μL 2× concentrated lysis buffer on ice, mix well, and keep on ice.

7. After every chase interval, collect a sample and add to 2× concentrated lysis buffer on ice, mix well, and keep on ice.
8. Spin the lysates at 16 000 g for 10 min at 4 °C to pellet the nuclei. The postnuclear lysate can be directly used for immunoprecipitations or transferred to a new microcentrifuge tube, snap frozen, and stored at −80 °C.

Semi-permeabilized Cell System (see Chapter 18) In vitro translation in the presence of semi-permeabilized cells [12] is a method with a complexity between the pulse-chase assay and the in vitro chase assay, since there is only one membrane between the protein and the experimenter, whereas there are two in the pulse-chase assay and none during the second part of the in vitro chase assay.

In the semi-permeabilized cell system, cells are treated with the detergent digitonin, which selectively permeabilizes the plasma membrane, leaving cellular organelles such as ER and Golgi complex intact. These semi-permeabilized cells (SP cells) are added to an in vitro translation system, where they act as a source of ER membranes. Only the mRNA of the protein of interest is present, which precludes the need for immunoprecipitations. This assay is especially useful when there are no antibodies available against the protein under study, for protease resistance studies, or when only the ER form of a protein is under study and ER-to-Golgi transport is undesirable.

In Vitro Chase Assay In the in vitro chase assay (Maggioni et al., manuscript submitted), a protein is pulse labeled/translated in vivo and chased in a detergent lysate. Since there are no membranes during the chase period, the direct environment of the folding protein can be manipulated. For example, ATP levels in the detergent lysate can be depleted, and chaperones and organellar cell fractions can be added.

After point 4 of the basic pulse-chase protocol (now using 10-cm dishes):

For the in vitro assay 10-cm dishes are used instead of 6-cm dishes. Therefore, all buffer volumes used in steps 1–4 of the basic pulse-chase protocol need to be multiplied by 2–2.5.

1. Wash cells twice with 4 mL ice-cold HBSS wash buffer, aspirate the buffer, and immediately lyse with 1.8 mL ice-cold lysis buffer without alkylating agent.
2. Scrape the cell lysate and nuclei off the dish with a cell scraper, transfer lysates to a microcentrifuge tube, and immediately take out 1/6 of the lysate and add NEM to this sample to a concentration of 20 mM (0 min sample in the in vitro chase assay).
3. Spin cell lysates for 10 min at 16 000 g at 4 °C to pellet nuclei. Transfer postnuclear cell lysate to a new tube. The supernatant of the lysate without NEM is used in the rest of the in vitro chase assay.

After transfer to a new tube, add 5 mM GSSG and transfer the tube immediately to a water bath at 30 °C.

4. After different times of in vitro chase, take 300 μL of the lysate and transfer it

into a new tube that contains NEM to a final concentration of 20 mM and immediately incubate on ice.
5. Use the supernatant directly for immunoprecipitations or snap freeze and store at −80 °C.

3.5.2
(Co)-immunoprecipitation and Accessory Protocols

3.5.2.1 Immunoprecipitation

1. Mix 50 µL of washed protein A-Sepharose beads and the optimal amount of antibody and shake in a shaker for 60 min at 4 °C. *Optimize the incubation time for each antibody. The optimal amount of antibody is the amount that precipitates all, or at least the maximum amount, of antigen from the solution. This needs to be tested for each antibody-antigen combination by re-incubation of the supernatant from the beads-antibody-antigen incubation (see step 3 below) with antibody and beads. If the antibody has a low affinity for protein A, protein G-Sepharose beads or a bridging antibody between the primary antibody and the protein A beads, such as a goat or rabbit anti-mouse Ig, can be used. Instead of Sepharose beads, heat-killed and fixed S. aureus cells can be used to immobilize the antibody. Because these cells are more difficult to resuspend than Sepharose beads, spin conditions should be 5 min at 1500 g rather than 1 min at max speed. Resuspension time will be longer and should be added to the wash time.*
2. Add 100 µL to 600 µL post-nuclear cell lysate and couple in the head-over-head rotator for at least 30 min (depending on the antibody) at 4 °C. *Test for each antibody the minimum incubation time needed (between 30 min and overnight). With one lysate, different immunoprecipitations (such as one with an antibody that recognizes all forms of the protein, a conformation-sensitive antibody, or a co-immunoprecipitation) can be done to obtain as much information as possible on the folding protein.*
3. Pellet beads by spinning 1 min at RT max, aspirate the supernatant, and use it for re-precipitation to test whether the antibody precipitated all antigen molecules in the solution (see step 1 above). Add 1 mL washing buffer and shake for at least 5 min in a shaker at RT. Repeat this washing procedure once. *Test different wash buffers with different concentrations of SDS, other detergents, combinations thereof, and salt. Also test different wash times and wash temperatures. It is not necessary to wash more than twice. Rather than increase the number of washes, conditions of the two washes should be changed. If this does not give the desired result, pre-clearing of the detergent cell lysate is the method of choice (see calnexin protocol).*
4. Pellet beads and aspirate supernatant. Add 20 µL TE to the beads, vortex to resuspend the beads, and add 20 µL 2× concentrated sample buffer and vortex again. *If an Endo H digestion is performed, add here the Endo H buffer instead of the TE. It is important to resuspend in TE before adding SDS, because addition of SDS to a pellet may induce or increase aggregation.*

5. Heat samples for 5 min at 95 °C, vortex when still hot, and pellet the beads. The supernatant is the non-reduced sample.
6. To make a reduced sample, transfer 18 µL supernatant to a new tube to which a defined drop of DTT solution has been added (final DTT concentration should be 20 mM) and vortex. Heat samples for 5 min at 95 °C. Centrifuge shortly to spin down the condensation fluid. The samples are now ready to be loaded onto an SDS-PA gel. *Addition of DTT to the bottom of the tube allows one to check whether DTT actually has been added to the sample. Re-oxidation during electrophoresis, which happens if proteins have many cysteines, is prevented by adding an excess alkylating agent, such as 100 mM NEM, to all samples before loading.*

Buffers for Immunoprecipitation
Protein A-Sepharose beads: 10% protein A-Sepharose beads in PBS or the same buffer in which the cells were lysed in, supplemented with 1/10 of the concentration of the same detergent used in the lysis buffer and 0.25% BSA. *The beads are washed twice in this solution to remove the buffer in which the Sepharose was stored. When using S. aureus cells, washing should be done twice as well, in the same centrifuge that will be used for the immunoprecipitation, to ensure that the same-sized particles will be pelleted during all washes.*

Wash buffer: PBS, pH 7.4, containing 0.5% Triton X-100 or 150 mM NaCl (representing some of the mildest conditions) or another buffer (such as a harsher one with SDS).

TE buffer: 10 mM Tris-HCl, pH 6.8, 1 mM EDTA.

2× concentrated sample buffer: 400 mM Tris-HCl, pH 6.8, 6% (w/v) SDS, 20% glycerol, 2 mM EDTA, 0.04% (w/v) bromophenol blue.

5× concentrated sample buffer: 1 M Tris-HCl, pH 6.8, 7.5% (w/v) SDS, 50% glycerol, 5 mM EDTA, 0.08% (w/v) bromophenol blue.

3.5.2.2 Co-precipitation with Calnexin ([30]; adapted from Ou et al. [105])
After step 7 from the pulse chase:

1. Aspirate the dish as dry as possible and add 600 µL ice-cold lysis buffer (2% CHAPS in 50 mM Na-HEPES pH 7.6 and 200 mM NaCl. *Do not add EDTA to the lysis buffer, since calcium is needed for the interaction.*
2. Scrape the cell lysate and nuclei from the dish and transfer into a microcentrifuge tube.
3. Use around 200 µL of the lysate for the co-immunoprecipitation and use a parallel amount for an immunoprecipitation using an antibody that recognizes all antigen that is present.
4. Add 0.1 mL of a 10% suspension of heat-killed and fixed S. aureus cells to the lysate to reduce background and rotate or shake for 30–60 min at 4 °C.

5. Pellet the fixed *S. aureus* and nuclei at max speed to remove the smallest particles and use the supernatant for the immunoprecipitation.
6. Couple the antibody to the protein A-Sepharose beads for 30 min at 4 °C in a shaker.
7. Add the pre-cleared lysate.
8. Couple at 4 °C for 1–16 h, depending on the antibody.
9. Pellet the beads by centrifuging for 2 min at 8000 g.
10. Wash twice with 0.5% CHAPS in 50 mM Na-HEPES pH 7.6, containing 200 mM NaCl.
11. After the last wash: continue the immunoprecipitation protocol at step 4.

3.5.2.3 Co-immunoprecipitation with Other Chaperones

The calnexin co-immunoprecipitation protocol can be used as a basic protocol when the interaction of a protein with another chaperone is to be studied. Since each protein-chaperone interaction is different, protocols should be optimized for each new co-immunoprecipitation. To stabilize the interaction of the molecular chaperone BiP with its substrate, adding an ATP-depleting agent to the lysis buffer is recommended, to lock the ATP cycle of the chaperone. When interactions with calcium-binding chaperones are studied, calcium chelators such as EDTA should be omitted from the lysis buffer.

3.5.2.4 Protease Resistance

Limited proteolysis can be used to investigate the conformation of proteins. However, most investigators often perform limited proteolysis on purified protein, which will represent only the steady-state conformation. To obtain more dynamic information on how proteins acquire their conformation during the folding process, one can combine in vivo pulse-chase analysis and immunoprecipitation, as described earlier in this chapter, with limited proteolysis. Important considerations should be made before using this technique. First, methionines and cysteines within the protein should be evenly distributed. Second, a highly specific antibody is needed to immunoprecipitate the protein of interest. Third, a mild denaturing buffer should be used for cell lysis and immunoprecipitation to prevent large conformational changes of the protein. For most proteins in our lab, we use MNT (20 mM MES, 100 mM NaCl, 30 mM Tris-HCl, pH 7.4) containing 0.5% Triton X-100 as a mild lysis buffer. It is important to omit protease inhibitors and EDTA during cell lysis. Therefore, perform the immunoprecipitation as shortly as possible to prevent protein degradation.

After the last wash of the immunoprecipitation protocol (step 3):

(Note: one protease concentration tested in a proteolysis experiment is one immunoprecipitation.)

1. Pellet the antigen-antibody-bead complexes and aspirate all wash buffer.

2. Add 10 μL MNT + 0.5% Triton X-100 to the beads and resuspend gently.
3. Add 1 μL protease from a 10× stock. Resuspend by vortexing and incubate for *exactly* 15 min on ice. Use a protease titration range (for example, 0.25, 1, 5, 25, 100, or 500 μg mL^{-1} final concentration) to investigate protease susceptibility of each protein. TPCK-trypsin, TLCK-chymotrypsin, proteinase K, and endoproteinase Glu-C V8 are suitable proteases (Sigma) for these studies.
4. After incubation, inhibit the protease by first adding 1 μL PMSF (phenylmethylsulfonyl fluoride) from a 10× stock before adding 10 μL 2× sample buffer. TPCK-trypsin is specifically inhibited by adding a fivefold excess of soybean trypsin inhibitor.
5. Heating the sample for 5 min at 95 °C will completely inactivate the protease.
6. Analyze the radio-labeled proteolytic pattern on a 12–15% SDS-PA gel and expose to film or phosphor screen.

3.5.2.5 Endo H Resistance
After the last wash of the immunoprecipitation protocol (step 3):

1. Pellet beads and aspirate all wash buffer from the protein A-Sepharose beads.
2. Add 15 μL 0.2% SDS in 100 mM sodium acetate, pH 5.5, and heat for 5 min at 95 °C to denature the protein.
3. Cool and add 15 μL 100-mM sodium acetate, pH 5.5, containing 2% Triton X-100 and protease inhibitors: 10 μg μL^{-1} each of chymostatin, leupeptin, antipain, and pepstatin and 1 mM PMSF. Add 0.0025 U Endo H (Roche).
4. Mix and incubate for 1.5–2 h at 37 °C. *Optimize incubation time for each protein.*
5. Spin down the fluid and add 7.5 μL 5× sample buffer and mix.
6. Heat for 5 min to 95 °C.
7. Spin down. At this moment a reducing sample can be prepared (see immunoprecipitation protocol).

3.5.2.6 Cell Surface Expression Tested by Protease
Arrival of a protein that followed the secretory pathway at the plasma membrane can be biochemically monitored. At the cell surface, the protein is accessible to various reagents, such as antibodies, proteases, and biotinylation agents. Below is a typical protocol for cell surface detection by proteases.

1. Wash the cells twice with 2 mL stop buffer.
2. Add 0.5 mL PBS containing 100 μg mL^{-1} trypsin or another protease and 2 mM CaCl$_2$ (if the protease needs calcium) at 4 °C to the cells.
3. Incubate for 30 min on ice. *At this step surface accessible protein either will be completely digested, or it will be cleaved specifically into a limited number of fragments.*
4. Collect fluid from cells and add 100 μg mL^{-1} soybean trypsin inhibitor, 1 mM PMSF, and 10 μg mL^{-1} each of chymostatin, leupeptin, antipain, and pepstatin (final concentrations).
5. Add to the cells 0.5 mL of 100 μg mL^{-1} soybean trypsin inhibitor, 1 mM PMSF,

and 10 μg mL^{-1} each of chymostatin, leupeptin, antipain, and pepstatin (final concentrations). Incubate on ice and repeat the incubation. *It is essential to properly inhibit the protease, because it will otherwise digest internal protein upon lysis.*
6. Lyse the cells in 600 μL lysis buffer and continue with the pulse-chase protocol at step 9. *The comparison between untreated and treated cells will show how much of the protein of interest disappeared because of the proteolysis, or the treated sample will contain a mixture of undigested (internal) and digested protein (cell surface–localized).*

3.5.3
SDS-PAGE [13]

1. Prepare 0.75-mm thick polyacrylamide separating and stacking gels. *The acrylamide percentage depends on the molecular weight of the studied protein and should position the protein just below the middle of the gel to allow maximal separation of different forms of the protein.*
2. Load 8 μL of each sample (and 3 μL of the total lysates on a separate gel to check labeling efficiency). Not using the outer lanes of the gel, load 1× sample buffer in the outside as well as empty lanes to prevent "smiling" of the bands. *When reduced and non-reduced samples are loaded in adjacent lanes in SDS-PAGE, DTT can diffuse into the non-reduced lanes. Either one or two lanes should be left open between the two samples, or an excess quencher should be added to all samples (including the non-reduced ones) just before loading. We frequently use 100 mM N-ethylmaleimide (final concentration in the sample) for this purpose.*
3. Run each gel at max speed at constant current (25 mA per gel for Hoefer minigels) until the dye front is at the bottom of the gel. *Maximum speed without overheating limits lateral diffusion and guarantees sharper bands.*
4. Stain the gel with Coomassie brilliant blue for 5 min and destain to visualize antibody bands.
5. Neutralize gels for 1–3 × 5 min. *If incubated too long in the absence of acid, bands will become diffuse. A minimum time of neutralization is needed to prevent precipitation of salicylic acid, if gels are too acidic.*
6. Treat gels with enhancer solution for 15 min. *Do not incubate longer than 15 min to prevent non-specific signal.*
7. Dry gels at 80 °C on 0.4-mm Schleicher & Schuell filter paper.
8. Expose to film at −80 °C or to a phosphor-imaging screen.

Solutions for SDS-PAGE

Coomassie stain: 0.25% (w/v) Coomassie brilliant blue in destain

Destain: 30% (v/v) methanol and 10% (v/v) acetic acid in H$_2$O

Neutralizer: 30% (v/v) methanol in PBS

Enhancer: 1.5 M sodium salicylate in 30% (v/v) methanol in H$_2$O

Various types of native gels can be used to study the conformation of proteins, oligomerization, aggregation, and the interactions between different proteins. The mobility depends on both the protein's charge and its hydrodynamic size. Blue native gels [106] are often are used for membrane proteins.

Two-dimensional SDS-PAGE

1. Load samples on a tube gel and run, or run on a slab gel with pre-stained markers on either side of the lane. *The first dimension can be a native gel or a non-reducing SDS-PA gel, an IEF gel, or virtually any other kind.*
2. Extract tube gel from capillary, or cut the lane of interest out of the first dimension slab gel.
3. Boil the tube or strip for 10 min in reducing sample buffer.
4. Place the tube or strip on an SDS-polyacrylamide slab gel.
5. A reduced sample can be run in a separate lane close to the edge of the slab gel.
6. Run in the second dimension as a regular slab gel.

Acknowledgements

We thank Claudia Maggioni for sharing the in vitro chase assay protocol and for assistance with the figures, and we thank other members of the Braakman lab, especially Ari Ora, Aafke Land, and Nicole Hafkemeijer, for critical reading of the manuscript.

References

1 ANFINSEN, C. B. (1973). Principles that govern the folding of protein chains. *Science* **181**, 223–230.
2 BERGMAN, L. W. & KUEHL, W. M. (1979). Formation of an intrachain disulfide bond on nascent immunoglobulin light chains. *J Biol Chem* **254**, 8869–8876.
3 CHEN, W., HELENIUS, J., BRAAKMAN, I. & HELENIUS, A. (1995). Cotranslational folding and calnexin binding during glycoprotein synthesis. *Proc Natl Acad Sci USA* **92**, 6229–6233.
4 FULTON, A. B. (1982). How crowded is the cytoplasm? *Cell* **30**, 345–347.
5 MINTON, A. P. (1983). The effect of volume occupancy upon the thermodynamic activity of proteins: some biochemical consequences. *Mol Cell Biochem* **55**, 119–140.
6 ELLIS, R. J. (2001). Macromolecular crowding: obvious but underappreciated. *Trends Biochem Sci* **26**, 597–604.
7 HAMMOND, C. & HELENIUS, A. (1995). Quality control in the secretory pathway. *Curr Opin Cell Biol* **7**, 523–529.
8 HARTL, F. U. & HAYER-HARTL, M. (2002). Molecular chaperones in the cytosol: from nascent chain to folded protein. *Science* **295**, 1852–1858.
9 ELLGAARD, L. & HELENIUS, A. (2003). Quality control in the endoplasmic reticulum. *Nat Rev Mol Cell Biol* **4**, 181–191.
10 GETHING, M. J. & SAMBROOK, J. (1992). Protein folding in the cell. *Nature* **355**, 33–45.
11 BRAAKMAN, I., HOOVER-LITTY, H.,

Wagner, K. R. & Helenius, A. (1991). Folding of influenza hemagglutinin in the endoplasmic reticulum. *J Cell Biol* **114**, 401–411.
12 Wilson, R., Allen, A. J., Oliver, J., Brookman, J. L., High, S. & Bulleid, N. J. (1995). The translocation, folding, assembly and redox-dependent degradation of secretory and membrane proteins in semi-permeabilized mammalian cells. *Biochem J* **307**, 679–687.
13 Laemmli, U. K. (1970). Cleavage of structural proteins during the assembly of the head of bacteriophage T4. *Nature* **227**, 680–685.
14 Porter, R. R. (1959). The hydrolysis of rabbit y-globulin and antibodies with crystalline papain. *Biochem J* **73**, 119–126.
15 Grassetti, D. R. & Murray, J. F., Jr. (1967). Determination of sulfhydryl groups with 2,2'- or 4,4'-dithio-dipyridine. *Arch Biochem Biophys* **119**, 41–49.
16 Ostergaard, H., Henriksen, A., Hansen, F. G. & Winther, J. R. (2001). Shedding light on disulfide bond formation: engineering a redox switch in green fluorescent protein. *EMBO J* **20**, 5853–5862.
17 Kosower, E. M. & Kosower, N. S. (1969). Lest I forget thee, glutathione. *Nature* **224**, 117–120.
18 Kosower, N. S. & Kosower, E. M. (1995). Diamide: an oxidant probe for thiols. *Methods Enzymol* **251**, 123–133.
19 Tortorella, D., Story, C. M., Huppa, J. B., Wiertz, E. J., Jones, T. R., Bacik, I., Bennink, J. R., Yewdell, J. W. & Ploegh, H. L. (1998). Dislocation of type I membrane proteins from the ER to the cytosol is sensitive to changes in redox potential. *J Cell Biol* **142**, 365–376.
20 Braakman, I., Helenius, J. & Helenius, A. (1992). Manipulating disulfide bond formation and protein folding in the endoplasmic reticulum. *EMBO J* **11**, 1717–1722.
21 Tatu, U., Braakman, I. & Helenius, A. (1993). Membrane glycoprotein folding, oligomerization and intracellular transport: effects of dithiothreitol in living cells. *EMBO J* **12**, 2151–2157.
22 de Silva, A., Braakman, I. & Helenius, A. (1993). Posttranslational folding of vesicular stomatitis virus G protein in the ER: involvement of noncovalent and covalent complexes. *J Cell Biol* **120**, 647–655.
23 Lodish, H. F., Kong, N. & Wikstrom, L. (1992). Calcium is required for folding of newly made subunits of the asialoglycoprotein receptor within the endoplasmic reticulum. *J Biol Chem* **267**, 12753–12760.
24 Ellis, R. J. (1990). The molecular chaperone concept. *Semin Cell Biol* **1**, 1–9.
25 Hartl, F. U. & Martin, J. (1995). Molecular chaperones in cellular protein folding. *Curr Opin Struct Biol* **5**, 92–102.
26 Jaenicke, R. & Lilie, H. (2000). Folding and association of oligomeric and multimeric proteins. *Adv Protein Chem* **53**, 329–401.
27 Haas, I. G. & Wabl, M. (1983). Immunoglobulin heavy chain binding protein. *Nature* **306**, 387–389.
28 Flynn, G. C., Pohl, J., Flocco, M. T. & Rothman, J. E. (1991). Peptide-binding specificity of the molecular chaperone BiP. *Nature* **353**, 726–730.
29 Blond-Elguindi, S., Cwirla, S. E., Dower, W. J., Lipshutz, R. J., Sprang, S. R., Sambrook, J. F. & Gething, M. J. (1993). Affinity panning of a library of peptides displayed on bacteriophages reveals the binding specificity of BiP. *Cell* **75**, 717–728.
30 Hammond, C., Braakman, I. & Helenius, A. (1994). Role of N-linked oligosaccharide recognition, glucose trimming, and calnexin in glycoprotein folding and quality control. *Proc Natl Acad Sci USA* **91**, 913–917.
31 Kassenbrock, C. K. & Kelly, R. B. (1989). Interaction of heavy chain binding protein (BiP/GRP78) with adenine nucleotides. *EMBO J* **8**, 1461–1467.
32 Wei, J. & Hendershot, L. M. (1995).

Characterization of the nucleotide binding properties and ATPase activity of recombinant hamster BiP purified from bacteria. *J Biol Chem* **270**, 26670–26676.

33 MOLINARI, M. & HELENIUS, A. (2000). Chaperone selection during glycoprotein translocation into the endoplasmic reticulum. *Science* **288**, 331–333.

34 SWANTON, E., HIGH, S. & WOODMAN, P. (2003). Role of calnexin in the glycan-independent quality control of proteolipid protein. *EMBO J* **22**, 2948–2958.

35 IHARA, Y., COHEN-DOYLE, M. F., SAITO, Y. & WILLIAMS, D. B. (1999). Calnexin discriminates between protein conformational states and functions as a molecular chaperone in vitro. *Mol Cell* **4**, 331–341.

36 BULLEID, N. J. & FREEDMAN, R. B. (1988). Defective co-translational formation of disulphide bonds in protein disulphide-isomerase-deficient microsomes. *Nature* **335**, 649–651.

37 MOLINARI, M. & HELENIUS, A. (1999). Glycoproteins form mixed disulphides with oxidoreductases during folding in living cells. *Nature* **402**, 90–93.

38 OLIVER, J. D., VAN DER WAL, F. J., BULLEID, N. J. & HIGH, S. (1997). Interaction of the thiol-dependent reductase ERp57 with nascent glycoproteins. *Science* **275**, 86–88.

39 FISCHER, G., BANG, H. & MECH, C. (1984). [Determination of enzymatic catalysis for the cis-trans-isomerization of peptide binding in proline-containing peptides]. *Biomed Biochim Acta* **43**, 1101–1111.

40 SCHMID, F. X., MAYR, L. M., MUCKE, M. & SCHONBRUNNER, E. R. (1993). Prolyl isomerases: role in protein folding. *Adv Protein Chem* **44**, 25–66.

41 LANG, K., SCHMID, F. X. & FISCHER, G. (1987). Catalysis of protein folding by prolyl isomerase. *Nature* **329**, 268–270.

42 STEINMANN, B., BRUCKNER, P. & SUPERTI-FURGA, A. (1991). Cyclosporin A slows collagen triple-helix formation in vivo: indirect evidence for a physiologic role of peptidyl-prolyl cis-trans-isomerase. *J Biol Chem* **266**, 1299–1303.

43 LODISH, H. F. & KONG, N. (1991). Cyclosporin A inhibits an initial step in folding of transferrin within the endoplasmic reticulum. *J Biol Chem* **266**, 14835–14838.

44 LOO, T. W. & CLARKE, D. M. (1994). Prolonged association of temperature-sensitive mutants of human P-glycoprotein with calnexin during biogenesis. *J Biol Chem* **269**, 28683–28689.

45 PIND, S., RIORDAN, J. R. & WILLIAMS, D. B. (1994). Participation of the endoplasmic reticulum chaperone calnexin (p88, IP90) in the biogenesis of the cystic fibrosis transmembrane conductance regulator. *J Biol Chem* **269**, 12784–12788.

46 LOO, T. W. & CLARKE, D. M. (1995). P-glycoprotein. Associations between domains and between domains and molecular chaperones. *J Biol Chem* **270**, 21839–21844.

47 YANG, Y., JANICH, S., COHN, J. A. & WILSON, J. M. (1993). The common variant of cystic fibrosis transmembrane conductance regulator is recognized by hsp70 and degraded in a pre-Golgi nonlysosomal compartment. *Proc Natl Acad Sci USA* **90**, 9480–9484.

48 MEACHAM, G. C., PATTERSON, C., ZHANG, W., YOUNGER, J. M. & CYR, D. M. (2001). The Hsc70 co-chaperone CHIP targets immature CFTR for proteasomal degradation. *Nat Cell Biol* **3**, 100–105.

49 LOO, M. A., JENSEN, T. J., CUI, L., HOU, Y., CHANG, X. B. & RIORDAN, J. R. (1998). Perturbation of Hsp90 interaction with nascent CFTR prevents its maturation and accelerates its degradation by the proteasome. *EMBO J* **17**, 6879–6887.

50 TRUSCOTT, K. N., BRANDNER, K. & PFANNER, N. (2003). Mechanisms of protein import into mitochondria. *Curr Biol* **13**, R326–337.

51 SOLL, J. (2002). Protein import into

chloroplasts. *Curr Opin Plant Biol* **5**, 529–535.

52 WALTER, P. & BLOBEL, G. (1980). Purification of a membrane-associated protein complex required for protein translocation across the endoplasmic reticulum. *Proc Natl Acad Sci USA* **77**, 7112–7116.

53 LI, Y., LUO, L., THOMAS, D. Y. & KANG, C. Y. (1994). Control of expression, glycosylation, and secretion of HIV-1 gp120 by homologous and heterologous signal sequences. *Virology* **204**, 266–278.

54 REHM, A., STERN, P., PLOEGH, H. L. & TORTORELLA, D. (2001). Signal peptide cleavage of a type I membrane protein, HCMV US11, is dependent on its membrane anchor. *EMBO J* **20**, 1573–1582.

55 JOSEFSSON, L. G. & RANDALL, L. L. (1983). Analysis of cotranslational proteolytic processing of nascent chains using two-dimensional gel electrophoresis. *Methods Enzymol* **97**, 77–85.

56 LAND, A., ZONNEVELD, D. & BRAAKMAN, I. (2003). Folding of HIV-1 envelope glycoprotein involves extensive isomerization of disulfide bonds and conformation-dependent leader peptide cleavage. *FASEB J* **17**, 1058–1067.

57 LI, Y., BERGERON, J. J., LUO, L., OU, W. J., THOMAS, D. Y. & KANG, C. Y. (1996). Effects of inefficient cleavage of the signal sequence of HIV-1 gp 120 on its association with calnexin, folding, and intracellular transport. *Proc Natl Acad Sci USA* **93**, 9606–9611.

58 GAVEL, Y. & VON HEIJNE, G. (1990). Sequence differences between glycosylated and non-glycosylated Asn-X-Thr/Ser acceptor sites: implications for protein engineering. *Protein Eng* **3**, 433–442.

59 SOUSA, M. C., FERRERO-GARCIA, M. A. & PARODI, A. J. (1992). Recognition of the oligosaccharide and protein moieties of glycoproteins by the UDP-Glc:glycoprotein glucosyltransferase. *Biochemistry* **31**, 97–105.

60 TROMBETTA, S. E. & PARODI, A. J. (1992). Purification to apparent homogeneity and partial characterization of rat liver UDP-glucose:glycoprotein glucosyltransferase. *J Biol Chem* **267**, 9236–9240.

61 RITTER, C. & HELENIUS, A. (2000). Recognition of local glycoprotein misfolding by the ER folding sensor UDP-glucose:glycoprotein glucosyltransferase. *Nat Struct Biol* **7**, 278–280.

62 HWANG, C., SINSKEY, A. J. & LODISH, H. F. (1992). Oxidized redox state of glutathione in the endoplasmic reticulum. *Science* **257**, 1496–1502.

63 TU, B. P., HO-SCHLEYER, S. C., TRAVERS, K. J. & WEISSMAN, J. S. (2000). Biochemical basis of oxidative protein folding in the endoplasmic reticulum. *Science* **290**, 1571–4.

64 FRAND, A. R. & KAISER, C. A. (1998). The ERO1 gene of yeast is required for oxidation of protein dithiols in the endoplasmic reticulum. *Mol Cell* **1**, 161–170.

65 POLLARD, M. G., TRAVERS, K. J. & WEISSMAN, J. S. (1998). Ero1p: a novel and ubiquitous protein with an essential role in oxidative protein folding in the endoplasmic reticulum. *Mol Cell* **1**, 171–182.

66 CABIBBO, A., PAGANI, M., FABBRI, M., ROCCHI, M., FARMERY, M. R., BULLEID, N. J. & SITIA, R. (2000). ERO1-L, a human protein that favors disulfide bond formation in the endoplasmic reticulum. *J Biol Chem* **275**, 4827–4833.

67 PAGANI, M., FABBRI, M., BENEDETTI, C., FASSIO, A., PILATI, S., BULLEID, N. J., CABIBBO, A. & SITIA, R. (2000). Endoplasmic reticulum oxidoreductin 1-lbeta (ERO1-Lbeta), a human gene induced in the course of the unfolded protein response. *J Biol Chem* **275**, 23685–23692.

68 FRAND, A. R. & KAISER, C. A. (1999). Ero1p oxidizes protein disulfide isomerase in a pathway for disulfide bond formation in the endoplasmic reticulum. *Mol Cell* **4**, 469–477.

69 SEVIER, C. S., CUOZZO, J. W., VALA, A., ASLUND, F. & KAISER, C. A. (2001). A flavoprotein oxidase defines a new endoplasmic reticulum pathway for biosynthetic disulphide bond formation. *Nat Cell Biol* **3**, 874–882.

70 TU, B. P. & WEISSMAN, J. S. (2004). Oxidative protein folding in eukaryotes: mechanisms and consequences. *J Cell Biol* **164**, 341–346.

71 CREIGHTON, T. E. (1978). Experimental studies of protein folding and unfolding. *Prog Biophys Mol Biol* **33**, 231–297.

72 CREIGHTON, T. E. (1990). Disulphide bonds between cysteine residues. In *Protein structure. A practical approach.* (CREIGHTON, T. E., ed.), pp. 155–168. IRL Press, Oxford.

73 SOMMER, T. & WOLF, D. H. (1997). Endoplasmic reticulum degradation: reverse protein flow of no return. *FASEB J* **11**, 1227–1233.

74 HOSOKAWA, N., WADA, I., HASEGAWA, K., YORIHUZI, T., TREMBLAY, L. O., HERSCOVICS, A. & NAGATA, K. (2001). A novel ER alpha-mannosidase-like protein accelerates ER-associated degradation. *EMBO Rep* **2**, 415–422.

75 JAKOB, C. A., BODMER, D., SPIRIG, U., BATTIG, P., MARCIL, A., DIGNARD, D., BERGERON, J. J., THOMAS, D. Y. & AEBI, M. (2001). Htm1p, a mannosidase-like protein, is involved in glycoprotein degradation in yeast. *EMBO Rep* **2**, 423–430.

76 ODA, Y., HOSOKAWA, N., WADA, I. & NAGATA, K. (2003). EDEM as an acceptor of terminally misfolded glycoproteins released from calnexin. *Science* **299**, 1394–1397.

77 MOLINARI, M., CALANCA, V., GALLI, C., LUCCA, P. & PAGANETTI, P. (2003). Role of EDEM in the release of misfolded glycoproteins from the calnexin cycle. *Science* **299**, 1397–1400.

78 CUMMINGS, R. D., KORNFELD, S., SCHNEIDER, W. J., HOBGOOD, K. K., TOLLESHAUG, H., BROWN, M. S. & GOLDSTEIN, J. L. (1983). Biosynthesis of N- and O-linked oligosaccharides of the low density lipoprotein receptor. *J Biol Chem* **258**, 15261–15273.

79 LEONARD, C. K., SPELLMAN, M. W., RIDDLE, L., HARRIS, R. J., THOMAS, J. N. & GREGORY, T. J. (1990). Assignment of intrachain disulfide bonds and characterization of potential glycosylation sites of the type 1 recombinant human immunodeficiency virus envelope glycoprotein (gp120) expressed in Chinese hamster ovary cells. *J Biol Chem* **265**, 10373–10382.

80 PELLETIER, L., JOKITALO, E. and WARREN, G. (2000). The effect of Golgi depletion on exocytic transport. *Nature Cell Biology* **2**, 840–846.

81 STEVENS, T., ESMON, B. & SCHEKMAN, R. (1982). Early stages in the yeast secretory pathway are required for transport of carboxypeptidase Y to the vacuole. *Cell* **30**, 439–448.

82 CAMPBELL, K. P., MACLENNAN, D. H., JORGENSEN, A. O. & MINTZER, M. C. (1983). Purification and characterization of calsequestrin from canine cardiac sarcoplasmic reticulum and identification of the 53,000 dalton glycoprotein. *J Biol Chem* **258**, 1197–1204.

83 LODISH, H. F. & KONG, N. (1990). Perturbation of cellular calcium blocks exit of secretory proteins from the rough endoplasmic reticulum. *J Biol Chem* **265**, 10893–10899.

84 KLENK, H. D., GARTEN, W. & ROTT, R. (1984). Inhibition of proteolytic cleavage of the hemagglutinin of influenza virus by the calcium-specific ionophore A23187. *EMBO J* **3**, 2911–2915.

85 THASTRUP, O., CULLEN, P. J., DROBAK, B. K., HANLEY, M. R. & DAWSON, A. P. (1990). Thapsigargin, a tumor promoter, discharges intracellular Ca2+ stores by specific inhibition of the endoplasmic reticulum Ca2(+)-ATPase. *Proc Natl Acad Sci USA* **87**, 2466–2470.

86 LYTTON, J., WESTLIN, M. & HANLEY, M. R. (1991). Thapsigargin inhibits the sarcoplasmic or endoplasmic reticulum Ca-ATPase family of calcium pumps. *J Biol Chem* **266**, 17067–17071.

87 Braakman, I., Helenius, J. & Helenius, A. (1992). Role of ATP and disulphide bonds during protein folding in the endoplasmic reticulum. *Nature* **356**, 260–262.
88 Datema, R. & Schwarz, R. T. (1979). Interference with glycosylation of glycoproteins. Inhibition of formation of lipid-linked oligosaccharides in vivo. *Biochem J* **184**, 113–123.
89 Dorner, A. J., Wasley, L. C. & Kaufman, R. J. (1990). Protein dissociation from GRP78 and secretion are blocked by depletion of cellular ATP levels. *Proc Natl Acad Sci USA* **87**, 7429–7432.
90 Lundin, A. & Thore, A. (1975). Analytical information obtainable by evaluation of the time course of firefly bioluminescence in the assay of ATP. *Anal Biochem* **66**, 47–63.
91 Joshi, S. & Burrows, R. (1990). ATP synthase complex from bovine heart mitochondria. Subunit arrangement as revealed by nearest neighbor analysis and susceptibility to trypsin. *J Biol Chem* **265**, 14518–14525.
92 Chen, L. L., Rosa, J. J., Turner, S. & Pepinsky, R. B. (1991). Production of multimeric forms of CD4 through a sugar-based cross-linking strategy. *J Biol Chem* **266**, 18237–18243.
93 Oliver, J. D., Roderick, H. L., Llewellyn, D. H. & High, S. (1999). ERp57 functions as a subunit of specific complexes formed with the ER lectins calreticulin and calnexin. *Mol Biol Cell* **10**, 2573–2582.
94 Takatsuki, A., Arima, K. & Tamura, G. (1971). Tunicamycin, a new antibiotic. I. Isolation and characterization of tunicamycin. *J Antibiot (Tokyo)* **24**, 215–223.
95 Patil, C. & Walter, P. (2001). Intracellular signaling from the endoplasmic reticulum to the nucleus: the unfolded protein response in yeast and mammals. *Curr Opin Cell Biol* **13**, 349–355.
96 Chang, X. B., Hou, Y. X., Jensen, T. J. & Riordan, J. R. (1994). Mapping of cystic fibrosis transmembrane conductance regulator membrane topology by glycosylation site insertion. *J Biol Chem* **269**, 18572–18575.
97 Dorner, A. J., Wasley, L. C., Raney, P., Haugejorden, S., Green, M. & Kaufman, R. J. (1990). The stress response in Chinese hamster ovary cells. Regulation of ERp72 and protein disulfide isomerase expression and secretion. *J Biol Chem* **265**, 22029–22034.
98 Rye, H. S., Burston, S. G., Fenton, W. A., Beechem, J. M., Xu, Z., Sigler, P. B. & Horwich, A. L. (1997). Distinct actions of cis and trans ATP within the double ring of the chaperonin GroEL. *Nature* **388**, 792–798.
99 Hendershot, L., Wei, J., Gaut, J., Melnick, J., Aviel, S. & Argon, Y. (1996). Inhibition of immunoglobulin folding and secretion by dominant negative BiP ATPase mutants. *Proc Natl Acad Sci USA* **93**, 5269–5274.
100 Stanley, P., Sallustio, S., Krag, S. S. & Dunn, B. (1990). Lectin-resistant CHO cells: selection of seven new mutants resistant to ricin. *Somat Cell Mol Genet* **16**, 211–223.
101 Ray, M. K., Yang, J., Sundaram, S. & Stanley, P. (1991). A novel glycosylation phenotype expressed by Lec23, a Chinese hamster ovary mutant deficient in alpha-glucosidase I. *J Biol Chem* **266**, 22818–22825.
102 Reitman, M. L., Trowbridge, I. S. & Kornfeld, S. (1982). A lectin-resistant mouse lymphoma cell line is deficient in glucosidase II, a glycoprotein-processing enzyme. *J Biol Chem* **257**, 10357–10363.
103 Howell, D. N., Andreotti, P. E., Dawson, J. R. & Cresswell, P. (1985). Natural killing target antigens as inducers of interferon: studies with an immunoselected, natural killing-resistant human T lymphoblastoid cell line. *J Immunol* **134**, 971–976.
104 Fire, A., Xu, S., Montgomery, M. K., Kostas, S. A., Driver, S. E. & Mello, C. C. (1998). Potent and specific

genetic interference by double-stranded RNA in Caenorhabditis elegans. *Nature* **391**, 806–811.

105 OU, W. J., CAMERON, P. H., THOMAS, D. Y. & BERGERON, J. J. (1993). Association of folding intermediates of glycoproteins with calnexin during protein maturation. *Nature* **364**, 771–776.

106 SCHAGGER, H. & VON JAGOW, G. (1991). Blue native electrophoresis for isolation of membrane protein complexes in enzymatically active form. *Anal Biochem* **199**, 223–231.

4
Characterization of ATPase Cycles of Molecular Chaperones by Fluorescence and Transient Kinetic Methods

Sandra Schlee and Jochen Reinstein

4.1
Introduction

4.1.1
Characterization of ATPase Cycles of Energy-transducing Systems

ATP has long been known to be involved in many important cellular processes such as energy transduction in molecular motors, ion transport and signal transmission, topological processing of nucleic acids and the fidelity of protein synthesis. In spite of the diverse functions of ATPases, their mechanism includes some universal steps: the specific recognition and binding of the nucleotide, a conformational change induced by nucleotide binding, phosphoryl transfer (chemical reaction), adjustment of the conformation to the product state, and the release of products, which is accompanied by the regeneration of the initial state of the enzyme.

Although ATPases in principle employ the same reaction steps, they have to cope with entirely different tasks. Therefore, the consequence of a single reaction step on a molecular level depends on the respective system. The ATP state of DnaK, for example, represents a form that rapidly traps and releases potential substrate proteins [1, 2]. In the context of molecular motor proteins, ATP binding causes the forward movement of conventional kinesin (power stroke), but ATP binding to myosin, on the other hand, leads to dissociation of myosin from actin and is accompanied by the recovery stroke of its lever arm [3].

With some exceptions, an exact definition of the coupling between ATP hydrolysis and enzymatic action is still missing for many of the systems. History has revealed repeatedly that mechanisms remain a matter of speculation until equilibrium constants and reaction rates of all relevant intermediates (functional states) have been identified. In addition, the stability of enzyme-bound nucleotide as well as the velocity of product release represent molecular timers that are carefully regulated by accessory proteins in many ATPases. Determination of kinetics and measurements of force and motion of single molecules are therefore essential for a comprehensive understanding of the respective enzymatic mechanism.

Protein Folding Handbook. Part II. Edited by J. Buchner and T. Kiefhaber
Copyright © 2005 WILEY-VCH Verlag GmbH & Co. KGaA, Weinheim
ISBN: 3-527-30784-2

This review summarizes various aspects of fluorescent nucleotide analogues that were used to determine the ATPase cycles of the molecular chaperones DnaK, Hsp90, and ClpB and their regulation by accessory co-chaperones. Experimental protocols and conceptual considerations regarding nucleotide-binding and ATPase reactions of these chaperone systems can be found in the Section 4.3 of this review. The particular details of roles and properties of DnaK/Hsc70 or Hsp90 are not described here since they are well covered by other chapters in this volume.

4.1.2
The Use of Fluorescent Nucleotide Analogues

4.1.2.1 Fluorescent Modifications of Nucleotides

Numerous ATP analogues have been synthesized to report on the action of ATP in biosystems by means of a spectroscopic signal (absorption, fluorescence, and EPR and NMR probes) [4, 5]. Since a comprehensive intrinsic tryptophan fluorescence signal for nucleotide binding is rarely available, the use of fluorescent nucleotide analogues to study nucleotide-binding proteins prevails. Fluorescent nucleotide analogues combine the localization to a defined position with the inherent high sensitivity of the method: fluorophores with high quantum yields (i.e., >0.1) can be readily detected at submicromolar concentrations. Fluorescence can provide information about the size of proteins and structural features of the nucleotide-binding site and allows quantification of the kinetic and equilibrium constants describing the system. The intrinsic fluorescence of the common nucleotides, at ambient temperatures and neutral pH, is far too low (quantum yields on the order of 10^{-4}) to be of general use in the investigation of nucleotide-binding proteins, especially since the furthest reaching absorption maxima of nucleotides are near 260 nm, which overlaps with intrinsic protein absorption. To improve the utility of nucleotides as fluorescent probes, modifications have been introduced to different positions of the nucleotide moiety: the adenosine ring, the ribose groups, and the phosphate moiety [4].

Among the most conservative changes is the substitution of an adenosine with a fluorescent formycin or aminopurine ring (Figure 4.1). These analogues have been used to study the ATPase kinetics of $(Na^+ + K^+)$-ATPase and myosin [6–8]. It was shown that C8-MABA-ADP, with a substitution in the 8 position of adenine, was a useful probe for the nucleotide-binding site of the molecular chaperone DnaK [9–11]. However, one limitation is that the substitutions tend to stabilize the base in the unusual *syn* conformation rather than in the extended *anti* conformation of unmodified nucleotides. As a consequence, binding parameters, specifically kinetic constants, may be altered.

Another fluorescent analogue, etheno-ATP (ε-ATP), includes the addition of an aromatic ring to the adenine moiety [12]. This modification and the addition of different aromatic rings result in bulky fluorophores that are not readily tolerated by many nucleotide-binding sites.

Modifications that do not alter the purine ring systems offer an alternative approach to fluorescent nucleotide analogues. For example, an ATP analogue with

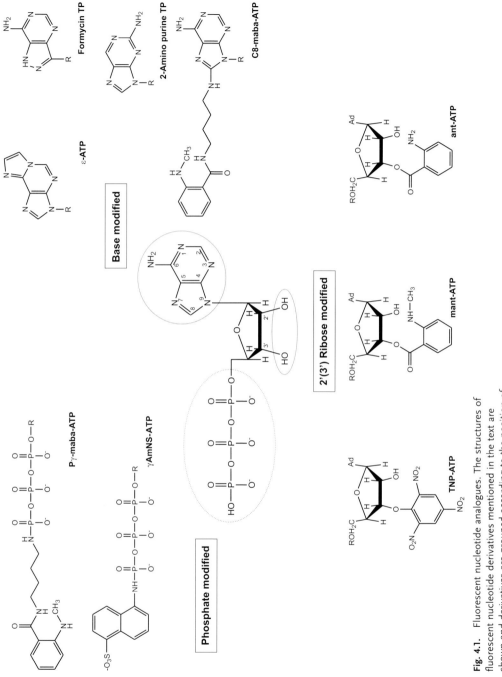

Fig. 4.1. Fluorescent nucleotide analogues. The structures of fluorescent nucleotide derivatives mentioned in the text are shown and derivatives are grouped according to the position of modifications.

altered phosphoryl structure, namely, adenosine-5′-triphospho-γ-1-(5-sulfonicacid)-naphtylamine (γAmNS-ATP), which is a good substrate for *E.coli* RNA polymerase, has been described [13]. By coupling a fluorescent *N*-methylanthraniloyl-group to the phosphate moiety of nucleotides via a butyl linker, nucleotide analogues were obtained that allow the direct determination of binding constants. (P_γ)MABA-ATP (adenosine-5′-triphospho-γ-(*N*′-methylanthraniloyl-aminobutyl)-phosphoamidate) was used for the kinetic characterization of the ATPase cycle of the molecular chaperone Hsp90 [14]. The fluorescent CDP analogue (P_β)MABA-CDP proved to be a specific probe for the NMP-binding site of UMP/CMP kinase [15].

In many cases modification of the ribose moiety of nucleotides offers a satisfactory approach in that the resultant analogues closely mimic the behavior of their parent nucleotides. The advantage of such ribose modification can be understood by examination of the structures of nucleoside 5′-triphosphatases such as p21ras [16], in which case the 2′,3′-hydroxyl groups of the nucleotide project out of the binding domain.

Among the first ribose-modified nucleotides was 2′,3′-O-(2,4,6-trinitrocyclohexadienylidene) adenosine 5′-triphosphate (TNP-ATP), introduced by Hiratsuka and Uchida [17]. Ribose-modified analogues used extensively in studies of nucleoside 5′-triphosphatases are the 2′(3′)-O-anthraniloyl (Ant) and 2′(3′)-O-methylanthraniloyl (MANT) derivatives of adenine [18–20]. The motivations for this choice of fluorophores were their relatively small size, which implies that perturbation of binding properties would be minimal, and the fact that the fluorescence properties of these probes are environmentally sensitive. MANT-ADP has been applied, among others, to investigate the ATPase cycle of ClpB [21, 22].

It has to be kept in mind that one complication of derivatizing the ribose hydroxyls is that a mixture of 2′ and 3′ isomers will be generated, which may differ in their fluorescent and binding capacities. One way to avoid the complication of mixed isomers is to use nucleotide derivatives based on 2′ deoxyribose.

4.1.2.2 How to Find a Suitable Analogue for a Specific Protein

ATP-utilizing systems vary enormously in their specificity, with the consequence that what acts as a good analogue in one case may be a very poor probe in another case. Without structural information about the nucleotide-protein complex, it is very difficult to accurately predict the effectiveness of a fluorescent nucleotide analogue in a novel biological system. To facilitate the screening of potential candidates, use can be made of the binding of ATPases to ATP-agarose. ATP- and ADP-agaroses are available with their ATP moiety coupled chemically to the N6 or C8 position of the adenine ring or the ribose hydroxyls via linkers of variable length (Table 4.1). If binding of an ATPase to a specific kind of ATP-agarose is observed, chances are high that fluorescent nucleotide analogues, modified at the according position, represent specific probes for the respective protein.

A good indicator for the suitability of a fluorescent derivative as substrate for a specific system is the accordance of the binding parameters (K_d, k_{on}, k_{off}) [23] and kinetic parameters (e.g., K_m, V_{max}) of the ATPase reaction of the fluorescent derivative and the unmodified nucleotide. In addition, the modified nucleotide should support the enzymatic activity of a protein equally to the unmodified nucleotide.

Tab. 4.1. Commercially available AXP matrices for affinity chromatography with ATPases.

Matrix	Nucleotide	Matrix attachment	Matrix spacer
Agarose	5'ATP/ADP/AMP	N6	11 atoms
Agarose	5'AMP	N6	8 atoms
Agarose	5'ATP/AMP	C8	9 atoms
Agarose	5'ATP/AMP	Ribose hydroxyl	11 atoms
Agarose	5'ATP	Ribose hydroxyl	22 atoms
Agarose	2'AMP	C8	9 atoms
Agarose	2'5'ADP	N6	8 atoms
Agarose	3'5'ADP	N6	8 atoms
Sepharose	2'5'ADP	N6	8 atoms
Sepharose	5'AMP	N6	8 atoms

4.2 Characterization of ATPase Cycles of Molecular Chaperones

4.2.1 Biased View

The following description of the ATPase cycles of DnaK, Hsp90, and ClpB systems certainly does not claim to give a comprehensive overview of the many roles and aspects of these chaperones. For this purpose, other excellent reviews in this book and outside, as mentioned in the individual chapters, should be referred to. We are also aware that the selection of results and methods here is highly biased towards fluorescence, nucleotide analogues, and related methods, but we feel that the reader may benefit more from a description of the various aspects of these techniques than from a (bound-to-fail) attempt to cover all conceivable approaches.

4.2.2 The ATPase Cycle of DnaK

The ATPase cycles of DnaK and its eukaryotic ortholog Hsc70 are well characterized from a variety of methods at this point. Comprehensive overviews may be found in this volume and in other reviews (e.g., Refs. [24–27]).

Where applicable, references to particular aspects concerning ATPase activity or nucleotide binding are covered in Section 4.3.

4.2.3 The ATPase Cycle of the Chaperone Hsp90

Hsp90 has a central role in communicating numerous folding as well as other processes in the cell [28–32]. Despite great interest in its molecular mechanism, the involvement of ATP in its function remained a matter of debate for an extended time period. It was reported by Csermely and co-workers that Hsp90 has an ATP-binding site and autophosphorylation activity that is dependent on catalytic Mg^{2+} and stimulated by Ca^{2+} [33]. ATP-binding studies showed similar dependencies on these cations and were based on affinity labeling with azido-ATP and the ability of Hsp90 to bind to ATP-agarose.

Apparent ATPase activity of Hsp90 was also reported by other groups [34, 35], but considering the low enzymatic activity involved, it was questionable whether the activities observed were not the consequences of minor impurities, e.g., protein kinases that constitute one of the substrates for Hsp90 and may be dragged along the purification process.

Further studies observed the dependence of circular dichroism and Fourier infrared spectra of Hsp90 as a function of ATP concentration. Here ATP was shown to increase the content of β-plated sheet structures and increased interchain interactions. Also, a decrease in intrinsic tryptophan fluorescence and susceptibility to tryptic digestion were seen in the presence of ATP [36]. The conditions necessary to achieve these observed effects, however, still raised some concerns about specific binding of ATP to Hsp90. In a comparative study of highly purified Hsp90 with Hsc70 [37], Hsp90 failed to show any significant binding to immobilized ATP (ATP-agarose) or three different fluorescent nucleotide analogues, namely, $2'(3')$-O-(N-methylanthraniloyl)-adenosine 5'-diphosphate (MANT-ADP), N^8-(4-N'-methylanthr$\underline{a}$niloylamino$\underline{b}$utyl)-8-$\underline{a}$minoadenosine 5'-diphosphate (MABA-ADP), or $1,N^6$-ethenoadenosine diphosphate (see also Figure 4.1). A subsequent reinvestigation with electron paramagnetic resonance (EPR) spectroscopy and spin-labeled nucleotide analogues, however, detected weak binding of ATP ($K_d = 400$ μM) [38]. This weak binding of ATP remained enigmatic, however, since even BSA is capable of binding ATP with this affinity, again as determined by EPR [39]. This should not mean that weak binding of substrates in general has to be regarded with distrust, but usually weak binding goes hand in hand with high activity, whereas the ATPase activity of Hsp90 was not identified to be particularly high.

A major breakthrough concerning this question arrived with the X-ray structure of the N-terminal domain of Hsp90 in complex with ADP and Mg^{2+} [40]. This important structure showed that ADP and ATP bind in the same binding pocket as the antitumor drug geldanamycin [41], suggesting that geldanamycin acts by blocking nucleotide binding and not in direct competition with protein substrates.

The nucleotide binds to Hsp90 in a peculiar mode with a kinked phosphate chain. The ribose and base are not readily accessible, explaining why fluorescent nucleotide analogues with modifications at the ribose or base could not bind [42]. Based on previous experience with modifications at the terminal phosphate groups [15], fluorescent analogues of ADP and ATP were synthesized. Binding of these new compounds, namely, adenosinediphospho-β-(N'-methylanthraniloylaminobutyl)-phosphoramidate and adenosinetriphospho-γ-(N'-methylanthraniloylaminobutyl)-phosphoramidate (see also Table 4.1), to Hsp90 results in enhancement of MANT fluorescence. Although it is usually desirable that modified fluorescent nucleotide analogues behave as closely to their "native" siblings as possible, with Hsp90 it turned out that this is fortunately not the case. Here the fluorescent analogues bind much more tightly (K_d of 20 μM for Hsp90) than the unmodified nucleotide, with K_d around 500 µM [14]. This fortunate fact allowed us to determine many individual steps of the ATPase cycle of Hsp90 with stopped-flow and other methods. Once the basic properties of P(γ)-MABA-ATP binding were determined, they served as a diagnostic tool to help clarify the role of individual residues or domains [43, 44] and, similarly to studies concerning DnaK and its nucleotide-exchange factor GrpE, the effect of regulating co-chaperones such as Sti1 on the catalytic cycle [45, 46].

4.2.4
The ATPase Cycle of the Chaperone ClpB

4.2.4.1 ClpB, an Oligomeric ATPase With Two AAA Modules Per Protomer

The molecular chaperone ClpB belongs to the AAA superfamily of proteins that couple ATP binding and hydrolysis to the disruption of protein-protein interaction and the remodeling of protein complexes in an extraordinary variety of biological circumstances (for reviews, see Refs. [47, 48]). AAA proteins contain at least one ATPase module consisting of two physical domains with conserved sequence motifs (Walker A, Walker B, sensor-1) [49] and form ring-shaped oligomers. The conformation of the oligomer and in some cases oligomerization itself are controlled by the bound nucleotide [50, 51].

In contrast to the related proteins ClpA and ClpX, which unfold soluble protein substrates for proteolysis, ClpB and its *S. cerevisiae* homologue Hsp104 act directly on protein aggregates to return them to the soluble state [52–54]. Recovery of active proteins from the aggregated state by ClpB requires the assistance of the DnaK-DnaJ chaperone system [52, 55–57].

The molecular details of the mechanism linking ATP binding and hydrolysis to the disruption of interactions in protein aggregates remain largely undefined. In the case of ClpB, which comprises two nucleotide-binding domains (NBD), clarification of this mechanism is a particularly difficult problem. On the one hand, one has to obtain site-specific information regarding the contribution of each NBD to the activity (catalysis and binding), but, on the other hand, one has to consider that these properties are regulated by the activity of the other site. In addition, the accumulation of ATP-binding sites within the ClpB oligomer is suggestive of complex allosteric behavior, in which inter-ring communications might contribute to complexity.

4.2.4.2 Nucleotide-binding Properties of NBD1 and NBD2

To provide an initial, general sense of the nucleotide-binding properties of the two NBDs, the fluorescent nucleotide analogue MANT-ADP was titrated with increasing amounts of ClpB from *T. thermophilus* [21, 22]. The same experiment was executed with ClpB mutants (ClpB(K204Q), ClpB(K601Q)), which were selectively inactivated at one NBD by substituting a conserved lysine residue within the Walker A consensus sequence. As these conserved lysine residues interact directly with the β- and γ-phosphates of bound nucleotides, the introduction of an uncharged amino acid at this position impairs nucleotide binding. The fluorescence of MANT-ADP

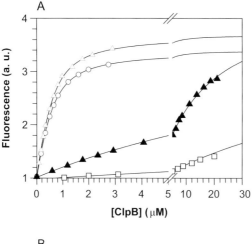

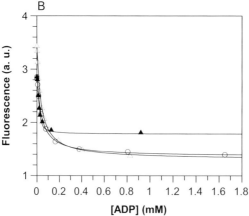

Fig. 4.2. Equilibrium titration experiments with ClpB$_{Tth}$ and MANT-ADP. (A) 0.5 µM MANT-ADP was titrated with ClpB$_{Tth}$(wt) (○), ClpB$_{Tth}$(K204Q) (△), ClpB$_{Tth}$(K601Q) (▲), and the double mutant ClpB$_{Tth}$(K204Q/K601Q) (□) at 25 °C. The excitation wavelength was 360 nm and emission was detected at 440 nm. Data were fitted according to the quadratic equation assuming a 1:1 binding stoichiometry. A K_d of 0.2 µM (±0.02) was observed for ClpB$_{Tth}$(wt) and ClpB$_{Tth}$(K204Q) and a K_d of 15.9 µM (±1.0) was observed for ClpB$_{Tth}$(K601Q). (B) Displacement titration. A preformed ClpB$_{Tth}$*MANT-ADP complex was titrated with ADP ($T = 25$ °C, $\lambda_{ex} = 360$ nm, $\lambda_{em} = 440$ nm). Fitting the curves to a cubic equation yields K_d values of 1.5 µM (±0.08) for ClpB$_{Tth}$(wt) (○), of 2.8 µM (±0.09) for ClpB$_{Tth}$(K204Q) (△), and of 2.3 µM (±0.8) for ClpB$_{Tth}$(K601Q) (▲).

changes in response to binding at both NBDs, and for both ClpB(wt) and the NBD mutants, a saturable increase in fluorescence was observed (Figure 4.2A). Fitting the data to the quadratic binding equation (see Section 4.3) yielded a K_d of 0.2 µM for ClpB(K204Q) and of 16 µM for ClpB(K601Q), indicating that the NBD2 represents a high-affinity and the NBD1 a low-affinity binding site. As the binding affinities differ by two orders of magnitude, MANT-ADP binding to the high-affinity site of ClpB(wt) masks binding to the low-affinity site. Active site titrations with MANT-ADP and ClpB(K204Q) showed that data are best described by a stoichiometry of 1:2 (one molecule of MANT-ADP binds to a dimer of ClpB), whereas binding to the low-affinity binding site NBD1 follows a 1:1 stoichiometry.

Determination of the affinities for ADP and ATP was based on displacement titrations with a preformed MANT-ADP–ClpB complex and the corresponding non-fluorescent nucleotides (Figure 4.2B). According to the K_d values obtained by fitting this data to the cubic equation (see Section 4.3), ATP binds with a ca. 10-fold lower affinity compared to ADP. (ATP hydrolysis might tamper with the apparent K_d calculated from displacement titrations with ATP. As ATPase activity especially of the NBD mutants is very low, this effect could be neglected.) Surprisingly, the displacement titrations did not confirm the assignment of the NBDs as low- and high-affinity binding sites. With regard to the displacement titrations, both NBDs bind nucleotides with approximately the same affinity (K_d (ADP) $\approx$ 2 µM; K_d (ATP) $\approx$ 30 µM).

In contrast to ClpB from *E.coli*, which is capable of hexamerization only in the presence of ATP [51, 58], ClpB$_{Tth}$ forms a hexamer in the absence of nucleotides and in the presence of ATP. ADP binding to ClpB$_{Tth}$ triggers dissociation of high-order oligomers and formation of dimers. Gel-filtration studies with ClpB$_{Tth}$(wt) and the NBD mutants indicated that ADP binding to the NBD1 is responsible for ClpB dissociation [21]. As the bound nucleotide determines the oligomeric state of ClpB, according to the rules of thermodynamics, the oligomeric state should influence nucleotide affinity. To investigate this, the ClpB concentration was kept constant during the fluorescence equilibrium titration with increasing amounts of MANT-ADP (Figure 4.3A). This experiment was repeated with ClpB concentrations ranging from 0.5 µM to 50 µM, thereby covering different distributions of oligomers. Analysis of the obtained data showed that the affinity of the NBD2 to MANT-ADP indeed depends on the oligomeric state of ClpB$_{Tth}$ (Figure 4.3B). The ClpB dimer represents the high-affinity state, while high-order oligomers bind with decreased affinity.

In the case of the chaperone ClpB, application of the fluorescent MANT nucleotide in combination with mutants partially deficient in nucleotide binding has some drawbacks. In the first place MANT-ATP does not meet the requirements of an ATP analogue. The rate of hydrolysis is 100-fold slower compared to ATP, and binding of MANT-ATP triggers oligomer dissociation in contrast to ATP, which stabilizes high-order oligomers. In addition, affinities for MANT-ADP and the unmodified nucleotide differ by a factor of 10, indicating altered kinetic parameters of the modified nucleotide. Notably, these effects are distinct from another technical problem: that of mutations in one active site, such as the K204Q mutation in the Walker A motif of ClpB$_{Tth}$ NBD1, which affects properties of the second site by altering the oligomeric state of the protein.

Because of such problems, we suggest that the site-specific modification of ClpB to introduce fluorescent reporter groups into the NBDs is more suitable to obtain information regarding the properties of each module. A single tryptophan substitution (Y819W) in the C-terminal domain of NBD2 of Hsp104, which has no other

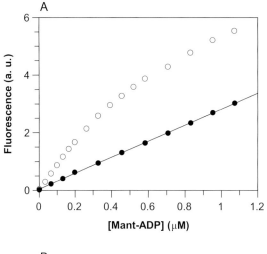

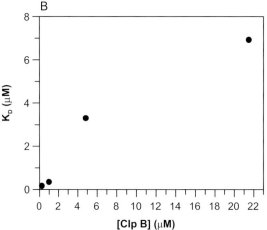

Fig. 4.3. Reverse equilibrium titration of ClpB$_{Tth}$. (A) 0.5 μM ClpB$_{Tth}$(wt) (○) or buffer (●) was titrated with MANT-ADP at 25 °C (λ_{ex} = 360 nm, λ_{em} = 440 nm). The titration curve was corrected for the contribution of the free fluorophore and fitted to a quadratic equation. (B) The plot shows the observed K_d dependent on the ClpB concentration applied in the titra-tion experiment.

tryptophan residues, could be used as an intrinsic fluorescent probe of binding to this site [59]. Site-specific labeling with IANBD of ClpB mutants, carrying Cys substitutions within the NBDs, also proved to be a practical method (S. Schlee and J. Reinstein, unpuplished data).

4.2.4.3 Cooperativity of ATP Hydrolysis and Interdomain Communication

Since ClpB is a hexamer with two binding sites per monomer, the ATP hydrolysis reaction is likely to be regulated at both the homotypic (between identical sites on

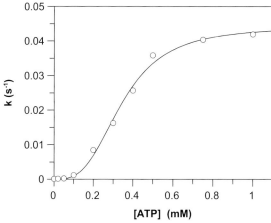

Fig. 4.4. Steady-state kinetic analysis of ATP hydrolysis by ClpB$_{Tth}$. The turnover rate k was determined in a coupled colorimetric assay by incubating 5 µM ClpB$_{Tth}$ with various amounts of Mg-ATP at 25 °C. A fit of the sigmoidal curve with the Hill equation (see "Coupled Enzymatic Assay") yields $n = 3.0$, $k_{cat} = 0.044$ s^{-1}, and $K_m = 0.34$ mM.

different subunits) and heterotypic (between NBD1 and NBD2) levels. Quantitative understanding of the catalytic properties of both ATPase domains is complicated by the fact that effects of mutations on the ATPase activity studied to date are not consistent for different ClpB/Hsp104 proteins [58, 60, 61]. Indeed, measurements of ATP hydrolysis, determined in a coupled colorimetric assay with ClpB$_{Tth}$ [21] or by quantifying released P$_i$ in the case of Hsp104/Hsp78 [61, 62], identified positive cooperativity (Figure 4.4). Mutations in both Walker A motifs of ClpB$_{Tth}$ caused a complete loss of cooperativity and a 100-fold reduction of the ATPase activity. However single-turnover experiments show that both NBDs retain a basal ATPase activity [21]. Remarkably, formation of mixed oligomers, consisting of subunits from ClpB(wt) and the Walker A mutants, leads to a decrease in the ATPase activity, indicating that cooperative effects are based on inter-ring interactions.

A very detailed characterization of the properties of the two AAA modules of Hsp104 could be attained by using [γ-^{32}P]ATP in combination with quantification of a molybdate–inorganic phosphate complex [59]. Analysis of the kinetics of ATP hydrolysis of NBD1 and NBD2 shows that one site (NBD1) is a low-affinity site with a relatively high turnover ($K_m1 = 170$ µM, $k_{cat}1 = 76$ min^{-1}) and that the second site is characterized by a much higher affinity and a 300-fold slower turnover ($K_m2 = 4.7$ µM, $k_{cat}2 = 0.27$ min^{-1}). Both sites show positive cooperativity, with Hill coefficients of 2.3 and 1.6 for the low- and high-affinity sites, respectively. Using mutations in the AAA sensor-1 motif of NBD1 and NBD2 that reduce the rate of ATP hydrolysis without affecting nucleotide binding, Lindquist and coworkers probed the communication between both sites [63]. Impairing ATP hydrolysis of the NBD2 significantly alters the steady-state kinetic behavior of NBD1. Thus, Hsp104 exhibits allosteric communication between the two sites in addition to homotypic cooperativity at both NBD1 and NBD2.

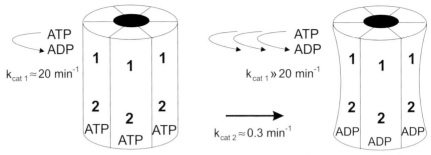

Fig. 4.5. Model for regulation of ATP hydrolysis at Hsp104 NBD1 by the hydrolysis cycle at NBD2 (adapted from Hattendorf et al. [63]). When ATP is bound to NBD2, hydrolysis at NBD1 is relatively slow. Upon ATP hydrolysis at NBD2, a presumed conformational change occurs, resulting in a state in which the rate at NBD1 is elevated.

Interpretation of this data led up to a model predicting that ATP hydrolysis at Hsp104 NBD1 is regulated by the nucleotide-bound state of NBD2 (Figure 4.5). Verification of this model will require transient kinetic measurement of the rate constants associated with ATP binding, ATP hydrolysis, and ADP release at NBD2. Because the oligomeric state stabilized by the bound nucleotide and the effect of mutations within the NBDs were not consistent for ClpB analogues from different organisms, it remains to be clarified whether this model can be generalized or is valid only for Hsp104. In addition, to gain a comprehension of the mechanism linking ATP binding and hydrolysis to the chaperone function of ClpB, the influence of substrates and chaperone cofactors on the ATPase cycle of ClpB has to be defined.

4.3
Experimental Protocols

4.3.1
Synthesis of Fluorescent Nucleotide Analogues

4.3.1.1 Synthesis and Characterization of (P_β)MABA-ADP and (P_γ)MABA-ATP

Synthesis of N-(4′-aminobutyl)-2-(methylamino)-benzoylamide N-(4′-aminobutyl)-2-(methylamino)-benzoylamide (referred to as tetralinker) was synthesized as follows. To 16.5 g (188 mmol) of 1,4-diaminobutane, 1.66 g (9.38 mmol) of N-methylisatoic anhydride was added in small portions. After 4 h of boiling under reflux, the diamine was distilled under vacuum, and the residual (1.86 g, 90%) was dissolved in CH_2Cl_2/MeOH/Et_3N (49:49:2). Chromatography over silica gel 60 (70–230 mesh; Merck, Germany) in the above solvent yielded 1.30 g (63% relative to N-methylisatoic acid) of tetralinker with $R_f = 0.14$ (silica gel; CH_2Cl_2/MeOH/Et_3N [49:49:2]) and a melting point of 90–91 °C.

^{1}H-NMR (DMSO-d$_6$, 500 MHz): $\delta = 1.57$ (m; 4H, (CH$_2$)$_2$), $\delta = 2.68$ (m; 4H, (NCH$_2$)$_2$), $\delta = 2.88$ (d; 3H, aryl-NCH$_3$), $\delta = 3.33$ (t; 2H, alkyl-NH$_2$), $\delta = 6.73$ (d; 1H, aryl-3-H), $\delta = 6.66$ (m; 1H, aryl-4-H), $\delta = 7.39$ (m; 1H, aryl-5-H), $\delta = 7.65$ (d; 1H, aryl-6-H), $\delta = 7.78$ (t; 1H, OCNH), $\delta = 8.45$ (q; 1H, aryl-NH).

Synthesis of (P$_\beta$)MABA-ADP [15] The water-soluble N-ethyl-N'-(3-dimethylaminopropyl)-carbodiimide (EDC) was used to condense the synthesized tetralinker with the β-phosphate of ADP to give adeninediphospho-β-(N'-methylanthraniloylaminobutyl)-phosphoramidate (referred to as (P$_\beta$)MABA-ADP) (Figure 4.6). 100 mg (450 µmol) of tetralinker, 19.1 mg (100 µmol) of EDC, and 100 mg (234 µmol) of ADP were dissolved in 1.0 mL 100-mM MES/NaOH pH 6.7 and heated to 50 °C for 35 min. The crude product was purified by HPLC (Waters, Massachusetts, USA) on a semi-preparative reversed-phase C$_{18}$ column (ODS Hypersil Reversed Phase C-18, 5 µm, 120 Å, 250 × 8 mm; Bischoff, Germany) in 0.1 M TEA/HOAc, 25% MeCN with absorption at 260 nm and 350 nm and fluorescence at 430 nm after excitation at 350 nm to identify the product that elutes between ADP and free tetralinker. (P$_\beta$)MABA-ADP–containing fractions were pooled, the solvent was distilled under vacuum, and the pure product was lyophilized. The final yield was 31% (relative to ADP) of (P$_\beta$)MABA-ADP; the integrity of (P$_\beta$)MABA-ADP was proved by ESI-MS and NMR.

^{1}H-NMR (D$_2$O, 500 MHz; TSP): $\delta = 1.29$ (t; 18H; CH$_3$; (CH$_3$CH$_2$)$_3$NH$^+$-counter-ion); 1.93 (m; 4H; (CH$_2$)$_2$); 2.78 (s; 3H; aryl-NCH$_3$); 2.91 (m; 4H; N(CH)$_2$); 3.21 (q; 12H; CH$_2$; (CH$_3$CH$_2$)$_3$NH$^+$-counter-ion); 3.74 (m; 1H; 4'-H); 4.23 (m; 2H; 5'-H); 4.37 (m; 1H; 3'-H); 4.49 (m; 1H; 2'-H); 6.08 (d; 1H; 1'-H); 6.72 (m; 1H; aryl-4-H); 6.79 (d; 1H; aryl-6-H); 7.30 (d; 1H; aryl-3-H); 7.37 (m; 1H; aryl-5-H); 8.16 (s; 1H; H-2); 8.47 (s; 1H; H-8).

^{31}P-NMR (D$_2$O; 500 MHz, PO$_4^{3-}$): $\delta = -1.24$ (d; α-phosphate); -11.29 (d; β-phosphate).

ESI-MS (negative ion mode): 629.1 g·mol^{-1}, 629.4 g·mol^{-1} were expected.

The extinction coefficient of (P$_\beta$)MABA-ADP was determined by comparison of the absorption at 325 nm of (P$_\beta$)MABA-ADP and that of the cleaved material. At this wavelength, only the tetralinker contributes to the absorption spectrum. The extinction coefficient of the tetralinker was determined to be 2700 M^{-1} cm^{-1} from a solution of known concentration. Cleavage of (P$_\beta$)MABA-ADP is complete after heating the compound for 15 min to 60 °C in 0.1 M HCl according to HPLC analysis. After neutralization, the absorption at 325 nm was unchanged, and the extinction coefficient of (P$_\beta$)MABA-ADP is therefore $\varepsilon_{325} = 2700$ M^{-1} cm^{-1}.

Synthesis of (P$_\gamma$)MABA-ATP EDC was used to condense the tetralinker with the γ-phosphate of ATP to give adeninetriphospho-γ-(N'-methylanthraniloylaminobutyl)-phosphoramidate (referred to as (P$_\gamma$)MABA-ATP). The synthesis was carried out as described for (P$_\beta$)MABA-ADP with 100 mg ATP instead of ADP.

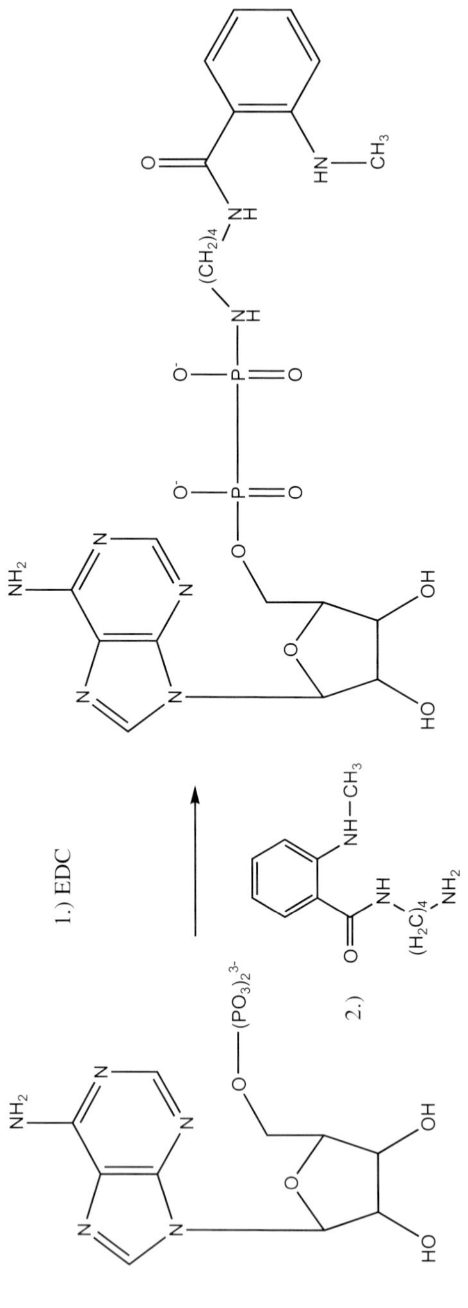

Fig. 4.6. Synthesis of P(β) MABA-ADP. Condensation of ADP with N-(4'-aminobutyl)-2-(methylamino)benzoylamide.

^{1}H-NMR (D$_2$O, 500 MHz; TSP): $\delta = 1.29$ (t; 27H; CH$_3$; (CH$_3$CH$_2$)$_3$NH$^+$-gegenion); 1.92 (m; 4H; (CH$_2$)$_2$); 2.78 (s; 3H; aryl-NCH$_3$); 2.91 (m; 4H; N(CH)$_2$); 3.21 (q; 18H; CH$_2$; (CH$_3$CH$_2$)$_3$NH$^+$-gegenion); 3.78 (m; 1H; 4'-H); 4.26 (m; 2H; 5'-H); 4.38 (m; 1H; 3'-H); 4.53 (m; 1H; 2'-H); 6.08 (d; 1H; 1'-H); 6.71 (m; 1H; aryl-4-H); 6.78 (d; 1H; aryl-6-H); 7.31 (d; 1H; aryl-3-H); 7.35 (m; 1H; aryl-5-H); 8.16 (s; 1H; H-2); 8.45 (s; 1H; H-8).

^{31}P-NMR (D$_2$O; 500 MHz, PO$_4^{3-}$): $\delta = -2.83$ (d; α-phosphate); -13.35 (d; γ-phosphate); -24.73 (t; β-phosphate).

ESI-MS (negative ion mode): 709.1 g·mol^{-1}, 708.4 g·mol^{-1} were expected.

4.3.1.2 Synthesis and Characterization of N8-MABA Nucleotides

Synthesis of MABA-AMP [11] The fluorophores N^8-(4-N'-methylanthraniloyl-aminobutyl)-8-aminoadenosine 5'-tri- and diphosphate (MABA-ATP and MABA-ADP) were synthesized in a four-step reaction. The starting material was AMP that was converted to N8-(4-aminobutyl)-8-aminoadenosine 5'-phosphate (ABA-AMP) [64]. ABA-AMP (50 µmol) was dissolved in 1 mL 0.5-M sodium carbonate (pH 9–10), and 600 µL of 200-mM succinimidyl-N-methylanthranylate (SMANT, Molecular Probes) in DMF was added. The mixture was agitated vigorously and left at 50 °C for about 4 h. To quench unreacted SMANT, 1.3 mL of concentrated ammonia (final concentration 8 M) was added, and after 30 min the reaction mixture was diluted with 60 mL water at 4 °C. To remove byproducts, the solution was applied to the quaternary anion-exchange resin QAE Sephadex A 25 (2.2 × 13 cm, Pharmacia). After a wash with water, the column was eluted with a linear gradient to 0.3 M triethyl ammonium bicarbonate (TEAB), pH 7.5. The effluent was monitored by UV absorbance at 260 nm; the product elutes between 0.2 M and 0.25 M TEAB in a single peak. The fractions estimated to contain pure MABA-AMP were pooled and lyophilized three times from water.

Purity of the product was analyzed by reverse phase HPLC with a C$_{18}$ column (ODS-Hypersil, 5 µM, 250 × 4.6 mm) and detection at 280 nm. The sample was eluted with 20% (v/v) acetonitrile in 50 mM potassium phosphate, pH 7.0, over 20 min at 2 mL min^{-1}. A single major peak was observed after 12 min.

The final concentration was determined by UV absorbance with $\varepsilon_{280} = 18\,000$ M^{-1} cm^{-1} at pH 7.5. The molar extinction coefficient was obtained from MABA-ATP by enzymatic ATP determination with hexokinase and glucose-6-phosphate-dehydrogenase [65]. The yield of MABA-AMP from ABA-AMP was ca. 50%.

Synthesis of MABA-ATP MABA-AMP (31 µmol) was dissolved in 0.6 mL water, and 0.23 mL (1 mol) tributylamine was added. The solution was concentrated under reduced pressure and the viscous residue was dried by repeated addition and evaporation of anhydrous pyridine (5 × 0.5 mL) followed by anhydrous DMF (3 × 0.3 mL). The anhydrous tributyl ammonium salt of MABA-AMP was dissolved in 0.5 mL DMF, and 24 mg (0.15 mmol) N,N'-carbonyldiimidazole (CDI) was added. The mixture was agitated vigorously for 20 min and left overnight at room temperature in a desiccator. In order to hydrolyze excess CDI, 7.5 μL methanol was added, and after 30 min the mixture was concentrated to dryness at 40 °C under reduced pressure. The imidazolide was dissolved in 0.3 mL DMF and stored under anhydrous conditions.

Pyrophosphoric acid (26.7 mg; 0.3 mmol) and 142 μL (0.6 mmol) tributylamine were dissolved in methanol, evaporated, and then evaporated three times with 0.5 mL pyridine and an additional 10 μL tributylamine (0.04 mmol). The precipitate was dissolved in 0.3 mL dry dimethyl formamide (DMF) and added to the imidazolide solution; after shaking vigorously for 30 min, the reaction mixture was left in a desiccator at room temperature for 1–2 days. The precipitate of imidazolium phosphate that formed during the reaction was removed by centrifugation and extracted three times with 0.5 mL DMF. The combined extracts were added to an equal volume of methanol, and the resulting solution was concentrated at 40 °C. The residue was dissolved in 50% (v/v) aqueous ethanol (20 mL) and applied to a QAE Sephadex A25 column (see above).

After washing with 100 mL of 50% (v/v) aqueous ethanol, the sample was eluted with a linear gradient to 1 M TEAB, pH 7.5. The eluent was monitored by UV absorbance at 260 nm. Three major peaks were observed: the first peak at about 0.25 M TEAB was unreacted MABA-AMP, the second at 0.4 M TEAB was MABA-ADP (presumably from phosphoric acid in the pyrophosphoric acid), and the last at 0.6 M TEAB contained MABA-ATP. These fractions were pooled and lyophilized three times from water. Purity was determined by reverse-phase HPLC as described for MABA-AMP. A single major peak was observed at 5 min. The final concentration was determined by enzymatic ATP determination [65]. The UV absorption properties are the same as for MABA-AMP. The yield of MABA-ATP was ca. 27%.

MABA-ADP can be prepared with tributyl-ammonium phosphate instead of the tributyl-ammonium pyrophosphate in the reaction sequence described above.

^{1}H-NMR of MABA-ATP (D$_2$O, 360 MHz) pH 7.0, 24 °C. The chemical shifts are relative to an external tetramethylsilane standard.

δ = 7.98 (s; 1H; H-2); 7.27 (t; 1H; CH3″) ^3J (7.63 Hz); 7.15 (d; 1H; CH5″) ^3J (7.83 Hz); 6.68 (d; 1H; CH2″) ^3J (8.41 Hz); 6.50 (t; 1H; CH4″) ^3J (7.46 Hz); 5.96 (d; 1H; 1′-H) $^3J1'2'$ (7.8 Hz); 4.65 (t; 1H; 2′-H) $^3J2'3'$ (6.74 Hz); 4.44 (m; 1H; 3′-H); 4.35 (s; 1H; 4′-H); 4.18 (m; 2H; 5′5″-H); 3.55 (m; 2H; (CH$_2$)$_4$); 3.38 (m; 2H; (CH$_2$)$_1$); 2.72 (s; 3H; Ph-N-CH$_3$); 1.79 (m, 2H, (CH$_2$)$_2$); 1.72 (m; 2H; (CH$_2$)$_3$); plus counter-ion (triethylammonium) peaks.

4.3.1.3 Synthesis of MANT Nucleotides

Synthesis of MANT-ADP and MANT-ATP [18] The nucleotide (0.5 mmol ADP, ATP) was dissolved in 7 mL water at 38 °C. The pH was adjusted to 9.6 with NaOH. To this solution a crystalline preparation of methylisatoic anhydride (2.5 mmol) was added in small portions with continuous stirring. The pH was maintained at 9.6 by titration with 2 N NaOH for 2 h. The reaction mixture was cooled and washed three times with 10 mL chloroform; the derivative remains in the aqueous phase. The reaction products can be monitored by TLC on silica gel. On a silica gel plate developed in 1-propanol/NH$_4$OH/water (6:3:1, v/v, containing 0.5 g L^{-1} of EDTA), the R_f values behave as follows: unmodified nucleotides < fluorescent nucleotides (R_f (MANT-ADP): 0.44, R_f (MANT-ATP): 0.21) < N-methylanthranilic acid. The fluorescent analogue can be identified under an ultraviolet lamp (366 nm) by its brilliant blue fluorescence, while N-methylanthranilic acid shows violet fluorescence.

The reaction mixture was then diluted to a volume of 100 mL and the pH adjusted to 7.5 with 1 M acetic acid. To remove byproducts, the solution was applied to the anion-exchange resin Q-Sepharose FF (3.0 × 25 cm, Pharmacia). After a wash with 50 mM triethyl ammonium bicarbonate (TEAB), pH 7.5 at a flow rate of 5 mL min^{-1}, the column was eluted with a linear gradient to 0.7 M TEAB. The effluent was monitored by UV absorbance at 260 nm and 366 nm; the fluorescent nucleotide analogues elute after the unreacted nucleotide and the fluorescent byproduct between 0.6 and 0.7 M TEAB in a single peak. MANT nucleotide-containing fractions were identified by TLC and the absorption profiles. Fractions were pooled, the solvent was distilled under vacuum, the residual was dissolved in methanol, and the pure product was lyophilized.

Purity of the product was analyzed by reverse-phase HPLC with a C$_{18}$ column (ODS-Hypersil, 5 μM 250 × 4.6 mm, Bischoff). The column was developed with a linear gradient to 50% (v/v) acetonitrile in 50 mM potassium phosphate, pH 6.5, over 20 min at 1.5 mL min^{-1} and elution products were monitored by absorption at 280 nm and fluorescence (excitation, 360 nm; emission, 440 nm). MANT-ADP elutes as a single peak after 12.5 min, while MANT-ATP elutes after 11.7 min.

The final concentration of MANT nucleotides was determined by UV absorbance with $\varepsilon_{255} = 23\,300$ M^{-1} cm^{-1} at pH 8.0. The yield of MANT nucleotide amounts to ∼50%. MANT nucleotides exist in solution as an equilibrium mixture of both the 2′ and 3′ isomers (60% 3′ isomer), which are in slow base-catalyzed exchange ($t_{1/2}$ ∼ 10 min, pH 7.5, 25 °C).

4.3.2
Preparation of Nucleotides and Proteins

4.3.2.1 Assessment of Quality of Nucleotide Stock Solution

It is important that fluorescent nucleotide analogues be fully characterized in terms of structure and purity. The latter should demonstrate the absence of unmodified nucleotides or non-nucleotide fluorescent species that could seriously affect binding studies. Checks on purity are most easily achieved by high-performance liquid chromatography (HPLC).

Unmodified Nucleotides In our experience, commercially available ATP contains a background level of ADP (typically 1–4%). Impurities in nucleotide stock solutions give rise to artifacts when determining rate constants of the ATPase cycle. Therefore, we advise controlling the purity of nucleotide stock solutions by HPLC analysis. The purity of ATP and ADP solutions can be analyzed by reverse-phase HPLC

Tab. 4.2. HPLC program for purity control of P-MABA nucleotides.

Time (min)	Flow rate (mL min^{-1})	A (%)	B (%)
Start	2	70	30
7	2	70	30
7.1	2	0	100
12	2	0	100
12.1	2	70	30
20	2	70	30

with a C_{18} column (ODS-Hypersil, 5 µM, 250 × 4.6 mm, Bischoff) and detection at 254 nm. The sample (usually 20 µL of a 10-µM solution) was eluted with 50 mM potassium phosphate (pH 6.8) over 10 min at 1.5 mL min^{-1}. Nucleotides elute as single peaks with retention times of 3.5 min (ATP), 4.1 min (ADP), and 5.0 min (AMP) at room temperature.

P(β)MABA-ADP and P(γ)MABA-ATP Purity of the phosphate-modified nucleotide analogues P(β)MABA-ADP and P(γ)MABA-ATP was determined by reverse-phase HPLC with a C_{18} column (ODS-Hypersil, 5 µM 250 × 8 mm, Bischoff) and monitoring UV absorption at 254 nm and 325 nm and fluorescence (excitation, 335 nm; emission, 440 nm). The sample was eluted with a step gradient at 2 mL min^{-1} (see Table 4.2) (buffer A: 100 mM triethylammonium acetate pH 7.0; buffer B: 100 mM triethylammonium acetate pH 7.0, 70% acetonitrile). Peaks were observed after 3.8 min (ADP, ATP), 5.7 min (P(γ)MABA-ATP), and 7.6 min (P(β)MABA-ADP).

N8-MABA-ADP and N8-MABA-ATP Purity of nucleotides modified at the N8 position was analyzed by reverse-phase HPLC with a C_{18} column (ODS-Hypersil, 5 µM, 250 × 4.6 mm) and detection at 280 nm [11]. The sample was eluted with 20% (v/v) acetonitrile in 50 mM potassium phosphate, pH 7.0, over 20 min at 2 mL min^{-1}. Peaks were observed after 5.3 min (N8-MABA-ATP), 9.1 min (N8-MABA-ADP), and 12.0 min (N8-MABA-AMP).

MANT-ADP and MANT-ATP The purity of MANT nucleotides can be examined by reverse-phase HPLC with a C_{18} column (ODS-Hypersil, 5 µM, 250 × 4.6 mm), and the effluent is monitored by measuring UV absorbance at 254 nm as well as fluorescence (excitation, 360; emission 440). The column was eluted with a linear gradient to 50% acetonitrile in 50 mM potassium phosphate (pH 6.5) at 1.5 mL min^{-1} (see Table 4.3). MANT-nucleotides elute as double peaks (2' and 3' derivatives) with retention times of 13.7 min (MANT-ATP) and 14.5 min (MANT-ADP).

4.3.2.2 Determination of the Nucleotide Content of Proteins

For the investigation of nucleotide-binding capacities of chaperones, the proteins had to be prepared in a nucleotide-free state. To determine the nucleotide content of a protein preparation, 16 µL of a 100-µM protein solution was mixed with 4 µL

Tab. 4.3. HPLC program for purity control of MANT nucleotides.

Time (min)	Flow rate (mL min^{-1})	A (%)	B (%)
0	1.5	100	0
2.0	1.5	100	0
22.0	1.5	0	100
23.0	1.5	100	0
30.0	1.5	100	0

50% (w/v) trichloroacetic acid and incubated for 10 min on ice. The precipitated protein was removed by centrifugation. 10 µL of the supernatant was neutralized with 20 µL of a 2-M solution of potassium acetate. The sample was diluted 1:10 in 50 mM potassium phosphate (pH 6.8) and 20 µL of this solution was analyzed by HPLC (see "Assessment of Quality of Nucleotide Stock Solution" for HPLC conditions). By comparing the integrated peaks with calibration peaks corresponding to nucleotide stock solutions of known concentrations, the nucleotide content of the probe can be calculated.

4.3.2.3 Nucleotide Depletion Methods

Preparation of nucleotide-free chaperones is crucial for the determination of nucleotide-binding kinetics. The nucleotide depletion method applied to a specific chaperone depends on the nucleotide affinity. To prepare ClpB, whose nucleotide affinity is comparatively low (micromolar range), in a nucleotide-free form, it is sufficient to remove Mg^{2+} before a gel-filtration step and thus weaken the binding of nucleotides [21]. Nucleotide depletion of DnaK, which binds nucleotides with dissociation constants in the nanomolar range [66], requires more thorough procedures, which are listed below.

K^+ and Mg^{2+} Depletion As two K^+ and one Mg^{2+} coordinate the bound nucleotide in the ATPase site of DnaK [67], removal of these ions should decrease nucleotide affinity. Therefore, the nucleotide depletion procedure included extensive dialysis over 2–3 days against a buffer containing 1 mM EDTA in the absence of K^+ and Mg^{2+} ions followed by size-exclusion chromatography [68]. A shortcoming of this method is that at high protein concentrations the dissociation of a very small percentage of the bound nucleotide provides enough free nucleotide to prevent further dissociation. When reducing the enzyme concentration to facilitate the dialysis of the bound nucleotide, protein denaturation occurs easily [69].

Exchange Against Low-affinity Nucleotide Analogues Nucleotide removal from DnaK preparations was accomplished by competing off bound nucleotide with excess AMP-PNP or AMP, followed by dialysis to remove less tightly bound AMP-PNP or AMP [68–70]. Nucleotide exchange can be accelerated by addition of catalytic amounts of GrpE to the sample. As for the K^+ and Mg^{2+} depletion procedure, extensive dialysis (72 h) is indispensable. Therefore, the procedure is not suitable with regard to aggregation-prone proteins. The final purification step is again a size-exclusion chromatography.

Incubation With Alkaline Phosphatase Entirely nucleotide-free DnaK$_{Tth}$ was obtained after treatment with 3–4 U alkaline phosphatase (Boehringer, Mannheim) for 4–6 h in a buffer without K$^+$ and Mg^{2+} and simultaneous removal of AMP by dialysis. Alkaline phosphatase was removed by a further gel-filtration step [9].

4.3.3
Steady-state ATPase Assays

Observing the ATPase reaction under steady-state conditions implies that the rate of reaction is constant for a relatively long period of time and that multiple turnovers take place. This is the case when the ATP concentration exceeds the chaperone concentration substantially and, ideally, the ATP concentration does not change during the experiment. This could be achieved by using an ATP-regenerating system. Steady-state ATPase assays allow determination of the enzymatic parameters K_m and V_{max}; however, it is not possible to define individual steps and rate constants of the ATPase mechanism. Procedures for determining the steady-state ATPase activity of chaperones are listed below.

4.3.3.1 Coupled Enzymatic Assay
In a colorimetric assay, the hydrolysis of ATP is coupled to the oxidation of NADH by using the enzymes pyruvate kinase and lactate dehydrogenase (Figure 4.7). Because NADH but not NAD$^+$ absorbs at 340 nm, ATP hydrolysis can be monitored by a decrease in A_{340} [71]. The steady-state ATPase rate of ClpB$_{Tth}$ has been investigated by the coupled enzymatic assay [21].

The assay is carried out in microtiter plates or plastic cuvettes. Measurements are initiated by the addition of an appropriate amount of chaperones (5 µM ClpB) to various concentrations of Mg*ATP and 20 µg mL^{-1} lactate dehydrogenase, 50 µg mL^{-1} pyruvate kinase in 50 mM Tris/HCl (pH 7.5), 100 mM KCl, 5 mM MgCl$_2$, 2 mM EDTA, 2 mM DTE, 0.4 mM PEP, 0.4 mM NADH. The ATP turnover rate in the presence of various ATP concentrations is calculated as follows:

$$k = \Delta A_{340}/(\varepsilon_{NADH} \cdot d \cdot c_p) \tag{1}$$

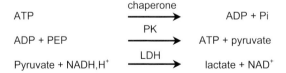

Fig. 4.7. The coupled colorimetric assay. ADP produced by the hydrolysis of ATP reacts with PEP (phosphoenolpyruvate) to form pyruvate in a reaction catalyzed by the PK (pyruvate kinase), whereby ATP is regenerated. Pyruvate is reduced to lactate by the LDH (lactate dehydrogenase), while NADH is oxidized. The decrease of the NADH absorption at 340 nm is a measure for hydrolytic activity.

k: ATP turnover rate in s^{-1}
ΔA_{340}: change of A_{340} in absorption units per s
ε_{NADH}: extinction coefficient of NADH $\varepsilon_{340} = 6220\ M^{-1}\ cm^{-1}$
d: path length in cm
c_p: protein concentration (M)

The k_{cat} and K_m of the ATPase reaction can be determined by plotting the ATP turnover rate k against the ATP concentration and by fitting the obtained curve to the Michaelis-Menten equation (Eq. (2)). If allosteric effects play a role in ATP hydrolysis, the Michaelis-Menten equation should be expanded by introducing the Hill coefficient n (Eq. (3)) (Figure 4.4).

$$k = k_{cat} \cdot S/(K_M + S) \tag{2}$$

$$k = k_{cat} \cdot S^n/(K_M' + S^n) \tag{3}$$

k: ATP turnover rate (s^{-1}) at specific ATP concentration
k_{cat}: maximal ATP turnover rate (s^{-1})
S: ATP concentration (M)
K_M: Michaelis-Menten constant (M)
n: Hill coefficient

4.3.3.2 Assays Based on [α-^{32}P]-ATP and TLC

The use of radioactive-labeled ATP facilitates quantification of hydrolytic products. Assays of this kind have been used extensively to characterize the ATPase activity of DnaK$_{Eco}$ [70, 72–75]. Exemplary reaction mixtures (75 μL) contained 50 μg mL^{-1} BSA, 1 nM DnaK, 3 μCi mL^{-1} of [α-^{32}P]ATP or [γ-^{32}P]ATP (1 nM), and 2–80 nM unlabeled ATP in buffer (40 mM HEPES/KOH, pH 7.6, 11 mM magnesium acetate, 200 mM potassium glutamate) [70]. DnaK was pre-incubated at the indicated reaction temperature for 5 min, and hydrolysis reactions were initiated with the addition of ATP. Samples from the reaction mixture (≥ 6 per reaction) were quenched at various times by adding 8 μL of the reaction mixture to 2 μL of 1-N HCl and placing the samples on ice. Each sample (2 μL) was spotted on polyethylenimine cellulase TLC sheets, which were developed in 1 M formic acid, 0.5 M LiCl. The fraction of nucleotide present as ADP could be determined using a phosphorimager system, or spots were cut out and the radioactivity was quantified by liquid scintillation counting. Hydrolysis of ATP should not exceed $\sim 15\%$, as formed ADP competes with ATP and decreases the hydrolytic rate.

At each concentration of ATP, v_0 was determined from a linear regression analysis. K_M and k_{cat} were determined by fitting a plot of $(v_0/[DnaK])$ against the concentration of ATP to the Michaelis-Menten equation.

4.3.3.3 Assays Based on Released P_i

Quantification of the P_i released during ATP hydrolysis represents a sensitive measure [76]. A modified method allowed the evaluation of cooperative effects in the

ATPase reaction of Hsp104 [63]. All measurements were performed in 20 mM HEPES pH 7.5, 20 mM NaCl, and 10 mM MgCl$_2$. Hydrolysis reactions were initiated by mixing 12.5 μL of an Hsp104 solution (final concentration 200 nM) and 12.5 μL of an ATP solution containing 10 μCi of [γ-^{32}P]ATP and 1–1000 μM ATP. At varying times reactions were stopped by addition of 175 μL of 1 M HClO$_4$ and 1 mM Na$_3$PO$_4$. Ammonium molybdate was added (400 μL of 20 mM) and the molybdate–inorganic phosphate complex was extracted by addition of 400 μL of isopropyl acetate followed by vortexing. The cold P$_i$ is present as carrier and facilitates extraction of the molybdate-phosphate complex into the organic phase.

The amounts of radioactivity in the organic phase and in the aqueous phase were determined by scintillation counting and used to determine the percentage of ATP hydrolysis. After subtraction of free phosphate in a blank reaction, the initial reaction rate was calculated. Plots of the initial reaction rates versus ATP concentration were fitted to a sum of two independent cooperative kinetic transitions [63].

Another possible way to monitor P$_i$ release in real time is the application of the fluorescent reporter PBP-MDCC (A197C mutant of phosphate-binding protein from E. coli with attached N-[2-(1-maleimidyl)ethyl]-7-(diethylamino)coumarin-3-carboxamide) [77].

4.3.4
Single-turnover ATPase Assays

For measurements of single-turnover ATPase activity, the chaperone has to be present in excess over ATP, so that saturation of the chaperone is achieved (depends on K$_d$!) and only one turnover of ATP takes place. Conversion of ATP to ADP as a function of time is described by a first-order exponential equation ($y = A \cdot e^{-kt}$).

Typically, ATP and the chaperone are mixed, the reaction is stopped at various times by addition of a quencher, and the fraction of nucleotide present as ADP is determined. As a rule of thumb, the reaction can be analyzed by "hand mixing" procedures if $t_{1/2}$ (ATP) > 10 s; faster ATPase rates have to be determined by quenched-flow procedures. If product release (ADP or P$_i$) is rate-limiting in the ATPase cycle, the rate constant determined under single-turnover conditions exceeds the one observed under steady-state conditions.

4.3.4.1 Manual Mixing Procedures
Reaction mixtures and reactions are essentially identical to those described for measurement of steady-state ATPase activity, except that the protein concentration exceeds ATP concentration and ATP hydrolysis is followed to completion. A protocol used for determining the single-turnover ATPase activity of ClpB$_{Tth}$ is given in ref. [21].

The reaction was started by adding 150 μM ClpB to 75 μM ATP in 50 mM Tris/HCl (pH 7.5), 5 mM MgCl$_2$, 2 mM EDTA, 2 mM DTE, 100 mM KCl. After incubation for appropriate periods (0, 10, 20, 30, ..., 120 min), 4 μL samples were drawn and the protein was precipitated by addition of 2 μL of 1-M HClO$_4$. After vortexing, the mixture was incubated for 5 min on ice, and 40 µL of 2-M potassium acetate was added for neutralization. After centrifugation, 20 µL of the supernatant was analyzed with reverse-phase HPLC on a C_{18} column (ODS-Hypersil, 5 µM 250 × 4.6 mm, Bischoff) and detection at 254 nm (for conditions, see 4.3.2.1 "Unmodified Nucleotides"). The relative amounts of ATP and ADP present in each sample were calculated by integrating peaks corresponding to ADP and ATP from the elution profile.

The single-turnover ATPase activity of DnaK has been characterized with radioactive [α-^{32}P]ATP [70, 78] and [γ-^{32}P]ATP [9] using procedures similar to those described for steady-state measurements.

4.3.4.2 Quenched Flow

DnaJ$_{Eco}$ is capable of stimulating the single-turnover ATPase activity of DnaK$_{Eco}$ ≥ 200-fold [78]. As the stimulated ATP-hydrolysis occurred too rapidly to measure by hand, it was monitored using a rapid quench-flow apparatus [79, 80]. Proteins and [α-^{32}P]ATP were loaded into separate 1-mL syringes that were used to fill the 15-µL sample loops. The drive syringes were pushed at a constant velocity, mixing the sample solutions in a 1:1 ratio. The reactions were automatically quenched after a defined reaction time (0.05–120 s) with 80 µL of a 0.3 N HCl solution present in a separate drive syringe [80]. To evaluate the influence of DnaJ on the ATPase activity of DnaK, equal volumes of ATP (8 µM) and nucleotide-free DnaK (9.6 µM) were pre-incubated for 10 s before 1:1 mixing with various concentrations of DnaJ in a quenched-flow apparatus. The reaction was quenched after 0.5 s with perchloric acid [79]. The ATP/ADP ratio in every sample is determined as described for steady-state measurements by HPLC analysis or, in the case of radioactive ATP, with TLC.

4.3.5
Nucleotide-binding Measurements

Techniques used to obtain equilibrium-binding constants (K_a and K_d) are manifold, and methods range from analytical ultracentrifugation to surface plasmon resonance and fluorescence titrations. Hereafter, the methods frequently used to characterize binding equilibria of nucleotides and chaperones are described.

4.3.5.1 Isothermal Titration Calorimetry

In isothermal titration calorimetry (ITC), the heat of interaction between two molecules is used to define the binding process. A syringe containing a "ligand", e.g., the nucleotide, is titrated into a cell containing a solution of the "macromolecule", mostly a protein. As the two elements interact, heat is released or absorbed in direct proportion to the amount of binding that occurs. When the macromolecule in the cell becomes saturated with added ligand, the heat signal diminishes until only the background heat of dilution is observed. The area underneath each injection peak is equal to the total heat released for that injection. When this is plotted against the molar ratio of ligand added to macromolecule in the cell, a complete

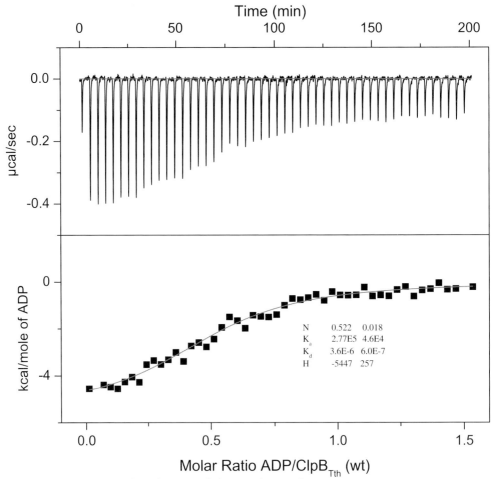

Fig. 4.8. Isothermal titration of ClpB$_{Tth}$ with ADP. ClpB$_{Tth}$ (40 µM) was titrated with 300 µM ADP at 25 °C in an isothermal titration calorimeter in 7-µL steps until saturation was achieved.

binding isotherm for the interaction is obtained (Figure 4.8). ITC measurements yield an entire set of experimental parameters for the binding process, including binding affinity (K_a or K_d), binding stoichiometry (n), heat (ΔH), and heat capacity (ΔCp). ITC measurements were used to identify the ATP-binding site of Hsp90 [81]. The protocol cited below was used for equilibrium studies of the DnaK-GrpE-ADP system from *T. thermophilus* [82].

A solution of 10 µM DnaK$_{Tth}$ in the presence or absence of 42 µM GrpE$_{Tth}$ in 25 mM HEPES-NaOH (pH 7.5), 100 mM KCl, 5 mM MgCl$_2$ (cell volume 1.3 mL), was titrated with 142 µM MABA-ADP in 6-µL steps until saturation was achieved. In the same way, a solution of 11 µM DnaK$_{Tth}$ or DnaK$_{Tth}$ pre-incubated with 20 µM

MABA-ADP was titrated with a solution of 162 µM GrpE$_{Tth}$ in 7-µL steps until saturation. Solutions were incubated for 5 h before measurements to allow for completion of the binary DnaK$_{Tth}$-GrpE$_{Tth}$ or DnaK$_{Tth}$–MABA-ADP complex formation. Data were analyzed by Eq. (4) using the program Microcal Origin for ITC to give the enthalpy, stoichiometry of binding, and association constant.

$$Q = \frac{n \cdot [E] \cdot \Delta H_B \cdot V}{2} \cdot \left(1 + \frac{[L]}{n \cdot [E]} + \frac{1}{n \cdot K_A \cdot [E]} - \sqrt{\left(1 + \frac{[L]}{n \cdot [E]} + \frac{1}{n \cdot K_A \cdot [E]}\right)^2 - \frac{4 \cdot [L]}{n \cdot [E]}}\right) \quad (4)$$

Q: heat in J
$[E]$: total concentration of the protein in M
$[L]$: total concentration of the ligand in M
V: volume of cell in liters
ΔH_B: molar enthalpy of the binding reaction in J mol^{-1}
n: number of active binding sites on the protein
K_A: association equilibrium constant in M^{-1}

4.3.5.2 Equilibrium Dialysis

Nucleotide-binding properties of Hsp70s have been studied extensively by equilibrium dialysis [66, 69]. In any mixture of the ligand and macromolecule, it is difficult to distinguish between bound and free ligand. If, however, the free ligand can be dialyzed through a membrane, until its concentration across the membrane is at equilibrium, free ligand concentration can be measured.

A typical experiment proceeds as follows [66]: [^{14}C]ATP or [^{14}C]ADP together with a regenerating system (creatine kinase/creatine phosphate for ATP, glucose/hexokinase for ADP) were added to both sides of the membrane, while nucleotide-free Hsp70 was added to only one side of the membrane at the beginning of the experiment. Then the dialysis chambers were put in a 4 °C cold room for 24 h with gentle agitation. The nucleotide contents on each side of the membrane were quantified by scintillation counting at the end of the dialysis and yielded the concentration of the free-nucleotide and total nucleotide concentration. Data obtained from several experiments and with varying initial concentrations of ligand can provide the association constant and stoichiometry of the binding reaction.

4.3.5.3 Filter Binding

Apparent rates of association and dissociation of nucleotides with Hsc70 as well as equilibrium dissociation constants were established by filter-binding experiments [83]. To determine the equilibrium-binding constant for MgADP-protein complexes, protein at concentrations from 20 nM to 2.56 µM was incubated with 1 nM [α-^{32}P]ADP at 25 °C. The solution was then filtered and the K_d and the filter efficiency E were determined by a least-squares fit of the equation

$$\theta = \frac{[\text{protein}]}{[\text{protein}] + K_D} E \qquad (5)$$

to the data on the fraction of background-corrected counts retained on the filter ($\theta = (C - C_B)/(C_T - C_B)$, where C is the total counts retained on the filter, C_B is the background counts retained when ATP is filtered without protein, and C_T is the total input counts) as a function of protein concentration.

4.3.5.4 Equilibrium Fluorescence Titration

There are two possible situations that can be exploited to study protein-ligand interaction by fluorescence [84]. Either the fluorescence of the protein changes on binding, or the fluorescence of the nucleotide ligand is altered. In general control of temperature and solution conditions is important in binding studies. The fluorescence of most fluorophores, as well as association constants, is sensitive to temperature. Also, care should be taken to control pH during a titration (check pH of nucleotide stock solution!). Dust particles should be removed from both ligand and protein solutions by Millipore filtration or by centrifugation, and buffers should be degassed thoroughly. It is necessary to have accurate values for the total concentrations of ligand ($[L]_0$) and enzyme ($[E]_0$), which may be the greatest limitation for the determination of accurate values of K_d.

Intrinsic Tryptophan Fluorescence When changes in the fluorescence of a protein are observed, the fluorophore may be one or more of the intrinsic tryptophan residues, because emission from this amino acid usually dominates the fluorescence of proteins. The λ_{max} of tryptophan emission ranges from 308 nm for deeply buried tryptophans, to 350 nm for solvent-exposed residues in unfolded proteins. Therefore, not only changes of the amplitude but also shifts of the emission maximum can be exploited as signal. Another advantage of using tryptophan fluorescence is that no chemical modification of either protein or ligand is needed.

The binding of a ligand to a protein may directly affect the fluorescence of a tryptophan residue by acting as a quencher (i.e., by a collisional or energy-transfer mechanism), as seen for ATP quenching the tryptophan fluorescence in human Hsp90 [81, 85], or by physically interacting with the fluorophore and thereby changing the polarity of its environment and/or its accessibility to solvent. Alternatively, a ligand may bind at a site that is remote from the tryptophan residue and may induce a change in conformation of the protein that alters the microenvironment of the tryptophan. These changes in microenvironment may result either in enhancement or quenching of fluorescence or in shifts in the spectrum to the red or blue. Such a fluorescent probe was reported for the NBD2 of Hsp104 [59], where a single tryptophan substitution (Y819W) in the C-terminal domain of NBD2 served as a probe for nucleotide binding.

Tryptophan fluorescence is measured with λ_{ex} between 290 nm and 295 nm and λ_{em} between 310 nm and 350 nm. For equilibrium fluorescence titrations, the applied concentration of the protein should be far below the K_d of the interaction. Nucleotide is added in sequential increments from highly concentrated stock solution (1–50 mM) until saturation of fluorescence emission is obtained. Fluorescence is corrected for dilution from nucleotide addition, and fluorescence is plotted as a function of nucleotide concentration. Data are usually fit to the quadratic equation (see 4.3.6.1 "Quadratic Equation" below).

Fluorescent Nucleotide Analogues The choice of the fluorescent probe depends on its ability to replace the native nucleotide (Is the modified nucleotide hydrolyzed? Does it bind like the unmodified nucleotide?) and its environmental sensitivity and thus the amplitude of the signal [5]. In addition, binding specificity should be probed before application. The presence of hydrophobic groups in the fluorescent nucleotide analogues presents the possibility that they may bind to proteins at sites other than the nucleotide-binding site. The occurrence of such nonspecific binding can be determined by carrying out the relevant experiment in the presence of a large excess of the physiological nucleotide, which should not give rise to changes in the fluorescent signal. Usually MANT nucleotides show little or no nonspecific binding to ATPases.

With regard to chaperones, examples for equilibrium fluorescence titrations with fluorescent nucleotide analogues are found in Ref. [21] (MANT-ADP and ClpB$_{Tth}$), Ref. [11] (N8-MABA-ADP and DnaK$_{Eco}$), and Ref. [14] ((Pγ)-MABA-ATP and yeast Hsp90). In a standard procedure, the fluorescent nucleotide is placed in a cuvette (0.5 µM in standard fluorescence buffer: 50 mM Tris/HCl (pH 7.5), 100 mM KCl, 5 mM MgCl$_2$, 2 mM EDTA, 2 mM DTE), aliquots of the chaperone solution are added (with mixing), and the signal after a brief period of time is measured (correcting the signal for dilution of the fluorescent species and for the inner-filter effect). Curves are fitted according to the quadratic equation (see below). MANT fluorescence is measured with λ_{ex} at 360 nm and λ_{em} at 440 nm. For a tightly binding system, one must be concerned that equilibrium is reached after each addition of reactant. Association is often limited only by diffusion, but dissociation rate constants are much smaller. In such a case, measurements can also be made with a series of prepared solutions, covering various ratios of protein and nucleotide. An example of a fluorescence-binding study is presented in Figure 4.2A.

Active Site Titration In order to perform proper analyses of ligand-binding data and to evaluate enzymatic mechanisms, it is desirable to know with which stoichiometry binding partners interact. For this purpose, a straightforward approach exists that can be performed with various methods; it is described here for the case of fluorescent nucleotide–protein interaction [86]. It is important to note that some constraints apply for successful application of this technique. Firstly, the initial concentration of the signaling ligand, in this case the fluorescent nucleotide, should be well above the estimated K_d (say 5-fold or, even better, 10-fold) such that addition of protein results in nearly complete binding until all nucleotide is bound (saturated). Secondly, the concentrations of the binding partners should be determined accurately, which is straightforward in the case of nucleotides since they are usually pure and have a well-defined extinction coefficient. In the case of protein, this could be more problematic since most common methods to deter-

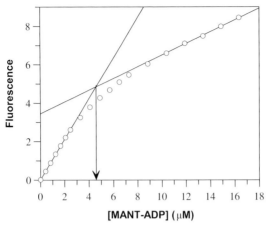

Fig. 4.9. Active site titration of ClpB$_{Tth}$(trunc) with MANT-ADP. Concentration of ClpB(trunc), a shortened ClpB variant lacking the NBD1, at start was 10 μM (measured according to Ref. [106]), and initial volume was 600 μL. Interpolation of the two phases gives an intersection at [MANT-ADP] = 4.8 μM. The volume-corrected concentration of ClpB(trunc) at this point is 9.8 μM. This indicates that one ClpB-dimer is able to bind one molecule of MANT-ADP.

mine protein concentration can deviate for certain proteins. Once these two prerequisites are met, the actual experiment/analysis is quite simple. Protein is added to the ligand in small steps, usually corresponding to concentration increases that amount to 1/20th of the initial ligand concentration until an appreciable saturation of the fluorescence signal is obtained. Now the protein may be added in larger steps until (after volume correction) no further change of signal is observed. The two resulting phases of this titration, initial nearly linear increase and saturation, are extrapolated by hand with linear slopes. The point of intersection then determines at which concentration of added protein the ligand is saturated and as such the stoichiometry of the binding isotherm.

An example for such an experiment is given in Figure 4.9, where MANT-ADP was titrated to ClpB$_{Tth}$(trunc), a shortened ClpB variant lacking a functional NBD1. Interpolation of the two phases gives an intersection at [MANT-ADP] = 4.8 μM, while the volume-corrected concentration of ClpB(trunc) at this point is 9.8 μM. This indicates that one molecule of MANT-ADP is bound per ClpB dimer. Adding ClpB$_{Tth}$(trunc) to MANT-ADP under similar conditions also results in a biphasic curve, where the slope of the second phase is close to zero. The intersection also points to a 2:1 stoichiometry (ClpB(trunc):MANT-ADP).

4.3.5.5 Competition Experiments

The affinities of the unmodified nucleotides ADP or ATP for DnaK or ClpB were determined by competition experiments [11, 21, 87]. Therefore, a preformed complex out of the chaperone and the fluorescent nucleotide was titrated with the unmodified nucleotide, and the decrease of the fluorescence signal caused by the displacement of the fluorophore by the non-fluorescent nucleotide was used together with the known K_d for the fluorescent nucleotide to obtain the affinity (see Figure 4.2B). The general mathematical description of this system is the solution of a cubic equation (see "Cubic Equation" below) [88]. This complete analysis is specifically required if the affinities of the two ligands are very high such that the corresponding K_d values are much lower than the initial concentration of the substance to be titrated.

4.3.6
Analytical Solutions of Equilibrium Systems

4.3.6.1 Quadratic Equation

$[A]$, $[B]$, and $[C]$ are the free concentrations of the compounds A, B, and C. $[A_o]$, $[B_o]$, and $[C_o]$ are the total concentrations of the compounds A, B, and C. $[AB]$ and $[AC]$ are complex concentrations of the complexes of A and B or A and C, respectively (e.g., with A being the protein and B a binding ligand).

$$A + B \underset{Kd_{AB}}{\rightleftharpoons} AB$$

Scheme 4.1

Many kinetic experiments require that the concentration of a ligand-protein complex be calculated as a function of any defined concentration of the two substances and a given dissociation constant Kd_{AB}. For a 1:1 stoichiometry (one ligand binds to one binding site of a protein), the dissociation constant is defined as:

$$Kd_{AB} = \frac{[A][B]}{[AB]} \tag{6}$$

Equations of mass conservation:

$$[A_o] = [A] + [AB]$$
$$[B_o] = [B] + [AB] \tag{7}, (8)$$

Inserting Eqs. (7) and (8) into Eq. (6) gives

$$\begin{aligned} Kd_{AB} &= \frac{[A][B]}{[AB]} \\ &= \frac{([A_o] - [AB])([B_o] - [AB])}{[AB]} \\ &= \frac{[A_o][B_o] - [AB][B_o] - [AB][A_o] + [AB]^2}{[AB]} \\ &\equiv [AB]^2 - [AB]([A_o] + [B_o] + Kd_{AB}) + [A_o][B_o] = 0 \end{aligned} \tag{9}$$

A quadratic equation usually gives two solutions, but here only one is biochemically meaningful, and the final solution to calculate the concentration of the complex AB as a function of the total concentrations of A and B and a given dissociation constant Kd_{AB} therefore is:

$$[AB] = \frac{[A_o] + [B_o] + Kd_{AB}}{2} - \sqrt{\left(\frac{[A_o] + [B_o] + Kd_{AB}}{2}\right)^2 - [A_o][B_o]} \qquad (10)$$

The measured quantity is usually the increase/decrease of the fluorescence emission ΔF in every titration step. This quantity is proportional to the binding ratio $[AB]/[B_o]$ and ΔF_{max}, the maximal emission change observed when all the ligand B is bound to the protein (saturation conditions).

$$\Delta F = \Delta F_{max} \cdot \frac{[AB]}{[B_o]} \qquad (11)$$

The observed total fluorescence F in every titration step is composed out of the fluorescence of the free ligand F_0 and the change in fluorescence ΔF. Substituting $[AB]$ with Eq. (10) yields an equation for the observed total fluorescence:

$$F = F_0 + \Delta F_{max} \cdot \frac{\frac{[A_0] + [B_0] + Kd_{AB}}{2} - \sqrt{\left(\frac{[A_0] + [B_0] + Kd_{AB}}{2}\right)^2 - [A_0][B_0]}}{B_0} \qquad (12)$$

4.3.6.2 Cubic Equation

If two ligands compete for one binding site, as it is the case for displacement titrations with a fluorescent and a non-fluorescent nucleotide, and a 1:1 binding stoichiometry holds, the cubic equation may be applied.

A kinetic system where the ligands B and C compete for binding to A is described in Scheme 4.2.

$$A + B \underset{Kd_{AB}}{\rightleftharpoons} AB \quad A + C \underset{Kd_{AC}}{\rightleftharpoons} AC$$

Scheme 4.2

The dissociation constants Kd_{AB} and Kd_{AC} are defined as follows:

$$Kd_{AB} = \frac{[A][B]}{[AB]}; \quad Kd_{AC} = \frac{[A][C]}{[AC]} \qquad (13), (14)$$

The corresponding mass conservation gives:

$$[A_o] = [A] + [AB] + [AC]$$
$$[B_o] = [B] + [AB] \quad\quad\quad (15), (16), (17)$$
$$[C_o] = [C] + [AC]$$

If the total concentrations of B and C are much higher than that of A, then the observed dissociation constant of the AC complex, $Kd_{AC}(obs)$, is defined by the simple relationship:

$$Kd_{AC}(obs) = Kd_{AC}\left(1 + \frac{[B]}{Kd_{AB}}\right) \approx Kd_{AC}\left(1 + \frac{B_o}{Kd_{AB}}\right)$$

If the total concentrations of B and C are close to that of A, however, the assumption that $[C] \approx [C_o]$ and $[B] \approx [B_o]$ no longer holds. The deviation from a simple hyperbolic binding is most evident if $[A_o] \gg Kd_{AB}$ and Kd_{AC}.

The general solution of the system in Scheme 4.2 is the solution of a cubic equation. This solution is not based on any assumptions with respect to the relationship of $[A_0]$, $[B_0]$, and $[C_0]$ to Kd_{AB} and Kd_{AC}.

Insert Eqs. (16) and (17) into Eq. (15):

$$[A_o] = [A] + ([B_o] - [B]) + ([C_o] - [C]) \quad\quad\quad (18)$$

Insert Eq. (13) into Eq. (16) and solve for $[B]$

$$[B] = \frac{[B_o]}{1 + \frac{[A]}{Kd_{AB}}} \quad\quad\quad (19)$$

Insert Eq. (14) into Eq. (17) and solve for $[C]$.

$$[C] = \frac{[C_o]}{1 + \frac{[A]}{Kd_{AC}}} \quad\quad\quad (20)$$

Now insert Eqs. (19) and (20) into Eq. (18)

$$[A_o] = [A] + \left([B_o] - \frac{[B_o]}{1 + \frac{[A]}{Kd_{AB}}}\right) + \left([C_o] - \frac{[C_o]}{1 + \frac{[A]}{Kd_{AC}}}\right) \quad\quad\quad (21)$$

$[A]$ has to be replaced by $[AB]$ since we want to calculate the concentration of the complex $[AB]$. Insert Eq. (19) into Eq. (18) and solve for $[A]$:

$$[AB] = \frac{[A]\dfrac{[B_o]}{1 + \dfrac{[A]}{Kd_{AB}}}}{Kd_{AB}}$$

rearrange,

$$[A] = [AB]\frac{Kd_{AB}}{\left(\dfrac{[B_o]}{1 + \dfrac{[A]}{Kd_{AB}}}\right)}$$

and finally

$$[A] = -\frac{[AB]Kd_{AB}}{[AB] - [B_o]} \tag{22}$$

Now insert Eq. (22) into Eq. (21)

$$[A_o] = -\frac{[AB]Kd_{AB}}{[AB] - [B_o]} + \left([B_o] - \frac{[B_o]}{1 + \dfrac{-\dfrac{[AB]Kd_{AB}}{[AB] - [B_o]}}{Kd_{AB}}}\right) + \left([C_o] - \frac{[C_o]}{1 + \dfrac{-\dfrac{[AB]Kd_{AB}}{[AB] - [B_o]}}{Kd_{AC}}}\right) \tag{23}$$

This equation now implicitly expresses $[AB]$ as a function of total concentrations of the three compounds and the two dissociation constants. Now we have to rearrange (factorize for $[AB]$) all terms to get the following cubic equation.

The general form of a cubic equation is:

$$[AB]^3 + a_1[AB]^2 + a_2[AB] + a_3 = 0 \tag{24}$$

and we find the corresponding coefficients to be

$$a_o = Kd_{AB} - Kd_{AC} \tag{25}$$

$$a_1 = \frac{[A_o](Kd_{AC} - Kd_{AB}) + [B_o](2Kd_{AC} - Kd_{AB}) + [C_o]Kd_{AB} - Kd_{AB}^2 + Kd_{AB}Kd_{AC}}{a_o} \tag{26}$$

$$a_2 = \frac{[A_o][B_o](Kd_{AB} - 2Kd_{AC}) - [B_o]^2 Kd_{AC} - [B_o]Kd_{AB}([C_o] + Kd_{AC})}{a_o} \tag{27}$$

$$a_3 = \frac{[A_o][B_o]^2 Kd_{AC}}{a_o} \tag{28}$$

The general solution of the cubic equation Eq. (24) may be obtained as described [89].

$$Q \equiv \frac{a_1^2 - 3a_2}{9}; \quad R \equiv \frac{2a_1^3 - 9a_1 a_2 + 27a_3}{54} \tag{29}$$

if $Q^3 - R^2 \geq 0$, then there are three solutions to the cubic equation:

$$\Theta = \arccos\left(\frac{R}{\sqrt{Q^3}}\right)$$

$$[AB]_1 = -2\sqrt{Q}\cos\left(\frac{\Theta}{3}\right) - \frac{a_1}{3}$$

$$[AB]_2 = -2\sqrt{Q}\cos\left(\frac{\Theta + 2\pi}{3}\right) - \frac{a_1}{3}$$

$$[AB]_3 = -2\sqrt{Q}\cos\left(\frac{\Theta + 4\pi}{3}\right) - \frac{a_1}{3}$$

However, only one out of the three solutions given above is meaningful and has to be selected.

The automatic selection is a bit tricky, but once it is implemented it works pretty well. Imagine there is no substance C present; then a quadratic solution would be sufficient to calculate the concentration of the complex $[AB]$. We can still apply the cubic solution since it is valid for any arbitrary situation. If we now compare the solutions from the quadratic equation with the three solutions of the cubic equation, we realize that only one of those three solutions is correct. Now we know which of the three solutions to use. It turns out that this will still hold even if $[C_0]$ is not zero.

A more comprehensive treatment of this problem revealed that it is always the first solution $[AB]_1$ that represents the physically meaningful solution [90].

If $R^2 - Q^3 > 0$, then there is only one solution:

$$[AB] = -\text{sign}(R)\left[\sqrt[3]{\sqrt{R^2 - Q^3} + |R|} + \frac{Q}{\sqrt[3]{\sqrt{R^2 - Q^3} + |R|}}\right] - \frac{a_1}{3}$$

This case rarely occurs with the usual values of concentrations and dissociation constants all being positive.

This approach to analyzing competition experiments is complicated and often unnecessary, but experimental constraints do not always allow choosing conditions such that a simple analysis is possible. The choice of appropriate experimental conditions, in conjunction with the complete analysis as described above, also allows gathering information about binding stoichiometries in a competition experiment.

4.3.6.3 Iterative Solutions

Four-component System as an Example: Newton-Raphson Root Search

$$A + C \rightleftharpoons AC; \quad A + D \rightleftharpoons AD; \quad B + C \rightleftharpoons BC; \quad B + D \rightleftharpoons BD$$

Scheme 4.3

In the examples given above, we searched for analytical solutions that allowed us to define the desired complex concentrations as a function of total concentrations and K_d values. The order of the corresponding equation was directly related to the number of different ligands present. It was a quadratic equation (order 2) for two compounds and a cubic equation (order 3) for three compounds. In principle we could also obtain an analytical solution for a four-component system since the resulting quadric equation would still be solvable. We will avoid this hassle and instead use an iterative method that searches for the roots of a coupled equilibrium system. This method is known as the Newton-Raphson algorithm to find roots in systems of nonlinear equations and is also used by most commercial software packages (e.g., Scientist). It should be noted that these computer programs solve implicit equilibrium systems with this numerical method even with simple two-component systems.

An early practical example of a four-component system was the competition of Mg^{2+} and Ca^{2+} for the metal chelators ATP and EDTA or EGTA in muscle research [91]. Solutions for multi-species equilibria are also regaining attention in chaperone research due to the multi-enzyme–multi-ligand complexes involved. The effects of the DnaK nucleotide-exchange factor GrpE as measured with fluorescent nucleotide in competition with its unlabeled form, e.g., constitute a quaternary system [10, 82, 92], with

$$A \equiv Ca^{2+}; \quad B \equiv Mg^{2+}; \quad C \equiv EGTA; \quad D \equiv ATP \tag{30}$$

Definition of equilibrium constants:

$$Kd_{AC} = \frac{[A][C]}{[AC]}; \quad Kd_{AD} = \frac{[A][D]}{[AD]} \tag{31, 32}$$

$$Kd_{BC} = \frac{[B][C]}{[BC]}; \quad Kd_{BD} = \frac{[B][D]}{[BD]} \tag{33, 34}$$

Conservation equations:

$$[A_o] = [A] + [AC] + [AD]$$
$$[B_o] = [B] + [BC] + [BD]$$
$$[C_o] = [C] + [AC] + [BC]$$
$$[D_o] = [D] + [AD] + [BD]$$

(35)–(38)

Rearrangement of Eqs. (35)–(38) and insertion into Eqs. (31)–(34) gives:

$$[A] + \frac{[A][C]}{Kd_{AC}} + \frac{[A][D]}{Kd_{AD}} - [A_o] = 0$$

$$[B] + \frac{[B][C]}{Kd_{BC}} + \frac{[B][D]}{Kd_{BD}} - [B_o] = 0$$

$$[C] + \frac{[B][C]}{Kd_{BC}} + \frac{[A][C]}{Kd_{AC}} - [C_o] = 0$$

$$[D] + \frac{[A][D]}{Kd_{AD}} + \frac{[B][D]}{Kd_{BD}} - [D_o] = 0$$

(39)–(42)

Eqs. (39)–(42) describe the system of equations that has to be solved. One step that is necessary to get the Newton-Raphson (NR) algorithm to work is to generate a matrix of partial derivatives where the species that are iterated (x) are freely selectable. For example:

- If the total concentrations are given, calculate free concentrations and concentrations of complexes.
- If free concentrations are fixed, calculate total concentrations and concentration of complexes.
- If complex concentrations are given, calculate free and total concentrations.

General case:

$$\begin{bmatrix} \frac{\partial y_1}{\partial x_1} & \cdots & \cdots & \frac{\partial y_1}{\partial x_4} \\ \cdot & \cdots & \cdots & \cdot \\ \cdot & \cdots & \cdots & \cdot \\ \frac{\partial y_4}{\partial x_1} & \cdots & \cdots & \frac{\partial y_4}{\partial x_4} \end{bmatrix}$$

Let us assume that we define the total concentrations and want to calculate the free concentrations; then the matrix of partial derivatives would read:

$$\begin{bmatrix} \frac{\partial\left([A] + \frac{[A][C]}{Kd_{AC}} + \frac{[A][D]}{Kd_{AD}} - [A_o]\right)}{\partial[A]} & \cdots & \cdots & \frac{\partial\left([A] + \frac{[A][C]}{Kd_{AC}} + \frac{[A][D]}{Kd_{AD}} - [A_o]\right)}{\partial[D]} \\ \frac{\partial\left([B] + \frac{[B][C]}{Kd_{BC}} + \frac{[B][D]}{Kd_{BD}} - [B_o]\right)}{\partial[A]} & \cdots & \cdots & \frac{\partial\left([B] + \frac{[B][C]}{Kd_{BC}} + \frac{[B][D]}{Kd_{BD}} - [B_o]\right)}{\partial[D]} \\ \frac{\partial\left([C] + \frac{[B][C]}{Kd_{BC}} + \frac{[A][C]}{Kd_{AC}} - [C_o]\right)}{\partial[A]} & \cdots & \cdots & \frac{\partial\left([C] + \frac{[B][C]}{Kd_{BC}} + \frac{[A][C]}{Kd_{AC}} - [C_o]\right)}{\partial[D]} \\ \frac{\partial\left([D] + \frac{[A][D]}{Kd_{AD}} + \frac{[B][D]}{Kd_{BD}} - [D_o]\right)}{\partial[A]} & \cdots & \cdots & \frac{\partial\left([D] + \frac{[A][D]}{Kd_{AD}} + \frac{[B][D]}{Kd_{BD}} - [D_o]\right)}{\partial[D]} \end{bmatrix}$$

(43)

4 Characterization of ATPase Cycles of Molecular Chaperones

The result is the following matrix:

$$\begin{bmatrix} 1 + \dfrac{[C]}{Kd_{AC}} + \dfrac{[D]}{Kd_{AD}} & 0 & \dfrac{[A]}{Kd_{AC}} & \dfrac{[A]}{Kd_{AD}} \\ 0 & 1 + \dfrac{[C]}{Kd_{BC}} + \dfrac{[D]}{Kd_{BD}} & \dfrac{[B]}{Kd_{BC}} & \dfrac{[B]}{Kd_{BD}} \\ \dfrac{[C]}{Kd_{AC}} & \dfrac{[C]}{Kd_{BC}} & 1 + \dfrac{[B]}{Kd_{BC}} + \dfrac{[A]}{Kd_{AC}} & 0 \\ \dfrac{[D]}{Kd_{AD}} & \dfrac{[D]}{Kd_{BD}} & 0 & 1 + \dfrac{[A]}{Kd_{AD}} + \dfrac{[B]}{Kd_{BD}} \end{bmatrix} \tag{44}$$

Why do we need this matrix of partial derivatives (see Ref. [89] p. 306)?

If $\bar{x}$ is the vector of all x_i, then each function f_i may be extended in a Taylor series around $\bar{x}$.

$$f_i(\bar{x} + \delta\bar{x}) = f_i(\bar{x}) + \sum_{j=1}^{N} \frac{\partial f_i}{\partial x_j} \delta x_j + O(\delta\bar{x}^2) \tag{45}$$

A system of linear equations for the correction factors $\delta\bar{x}$ is obtained if $\delta\bar{x}^2$ and all higher-order terms are neglected. All equations could thus be optimized simultaneously:

$$\sum_{j=1}^{N} \alpha_{ij}\delta x_j = \beta_i \quad \text{where } \alpha_{ij} \equiv \frac{\partial f_i}{\partial x_j} \text{ and } \beta_i = -f_i \tag{46}$$

α thus represents the matrix of partial derivatives. Since we reduced our problem to a system of linear equations, it may be solved by any standard numerical technique like partial pivoting or lower/upper triangular (LU) decomposition. This proceeds iteratively and usually needs some 2 to 10 iterations to converge; the computer simulates the approach to equilibrium!

Trial runs indicated that the numerical calculations became very unstable when the matrix of partial derivatives was generated by the approximation of finite differences, e.g.,

$$\frac{\partial f_i}{\partial x_j} \simeq \frac{f_i(x + \Delta x_j)}{f_i(x)} \tag{47}$$

The partial derivatives thus have to be generated analytically.

If the free concentrations of the complexes are defined and the corresponding total concentrations should be calculated, then the matrix of partial derivatives is:

$$\begin{bmatrix} \dfrac{\partial\left([A]+\dfrac{[A][C]}{Kd_{AC}}+\dfrac{[A][D]}{Kd_{AD}}-[A_o]\right)}{\partial[A_o]} & \cdots & \dfrac{\partial\left([A]+\dfrac{[A][C]}{Kd_{AC}}+\dfrac{[A][D]}{Kd_{AD}}-[A_o]\right)}{\partial[D_o]} \\[2ex] \dfrac{\partial\left([B]+\dfrac{[B][C]}{Kd_{BC}}+\dfrac{[B][D]}{Kd_{BD}}-[B_o]\right)}{\partial[A_o]} & \cdots & \dfrac{\partial\left([B]+\dfrac{[B][C]}{Kd_{BC}}+\dfrac{[B][D]}{Kd_{BD}}-[B_o]\right)}{\partial[D_o]} \\[2ex] \dfrac{\partial\left([C]+\dfrac{[B][C]}{Kd_{BC}}+\dfrac{[A][C]}{Kd_{AC}}-[C_o]\right)}{\partial[A_o]} & \cdots & \dfrac{\partial\left([C]+\dfrac{[B][C]}{Kd_{BC}}+\dfrac{[A][C]}{Kd_{AC}}-[C_o]\right)}{\partial[D_o]} \\[2ex] \dfrac{\partial\left([D]+\dfrac{[A][D]}{Kd_{AD}}+\dfrac{[B][D]}{Kd_{BD}}-[D_o]\right)}{\partial[A_o]} & \cdots & \dfrac{\partial\left([D]+\dfrac{[A][D]}{Kd_{AD}}+\dfrac{[B][D]}{Kd_{BD}}-[D_o]\right)}{\partial[D_o]} \end{bmatrix}$$

which is as simple as:

$$\begin{bmatrix} -1 & 0 & 0 & 0 \\ 0 & -1 & 0 & 0 \\ 0 & 0 & -1 & 0 \\ 0 & 0 & 0 & -1 \end{bmatrix}$$

All the examples presented above concern calculations of the same type: free concentrations, total concentrations, or complex concentrations. The question is now whether there is a way to calculate mixed species. It is obvious that the choice of species that may be defined or calculated is limited to a certain extent. For example, if the total amount of A_o and the complex concentration is specified, then the concentration of A_f cannot be specified since it has to be varied to give the AC desired. In this case, the system would be overdetermined. There could also be cases where the system is underdetermined; then the solution will not be unique or the system will not converge.

4.3.7
Time-resolved Binding Measurements

4.3.7.1 Introduction

Mechanistic questions and kinetic models are closely related, since a kinetic model is usually derived from hypotheses of the enzyme's mechanism. The model is then scrutinized by other appropriate kinetic measurements that should ideally be designed to falsify – not "prove" – the model. The link between the model and experimental data is provided by the equations that are obtained by solving the differential system determined by the model. Although it is becoming more popular to use computer programs that avoid the hassle of analytical solutions (through numerical integration), it is often still more instructive to use these solutions, and some examples of the most often needed types are shown below.

Differential equations systems could be solved in principle by three major methods: Laplace transformation, matrix method, and substitution method. Because a proper introduction to these mathematical approaches, as well as experimental methods to obtain relevant transient kinetic data, would go far beyond the scope of this chapter, the reader is kindly asked to refer to the following major references that provide fundamental and thorough introductions and overviews.

1) Laplace transformation: Refs. [93, 94]
2) Substitution: Ref. [94]
3) Matrix method: Ref. [95]

A "cookbook" that provides solutions to many systems [96] is unfortunately no longer commercially available but could be found in some libraries.

4.3.7.2 One-step Irreversible Process

A well-known example for one-step irreversible processes is radioactive decay of isotopes.

The according kinetic model represents the simplest scheme available:

$$A \xrightarrow{k} B$$

Under certain boundary conditions, this scheme can also be applied to enzymatic reactions, e.g., single-step hydrolysis of ATP where either ATP binding is not rate-limiting and saturating or the reaction is initiated with pre-bound ATP (e.g., by addition of catalytic Mg^{2+}). Another example is the displacement of fluorescently labeled, pre-bound nucleotide by an excess of unlabeled ligand (usually also nucleotide) to measure directly the rate constant for dissociation [10, 11, 97].

If the initial concentration of B at $t = 0$ (B_0) is 0, then the differential equations for that system read:

$$\frac{d[A]}{dt} = -k[A]$$

$$\frac{d[B]}{dt} = k[A]$$

(D1), (D2)

Converted to Laplace space (please refer to literature above for information on how to set up equations)

$$sa = -ka + a_0$$

$$sb = ka$$

(L1), (L2)

and solved for a and b, respectively:

$$a = \frac{a_0}{s+k}$$

$$b = \frac{ka}{s} = \frac{ka_0}{s(s+k)} \qquad \text{(S1), (S2)}$$

Looking in the table for the back transformation and setting k equal to α gives

$$L^{-1}\left(\frac{const}{s(s+\alpha)}\right) = \frac{const}{\alpha}(1 - e^{-\alpha t})$$

and

$$L^{-1}\left(\frac{const}{s+\alpha}\right) = e^{-\alpha t}.$$

Therefore, $[A](t) = [A_0] \cdot e^{-kt}$ and $[B](t) = [A_0](1 - e^{-kt})$.

4.3.7.3 One-step Reversible Process

First-order reversible processes are routinely encountered in protein folding and are discussed here in great detail. Related to that are structural changes that may be induced by ligands or other factors, e.g., light, pH, or temperature among others.

Model:

$$A \underset{k_{-1}}{\overset{k_1}{\rightleftharpoons}} B$$

$$\frac{d[A]}{dt} = -k_1[A] + k_{-1}[B]$$

$$\frac{d[B]}{dt} = k_1[A] - k_{-1}[B] \qquad \text{(D1), (D2)}$$

$$sa - a_0 = -k_1 a + k_{-1} b$$

$$sb = k_1 a + k_{-1} b \qquad \text{(L1), (L2)}$$

$$\Rightarrow a = \frac{a_0 + k_{-1} b}{s + k_1} \qquad \text{(S1)}$$

Inserting S1 into L2 and factorizing for b gives

$$b = \frac{k_1 a_0}{(s+k_1)\left(s + k_{-1} - \frac{k_1 k_{-1}}{s+k_1}\right)} = \frac{k_1 a_0}{s(s + k_1 + k_{-1})}$$

with $k_1 + k_{-1} \equiv \alpha$.

The transform table gives as the back transformation of

$$L^{-1}\frac{\text{const}}{s(s+\alpha)} = \frac{\text{const}}{\alpha}(1 - e^{-\alpha t}) \tag{S2}$$

and hence $B(t) = \dfrac{k_1[A_0]}{k_1 + k_{-1}}(1 - e^{-(k_1+k_{-1})\cdot t})$.

4.3.7.4 Reversible Second Order Reduced to Pseudo-first Order

Reversible second-order processes are mostly observed in simple ligand-binding processes where the binding of ligand does not induce further structural changes.

The corresponding model reads

$$A + B \underset{k_{-1}}{\overset{k_{+1}}{\rightleftharpoons}} C$$

with an according set of differential equations

$$\frac{d[A]}{dt} = -k_{+1}[A][B] + k_{-1}[C]$$

$$\frac{d[B]}{dt} = -k_{+1}[A][B] + k_{-1}[C] \tag{D1)-(D3}$$

$$\frac{d[C]}{dt} = k_{+1}[A][B] - k_{-1}[C]$$

If the boundary conditions are such that $B_o \gg A_o$ (that is, the species B at $t = 0$ is in excess) and $C_o = 0$, then the reaction becomes pseudo-first order since the concentration of B does not change significantly over time, $B(t) \approx B_o$. The set of (approximated) differential equations then is:

$$\frac{d[A]}{dt} \cong -k_{+1}[A][B_o] + k_{-1}[C]$$

$$\frac{d[B]}{dt} \cong 0 \tag{D4)-(D6}$$

$$\frac{d[C]}{dt} \cong k_{+1}[A][B_o] - k_{-1}[C]$$

The equations in Laplace space are:

$$sa - a_o \cong -k_{+1}ab_o + k_{-1}c$$

$$sc \cong k_{+1}ab_o - k_{-1}c \tag{L1), (L3}$$

Solve for a

$$a \cong \frac{a_o + k_{-1}c}{s + k_{+1}b_0} \qquad (S1)$$

Insert S1 into L3 and solve for c:

$$c \cong \frac{a_o b_o k_{+1}}{s(s + b_o k_{+1} + k_{-1})} \qquad (S2)$$

In the tables we find the transformation back to real space

$$L^{-1}\left(\frac{1}{s(s+\alpha)}\right) = \frac{1}{\alpha}(1 - e^{-\alpha t}) \quad \text{with } \alpha = b_o k_{+1} + k_{-1}$$

and therefore the solution is:

$$[C](t) \cong \frac{A_o B_o k_{+1}}{B_o k_{+1} + k_{-1}}(1 - e^{-(B_o k_{+1} + k_{-1})t})$$

Note that the amplitude is also an approximation, but in practice a three- to four-fold excess of B_o should be sufficient. To answer how close this approximation is to the real solution, we perform an exemplary calculation.

Assume: $A_0 = 1\ \mu M$; $B_0 = 10\ \mu M$; $k_{+1} = 1\ \mu M^{-1} \cdot s^{-1}$; $k_{-1} = 1\ s^{-1} \Rightarrow K_d = 1\ \mu M$

The approximation according to the kinetic solution is:

$$[C](t = \text{infinity}) = \frac{1 \cdot 10 \cdot 1}{10 \cdot 1 + 1} = \frac{10}{11} = 0.909\ \mu M$$

The exact solution as calculated with the quadratic binding equation is:

$$\frac{1 + 1 + 10}{2} - \sqrt{\left(-\frac{1 + 1 + 10}{2}\right)^2 - 10 \cdot 1} = 0.90098\ \mu M$$

We can conclude that the approximation of the amplitude is sufficiently good under these conditions.

If $B_0 = 2\ \mu M$, i.e., if the excess of ligand is just twofold over A_0, then the kinetic "solution" gives an amplitude of $\frac{1 \cdot 2 \cdot 1}{2 \cdot 1 + 1} = \frac{2}{3} = 0.666$ in comparison to the exact quadratic equation, which gives $2 - \sqrt{4 - 2} = 0.5857$ as a value for the final complex concentration after equilibration. This shows that, even being far from pseudo-first-order conditions, the amplitude derived under these conditions is still a fair approximation. We should be aware, however, that the shape of the time trace will be substantially different and thus large errors will be introduced for the resulting rate constant after a fitting procedure, including large visible systematic deviations that are clearly visible with a residual plot [95].

One should also note here that in general, without the assumption that one ligand is in excess, binding systems are rarely solvable – as are many other nonlinear systems.

4.3.7.5 Two Simultaneous Irreversible Pathways – Partitioning

Partitioning experiments are very useful and often are the only accessible way to obtain information about particular kinetic systems. One example where this technique was used successfully is in the case of Hsp90. Here a complex of radioactive ATP with Hsp90 was chased off with an excess of unlabeled ATP, and the yield of hydrolyzed, radioactive ADP compared to the yield without chase showed commitment to hydrolysis since the enzyme goes through a complete round of hydrolysis before the chase takes affect [14].

A complete treatment of this method follows the model in Scheme 4.4:

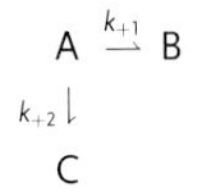

$$A \xrightleftharpoons{k_{+1}} B$$
$$\downarrow k_{+2}$$
$$C$$

Scheme 4.4

Again, the differential equation system (DES) is generated by summing up all steps that lead away from a certain species with the steps that lead to it.

$$\frac{d[A]}{dt} = -(k_1 + k_2)[A]$$

$$\frac{d[B]}{dt} = k_1[A] \qquad (D1)-(D3)$$

$$\frac{d[C]}{dt} = k_2[A]$$

Again, in Laplace space, we replace d/dt with the symbol s. Any species N that is not initially zero (at time 0, start of the reaction) gets the additional integration term n_o.

$$sa - a_o = -(k_1 + k_2)a$$
$$sb = k_1 a \qquad (L1)-(L3)$$
$$sc = k_2 a$$

Solving the resulting linear system in Laplace space:

Solve L1 for a to give S1

$$a = \frac{a_o}{s + k_1 + k_2} \tag{S1}$$

Now insert S1 into L2 and solve for b

$$sb = k_1 \frac{a_o}{s + k_1 + k_2} \Rightarrow b = \frac{k_1 a_o}{s(s + k_1 + k_2)} \tag{S2}$$

The table of Laplace back transformations tells us that the back transformation to real space (L^{-1}) of our system is:

$$L^{-1} \frac{1}{s(s + \alpha)} = \frac{1}{\alpha}(1 - e^{-\alpha t})$$

so that:

$$[B](t) = [A_o]\frac{k_1}{k_1 + k_2}(1 - e^{-(k_1 + k_2)t}) \tag{S3}$$

and accordingly:

$$[A](t) = [A_o]\left(1 - \frac{k_1}{k_1 + k_2}(1 - e^{-(k_1 + k_2)t})\right)$$

$$[C](t) = [A_o]\frac{k_2}{k_1 + k_2}(1 - e^{-(k_1 + k_2)t}) \tag{S3}, (S4)$$

The observed rate constant of $k_1 + k_2$ as well as the observed amplitude of $k_1/(k_1 + k_2)$ for the formation of B gives independent information about the rate constants and may be used to obtain k_1 and k_2 from one simple time trace. The example shown in Figure 4.10 demonstrates that for the same $k_{obs} = k_1 + k_2 = 10 + 1 = 11\ s^{-1}$, the different amplitudes allow us to unequivocally assign the individual rate constants. The same holds for equal amplitudes: now the observed rate constants allow us to assign the individual rate constants.

Some useful relationships are

$$k_{obs} = k_1 + k_2$$

$$B_{obs} = A_o \frac{k_1}{k_1 + k_2}$$

$$k_1 = \frac{A_{obs}}{A_o} k_{obs}$$

$$k_2 = k_{obs} - k_1$$

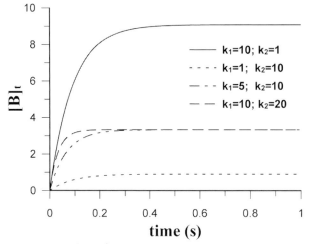

Fig. 4.10. Simulation for time traces according to partitioning scheme.

4.3.7.6 Two-step Consecutive (Sequential) Reaction
The following model

$$A \xrightarrow{k_1} B \xrightarrow{k_2} C$$

provides a simplified version of two-step irreversible mechanisms that can be applied under certain conditions e.g., binding of ATP (treated as a single step since a conformational change following initial encounter [to give complex A] is rate-limiting) followed by irreversible hydrolysis [79, 80]. In these literature examples, it was important to first generate a hydrolysis-competent DnaK-ATP state that could then be "chased" by ATPase-stimulating DnaJ in a second mixing reaction. The experiments therefore had to be designed to allow for complete ATP binding, including the conformational change it induces, without appreciable hydrolysis occurring before the DnaJ "chase". The amount of the "reaction" intermediate B (DnaK-ATP complex in the closed and DnaJ stimulation-competent conformation) thus had to be maximized prior to addition of DnaJ.

The **differential equation system** that describes this kinetic system is:

$$\frac{d[A]}{dt} = -k_1[A]$$

$$\frac{d[B]}{dt} = k_1[A] - k_2[B] \qquad \qquad \text{(D1)–(D3)}$$

$$\frac{d[C]}{dt} = k_2[B]$$

This system is transformed forward into Laplace space to give the following **linear equation system**: boundary conditions: $A(t = 0, A_0)$ and $B_0, C_0 = 0$

$$sa - a_o = -k_1 a$$
$$sb = -k_1 a - k_2 b \quad \text{(L1)–(L3)}$$
$$sc = k_2 b$$

To solve for c, we first solve for a to give:

$$a = \frac{a_o}{s + k_1} \quad \text{(S1)}$$

Then insert this expression into L2 and solve for b:

$$b = \frac{k_1 a_o}{s + k_1} \cdot \frac{1}{s + k_2} \quad \text{(S2)}$$

Finally, insert this expression into L3 and solve for c:

$$c = \frac{k_2}{s} \cdot \frac{k_1 a_o}{s + k_1} \cdot \frac{1}{s + k_2} \quad \text{(S3)}$$

Looking in the table of Laplace back transformations gives us:

$$L^{-1}\left(\frac{1}{s(s + \alpha_1)(s + \alpha_2)}\right) = \frac{1}{\alpha_1 \alpha_2}\left(1 + \frac{1}{\alpha_1 - \alpha_2}(\alpha_2 e^{-\alpha_1 t} - \alpha_1 e^{-\alpha_2 t})\right)$$

so that transforming back to real space gives:

$$[C](t) = [A_o]\left(1 + \frac{1}{k_1 - k_2}(k_2 e^{-k_1 t} - k_1 e^{-k_2 t})\right)$$

In summary, if the boundary conditions are chosen such that $A_0 \neq 0$, $B_0, C_0 = 0$, then the solutions are:

$$[A](t) = [A_o]e^{-k_1 t}$$

$$[B](t) = [A_o]\frac{k_1}{k_2 - k_1}(e^{-k_1 t} - e^{-k_2 t})$$

$$[C](t) = [A_o]\left(1 + \frac{1}{k_1 - k_2}(k_2 e^{-k_1 t} - k_1 e^{-k_2 t})\right)$$

Common questions that appear (e.g., in double-mixing or time-resolved structural measurements) are what the maximum of $[B]$ is that can be obtained and at which time this is reached.

$$\frac{d[B]}{dt} = 0 = \frac{d\left([A_o]\frac{k_1}{k_2 - k_1}(e^{-k_1 t} - e^{-k_2 t})\right)}{dt}$$

$$= [A_o]\frac{k_1}{k_2 - k_1}(-k_1 e^{-k_1 t_{B\,max}} + k_2 e^{-k_2 t_{B\,max}})$$

$$t_{B\,max} = \frac{\ln\frac{k_2}{k_1}}{k_2 - k_1}$$

Accordingly, the maximal amount of the intermediate B formed is:

$$[B](t_{B\,max}) = [A_o]\frac{k_1}{k_2 - k_1}(e^{-k_1 t_{B\,max}} - e^{-k_2 t_{B\,max}})$$

A plot of the maximal relative yield of the intermediate B as a function of the individual rate constants is simulated and shown in Figure 4.11. Additionally, a similar plot for the time when this maximum is reached is shown in Figure 4.12.

4.3.7.7 Two-step Binding Reactions

The following system represents a reversible two-step binding process that already results in substantial complexity of the according equations. This description was kindly provided by Roger S. Goody and was only slightly adjusted in minor parts.

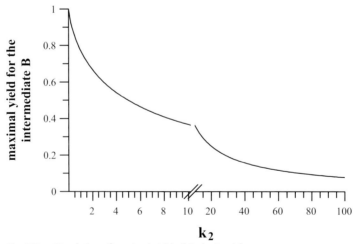

Fig. 4.11. Simulation of maximal yield of B obtained for different ratios of k_1 (10) and k_2 of a consecutive reaction ($A \xrightarrow{k_1} B \xrightarrow{k_2} C$).

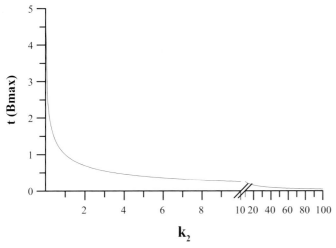

Fig. 4.12. Plot of time point where maximal yield of B is reached in consecutive reaction ($k_1 = 10$, k_2 as indicated on x-axis).

For an even advanced treatment, which is necessary for two-step binding mechanisms including competition, the reader is referred to Refs. [98, 99] and to some related "classics" that are concerned mostly with muscle research [100–104].

These mechanisms occur when, e.g., a fluorescent nucleotide serves as an established monitor for a two-step binding process, but of interest are the properties of the unmodified analogue that may also bind in two steps. An approach that uses numerical integration was also described for binding/competition of N^8-(4-N'-methylanthraniloylaminobutyl)-8-aminoadenosine 5'-diphosphate and ADP to DnaK from *Thermus thermophilus* [9].

Many binding interactions occur in two or more steps. How do we recognize this and what are the consequences? Two types of mechanism have often been observed. In principle, two different mechanisms are possible: initial "weak" binding followed by an isomerization reaction

$$E + L \rightleftharpoons EL \rightleftharpoons E^*L \quad \text{(Case 1)}$$

or pre-binding isomerization of one of the components

$$E \rightleftharpoons E^* + L \rightleftharpoons E^*L \quad \text{(Case 2)}$$

with the asterisk indicating a conformational change.

This problem was also discussed comprehensively for possible two-state binding mechanisms when ATP binding to Hsc70 was analyzed for the first time with stopped-flow measurements using the intrinsic single-tryptophan signal [105].

4 Characterization of ATPase Cycles of Molecular Chaperones

To derive the according kinetic equations in the most general fashion, we use the following model:

$$A \underset{k_{-1}}{\overset{k_1}{\rightleftharpoons}} B \underset{k_{-2}}{\overset{k_{+2}}{\rightleftharpoons}} C$$

$$\frac{dc}{dt} = k_{+2}b - k_{-2}c = k_{+2}(a_0 - a - c) - k_{-2}c$$

$$\frac{da}{dt} = -k_{+1}a + k_{-1}b = -k_{+1}(a_0 - b - c) + k_{-1}b$$

$$b = \frac{dc}{dt} \cdot \frac{1}{k_{+2}} + \frac{k_{-2}c}{k_{+2}}$$

$$\frac{da}{dt} = -\left(a_0 - \frac{dc}{dt} \cdot \frac{1}{k_{+2}} - \frac{k_{-2}c}{k_{+2}} - c\right)k_{+1} + \frac{dc}{dt} \cdot \frac{k_{-1}}{k_{+2}} + \frac{k_{-1}k_{-2}c}{k_{+2}}$$

$$\frac{d^2c}{dt^2} = -k_{+2}\frac{da}{dt} - k_{+2}\frac{dc}{dt} - k_{-2}\frac{dc}{dt}$$

$$= -k_{+2}\left[-a_0k_{+1} + \frac{dc}{dt} \cdot \frac{k_{+1}}{k_{+2}} + k_{+1}c\frac{k_{-2}}{k_{+2}} - ck_{+1} + \frac{dc}{dt} \cdot \frac{k_{-1}}{k_{+2}} + \frac{k_{-1}k_{-2}c}{k_{+2}}\right]$$

$$- k_{+2}\frac{dc}{dt} - k_{-2}\frac{dc}{dt}$$

$$\frac{d^2c}{dt} + \frac{dc}{dt}[k_{+1} + k_{-1} + k_{+2} + k_{-2}] + c[k_{+1}k_{-1} + k_{+2}k_{+1} + k_{-1}k_{-2}] = k_{+2}k_{+1}a_0$$

$$\lambda_1, \lambda_2 = \frac{-[k_{+1} + k_{-1} + k_{+2} + k_{-2}] \pm \sqrt{[k_{+1} + k_{-1} + k_{+2} + k_{-2}]^2 - 4[k_{+1}k_{-2} + k_{+2}k_{+1} + k_{-1}k_{-2}]}}{2}$$

For the case $1(E + L \rightleftharpoons EL \rightleftharpoons E^*L)$ and assuming that $l_0 \gg e_0$, the expression k_{+1} is replaced by $k_{+1}l_0$, or briefly k_{+1}', to indicate that the product of k_{+1} and l_o is virtually constant under these conditions:

$$\lambda_1\lambda_2 = \frac{-[k_{+1}' + k_{-1} + k_{+2} + k_{-2}] \pm \sqrt{[k_{+1}' + k_{-1} + k_{+2}]^2 - 4[k_{+1}'k_{-2} + k_{+2}k_{+1}' + k_{-1}k_{-2}]}}{2}$$

In the general case, this means complex kinetic behavior, with both solutions (λ_1 and λ_2) of the differential equation contributing to the transient observed. However, under certain conditions, simpler behavior is seen.

Examples are:

1. If k_{-1} is much larger than k_{+2} and k_{-2}, the second term under the square root bracket will always be much smaller than the first, and we can use the first two terms of the binomial expansion as a good approximation.

We can also simplify the first term by removing k_{+2} and k_{-2}.

[Binomial expansion of $a(a-b)^n = a^n - \dfrac{na^{n-1}b}{1}\ldots$]

Thus,

$$\lambda_1, \lambda_2 = -\dfrac{[k_{+1}' + k_{-1}] \pm \left[[k_{+1}' + k_{-1}] - \dfrac{1}{2} \cdot 4 \dfrac{[k_{+1}'k_{-2} + k_{+2}k_{+1}' + k_{-1}k_{-2}]}{[k_{+1}' + k_{-1}]}\right]}{2}$$

$$\approx [k_{+1}' + k_{-1}] \tag{48}$$

or

$$\dfrac{k_{+1}'k_{-2} + k_2 k_{+1}' + k_{-1}k_{-2}}{k_{+1}' + k_{-1}}$$

Note: The solutions of the equation are actually all negative, but the sign will be dropped at this point (terms in the solution are of the form $Ae^{-\lambda t}$).

$$k_{obs} = \dfrac{k_{-2}(k_{+1}' + k_{-1}) + k_2 k_{+1}'}{k_{+1}' + k_{-1}}$$

$$= k_{-2} + \dfrac{k_2 k_{+1}'}{k_{+1}' + k_{-1}}$$

$$= k_{-2} + \dfrac{k_2}{1 + \dfrac{k_{-1}}{k_{+1}'}} = k_{-2} + \dfrac{k_2}{1 + \dfrac{K_d'}{l_o}}$$

where K_d' is the dissociation constant for the first step. This behavior is shown in Figure 4.13 and was, e.g., observed for the binding of ATP to DnaK [11].

2. If k_{-1} is much smaller than k_{+2} and k_{-2}, we have the following situation when k_{+1}' is also small (i.e., at low $l = l_o$):

$$\lambda_{1,2} = \dfrac{-[k_{+2} + k_{-2}] \pm \sqrt{[k_{+2} + k_{-2}]^2 - 4[k_{+1}lk_{-2} + k_{+2}k_{+1}l + k_{-1}k_{+2}]}}{2} \tag{49}$$

Again, under the conditions chosen, the second term under the square root sign is small compared to the first and to a first approximation

$$\lambda_{1,2} = [k_{+2} + k_{-2}] \quad \text{or} \quad \dfrac{k_{+1}lk_{-2} + k_{+2}k_{+1}l + k_{-1}k_{+2}}{k_{+2} + k_{-2}} \tag{50}$$

large and of small amplitude if E^*L gives the signal.

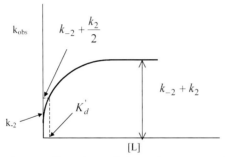

Fig. 4.13. Dependence of observed rate constant for large k_{-1} (fast first step).

The second solution simplifies to

$$k_{+1}l + \frac{k_{-1}k_{+2}}{k_{+2} + k_{-2}} \tag{51}$$

$$k_{+1}l + \frac{k_{-1}}{1 + \frac{k_{-2}}{k_{+2}}} = k_{+1}l + \frac{k_{-1}}{1 + \frac{1}{K_2}} = k_{obs} \tag{52}$$

A corresponding plot is shown in Figure 4.14.

At intermediate concentrations of L (when $k_{+1} \approx k_{+2}l + k_{-2}$), this system will show a lag phase in the kinetics of production of E^*L (comparable to the simple case of $A \rightarrow B \rightarrow C$ when the rate constants are of similar magnitude).

At concentrations at which $k_{+1}l$ is much greater than all the other rate constants, k_{obs} simplifies to:

$$k_{obs} = k_{+2} + k_{-2} \tag{53}$$

(i.e., the rate constant for production of E^*L can never be greater than the sum of k_{+2} and k_{-2}. This essentially simplifies the system to be analogous to the situation discussed previously for the rate of equilibration of $A \rightleftarrows B$).

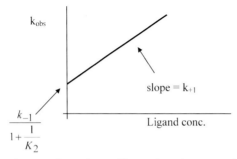

Fig. 4.14. Dependence of k_{obs} on ligand concentration for fast second step in two-step binding mechanism.

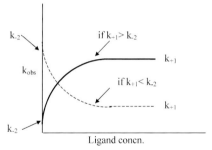

Fig. 4.15. Pre-isomerization dependence of k_{obs} on ligand concentration for large k_{-1}.

In the second possible two-step binding mechanism, a pre-binding isomerization occurs:

$$E \rightleftharpoons E^* + L \rightleftharpoons E^*L$$

In this case, k_{+2} in Eq. (49) is replaced with $k_{+2}l$:

$$\lambda_{1,2} = \frac{-[k_{+1} + k + k_{+2}l + k_{-2}] \pm \sqrt{[k_{+1} + k_{-1} + k_{+2}l + k_{-2}]^2 - 4[k_{+1}k_{-2} + k_{+2}l + k_{-1}k_{-2}]}}{2} \quad (54)$$

If E is mainly in the form without a star in the absence of the ligand, then k_{-1} must be much larger than k_{+1}.

If we also assume that $k_{-1} \gg k_{-2}$, the equation simplifies to:

$$\lambda_{1,2} = \frac{-[k_{-1} + k_{+2}l] \pm \sqrt{[k_{+1} + k_{+2}l]^2 - 4[k_{+2}lk_{+1} + k_{-1}k_{-2}]}}{2} \quad (55)$$

Again, the second term under the square root sign is much smaller than the first, so that:

$$\lambda_{1,2} = [k_{-1} + k_{+2}l] \pm [k_{-1} + k_{+2}l] - \frac{1}{2} \cdot \frac{4[k_{+2}lk_{+1} + k_{-1}k_{-2}]}{[k_{-1} + k_{+2}l]}$$

$$= [k_{-1} + k_{+2}l] \quad \text{(always large and of small amplitude)}$$

or

$$\frac{k_{+2}lk_{+1} + k_{-1}k_{-2}}{k_{-1} + k_{+2}l} = \frac{k_{+1}}{1 + \frac{k_{-1}}{k_{+2}l}} + \frac{k_{-2}}{1 + \frac{k_{+2}l}{k_{-1}}}$$

$$k_{obs} = \frac{k_{+1}}{1 + \frac{k_{-1}}{k_{+2}l}} + \frac{k_{-2}}{1 + \frac{k_{+2}l}{k_{-1}}}$$

The first term tends to 0 as l tends to 0 and to k_{+1} as l tends to ∞. The second term tends to k_{-2} as l tends to 0 and to 0 as l tends to ∞. The corresponding plot is shown in Fig. 4.15.

References

1 BUCHBERGER, A., THEYSSEN, H., SCHRÖDER, H., MCCARTY, J. S., VIRGALLITA, G., MILKEREIT, P., REINSTEIN, J., & BUKAU, B. (1995). Nucleotide-induced conformational Changes in the ATPase and Substrate Binding Domains of the DnaK Chaperone Provide Evidence for Interdomain Communication. *J. Biol. Chem.* **270**, 16903–16910.

2 MCCARTY, J. S., BUCHBERGER, A., REINSTEIN, J., & BUKAU, B. (1995). The Role of ATP in the Functional Cycle of the DnaK Chaperone System. *J. Mol. Biol.* **249**, 126–137.

3 VALE, R. D. & MILLIGAN, R. A. (2000). The way things move: looking under the hood of molecular motor proteins. *Science* **288**, 88–95.

4 BAGSHAW, C. R. (2001). ATP analogues at a glance. *J. Cell Sci.* **114**, 459–460.

5 JAMESON, D. M. & ECCLESTON, J. F. (1997). Fluorescent nucleotide analogues: synthesis and applications. *Methods Enzymol.* **278**, 363–390.

6 CROSS, R. A., JACKSON, A. P., CITI, S., KENDRICK-JONES, J., & BAGSHAW, C. R. (1988). Active site trapping of nucleotide by smooth and non-muscle myosins. *J. Mol. Biol.* **203**, 173–181.

7 KARLISH, S. J., YATES, D. W., & GLYNN, I. M. (1978). Conformational transitions between Na+-bound and K+-bound forms of (Na+ + K+)-ATPase, studied with formycin nucleotides. *Biochim. Biophys. Acta* **525**, 252–264.

8 WHITE, D. C., ZIMMERMAN, R. W., & TRENTHAM, D. R. (1986). The ATPase kinetics of insect fibrillar flight muscle myosin subfragment-1. *J. Muscle Res. Cell Motil.* **7**, 179–192.

9 KLOSTERMEIER, D., SEIDEL, R., & REINSTEIN, J. (1998). Functional Properties of the Molecular Chaperone DnaK from *Thermus thermophilus*. *J. Mol. Biol.* **279**, 841–853.

10 PACKSCHIES, L., THEYSSEN, H., BUCHBERGER, A., BUKAU, B., GOODY, R. S., & REINSTEIN, J. (1997). GrpE Accelerates Nucleotide Exchange of the Molecular Chaperone DnaK with an Associative Displacement Mechanism. *Biochemistry* **36**, 3417–3422.

11 THEYSSEN, H., SCHUSTER, H.-P., PACKSCHIES, L., BUKAU, B., & REINSTEIN, J. (1996). The Second Step of ATP Binding to DnaK Induces Peptide Release. *J. Mol. Biol.* **263**, 657–670.

12 LEONARD, N. J. (1993). Etheno-bridged nucleotides in enzyme reactions and protein binding. *Chemtracts Biochem. Mol. Biol.* **4**, 251–284.

13 YARBROUGH, L. R., SCHLAGECK, J. G., & BAUGHMAN, M. (1979). Synthesis and properties of fluorescent nucleotide substrates for DNA-dependent RNA polymerases. *J. Biol. Chem.* **254**, 12069–12073.

14 WEIKL, T., MUSCHLER, P., RICHTER, K., VEIT, T., REINSTEIN, J., & BUCHNER, J. (2000). C-terminal regions of hsp90 are important for trapping the nucleotide during the ATPase cycle. *J. Mol. Biol.* **303**, 583–592.

15 RUDOLPH, M. G., VEIT, T. J. H., & REINSTEIN, J. (1999). The novel fluorescent CDP-analogue (P_b)MABA-CDP is a specific probe for the NMP binding site of UMP/CMP-kinase. *Protein Science* **8**, 2697–2704.

16 SCHEIDIG, A. J., FRANKEN, S. M., CORRIE, J. E. T., REID, G. P., WITTINGHOFER, A., PAI, E. F., & GOODY, R. S. (1995). X-ray crystal-structure analysis of the catalytic domain of the oncogene product p21(h-ras) complexed with caged gtp and mant dgppnhp. *J. Mol. Biol.* **253**, 132–150.

17 HIRATSUKA, T. & UCHIDA, K. (1973). Preparation and properties of 2'(or

3′)-O-(2,4,6-trinitrophenyl) adenosine 5′-triphosphate, an analogue of adenosine triphosphate. *Biochim. Biophys. Acta* **320**, 635–647.

18 HIRATSUKA, T. (1983). New ribose-modified fluorescent analogues of adenine and guanine nucleotides available as substrates for various enzymes. *Biochim. Biophys. Acta* **734**, 496–508.

19 MOORE, K. J. M. & LOHMAN, T. M. (1994). Kinetic Mechanism of Adenine Nucleotide Binding to and Hydrolysis by the *Escherichia coli* Rep Monomer. 1. Use of Fluorescent Nucleotide Analogues. *Biochemistry* **33**, 14550–14564.

20 WOODWARD, S. K., ECCLESTON, J. F., & GEEVES, M. A. (1991). Kinetics of the interaction of 2′(3′)-O-(N-methyl-anthraniloyl)-ATP with myosin subfragment 1 and actomyosin subfragment 1: characterization of two acto-S1-ADP complexes. *Biochemistry* **30**, 422–430.

21 SCHLEE, S., GROEMPING, Y., HERDE, P., SEIDEL, R., & REINSTEIN, J. (2001). The Chaperone Function of ClpB from *Thermus thermophilus* Depends on Allosteric Interactions of its Two ATP-binding Sites. *J. Mol. Biol.* **306**, 889–899.

22 WATANABE, Y. Y., MOTOHASHI, K., & YOSHIDA, M. (2002). Roles of the Two ATP Binding Sites of ClpB from *Thermus thermophilus*. *J. Biol. Chem.* **277**, 5804–5809.

23 BUJALOWSKI, W. & JEZEWSKA, M. J. (2000). Kinetic mechanism of nucleotide cofactor binding to Escherichia coli replicative helicase DnaB protein. stopped-flow kinetic studies using fluorescent, ribose-, and base-modified nucleotide analogues. *Biochemistry* **39**, 2106–2122.

24 WITT, S. N. & SLEPENKOV, S. V. (1999). Unraveling the Kinetic Mechanism of the 70-kDa Molecular Chaperones Using Fluorescence Spectroscopic Methods. *Journal of Fluorescence* **9**, 281–293.

25 BUCHBERGER, A., REINSTEIN, J., & BUKAU, B. (1999) in *Molecular Chaperones and Folding Catalysts: Regulation, Cellular Functions and Mechanisms* (BUKAU, B., Ed.) pp 573–635, Harwood Academic Publishers, Amsterdam.

26 BUKAU, B. & HORWICH, A. L. (1998). The hsp70 and hsp60 chaperone machines. *Cell* **92**, 351–366.

27 SCHLEE, S. & REINSTEIN, J. (2002). The DnaK/ClpB chaperone system from Thermus thermophilus. *Cellular and Molecular Life Sciences* **59**, 1598–1606.

28 PEARL, L. H. & PRODROMOU, C. (2002). Structure, function, and mechanism of the Hsp90 molecular chaperone. *Adv. Protein Chem.* **59**, 157–186.

29 RICHTER, K. & BUCHNER, J. (2001). Hsp90: chaperoning signal transduction. *J. Cell Physiol* **188**, 281–290.

30 MAYER, M. P. & BUKAU, B. (1999). Molecular chaperones: the busy life of Hsp90. *Curr. Biol.* **9**, R322–R325.

31 BUCHNER, J. (1999). Hsp90 & Co. – a holding for folding. *Trends Biochem. Sci.* **24**, 136–141.

32 PRATT, W. B. & TOFT, D. O. (2003). Regulation of signaling protein function and trafficking by the hsp90/hsp70-based chaperone machinery. *Exp. Biol. Med. (Maywood.)* **228**, 111–133.

33 CSERMELY, P. & KAHN, C. R. (1991). The 90-kDa heat shock protein (hsp-90) possesses an ATP binding site and autophosphorylating activity. *J. Biol. Chem.* **266**, 4943–4950.

34 NADEAU, K., DAS, A., & WALSH, C. T. (1993). Hsp90 chaperonins possess ATPase activity and bind heat shock transcription factors and peptidyl prolyl isomerases. *J. Biol. Chem.* **268**, 1479–1487.

35 NADEAU, K., SULLIVAN, M. A., BRADLEY, M., ENGMAN, D. M., & WALSH, C. T. (1992). 83-kilodalton heat shock proteins of trypanosomes are potent peptide-stimulated ATPases. *Protein Sci.* **1**, 970–979.

36 CSERMELY, P., KAJTAR, J., HOLLOSI, M., JALSOVSZKY, G., HOLLY, S., KHAN, C. R., GERGELY, P., SÖTI, C., MIHALY, K., & SOMOGYI, J. (1993). ATP Induces a Conformational Change of the 90-kDa Heat Shock Protein (hsp90). *J. Biol. Chem.* **268**, 1901–1907.

37 Jakob, U., Scheibel, T., Bose, S., Reinstein, J., & Buchner, J. (1996). Assessment of the ATP Binding Properties of Hsp90. *J. Biol. Chem.* **271**, 10035–10041.

38 Scheibel, T., Neuhofen, S., Weikl, T., Mayr, C., Reinstein, J., Vogel, P. D., & Buchner, J. (1997). ATP-binding Properties of Human Hsp90. *J. Biol. Chem.* **272**, 18608–18613.

39 Bauer, M., Baumann, J., & Trommer, W. E. (1992). ATP binding to bovine serum albumin. *FEBS Lett.* **313**, 288–290.

40 Prodromou, C., Roe, S. M., O'Brien, R., Ladbury, J. E., Piper, P. W., & Pearl, L. H. (1997). Identification and Structural Characterization of the ATP/ADP-Binding Site in the HSP90 Molecular Chaperone. *Cell* **90**, 65–75.

41 Stebbins, C. E., Russo, A. A., Schneider, C., Rosen, N., Hartl, F. U., & Pavletich, N. P. (1997). Crystal structure of an Hsp90-geldanamycin complex: targeting of a protein chaperone by an antitumor agent. *Cell* **89**, 239–250.

42 Buchner, J. (1996). Supervising the fold: functional principles of molecular chaperones. *FASEB Journal* **10**, 10–19.

43 Wegele, H., Muschler, P., Bunck, M., Reinstein, J., & Buchner, J. (2003). Dissection of the contribution of individual domains to the ATPase mechanism of Hsp90. *J. Biol. Chem.* **78**, 39303–39310.

44 Richter, K., Reinstein, J., & Buchner, J. (2002). N-terminal residues regulate the catalytic efficiency of the Hsp90 ATPase cycle. *J. Biol. Chem.* **277**, 44905–44910.

45 Richter, K., Muschler, P., Hainzl, O., Reinstein, J., & Buchner, J. (2003). Sti1 Is a Non-competitive Inhibitor of the Hsp90 ATPase. Binding prevents the N-terminal dimerization reaction during the ATPase cycle. *J. Biol. Chem.* **278**, 10328–10333.

46 Wegele, H., Haslbeck, M., Reinstein, J., & Buchner, J. (2003). Sti1 Is a Novel Activator of the Ssa Proteins. *J. Biol. Chem.* **278**, 25970–25976.

47 Vale, R. D. (2000). AAA proteins. Lords of the ring. *J. Cell Biol.* **150**, F13–F19.

48 Ogura, T. & Wilkinson, A. J. (2001). AAA+ superfamily ATPases: common structure–diverse function. *Genes Cells* **6**, 575–597.

49 Neuwald, A. F., Aravind, L., Spouge, J. L., & Koonin, E. V. (1999). AAA+: A class of chaperone-like ATPases associated with the assembly, operation, and disassembly of protein complexes. *Genome Res.* **9**, 27–43.

50 Parsell, D. A., Kowal, A. S., & Lindquist, S. (1994). *Saccharomyces cerevisiae* Hsp104 Protein: Purification and characterization of ATP-induced structural changes. *J. Biol. Chem.* **269**, 4480–4487.

51 Zolkiewski, M., Kessel, M., Ginsburg, A., & Maurizi, M. R. (1999). Nucleotide-dependent oligomerization of ClpB from *Escherichia coli*. *Protein Sci.* **8**, 1899–1903.

52 Glover, J. R. & Lindquist, S. (1998). Hsp104, Hsp70, and Hsp40: A Novel Chaperone System that Rescues Previously Aggregated Proteins. *Cell* **94**, 73–82.

53 Motohashi, K., Watanabe, Y., Yohda, M., & Yoshida, M. (1999). Heat-inactivated proteins are rescued by the DnaK·J·GrpE set and ClpB chaperones. *Proc. Natl. Acad. Sci. U.S.A.* **96**, 7184–7189.

54 Parsell, D. A., Kowal, A. S., Singer, M. A., & Lindquist, S. (1994). Protein disaggregation mediated by heat-shock protein Hsp104. *Nature* **372**, 475–478.

55 Goloubinoff, P., Mogk, A., Zvi, A. P., Tomoyasu, T., & Bukau, B. (1999). Sequential mechanism of solubilization and refolding of stable protein aggregates by a bichaperone network. *Proc. Natl. Acad. Sci. U.S.A.* **96**, 13732–13737.

56 Mogk, A., Tomoyasu, T., Goloubinoff, P., Rüdiger, S., Röder, D., Langen, H., & Bukau, B. (1999). Identification of thermolabile *Escherichia coli* proteins: prevention

and reversion of aggregation by DnaK and ClpB. *EMBO J.* **18**, 6934–6949.

57 ZOLKIEWSKI, M. (1999). ClpB Cooperates with DnaK, DnaJ, and GrpE in Suppressing Protein Aggregation: A NOVEL MULTI-CHAPERONE SYSTEM FROM *ESCHERICHIA COLI*. *J. Biol. Chem.* **274**, 28083–28086.

58 MOGK, A., SCHLIEKER, C., STRUB, C., RIST, W., WEIBEZAHN, J., & BUKAU, B. (2003). Roles of individual domains and conserved motifs of the AAA+ chaperone ClpB in oligomerization, ATP hydrolysis, and chaperone activity. *J. Biol. Chem.* **278**, 17615–17624.

59 HATTENDORF, D. A. & LINDQUIST, S. L. (2002). Analysis of the AAA sensor-2 motif in the C-terminal ATPase domain of Hsp104 with a site-specific fluorescent probe of nucleotide binding. *Proc. Natl. Acad. Sci. U.S.A.*

60 KIM, K. I., WOO, K. M., SEONG, I. S., LEE, Z.-W., BAEK, S. H., & CHUNG, C. H. (1998). Mutational analysis of the two ATP-binding sites in ClpB, a heat shock protein with protein-activated ATPase activity in *Escherichia coli*. *Biochem. J.* **333**, 671–676.

61 SCHIRMER, E. C., QUEITSCH, C., KOWAL, A. S., PARSELL, D. A., & LINDQUIST, S. (1998). The ATPase activity of Hsp104, effects of environmental conditions and mutations [published erratum appears in *J Biol Chem* 1998 Jul 31; 273(31):19922]. *J. Biol. Chem.* **273**, 15546–15552.

62 KRZEWSKA, J., KONOPA, G., & LIBEREK, K. (2001). Importance of Two ATP-binding Sites for Oligomerization, ATPase Activity and Chaperone Function of Mitochondrial Hsp78 Protein. *J. Mol. Biol.* **314**, 901–910.

63 HATTENDORF, D. A. & LINDQUIST, S. L. (2002). Cooperative kinetics of both Hsp104 ATPase domains and interdomain communication revealed by AAA sensor-1 mutants. *EMBO J.* **21**, 12–21.

64 BARKER, R., TRAYER, I. P., & HILL, R. L. (1974). Nucleoside Phosphates Attached to Agarose. *Methods Enzymol.* **34B**, 479–491.

65 LAMBRECHT, W. & TRAUTSCHOLD, I. (1974) in *Methoden der enzymatischen Analyse II* (Anonymouspp 2151–2160, Verlag Chemie, Weinheim.

66 GAO, B., EMOTO, Y., GREENE, L., & EISENBERG, E. (1993). Nucleotide Binding Properties of Bovine Brain Uncoating ATPase. *J. Biol. Chem.* **268**, 8507–8513.

67 WILBANKS, S. M. & MCKAY, D. B. (1995). How Potassium Affects the Activity of the Molecular Chaperone Hsc70: II. POTASSIUM BINDS SPECIFICALLY IN THE ATPase ACTIVE SITE. *J. Biol. Chem.* **270**, 2251–2257.

68 FEIFEL, B., SANDMEIER, E., SCHÖNFELD, H.-J., & CHRISTEN, P. (1996). Potassium ions and the molecular-chaperone activity of DnaK. *Eur. J. Biochem.* **237**, 318–321.

69 GAO, B., GREENE, L., & EISENBERG, E. (1994). Characterization of Nucleotide-Free Uncoating ATPase and Its Binding to ATP, ADP, and ATP Analogues. *Biochemistry* **33**, 2048–2054.

70 RUSSELL, R., JORDAN, R., & MCMACKEN, R. (1998). Kinetic Characterization of the ATPase Cycle of the DnaK Molecular Chaperone. *Biochemistry* **37**, 596–607.

71 ADAM, H. (1962) in *Methoden der enzymatischen Analyse* (BERGMEYER, H. U., Ed.) pp 573–577, Verlag Chemie, Weinheim.

72 BUCHBERGER, A., VALENCIA, A., MCMACKEN, R., SANDER, C., & BUKAU, B. (1994). The chaperone function of DnaK requires the coupling of ATPase activity with substrate binding through residue E171. *EMBO J.* **13**, 1687–1695.

73 LIBEREK, K., MARSZALEK, J., ANG, D., GEORGOPOULOS, C., & ZYLICZ, M. (1991). *Escherichia coli* DnaJ and GrpE heat shock proteins jointly stimulate ATPase activity of DnaK. *Proc. Natl. Acad. Sci. U.S.A.* **88**, 2874–2878.

74 MCCARTY, J. S. & WALKER, G. C. (1991). DnaK as a thermometer: Threonine-199 is site of autophosphorylation and is critical for ATPase activity. *Proc. Natl. Acad. Sci. U.S.A.* **88**, 9513–9517.

75 O'BRIEN, M. C. & McKAY, D. B. (1995). How Potassium Affects the Activity of the Molecular Chaperone Hsc70: I. POTASSIUM IS REQUIRED FOR OPTIMAL ATPase ACTIVITY. *J. Biol. Chem.* **270**, 2247–2250.

76 CHOCK, S. P., CHOCK, P. B., & EISENBERG, E. (1979). The mechanism of the skeletal muscle myosin ATPase. II. Relationship between the fluorescence enhancement induced by ATP and the initial Pi burst. *J. Biol. Chem.* **254**, 3236–3243.

77 BRUNE, M., HUNTER, J. L., CORRIE, J. E. T., & WEBB, M. R. (1994). Direct, Real-Time Measurement of Rapid Inorganic Phosphate Release Using a Novel Fluorescent Probe and Its Application to Actomyosin Subfragment 1 ATPase. *Biochemistry* **33**, 8262–8271.

78 KARZAI, A. W. & McMACKEN, R. (1996). A bipartite signaling mechanism involved in DnaJ-mediated activation of the *Escherichia coli* DnaK protein. *J. Biol. Chem.* **271**, 11236–11246.

79 LAUFEN, T., MAYER, M. P., BEISEL, C., KLOSTERMEIER, D., MOGK, A., REINSTEIN, J., & BUKAU, B. (1999). Mechanism of regulation of Hsp70 chaperones by DnaJ cochaperones. *Proc. Natl. Acad. Sci. U.S.A* **96**, 5452–5457.

80 RUSSELL, R., KARZAI, A. W., MEHL, A. F., & McMACKEN, R. (1999). DnaJ Dramatically Stimulates ATP Hydrolysis by DnaK: Insight into Targeting of Hsp70 Proteins to Polypeptide Substrates. *Biochemistry* **38**, 4165–4176.

81 GARNIER, C., LAFITTE, D., TSVETKOV, P. O., BARBIER, P., LECLERC-DEVIN, J., MILLOT, J. M., BRIAND, C., MAKAROV, A. A., CATELLI, M. G., & PEYROT, V. (2002). Binding of ATP to heat shock protein 90: evidence for an ATP-binding site in the C-terminal domain. *J. Biol. Chem.* **277**, 12208–12214.

82 GROEMPING, Y., KLOSTERMEIER, D., HERRMANN, C., VEIT, T., SEIDEL, R., & REINSTEIN, J. (2001). Regulation of ATPase and Chaperone Cycle of DnaK from *Thermus thermophilus* by the Nucleotide Exchange Factor GrpE. *J. Mol. Biol.* **305**, 1173–1183.

83 HA, J.-H. & McKAY, D. B. (1994). ATPase Kinetics of Recombinant Bovine 70 kDa Heat Shock Cognate Protein and Its Amino-Terminal ATPase Domain. *Biochemistry* **33**, 14625–14635.

84 EFTINK, M. R. (1997). Fluorescence Methods for Studying Equilibrium Macromolecule-Ligand Interactions. *Methods Enzymol.* **278**, 221–257.

85 BARTHA, B. B., AJTAI, K., TOFT, D. O., & BURGHARDT, T. P. (1998). ATP sensitive tryptophans of hsp90. *Biophys. Chem.* **72**, 313–321.

86 REINSTEIN, J., VETTER, I. R., SCHLICHTING, I., RÖSCH, P., WITTINGHOFER, A., & GOODY, R. S. (1990). Fluorescence and NMR Investigations on the Ligand Binding Properties of Adenylate Kinases. *Biochemistry* **29**, 7440–7450.

87 KLOSTERMEIER, D., SEIDEL, R., & REINSTEIN, J. (1999). The Functional Cycle and Regulation of the *Thermus thermophilus* DnaK Chaperone System. *J. Mol. Biol.* **287**, 511–525.

88 THRALL, S. H., REINSTEIN, J., WÖHRL, B. M., & GOODY, R. S. (1996). Evaluation of human immunodeficiency virus type 1 reverse transcriptase primer tRNA binding by fluorescence spectroscopy: specificity and comparison to primer/template binding. *Biochemistry* **35**, 4609–4618.

89 PRESS, W. H., FLANNERY, B. P., TEUKOLSKY, S. A., & VETTERLING, W. T. (1989) in *Numerical recipes in Pascal; The art of scientific computing*, Cambridge University Press, Cambridge, USA.

90 WANG, Z. X. (1995). An exact mathematical expression for describing competitive binding of two different ligands to a protein molecule. *FEBS Lett.* **360**, 111–114.

91 FABIATO, A. & FABIATO, F. (1987). Multiligand program and constants. *Eur. J. Biochem.* **162**, 357–363.

92 BREHMER, D., RÜDIGER, S., GASSLER, C. S., KLOSTERMEIER, D., PACKSCHIES, L., REINSTEIN, J., MAYER, M. P., &

Bukau, B. (2001). Tuning of chaperone activity of Hsp70 proteins by modulation of nucleotide exchange. *Nat. Struct. Biol.* **8**, 427–432.

93. Steinfeld, J. I., Francisco, J. S., & Hase, W. L. (1989) in *Chemical Kinetics and Dynamics*, Prentice-Hall, Inc., New Jersey 07632.

94. Roberts, D. V. (1977) in *Enzyme kinetics*, Cambridge University Press, Cambridge.

95. Gutfreund, H. (1995) in *Kinetics for the Life Sciences: Receptors, transmitters and catalysts*, University Press, Cambridge.

96. Bernasconi, C. F. (1976) in *Relaxation kinetics*, AcademicPress, New York.

97. Schlee, S., Beinker, P., Akhrymuk, A., & Reinstein, J. (2004). A chaperone network for the resolubilization of protein aggregates: direct interaction of ClpB and DnaK. *J. Mol. Biol.* **336**, 275–285.

98. Wu, X., Gutfreund, H., & Chock, P. B. (1992). Kinetic Method for Differentiating Mechanisms for Ligand Exchange Reactions: Application To Test for Substrate Channeling in Glycolysis. *Biochemistry* **31**, 2123–2128.

99. Johnson, K. A. (2003) in *Kinetic Analysis of Macromolecules*, Oxford University Press.

100. Bagshaw, C. R., Eccleston, J. F., Eckstein, F., Goody, R. S., Gutfreund, H., & Trentham, D. R. (1974). The Magnesium Ion-Dependent Adenosine Triphosphatase of Myosin: Two-Step processes of adenosine triphospahte association and adenosine dophosphate dissociation. *Biochem. J.* **141**, 351–364.

101. Nowak, E., Strzelecka-Golaszewska, H., & Goody, R. S. (1988). Kinetics of Nucleotide and Metal Ion Interaction with G-Actin. *Biochemistry* **27**, 1785–1792.

102. Geeves, M. A., Branson, J. P., & Attwood, P. V. (1995). Kinetics of nucleotide binding to pyruvate carboxylase. *Biochemistry* **34**, 11846–11854.

103. Trentham, D. R., Eccleston, J. F., & Bagshaw, C. R. (1976). Kinetic analysis of ATPase mechanisms. *Q. Rev. Biophys.* **9**, 217–281.

104. Eccleston, J. F. & Trentham, D. R. (1979). Magnesium ion dependent rabbit skeletal muscle myosin guanosine and thioguanosine triphosphatase mechanism and a novel guanosine diphosphatase reaction. *Biochemistry* **18**, 2896–2904.

105. Ha, J.-H. & McKay, D. B. (1995). Kinetics of Nucleotide-Induced Changes in the Tryptophan Fluorescence of the Molecular Chaperone Hsc70 and Its Subfragments Suggest the ATP-Induced Conformational Change Follows Initial ATP Binding. *Biochemistry* **34**, 11635–11644.

106. Ehresmann, B., Imbault, P., & Weil, J. H. (1973). Spectrophotometric Determination of Protein Concentration in Cell Extracts Containing tRNA's and rRNA's. *Anal. Biochem.* **54**, 454–463.

5
Analysis of Chaperone Function in Vitro

Johannes Buchner and Stefan Walter

5.1
Introduction

The intention of this chapter is to provide a manual of how to identify a protein as a molecular chaperone and to carry out the experiments required for its basic characterization.

Originally, the term *molecular chaperone* had been coined by Laskey to describe the role of nucleoplasmin in the association of histones (1, 2). Later, the concept was extended to heat shock proteins by John Ellis (3, 4). He also invented the term "chaperonin" for GroE (Hsp60) and related proteins. Over the years, *molecular chaperone* has acquired additional meanings, some of which represent useful extensions of the original definition, whereas others, such as "chemical chaperones", may cause some confusion. In the first section of this chapter, we will discuss the distinct features of a molecular chaperone. We will then describe the types of reactions that can be facilitated by molecular chaperones in general and present some important functional aspects. The theoretical section concludes with a discussion of the principal assays that can be used to investigate chaperone function in vitro. In the second part of this chapter, we provide selected protocols to measure and characterize chaperone function.

Molecular chaperones are a functionally related family of proteins assisting conformational changes in proteins without becoming part of the final structure of the client protein [5–9]. Originally, the focus of the concept was on the correct association of proteins and the prevention of unspecific aggregation of polypeptide chains during protein folding. Interestingly, all major chaperones belong to the class of heat shock or stress proteins, i.e., they are synthesized in response to stress situations such as elevated temperatures. Under these conditions, proteins tend to aggregate in vivo. This was demonstrated for *E. coli* strains lacking the heat shock response [10]. In these bacteria, elevated temperature caused the accumulation of proteins in large insoluble aggregates, which resemble the so-called inclusion bodies sometimes obtained upon overexpression of recombinant proteins [11]. These experiments showed that under stress there is a strong influence of molecular chaperones on protein homeostasis. However, this seems to hold true also for physiological conditions: many molecular chaperones have homologues that are

Protein Folding Handbook. Part II. Edited by J. Buchner and T. Kiefhaber
Copyright © 2005 WILEY-VCH Verlag GmbH & Co. KGaA, Weinheim
ISBN: 3-527-30784-2

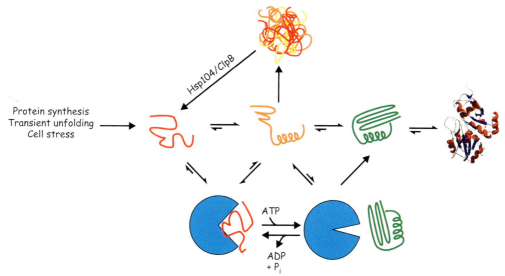

Fig. 5.1. Model for assisted protein folding by molecular chaperones. Several processes generate unfolded polypeptide chains (red), which can fold via partially structured intermediates (orange, green) to the native, biologically active state (right). Depending on the conditions and the nature of the polypeptide, a significant fraction of the molecules may form aggregates (top). The yield of cellular protein folding is significantly improved by the action of general molecular chaperones (bottom, blue). They bind to nonnative polypeptides and thus prevent their aggregation. Further, they may alter the conformation of the bound polypeptide so that once it dissociates, it is less prone to aggregation. Often, release is triggered by a change in the nucleotide state of the chaperone or by ATP hydrolysis. Recently, it was shown that some chaperones of the Hsp100 class such as ClpB and Hsp104 are able to reverse aggregation (top).

constitutively expressed [12], and some of them are essential for viability [13, 14]. This finding suggests that these helper proteins are of general importance for the folding and assembly of newly synthesized polypeptide chains. Furthermore, the observation of strong aggregation of overexpressed proteins under otherwise physiological conditions is consistent with the notion of a balance between protein synthesis/folding and chaperone expression, which has been developed through evolution and can be modulated only within a certain range.

At first glance, the concept outlined above is in conflict with the paradigm of autonomous, spontaneous protein folding established by Anfinsen and many others [15, 16]. The apparent contradiction is that if the acquisition of the unique three-dimensional structure of a protein is governed by its amino acid sequence and the resulting interactions between amino acid side chains, there should be no need for molecular chaperones (Figure 5.1). However, additional factors have to be taken into account. The most important is the possibility of unspecific intermolecular interactions during the folding process of proteins [17, 18]. This reaction depends on the concentration of unfolded or partially folded polypeptide chains and their folding kinetics. Thus, either conditions under which proteins fold slowly or high concentrations of folding proteins will favor aggregation [19]. In consequence, the fraction of molecules reaching the native state will decrease. The basic function of molecular chaperones is to counteract this nonproductive side reaction and thereby

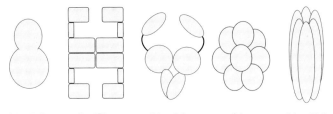

Fig. 5.2. Classes of general molecular chaperones. From left to right: Hsp70 is a monomeric chaperone consisting of an ATPase domain and a peptide-binding domain. The bacterial chaperone GroEL forms two rings comprised of seven subunits each. In the cross-section shown here, two subunits of each ring are depicted. Interactions between the subunits are mediated by the equatorial domains in the center of the molecule, which also contain the ATPase site. The dimeric Hsp90 chaperone consists of three domains. While the C-terminal domains are responsible for dimerization, the N-terminal domains bind and hydrolyze ATP. The small Hsps form large oligomeric structures that contain 24–32 identical subunits. They are the only general chaperones without ATPase function. Chaperones of the Hsp100 class form ring-shaped homohexamers. Each subunit contains two similar, but not identical, ATPase sites.

increase the efficiency of folding (Figure 5.1). Molecular chaperones do not provide steric information for the folding process and thus do not violate the concept of autonomous folding. (The information for folding is encoded solely in the amino acid sequence.) Rather, molecular chaperones assist the spontaneous folding of proteins. Interestingly, the principal possibility of assisted folding had already been hypothesized by Anfinsen in 1963 [20].

It is difficult to assess what fraction of the cellular proteome depends on molecular chaperones to reach its active conformation. But clearly not all newly synthesized proteins need their assistance. While for the chaperone role model GroE (see Chapter 20) it was shown that 40% of the *E. coli* proteins can be influenced in their folding [21], in the cell only about 10–15% may need this assistance under physiological conditions [22, 23]. Similar values were reported for other chaperones [24, 25]. Under conditions where proteins are destabilized (such as heat shock), the requirement for chaperone assistance is clearly increased.

There are only a few classes of general molecular chaperones (Figure 5.2) that have to take care of a large number of different proteins, implying that chaperones have to be promiscuous in their interaction with nonnative proteins, i.e., they must be able to recognize and handle a large variety of different proteins. With the exception of small Hsps, all of them possess an intrinsic ATPase activity (Table 5.1). ATP-dependent structural changes in molecular chaperones are essential for the conformational processing of client proteins. While some basic functional features seem to be conserved between different classes of chaperones, the structure and mode of action are completely different among individual chaperones.

5.2
Basic Functional Principles of Molecular Chaperones

The classification of an enzyme is usually based on the type of reaction it catalyzes and the substrate(s) it turns over. In the field of protein folding and molecular

Tab. 5.1. Properties of general chaperones.

	Hsp60	Hsp70	Hsp90	Hsp100	sHsp
Representatives	GroEL (eubacteria) CCT/TRiC (eukaryotes)	DnaK (E. coli) Ssa1 (S. cerevisiae) Hsp72 (higher eukaryotes)	HtpG (E. coli) Hsp82 (S. cerevisiae) Hsp90 (human)	Hsp104 (S. cerevisiae) ClpB (bacteria) ClpA (bacteria)	Hsp16.5 (M. jannashii) Hsp26 (S. cerevisiae) α-crystallin
Structure: PDB accession number	1AON (GroEL) [40] 1A6D (thermosome) [125]	1DKZ (DnaK peptide-binding domain) [33] 1S3X(Hsp72 ATPase domain) [150]	1AM1 (Hsp82 ATPase domain) [126] 1HK7 (Hsp82 middle domain) [127]	1QVR (ClpB T. thermophilus) [128]	1SHS (Hsp16.5) [129] 1GME (Hsp16.9) [151]
Molecular mass of the monomer (kDa)	57 (GroEL) 57–60 (human CCT-α)	69 (DnaK) 72 (human Hsp72)	81 (Hsp82) 85 (human Hsp90α)	102 (Hsp104) 96 (ClpB) 84 (ClpA)	16.5 (Hsp16.5) 24 (Hsp26) 20 (bovine α-crystallin)
Co-chaperones	GroES (E. coli) Prefoldin (human)	Hsp40 GrpE (E. coli) Hop/Sti1	Hop/Sti1 p23/Sba1 Cdc37	Hsp70/Hsp40 (Hsp104) DnaK/DnaJ/GrpE (ClpB)	Not known
Quaternary structure	Homo 14mer (GroEL) Hetero 16mer (archaea/eukaryotes)	Monomer[1]	Dimer	Dynamic hexamer	Dynamic 24mer (Hsp26) 32mer (α-crystallin)
ATPases per subunit	1	1	1	2[2]	–
ATPase rate per subunit (min^{-1})	3 (GroEL) [67] 1–7 (bovine CCT) [130]	0.02–0.04 (DnaK) [131] 0.04 (Ssa1) [132]	1.0 (Hsp82) [54] 0.05 (Hsp90) [60]	70 (Hsp104) [93] 10–30 (ClpB) [133]	–
T_M (°C)	67 (GroEL) [134]	50 (Ssa1)[3]	62 (Hsp82)[3]	48 (Hsp104)[3]	75 (Hsp26) [135]

[1] Can form oligomers in the absence of ATP.
[2] Some Hsp100s contain only one ATPase/monomer.
[3] Unpublished data from our labs.

chaperones, applying this concept is rather difficult for three reasons. First, though we currently lack a detailed understanding of the mechanisms of protein folding, especially in the context of a living cell, we know it is a rather complex process involving multiple pathways, intermediates, and potential side reactions. As a consequence, molecular chaperones can interfere with and assist protein folding in many different ways, and in most cases our knowledge of how chaperones influence protein folding on a molecular level is very limited. A second problem arises from the heterogeneity of the substrates. As already mentioned, the in vivo substrates of most chaperones are not known in detail. However, it is clear that they include a substantial fraction of the proteome. Finally, it is very difficult to assess which conformational state(s) of these client proteins interact with a certain chaperone. In classic enzymology, reaction intermediates can often be trapped by changing the solvent conditions or temperature. Due to their low intrinsic stability, this is in general not possible for nonnative proteins. Here, a change in solvent conditions will almost always result in a change in the conformational state of the client protein. Under "folding" conditions, the nonnative protein will continue to "react" until a final state (e.g., native structure or aggregate) is reached. Therefore, great care must be taken when the structure of these intermediates is studied. Otherwise, the result will most likely not reflect the conformation of the protein at the time the sample was taken [26–29].

In the following sections, we present key features that many chaperones have in common and that we consider as the functionally most relevant.

5.2.1
Recognition of Nonnative Proteins

The most fascinating trait of a molecular chaperone is its ability to discriminate between the folded and nonnative states of a protein. It would be deleterious for the cell if chaperones would bind to native proteins or fail to recognize the accumulation of the nonnative protein species. Thus, the promiscuous general chaperones have to use a property for recognition and binding that is common to all nonnative proteins. This is the exposure of hydrophobic amino acid side chains, the hallmark of unfolded or partially folded protein structures. In the unfolded state, extended stretches of polypeptide chains containing hydrophobic residues can be recognized by chaperones. This is well documented for proteins of the Hsp70 family [30, 31]. In partially folded proteins, these residues are often clustered in areas that are part of a domain/subunit interface in the folded protein, and they will become buried in subsequent folding/association events. These regions can thus be utilized by molecular chaperones to identify potential client polypeptides, i.e., to distinguish a nonnative conformation from a native protein. The respective binding site of the chaperone usually contains a number of hydrophobic residues that interact with the substrate [32, 33]. The low specificity of hydrophobic interactions is the basis for the substrate promiscuity of molecular chaperones. Since the extended exposure of hydrophobic surfaces in an unfolded or partially folded state is an important determinant for its probability to aggregate, the shielding of these surfaces upon binding to a chaperone is a very efficient means to prevent aggregation.

Besides hydrophobicity, a polypeptide may have to meet additional requirements in order to bind to a chaperone. It was shown for GroEL that negative charges on the substrate significantly weaken interaction with the chaperone due to electrostatic repulsion [34]. In the case of DnaK (Hsp70), the preferred substrate is a stretch of hydrophobic residues flanked by basic amino acids [35, 36]. Another important aspect that can influence the affinity of a polypeptide towards a chaperone is its conformational plasticity. Due to their promiscuity, chaperones cannot provide an optimal binding surface for each of their client proteins. Consequently, the polypeptide, and to some extent the chaperone itself, may have to adjust its conformation to establish tight binding [37]. Because this process requires a certain degree of flexibility, it may serve as a second checkpoint to discriminate between native and nonnative conformations.

An important consequence of the unspecific recognition of exposed hydrophobic residues is that chaperones will not bind just one specific nonnative conformation of a protein. In contrast, it has to be assumed that all conformations exhibiting a certain degree of hydrophobicity will be recognized. In the case of GroE, the conformations that are stably bound range from completely unfolded to a native-like dimer [38, 39].

The sites of interaction with nonnative proteins are known for Hsp70 and GroEL. In the crystal structure of the protein-binding domain of DnaK (Hsp70) in complex with a short peptide [33], the peptide adopts an extended conformation and the hydrophobic side chains of the peptide make contact with the binding groove, but there are also polar interactions between Hsp70 and the peptide backbone. In the case of GroEL, the analysis of numerous point mutations together with the crystal structure indicated that the nonnative protein binds to hydrophobic residues of the apical domains, which are located on the top of the inner rim of the GroEL cylinder [32, 40]. This conclusion was confirmed several years later by the X-ray structure of a GroEL-peptide complex [41, 42]. Detailed descriptions of the substrate-binding sites and their properties can be found in the chapters dedicated to the respective molecular chaperones.

A notable exception from the theme of hydrophobic binding motifs is chaperones involved in the unfolding of proteins. Many of their client proteins adopt a fully folded conformation and thus do not expose hydrophobic surfaces. At the same time, an unambiguous signal is required to tell the chaperone which protein should be unfolded for subsequent degradation. Evolution has solved this problem by attaching specific tags, such as ubiquitin [43, 44] or the ssrA peptide [45], to these proteins, which serve as recognition sites for the respective degradation systems [46]. Little is known about the structure/sequence motifs recognized by chaperones involved in disaggregation, such as ClpB or Hsp104, as it has proved difficult to obtain stable chaperone-substrate complexes in these cases.

5.2.2
Induction of Conformational Changes in the Substrate

Prevention of aggregation by (transient) binding of partially folded proteins clearly is an important part of chaperone function in vivo. But in a number of cases, this "buffering capacity" appears not to be the primary task. This is especially true for the ATP-consuming chaperones. An increasing amount of evidence suggests that the energy provided by ATP hydrolysis is used to induce conformational changes in the bound polypeptide. For the ClpA chaperone, Horwich and coworkers demonstrated that it can actively unfold ssrA-tagged, native GFP [47]. In this case, ATP hydrolysis is required to overcome the free energy of folding. In the case of the Hsp90 system, an immature conformation of a bound steroid hormone receptor (a client protein) is converted into a conformation that is capable of binding the hormone ligand [48, 49]. ATP binding and hydrolysis are essential for this process, as they seem to control the association of the chaperone with its various cofactors [50] (see also Chapter 23). During the whole sequence of reactions, the receptor is assumed to remain bound to the Hsp90 chaperone.

For the general chaperones, we know that conformational changes occur in the chaperone upon both ATP binding and hydrolysis [51–54]. However, much less is known about how and to what extent these changes result in conformational rearrangements in substrate proteins associated with the chaperone. As mentioned above, this is mainly because the species involved are often only transiently populated and very labile and thus are not easily susceptible to methods of structure determination. Also, the bound target protein may represent an ensemble of conformations [39].

Two mechanisms can be envisaged as to how a chaperone causes a structural change in a client protein. In both cases, a conformational change in the chaperone is induced by the binding of a ligand (ATP or a cofactor) or by ATP hydrolysis. This in turn results in a change in the nature/geometry of the substrate-binding site that weakens the interaction with the bound polypeptide. As the system ligand/chaperone tends to assume a state of minimum free energy, the substrate has two choices: (1) it may undergo structural rearrangements to adopt a conformation that has a higher affinity towards the binding site or (2) it may start to fold. Both processes will lower the free energy of the system, but in the first case the energy drop primarily results from interactions between chaperone and substrate, whereas in the second case it results solely from interactions within the client polypeptide. Although these rearrangements are driven by thermodynamics, kinetics may be important as well in defining whether binding or folding dominates for a given polypeptide conformation [55–57].

An alternative to this "passive" model is that the conformational change in the chaperone generates a "power stroke" that is used to perform work on the substrate, e.g., by moving apart two sections of the bound protein. If the substrate is a single polypeptide, this could result in a forced unfolding reaction [58]. If the substrate is an oligomeric species, e.g., a small aggregate, the power stroke may cause its dissociation into individual molecules. This "active" mechanism requires that the substrate remains associated with the chaperone during the conformational switch, which is not required in the first model. After the power stroke, the substrate will probably undergo a series of conformational relaxation reactions.

For both models, the concept of general chaperones requires that the conformational change in the substrate be encoded by its amino acid sequence rather than being forced onto the substrate by the chaperone. The chaperone merely enables the polypeptide to use its inherent folding information.

5.2.3
Energy Consumption and Regulation of Chaperone Function

Most of the general chaperones require ATP to carry out their task (Table 5.1). Examples of ATP-hydrolyzing chaperones are GroEL; the Hsp104/ClpB proteins, which participate in the dissociation of protein aggregates; Hsp90; and Hp70, which is involved in a large variety of reactions ranging from the general folding of nascent polypeptide chains to specific functions. As stated above, the cycle of ATP binding and hydrolysis drives an iterative sequence of conformational changes in the chaperone, which may induce structural rearrangements in a bound polypeptide or trigger binding/release of client proteins and co-chaperones. In contrast to the conformational changes that occur within both the chaperone and a bound substrate protein, the ATPase activity of molecular chaperones is experimentally very easily accessible. Because of its crucial functional importance, ATP turnover is usually modulated by both the substrate and helper chaperones.

Stimulation of ATP hydrolysis in the presence of unfolded proteins or peptides has been observed, e.g., for Hsp70, Hsp90, and Hsp104 [30, 59–61]. Apparently, the binding of the substrate induces a conformational change in these chaperones that results in a state with a higher intrinsic ATPase activity. This "activation on demand" may serve as a means to reduce the energy consumption by the empty chaperones. In contrast, ATP hydrolysis by GroEL is only marginally affected by the presence of protein substrates.

A well-characterized example of the regulation of ATP hydrolysis by co-chaperones is the DnaK system of *E. coli* (see Chapter 15). The ATPase cycle, and thus the cycle of substrate binding and release, is governed by two potential rate-limiting steps: the hydrolysis reaction and the subsequent exchange of ADP by ATP [62]. Both steps can be accelerated by specific cofactors, namely DnaJ and GrpE [59, 63, 64]. The rates of nucleotide exchange and hydrolysis determine how fast a substrate is processed and how long it is on average associated with the ADP and ATP states of DnaK, respectively.

While Hsp70 is subject mainly to positive regulation, other chaperone systems appear to include negative regulators as well. Hsp90, which plays an essential role in the activation of many regulatory proteins such as hormone receptors and kinases, is controlled by a large number of cofactors (see Chapter 23). Two of these co-chaperones, p23 and Hop/Sti1, were found to decrease the ATPase activity by stabilizing specific conformational states of Hsp90 [50, 65]. Intriguingly, Hop/Sti1 from *S. cerevisiae* seems to have a dual regulatory function: besides its inhibitory effect on Hsp90, Sti1 was also shown to be a very potent activator of Hsp70 [66]. Another example of a negative regulator is GroES, which reduces the ATP hydrolysis of its partner protein GroEL to 10–50% depending on the conditions used [67, 68]. The role of GroES in the functional cycle of the GroE chaperone, however, goes far beyond its modulation of ATP turnover, as it is essential for the encapsulation of a polypeptide in the folding compartment of the GroE particle (see Chapter 20). The example of GroES is instructive, as it demonstrates that regulation of ATP turnover may not be the most important function of a cofactor. Nevertheless, the combination of experimental simplicity and functional relevance makes the ATPase of a general chaperone a preferred choice for mechanistic studies.

Some chaperones are regulated by temperature. Hsp26, a small heat shock protein from *S. cerevisiae*, has no chaperone activity at 30 °C, the physiologic temperature for yeast. At this temperature, Hsp26 forms hollow spherical particles consisting of 24 subunits [69]. Under heat stress, e.g., at 42 °C, these large complexes dissociate into dimers that exhibit chaperone activity and bind to nonnative polypeptides. Another interesting example of temperature regulation is the hexameric chaperone/protease degP of the *E. coli* periplasm [70]. At low temperature the protein acts as a chaperone, whereas at elevated temperature it possesses proteolytic activity (see Chapter 28). The regulatory switch is thought to involve the reorientation of inhibitory domains, which block access to the active site of the protease at low temperature [71]. Interestingly, switching can also be achieved by the addition of certain peptides.

A further variation of the scheme of activating chaperones by stress conditions is the regulation of Hsp33. Here, oxidative stress leads to the formation of disulfide bonds within the chaperone and its subsequent dimerization, which is a prerequisite for efficient substrate binding [72] (see also Chapter 19).

5.3
Limits and Extensions of the Chaperone Concept

The term chaperone and the underlying concept are attractive. Therefore, it is not surprising that the term itself has been proved to be rather promiscuous, as it is increasingly used for different proteins and even non-proteinaceous molecules. In its most basic interpretation, it has been applied to all proteins that have an effect on the aggregation of another protein or that influence the yield of correct folding. Most of these proteins have hydrophobic patches on their surface and thus, irrespective of their biological function, will associate with nonnative proteins. However, to qualify for a molecular chaperone, additional requirements should be met. This includes the defined stoichiometry of the binding reaction and the regulated release of the bound polypeptide (e.g., triggered by binding of ATP).

An extension of the term chaperone are the so-called intramolecular chaperones. These are regions of proteins that are required for their productive folding without being part of the mature protein because they are cleaved off after folding is complete. Well-investigated examples include the pro-regions of subtilisin E and α-lytic protease [73, 74]. The variation of the chaperone concept is that an intramolecular chaperone is specific for one protein and is covalently attached to its target. Furthermore, these chaperones act only once: after completing their job, they are cleaved off and degraded. These properties do not agree with the basic definition of a chaperone.

Chemical chaperone is a term that has become fashionable in recent years. It refers to small molecules (not proteins or peptides) that are added to refolding reactions to increase the efficiency of protein folding in vitro. For instance, the folding yields for many enzymes can be improved by adding a substrate to the refolding buffer [75–77]. These compounds are thought to stabilize critical folding intermediates by binding to the active site, thereby enhancing folding. Besides these rather specific ligands, a number of substances, the most prominent of them being Tris and arginine, were found to promote the yield of folding in many cases when present in large excess over the folding proteins (see Chapter 38). In contrast to molecular chaperones, both chemical chaperones and intramolecular chaperones cannot be regulated in their activity.

5.3.1
Co-chaperones

For many chaperones, cofactors/co-chaperones exist that seem to directly influence the function of the chaperone. Some of them are essential constituents of their respective chaperone system, while others only improve its efficiency. Often, the role of cofactors is not limited to the regulation of the ATPase function but involves the recognition of nonnative proteins. These proteins thus have features of a chaperone. The DnaJ/Hsp40 co-chaperones seem to bind unfolded proteins and transfer them to Hsp70 [62, 78]. At the same time, as mentioned above, binding of DnaJ/Hsp40 stimulates ATP hydrolysis by Hsp70. Far more complex is the situation for the Hsp90 system. Here a plethora of cofactors exist, many of which seem to be able to recognize and bind nonnative proteins as demonstrated by in vitro chaperone assays [79, 80]. There is good reason to assume that they exert this function also within the Hsp90 chaperone complex, i.e., they interact with the substrate when it is associated with Hsp90.

The term co-chaperone implies that the function of these proteins is restricted to the context of their respective chaperones. However, recent evidence suggests that this may not be true in all cases. One example is the Hsp90 co-chaperone p23, which clearly has a distinct function in the Hsp90 system but also seems to be able to disassemble transcription factor complexes bound to DNA [81]. Similarly, Cdc37 was reported to exhibit an autonomous chaperone function [82].

It is important to remember that while in vitro chaperone assays can be used to detect chaperone properties in a cofactor, their functional relevance in the context of a chaperone complex has to be addressed in subsequent experiments.

5.3.2
Specific Chaperones

In addition to the general chaperones, a number of proteins exist for which the basic definition of chaperone function applies. However, these chaperones are dedicated to one or a few target proteins. In these cases, recognition is highly specific and does not necessarily involve hydrophobic interactions. An example for this class of chaperones is Hsp47, a resident protein of the endoplasmic reticulum that is a dedicated folding factor for collagen [83]. It interacts with immature collagen, is required for its correct folding and association, and is not part of the final collagen trimer. Thus, all the criteria for a molecular chaperone are fulfilled. Collagen release is not triggered by ATP hydrolysis but by changes in the pH upon transit from the ER to the Golgi [84, 85].

Another well-studied example is the pili assembly proteins of gram-negative bacteria (see Chapter 29). Here, unassembled pili subunits are bound by specific chaperones that prevent their premature association in the periplasmic space. These chaperones bind very specifically and can be considered as surrogate subunits.

5.4
Working with Molecular Chaperones

5.4.1
Natural versus Artificial Substrate Proteins

An important point in studying molecular chaperones is the choice of substrate protein. Section 5.6 of this chapter will describe several of the most commonly used chaperone substrate proteins in detail. For many of them it is clear that they do not represent authentic in vivo substrates for the chaperone studied. The question is whether this poses a limitation on the validity of the results obtained with these proteins.

For the promiscuous general chaperones and co-chaperones, the interaction with the substrate protein is relatively unspecific. As a consequence, any unfolded protein can be regarded as a bona fide substrate. Interestingly, almost all mechanistic studies on GroE have been performed with "artificial" substrate proteins. Using a standard set of model substrates has a number of important advantages: (1) they are commercially available at a fairly low cost, (2) they are very well characterized in their folding and aggregation properties, and (3) they allow a direct comparison among chaperones to assess functional differences (Table 5.2). Moreover, they likely reflect the spectrum of polypeptides that chaperones may encounter in the cell. The situation is different for chaperones with high substrate specificity. Here, assays with model substrate proteins may allow analysis of certain aspects of their mechanisms. However, it is desirable to use authentic substrate proteins whenever possible.

5.4.2
Stability of Chaperones

There seems to be a common misconception that chaperones are exceptionally stable proteins because they are active under conditions, such as heat shock, that lead to the unfolding of many proteins. However, it has to be considered that the heat shock response is able to protect organisms only a few degrees above their optimum growth temperature. In general, there is no selection pressure on any intracellular protein to be stable above the maximum temperature tolerated by the host. As outlined above, many chaperones are functionally important under both physiological and stress conditions, i.e., they have to be active over a broad temperature range. From a mechanistic point of view, flexibility and conformational changes are of exceptional importance for molecular chaperones since they have to interact

Tab. 5.2. Protein substrates for molecular chaperones.

	Organism	Molecular mass (kDa)	$A_{280\ nm}$ (1 mg mL^{-1}, 1 cm)	Commercially available	Activity assay available	General handling	Remarks	References
Alpha-glucosidase	S. cerevisiae	68	2.61	Yes	Yes	Good		115, 136
Alpha-lactalbumin	Bos taurus	14	1.7	Yes	Yes, but awkward	Very good	Permanently unfolded derivative (RCMLA)	137, 138
Citrate synthase	Sus scrofa	2 × 49	1.56	Yes	Yes (expensive)	Good	Model substrate	87, 107, 139
Luciferase	Photinus pyralis	60.7	0.59	Yes	Yes (expensive)	Very poor	Very sensitive	140–143
Insulin	Bos taurus	5.8	1.06	Yes	No	Good	Useful only for aggregation assays	114, 115, 144
Malate dehydrogenase	Sus scrofa	2 × 33	0.2	Yes	Yes	Good	Model substrate	99, 145
Maltose binding protein	E. coli	40.7	1.6	No	No (but change in Trp fluorescence upon folding)	Good	Spontaneous folding is decelerated by chaperones	55, 146
Rhodanese	Bos taurus	33	1.8	Yes	Yes (toxic)	Good	Model substrate	56, 97, 147
Rubisco	R. rubrum	2 × 50.5	1.11	No	Yes, but uses $^{14}CO_2$	Good	First substrate ever used for in vitro assays	139, 148, 149

with a large variety of substrate proteins. For the reasons given, it is therefore plausible that chaperones cannot be extremely stable and rigid proteins (Table 5.1). This imposes certain restrictions on the design of in vitro chaperone assays (see experimental Section).

5.5
Assays to Assess and Characterize Chaperone Function

The first in vitro assays for a molecular chaperone using purified components were presented by Lorimer and coworkers in 1989 [86]. They analyzed and dissected the function of the two components of the GroE system (GroEL and GroES) using unfolded bacterial Rubisco as a substrate protein. The methods employed included activity assays to monitor the refolding to the native state, CD spectroscopy to analyze the conformation of the nonnative substrate protein at the beginning of the reaction, and native PAGE to demonstrate a direct physical interaction between the chaperone and Rubisco. This landmark paper became a role model for numerous studies to follow. Among other things, it showed that artificially denatured proteins can serve as substrates for molecular chaperones and that they can be refolded in a strictly ATP-dependent manner. In many cases, the increase in the yield of refolded protein is due to the suppression of aggregation during refolding. This was first shown for GroE using light scattering [87].

5.5.1
Generating Nonnative Conformations of Proteins

To produce the nonnative conformation of a substrate protein, all the conditions and substances known to denature proteins can be employed. Chaotropic compounds such as guanidinium chloride (GdmCl) and urea are used if a completely unfolded state of the substrate protein is desirable. This is referred to as "chemical unfolding." In these cases, the substrate protein is diluted from a high concentration of denaturant into a solution containing the chaperone. Note that for most chaperone substrates there is competition between binding to the chaperone, folding, and aggregation. Alternatively, the test protein can be unfolded at elevated temperature. In this case, the unfolded protein cannot be stored because aggregation starts shortly after the temperature increase. In assays that employ thermally unfolded model substrates, the chaperone is therefore added before the heating step. Unfolding of proteins can also be achieved at low pH (acid denaturation) or, in the case of disulfide-containing polypeptides, by reduction.

5.5.2
Aggregation Assays

Since the prevention of aggregation is one of the key biological functions of general chaperones, this test can be considered an "activity assay." In the simplest case, a temperature-labile protein, such as citrate synthase, is incubated at a temperature at which it starts to unfold. The subsequent aggregation process can be monitored by the extent of light scattering shown by the sample, preferably in a fluorescence spectrometer or a regular UV/VIS photometer. In the presence of a chaperone, a decrease in the scattering signal at the end of the reaction should be observed. When an excess of chaperone is used, aggregation may be suppressed completely.

However, one should be cautious not to over-interpret results obtained from aggregation experiments. The intrinsic problem with aggregation is that, as a stochastic process, it is of limited reproducibility. Moreover, it is notoriously sensitive to small changes in conditions. Therefore, it is essential to include appropriate controls. Further, it yields only qualitative data: during the time course of aggregation, there is a concomitant increase in the number of aggregates as well as in the size of the aggregates. The contribution of the two processes will be different at different time points during the kinetics. Lastly, some proteins may suppress the aggregation of a test substrate without being chaperones.

5.5.3
Detection of Complexes Between Chaperone and Substrate

Many chaperones form remarkably stable complexes with unfolded polypeptides [48, 88, 89]. As described above, complex formation is due mainly to hydrophobic interactions between the binding site on the molecular chaperone and hydrophobic patches on the surface of the substrate protein.

To obtain complexes of chaperones with nonnative proteins, a concentrated solution of a chemically unfolded test protein is added to the chaperone. Alternatively, a mixture of a thermolabile protein and a chaperone is incubated at an elevated temperature. Depending on their stability, the complexes can be identified by several methods. Usually, unbound substrate is first separated from the complexes by size-exclusion chromatography, native PAGE, centrifugation, or other methods. The presence of the test substrate in the purified complexes is then shown on SDS-PAGE gels or by immunological methods if an antibody against the substrate is available. Alternatively, the test protein can be tagged for detection with, e.g., biotin or labeled with a fluorescent marker. Labile chaperone-substrate complexes must be trapped by either chemical- or photo-cross-linking before further analyses can be carried out. An extension of this type of experiment is the identification of complexes between a chaperone and its cofactors [90] (see also Chapter 23).

5.5.4
Refolding of Denatured Substrates

This type of assay is more complex than the ones described above because (1) it requires some knowledge about the mechanism of substrate release and (2) successful refolding may depend on the presence of (co-)chaperones or other factors. The most convenient refolding assays use test substrates whose enzymatic activity can be measured without interference from the chaperone. The use of classical spectroscopic probes for folding such as fluorescence or far-UV CD may be difficult because of the high background signal from the chaperone and (if present) ATP. However, some chaperones such as GroEL, GroES, and Hsp104 lack intrinsic tryptophans, and in these cases one may use fluorescence to directly monitor changes in the structure of Trp-containing test proteins.

The general procedure for a refolding assay is as follows. First, a complex between the chaperone and the test substrate is formed (see Section 5.5.3). Usually, it is not necessary to remove molecules that have not bound to the chaperone. Afterwards, substrate release and refolding are triggered by adding the appropriate components, e.g., GroES and ATP in the case of GroEL. At various time points, aliquots are withdrawn from the sample and the enzyme activity is determined.

5.5.5
ATPase Activity and Effect of Substrate and Cofactors

Many general chaperones have an intrinsic ATPase activity, though their rates of hydrolysis can differ significantly (Table 5.1). ATP hydrolysis by a chaperone can be measured using several assay systems (see Section 5.6). Once the basal ATPase activity has been determined, the effect of substrates and cofactors in various combinations can be tested. Concerning the substrate, a polypeptide is preferred that adopts a stable, nonnative conformation. Otherwise, its conformation and thus its effect on ATP hydrolysis may change during the experiment. Alternatively, a small peptide may be used.

A more extensive characterization of the ATPase may include the determination of enzymatic parameters such as K_M or k_{cat}, in both the absence and the presence of effectors (see Chapter 4). However, one should be aware that in the case of oligomeric chaperones, a detailed analysis can be very challenging due to homotropic and heterotropic allosteric effects [91].

5.6
Experimental Protocols

5.6.1
General Considerations

5.6.1.1 Analysis of Chaperone Stability
When setting out to analyze the function of molecular chaperones, it is highly recommended to determine the intrinsic stability of the chaperone as a first step. This enables the experimenter to choose the appropriate reaction conditions for the chaperone assays. The conformational stability against urea or GdmCl should be measured using the protocols provided in the first volume of this book (see Chapter 3 in Part I). In the case of ATP-dependent chaperones, the GdmCl- or urea-induced unfolding transitions may be recorded in the presence and absence of ATP to determine whether the ligand has an effect on protein stability. In addition to the conventional assays for determining protein stability, for molecular

chaperones it is useful to specifically determine the stability against thermal unfolding because assays at heat shock temperatures are convenient. CD spectroscopy is frequently used to monitor changes in the secondary structure with increasing temperature.

5.6.1.2 Generation of Nonnative Proteins

Because most substrates are stable proteins at ambient temperature, they have to be unfolded prior to complex formation. This can be achieved by chemical denaturation, thermal unfolding, or – in the case of some disulfide-containing proteins – reduction. Each of these methods has its own merits but may not be applicable for a certain combination of substrate and chaperone. Chemical unfolding is a very reproducible method to generate unfolded protein substrates, and the product can usually be stored for several hours. However, one has to be aware that urea slowly decomposes into cyanate, which is reactive and can covalently modify amino acid side chains [92]. Further, it has been shown for both Hsp104 and GroE that GdmCl can severely interfere with chaperone function [93, 94]. To keep the residual concentration of denaturant in the chaperone assay below 0.1 M (urea) and 0.05 M (GdmCl), respectively, one should use fairly concentrated stock solutions of unfolded polypeptide.

Unfolding substrates at elevated temperature mimics conditions of heat shock and thus could be considered the most physiologic method to generate unfolded polypeptides. Also, there are no problems arising from the addition of problematic chemicals. The main disadvantage is that thermally unfolded proteins cannot be stored, as they will start to aggregate quickly. Thus, the thermal inactivation has to take place in the presence of chaperone. This implies that the chaperone itself has to be stable under the conditions used to unfold the substrate. Also, the reproducibility is not as high as when chemical denaturation is used. In the case of proteins that contain disulfide bonds, these can be broken by the addition of reducing agents such as glutathione or DTT. This may destabilize the protein enough so that it will unfold either spontaneously or at mildly denaturing conditions. As for the previous procedures, it should be verified that the chaperone is not compromised by the reducing agent. While reducing conditions may not be a problem for cytosolic chaperones, proteins operating in a more oxidative environment such as the ER or the periplasm may not tolerate low redox potentials.

5.6.1.3 Model Substrates for Chaperone Assays

Numerous proteins have been used as substrates to investigate assisted protein folding. Due to features such as availability, ease of handling and detection, and general suitability as chaperone substrates, some of these proteins were repeatedly used and eventually became model substrates. In the following sections, we will briefly describe the most important ones. Additionally, Table 5.2 includes a number of less commonly used substrates along with selected references.

Rhodanese (Bovine Liver) Rhodanese from liver mitochondria is a small (33 kDa) monomeric enzyme that catalyzes sulfur transfer reactions [95]. It is commercially available from several sources, but batches may differ in the degree of purity. The structure of rhodanese has been determined [96]. The reactivation of rhodanese is dependent on chaperones [97], although spontaneous refolding has been observed in the presence of lauryl maltoside, β-mercaptoethanol, and thiosulfate [75]. Rhodanese has been extensively used over the years to study assisted protein folding. One disadvantage of rhodanese is that the activity assay uses the toxic compound KCN.

Mitochondrial Malate Dehydrogenase MDH (Porcine Heart) Malate dehydrogenase is an enzyme of the mitochondrial Krebs cycle. The enzyme from *Sus scrofa* consists of two subunits of $\sim$33 kDa each and is commercially available. Its reactivation is highly dependent on the assistance of chaperones, although spontaneous refolding occurs in the presence of high phosphate concentrations [98, 99]. The crystal structure of MDH has been solved [100]. This model substrate was extensively used to investigate the mechanism of GroEL [29, 101]. MDH can be unfolded with both urea and GdmCl. Moreover, it starts to unfold (and aggregate) at $\sim$45 °C in Tris buffer at pH 8 [102]. MDH has been shown to form stable complexes with a number of chaperones, including sHsps (Hsp18.1) and GroEL. In addition, aggregated MDH has been used to study the disaggregation properties of the chaperone ClpB [103]. Finally, the regaining of MDH activity during refolding is easy to monitor (see Chapter 20).

Mitochondrial Citrate Synthase CS (Porcine Heart) Citrate synthase, a mitochondrial enzyme like MDH, is a homodimeric protein of two 49-kDa subunits. It is commercially available, and its crystal structure is solved at a resolution of 2.7 Å [104]. Spontaneous reactivation after denaturation by urea or GdmCl is possible only at low protein concentrations [87]. Folding of CS can be assisted by GroE in vitro and in *E. coli* [87, 105]. A convenient enzymatic assay exists that can be used to monitor the regaining of activity [106, 107].

CS is temperature-sensitive and starts to unfold above $\sim$40 °C. Because of its low stability, CS is the substrate of choice to measure temperature-induced aggregation and its suppression by molecular chaperones. Thermal aggregation is preceded by the loss of enzymatic activity. Several molecular chaperones have been tested for their influence on the thermal aggregation and inactivation of CS [87, 107–109].

Reduced Carboxymethylated Alpha-lactalbumin RCMLA (Bovine Milk) The disadvantage of all substrates mentioned so far is that once the unfolded protein is added to the chaperone solution, its conformation is no longer defined. As a consequence, these types of substrates are not suited to carry out equilibrium studies, e.g., to determine binding constants. For these purposes, one requires a permanently unfolded polypeptide that does not aggregate. Reduced, carboxymethylated alpha-lactalbumin has been successfully employed in this context [110, 111]. It was also used as a competitive inhibitor for the binding of other nonnative proteins [112]. Its only drawback is that it is not very hydrophobic, and the resulting chaperone complexes are often rather labile.

Alpha-lactalbumin is a small protein (14.2 kDa) from bovine milk and can be purchased from a number of suppliers. Its conformational stability is highly dependent on the presence of the four disulfide bridges. To obtain a permanently unfolded version of lactalbumin, the disulfides first have to be disrupted by DTT in the presence of GdmCl (Protocol 1). This unfolded conformation can be maintained by irreversible alkylation of the eight cysteine residues with iodoacetic acid. Compared with unmodified lactalbumin, this derivative carries eight additional negative charges that increase its solubility but may interfere with the binding to certain chaperones. An alternative form can be obtained by using iodoacetamide in the alkylation step. The resulting derivative carries no additional charges but is prone to aggregation at high concentration.

Protocol 1: Preparation of RCMLA

1. Dissolve 5.6 mg alpha-lactalbumin in 305 µL 200 mM Tris/HCl pH 8.7, 7 M GdmCl, 2 mM EDTA, 20 mM DTT, and incubate for 90 min at room temperature.
2. Add 60 µL of 0.6 M iodoacetic acid in 200 mM Tris/HCl pH 7.5 (check pH!) under dark room conditions and incubate for 10 min.
3. Stop the reaction by adding 100 µL of 0.5 M reduced glutathione in 200 mM Tris/HCl pH 7.5.
4. The derivatized protein is purified via a Sephadex G25 desalting column equilibrated with 200 mM Tris/HCl pH 7.5.

5.6.2
Suppression of Aggregation

The simplest type of activity test for a chaperone determines its capability to prevent the aggregation of a model substrate. An unfolded conformation of the test substrate is first prepared. Then the aggregation of this protein is measured, in both the presence and the absence of the potential chaperone. A number of controls have to be carried out to verify that the chaperone does not aggregate itself and that the buffer in which the chaperone is stored does not change aggregation. The process of aggregation is usually monitored by the increase in light scattering. For detection, a thermostated photometer or fluorescence spectrometer is required. Aggregates can be monitored at wavelengths between 320 nm and 500 nm (no interference from aromatic residues) depending on the size of the aggregates. The scattering intensity increases with decreasing wavelengths. Since there is no linear relationship between the extent of aggregation and the light-scattering signal, we advise against using a percent scale when plotting aggregation data.

Protocol 2, in which the aggregation of CS at 43 °C is measured, can be performed only for chaperones that are stable at this temperature. If this is not the case, chemical denaturants have to be used to destabilize the protein whose aggregation will be monitored. Two different approaches can be applied. In the cases of insulin (Protocol 3) and alpha-lactalbumin (Protocol 4), aggregation can be induced by the addition of a reducing agent such as DTT. The B chain of reduced insulin will start to aggregate within minutes at room temperature [113]. For lactalbumin, efficient aggregation requires reduction at higher temperatures. Additionally, aggregation can be induced by diluting a concentrated solution of an unfolded test protein, e.g., in GdmCl or urea, into buffer without denaturant. In the absence of chaperones, the polypeptide will aggregate. This process can be enhanced by increasing the temperature.

As a rule of thumb, complete suppression of aggregation requires that the amount of chaperone – or, more precisely, the concentration of protein-binding sites – be at least as high as that of the aggregating polypeptide. Thus, chaperones act stoichiometrically and not catalytically. As some of the assays, especially the ones using insulin and alpha-lactalbumin, require a high concentration of the test protein, the consumption of chaperone can be considerable. Secondly, the interaction between the chaperone and the polypeptide has to be tight (see Figure 5.3, left panel). Otherwise, the irreversible process of aggregation will eventually pull all polypeptide molecules from the chaperone. In this case, a deceleration of the aggregation process may be observed in the presence of the chaperone, while the final amplitude remains unchanged (Figure 5.3, middle panel).

It may be worthwhile to check whether the test protein is indeed associated with the chaperone at the end of the aggregation assay by employing one of the methods described in the following section. If a refolding assay for the test protein is available, one can also examine whether the protein can be recovered from the chaperone complexes by transferring the sample to refolding conditions, e.g., by lowering the temperature (see Protocol 7). Finally, one should note that a positive result from an aggregation assay alone is not sufficient to classify the protein under investigation as a molecular chaperone.

Protocol 2: Thermal Aggregation of Citrate Synthase The major advantage of this assay is that it requires only small amounts of chaperone. However, the chaperone must withstand a temperature of 43 °C for at least 1 h (Figure 5.3). Tris and a number of salts such as KCl have been observed to stabilize CS and should be avoided.

1. Prepare a stock solution of 65 µM CS (monomers) in 50 mM Tris/HCl pH 8, 2 mM EDTA (if an ammonium sulfate suspension is used, the protein has to be extensively dialyzed first).
2. Remove aggregates by centrifugation at 12 000 g in a bench-top centrifuge.
3. Assay buffer: 50 mM HEPES/KOH pH 7.5, 150 mM KCl, 10 mM $MgCl_2$ (the buffer must be degassed and filtered through a 0.2-µm membrane before use).
4. Add 2494 µL assay buffer to a 1 × 1 cm fluorescence cuvette equipped with a stirring bar.
5. Incubate the cuvette for >5 min at 43 °C in a thermostated fluorescence spectrometer.
6. To start aggregation, add 6 µL CS solution and follow the increase in light scattering for 60 min (set both excitation and emission to 360 nm; adjust the slits to obtain a good signal/noise ratio).

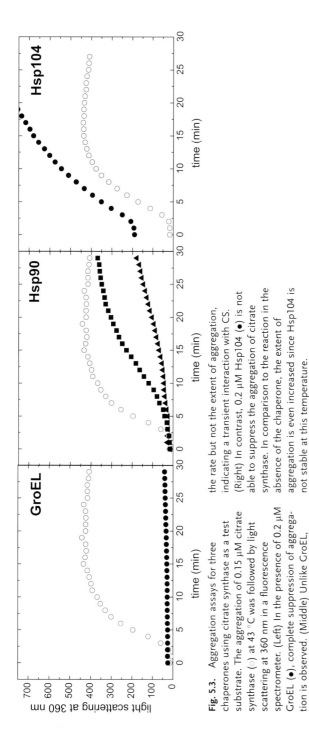

Fig. 5.3. Aggregation assays for three chaperones using citrate synthase as a test substrate. The aggregation of 0.15 μM citrate synthase (○) at 43 °C was followed by light scattering at 360 nm in a fluorescence spectrometer. (Left) In the presence of 0.2 μM GroEL (●), complete suppression of aggregation is observed. (Middle) Unlike GroEL, Hsp90 (0.1 μM ■ and 0.2 μM ▲) decreases the rate but not the extent of aggregation, indicating a transient interaction with CS. (Right) In contrast, 0.2 μM Hsp104 (●) is not able to suppress the aggregation of citrate synthase. In comparison to the reaction in the absence of the chaperone, the extent of aggregation is even increased since Hsp104 is not stable at this temperature.

7. Start a new experiment in which some of the buffer in the sample cuvette is replaced by a stock solution of your chaperone before the addition of CS; use the following chaperone concentrations: 0.1, 0.2, and 0.5 µM (the final concentration of CS in the cuvette is 0.15 µM).
8. To test the influence of the buffer in which the chaperone is stored, carry out a set of control experiments in which the chaperone is replaced by equal volumes of this buffer.
9. If an absorbance photometer is used, concentrations may have to be increased. Set the wavelength to 360 nm.

Protocol 3: Aggregation of Insulin B Chain In the presence of DTT, the three disulfide bridges within the insulin molecule are broken, and the two chains dissociate. While the A chain remains soluble, the B chain will readily start to aggregate [113, 114]. This process can be suppressed by molecular chaperones. A major disadvantage of this assay is that it consumes substantial amounts of chaperone, of which a highly concentrated stock solution is needed. The assay can be performed at different temperatures [69] and salt concentrations, but note that the extent of aggregation is very sensitive to changes in pH.

1. Dissolve 4.06 mg insulin (bovine pancreas) in 1 mL 10 mM sodium phosphate buffer pH 7 (if the protein does not dissolve, add concentrated HCl until the solution is clear, then add 10 M NaOH and adjust the pH to ∼7).
2. Keep the stock solution (∼ 700 µM) on ice.
3. Assay buffer: 50 mM Tris/HCl pH 8, 2 mM EDTA.
4. To a 120-µL UV cuvette, add 110 µL of assay buffer and 10 µL of insulin (the small sample volume minimizes chaperone consumption, but larger cuvettes may be used as well).
5. Incubate the cuvette for >5 min at 25 °C in a thermostated photometer.
6. Start the aggregation by adding 20 µL of 0.35 M DTT in assay buffer.
7. Follow the aggregation by measuring the extinction at 400 nm.
8. Start a new experiment in which some of the buffer in the sample cuvette is replaced by a stock solution of your chaperone before the addition of insulin; use the following chaperone concentrations: 25, 50, and 100 µM (the final concentration of insulin in the cuvette is 50 µM).
9. To test the influence of the buffer in which the chaperone is stored, carry out a set of control experiments in which the chaperone is replaced by equal volumes of this buffer.

Protocol 4: Aggregation of Alpha-lactalbumin The conformational stability of alpha-lactalbumin is highly dependent on the presence of the four disulfide bridges and the bound Ca^{2+}. Thus, the addition of a reducing agent such as DTT in the presence of EDTA causes the protein to unfold and aggregate [114]. This process is quite slow at 25 °C but can be significantly accelerated by raising the temperature. The best results are obtained between 25 °C and 45 °C.

1. Assay buffer: 50 mM sodium phosphate pH 7.0, 0.2 M NaCl, 2 mM EDTA.
2. 300 µM alpha-lactalbumin (bovine) in assay buffer. (Note: Aggregation is significantly inhibited by Tris and HEPES. It is therefore recommended to dialyze the chaperone against the assay buffer.)
3. Add 100 µL buffer and 20 µL alpha-lactalbumin to a 120-µL UV cuvette (the small sample volume minimizes chaperone consumption, but larger cuvettes may be used as well).
4. Incubate the cuvette for >5 min at the chosen temperature (25–45 °C) in a thermostated photometer.
5. Start the reaction with 20 µL of 0.35 M DTT in assay buffer and follow the aggregation by measuring the extinction at 500 nm.
6. Start a new experiment in which some of the buffer in the sample cuvette is replaced by a stock solution of your chaperone before the addition of alpha-lactalbumin; use the following chaperone concentrations: 25, 50, and 100 µM (the final concentration of lactalbumin in the cuvette is 43 µM).
7. To test the influence of the buffer in which the chaperone is stored, carry out a set of control experiments in which the chaperone is replaced by equal volumes of this buffer.

The kinetics of the aggregation reaction depends on the DTT concentration and is considerably faster at higher temperature. At 25 °C, complete aggregation may take several hours.

5.6.3
Complex Formation between Chaperones and Polypeptide Substrates

For the preparation of complexes with a nonnative polypeptide, an unfolded test substrate must be generated first. As mentioned in "Generation of Nonnative Proteins" above, it is very important to verify that the chaperone under investigation is compatible with the selected methods of protein denaturation.

Protocol 5: Preparation of Chaperone-Substrate Complexes Using Urea-denatured MDH

1. Remove stabilizing agents such as glycerol or $(NH_4)_2SO_4$ by dialysis or by passage through a desalting column (buffer: 50 mM Tris/HCl pH 7.5, 2 mM DTT); the concentration of MDH monomers after this step should be at least 150 µM (5 mg mL^{-1}).
2. To 3 µL of mMDH (150 µM), add 6 µL of 9 M urea, 50 mM Tris/HCl pH 7.5, 10 mM DTT; allow the protein to unfold for 30 min at 25 °C.
3. Prepare a solution of your chaperone (1 µM) in 198 µL 50 mM Tris/HCl pH 7.5, 10 mM DTT, 50 mM KCl and incubate at 25 °C for 5 min.
4. Add 2 µL of unfolded MDH, mix rapidly, and incubate for 15 min.
5. Spin down the sample in a bench-top centrifuge (12 000 g) to remove aggregated MDH.

Protocol 6: Formation of Complexes Between GroEL and Thermally Denatured CS

1. Prepare a stock solution of 65 µM CS (monomers) in 50 mM Tris/HCl pH 8, 2 mM EDTA (if an ammonium sulfate suspension is used, the protein has to be extensively dialyzed first).
2. Remove aggregates by centrifugation at 12 000 g in a bench-top centrifuge.
3. Aliquots are prepared, frozen in liquid nitrogen, and stored at −20 °C; once thawed, the aliquots should be used up.
4. Prepare a solution of your chaperone (1 µM) in 197 µL 50 mM HEPES/KOH pH 7.5, 50 mM KCl and incubate at 43 °C for 5 min.
5. Add 3 µL of CS and incubate for additional 30 min.
6. Remove aggregates by centrifugation at 12 000 g in a bench-top centrifuge.

Some applications may require concentrations of chaperone-CS complexes higher than 1 µM, as given in the protocol. This can be achieved by increasing the concentration of both CS and chaperone in the mixture. However, there may be a significant loss of CS due to aggregation.

Once the complexes between the polypeptide and the chaperone have been prepared, their composition can be examined by various methods. If binding is strong, analytical size-exclusion chromatography can be used in combination with a UV or fluorescence detection unit. In addition, the substrate polypeptide may be modified to carry a fluorescent label (Figure 5.4). This allows the direct analysis of different components in one peak. Otherwise, the elution profile may be analyzed with SDS-PAGE [115]. If complexes are only transient, as in the case of Hsp104 or Hsp70-ATP, chemical cross-linking with glutaraldehyde in combination with SDS-PAGE can be used. Other methods that have been successfully used to demonstrate physical interaction include immunoprecipitation [66], native PAGE [86, 90], and analytical ultracentrifugation [116].

5.6.4
Identification of Chaperone-binding Sites

The identification of binding sites for chaperones in a given substrate is an important issue in the context of characterizing a molecular chaperone. The experimental approaches that have been developed to address this kind of problem are rather advanced, and thus we will give only a brief outline.

For chaperones known to bind to extended peptide sequences such as the Hsp70 family members, early studies used phage display to analyze the interaction with the chaperone [117, 118]. The analysis of this large dataset of binding and nonbinding peptides led to an algorithm for the prediction of BiP-binding sites in the primary sequences of proteins. The predictive power of this program was found to be about 50% for binding sequences as tested by a survey of synthetic peptides [119].

A related approach is to screen for chaperone-binding sites in proteins using sets

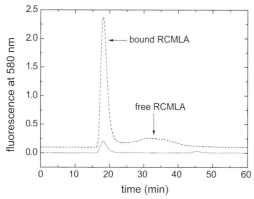

Fig. 5.4. Size-exclusion chromatography of a complex between GroEL and RCMLA (dashed line). Complexes were formed by incubating 3 µM RCMLA (labeled with tetramethylrhodamin) with 2 µM GroEL$_{14}$ for 5 min at 30 °C in 50 mM HEPES/KOH pH 7.5, 150 mM KCl, 10 mM MgCl$_2$. 100 µL of this mixture was applied to a Superdex-200 HR 10/30 column equilibrated in the same buffer. Complexes were detected by following the fluorescence of the labeled substrate. In a control experiment, GroEL alone was injected (solid line). Note that the position of the GroEL peak at 18 min is not significantly shifted upon binding of RCMLA, as the apparent molecular mass increases only marginally from ~800 kDa to ~820 kDa.

of overlapping synthetic peptides that represent the entire primary structure of the respective proteins. The peptides are immobilized on filters. This allows a correlation between sequence and signals. Binding is usually detected by antibodies against the respective chaperone in a Western blot–like manner. This technique has been successfully applied for several chaperones and target proteins [120, 121].

A caveat is that these approaches will provide information on all potential binding sites in a protein. Which of these will actually be used in vivo may depend on the kinetics of folding or on the degree of unfolding under stress conditions, similar to N-glycosylation sites in proteins. Moreover, this technique is unable to detect binding sites that are formed by residues distant in sequence but close in space, e.g., in a partially folded intermediate. Thus, its value may be restricted to identifying binding sites for chaperones such as Hsp70, which are known to bind consecutive stretches of amino acids.

Identifying the binding site(s) for nonnative proteins in chaperones is even more challenging. Here, chemical cross-linking of chaperone-substrate complexes and subsequent analysis of the site of interaction by protease digestion and mass spectrometry is the method of choice. Mutational analyses using deletion fragments to identify protein-binding sites should be considered with great caution because they have repeatedly yielded false or contradictory results. However, after a potential binding site has been identified biochemically, site-specific mutagenesis of residues in the respective region may be employed to further corroborate the result.

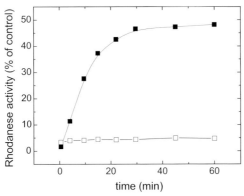

Fig. 5.5. GroE-assisted reactivation of rhodanese. Urea-denatured rhodanese (0.5 µM) was refolded in the presence of 1 µM GroEL/2 µM GroES/2 mM ATP (■) or in the absence of chaperones (□). At the indicated time points, aliquots were withdrawn from the refolding reaction, and the rhodanese activity was determined as described in Protocol 7.

5.6.5
Chaperone-mediated Refolding of Test Proteins

Usually, refolding is carried out by diluting a solution of a nonnative test enzyme, e.g., unfolded in 6 M GdmCl, into a buffer that contains the chaperone under investigation. At various time points, aliquots are withdrawn and the enzyme activity of the test protein is determined (Figure 5.5). In a control experiment, the assay is carried out with the native test enzyme instead of the unfolded protein. Concentrations, pre-incubation times, buffer conditions, etc., should be identical to the ones used in the refolding experiment. If a denaturant was used to prepare the unfolded test protein, an amount of denaturant equivalent to its residual concentration in the refolding assay should be added. This experiment yields the 100% value for the activity of the substrate, as it accounts for any changes in the enzyme activity that are due to its incubation in the refolding solution. In a further control experiment, one has to determine to what extent the enzyme refolds spontaneously. This is simply done by carrying out the refolding experiment in the absence of the chaperone but under otherwise unchanged conditions. Again, it is essential to include all additional factors such as ATP in this control, as they may stabilize or destabilize the enzyme and thus influence spontaneous refolding. Furthermore, an additional control reaction should include substantial amounts of a non-chaperone protein such as BSA, e.g., a fivefold excess over the substrate protein, to account for unspecific protein effects on refolding.

For many test proteins, spontaneous refolding can be quite efficient at certain conditions, e.g., at low protein concentration or temperature or in the presence of stabilizing reagents. To identify potential chaperones, refolding experiments should be carried out under stringent conditions where no refolding occurs in the absence of chaperones. As already mentioned, efficient refolding by the chaperone may require further components such as ATP, cofactors, or other chaperones. Therefore, some knowledge about the chaperone under investigation has to be acquired beforehand. Once the conditions for successful reactivation of a substrate have been established, parameters such as the type or concentration of cofactors can be varied to obtain further insight into the mechanism of the chaperone.

Protocol 7: GroE-assisted Refolding of Rhodanese

Unfolding of Rhodanese

In this protocol urea is used to denature rhodanese, but 6 M GdmCl will work as well. Depending on your rhodanese stock, the protein solution may have to be concentrated by ultrafiltration prior to unfolding.

1. Remove stabilizing agents such as glycerol or $(NH_4)_2SO_4$ by dialysis or by using a desalting column (buffer: 50 mM Tris/HCl, 2 mM DTT); the concentration of rhodanese after this step should be at least 300 µM (10 mg mL^{-1}).
2. To 3.3 µL of rhodanese (300 µM), add 17 µL of 10 M urea, 50 mM Tris/HCl pH 7.5, 10 mM DTT; allow the protein to unfold for 60 min at 25 °C.

Refolding of Rhodanese by the GroE System

The concentration of rhodanese in the refolding reaction is 0.5 µM. A twofold excess of GroEL and GroES is used (Figure 5.5).

1. Prepare a solution of 1 µM GroEL$_{14}$ and 2 µM GroES$_7$ in 50 mM Tris/HCl pH 7.5, 1 mM DTT, 50 mM KCl, 10 mM MgCl$_2$, 20 mM Na$_2$S$_2$O$_3$.
2. Add 2.5 µL of unfolded rhodanese, mix rapidly, and incubate at 25 °C for 5 min.
3. Start reactivation by adding 1 mM ATP (total volume of refolding reaction: ~250 µL).
4. Withdraw aliquots of 24 µL every 4 min and measure rhodanese activity (see next section).

Note that in the refolding assay 20 mM $S_2O_3^{2-}$ is present to stabilize refolded rhodanese. To determine the extent of spontaneous refolding, conduct a control experiment without GroEL.

Enzyme Assay for Rhodanese

The enzyme test is based on a sulfur transfer reaction catalyzed by rhodanese [95]. Sulfur atoms of the substrate thiosulfate are passed onto cyanide anions, generating SCN^-. The reaction is stopped at timed intervals with a solution of iron(III)-nitrate. The concentration of the resulting red complex [Fe(SCN)$_3$] is determined by measuring the absorbance of the samples at $\lambda = 470$ nm using a standard spectrophotometer.

EDTA is important to complex magnesia ions. This blocks the ATPase activity of residual chaperones. Otherwise, refolding would continue during the activity assay.

Caution: KCN is toxic. Work under a fume hood and wear gloves at all times.
1. Premix: 9% (v/v) formaldehyde.
2. Staining solution: 16% (w/v) $Fe(NO_3)_3$, 1 M HNO_3.
3. Prepare the number of activity tests tubes you will require (one for every time point of the refolding kinetics): 24 µL KH_2PO_4 (1 M), 12 µL EDTA (0.5 M, pH 9), 30 µL $Na_2S_2O_3$ (1 M), 30 µL KCN (1 M), 480 µL H_2O, total volume 576 µL each.
4. For each activity test tube, prepare six disposable cuvettes with 575 µL of premix solution.
5. Incubate test tubes at 25 °C.
6. Start the activity test by adding 24 µL of rhodanese-containing solution from the refolding reaction to the test tube. The final conditions of the activity tests are: 25 °C, 40 mM KH_2PO_4, 10 mM EDTA, 30 mM $S_2O_3^{2-}$, 30 mM CN^-, 20 nM rhodanese (from refolding reaction), total volume = 600 µL.
7. Every 2 min, withdraw 90 µL from the activity test tube and add it to the premix solution in the cuvette.
8. Quench sample with 325 µL staining solution and mix rapidly.
9. Measure the absorbance at 470 nm (the orange color is stable for at least 2 h).
10. Blank: cuvette with premix solution.
11. Draw a straight line by plotting absorbance at 470 nm vs. time.
12. Use slope $dA_{470\,nm}/dt$ to calculate enzymatic activity.

5.6.6
ATPase Activity

Several assay systems have been developed to monitor ATP hydrolysis. The most sensitive technique uses either α-[^{32}P]-ATP or γ-[^{32}P]-ATP. Upon hydrolysis, either radioactive phosphate or ADP is released and can be detected and quantified [122]. Another method uses the dye malachite green to determine the amount of phosphate produced by hydrolysis of ATP [123]. A more elegant approach employs a coupled enzyme system [124]. In the first step, pyruvate kinase regenerates ATP from ADP and phosphoenol pyruvate. This reaction also produces pyruvate, which is reduced in a second step catalyzed by lactate dehydrogenase under the consumption of NADH, a reaction that can be followed at 340 nm in a spectrometer. Detailed information on the experimental design and evaluation of ATPase experiments is provided in Chapter 4 of this volume.

Acknowledgments

We would like to thank V. Grimminger, B. Bösl, and K. Ruth for contributing to this chapter and M. Haslbeck and H. Wegele for critically reading the manuscript. Work in the authors' labs was supported by grants from the DFG to J.B. and S.W. and by grants from the Fonds der Chemischen Industrie to J.B.

References

1. EARNSHAW, W. C., HONDA, B. M., LASKEY, R. A., & THOMAS, J. O. (1980). Assembly of nucleosomes: the reaction involving X. laevis nucleoplasmin. *Cell* **21**, 373–383.
2. DINGWALL, C. & LASKEY, R. A. (1990). Nucleoplasmin: the archetypal molecular chaperone. *Semin. Cell Biol.* **1**, 11–17.
3. ELLIS, R. J. & HEMMINGSEN, S. M. (1989). Molecular chaperones: proteins essential for the biogenesis of some macromolecular structures. *Trends Biochem. Sci.* **14**, 339–342.
4. ELLIS, R. J. & VAN DER VIES, S. M. (1991). Molecular chaperones. *Annu. Rev. Biochem.* **60**, 321–347.
5. GETHING, M. J. & SAMBROOK, J. (1992). Protein folding in the cell. *Nature* **355**, 33–45.
6. HARTL, F. U. (1996). Molecular chaperones in cellular protein folding. *Nature* **381**, 571–579.
7. BUKAU, B. & HORWICH, A. L. (1998). The Hsp70 and Hsp60 chaperone machines. *Cell* **92**, 351–366.
8. HARTL, F. U. & HAYER-HARTL, M. (2002). Molecular chaperones in the cytosol: from nascent chain to folded protein. *Science* **295**, 1852–1858.
9. WALTER, S. & BUCHNER, J. (2002). Molecular chaperones–cellular machines for protein folding. *Angew. Chem. Int. Ed Engl.* **41**, 1098–1113.
10. GRAGEROV, A. I., MARTIN, E. S., KRUPENKO, M. A., KASHLEV, M. V., & NIKIFOROV, V. G. (1991). Protein aggregation and inclusion body formation in *Escherichia coli* rpoH mutant defective in heat shock protein induction. *FEBS Lett.* **291**, 222–224.
11. SCHOEMAKER, J. M., BRASNETT, A. H., & MARSTON, F. A. (1985). Examination of calf prochymosin accumulation in *Escherichia coli*: disulphide linkages are a structural component of prochymosin-containing inclusion bodies. *EMBO J.* **4**, 775–780.
12. INGOLIA, T. D. & CRAIG, E. A. (1982). Drosophila gene related to the major heat shock-induced gene is transcribed at normal temperatures and not induced by heat shock. *Proc. Natl. Acad. Sci. U.S.A* **79**, 525–529.
13. FAYET, O., ZIEGELHOFFER, T., & GEORGOPOULOS, C. (1989). The groES and groEL heat shock gene products of *Escherichia coli* are essential for bacterial growth at all temperatures. *J. Bacteriol.* **171**, 1379–1385.
14. BORKOVICH, K. A., FARRELLY, F. W., FINKELSTEIN, D. B., TAULIEN, J., & LINDQUIST, S. (1989). hsp82 is an essential protein that is required in higher concentrations for growth of cells at higher temperatures. *Mol. Cell Biol.* **9**, 3919–3930.
15. ANFINSEN, C. B., HABER, E., SELA, M., & WHITE, F. H. (1961). The kinetics of formation of native ribonuclease during oxidation of the reduced polypetide chain. *Proc. Natl. Acad. Sci. U.S.A.* **47**, 1309–1314.
16. ANFINSEN, C. B. (1973). Principles that govern the folding of protein chains. *Science* **181**, 223–230.
17. JAENICKE, R. (1987). Folding and association of proteins. *Prog. Biophys. Mol. Biol.* **49**, 117–237.
18. SECKLER, R. & JAENICKE, R. (1992). Protein folding and protein refolding. *FASEB J.* **6**, 2545–2552.
19. KIEFHABER, T., RUDOLPH, R., KOHLER, H. H., & BUCHNER, J. (1991). Protein aggregation in vitro and in vivo: a quantitative model of the kinetic competition between folding and aggregation. *Biotechnology* **9**, 825–829.
20. EPSTEIN, C. J., GOLDBERGER, R. F., & ANFINSEN, C. B. (1963). Genetic Control of Tertiary Protein Structure: Studies with Model Systems. *Cold Spring Harbor Symp. Quant. Biol.* **28**, 439–449.
21. VIITANEN, P. V., GATENBY, A. A., & LORIMER, G. H. (1992). Purified chaperonin 60 (groEL) interacts with the nonnative states of a multitude of *Escherichia coli* proteins. *Protein Sci.* **1**, 363–369.
22. HORWICH, A. L., LOW, K. B., FENTON, W. A., HIRSHFIELD, I. N., & FURTAK, K. (1993). Folding in vivo of bacterial

cytoplasmic proteins: role of GroEL. *Cell* **74**, 909–917.

23 HOURY, W. A., FRISHMAN, D., ECKERSKORN, C., LOTTSPEICH, F., & HARTL, F. U. (1999). Identification of in vivo substrates of the chaperonin GroEL. *Nature* **402**, 147–154.

24 TETER, S. A., HOURY, W. A., ANG, D., TRADLER, T., ROCKABRAND, D., FISCHER, G., BLUM, P., GEORGOPOULOS, C., & HARTL, F. U. (1999). Polypeptide flux through bacterial Hsp70: DnaK cooperates with trigger factor in chaperoning nascent chains. *Cell* **97**, 755–765.

25 HASLBECK, M., BRAUN, N., STROMER, T., RICHTER, B., MODEL, N., WEINKAUF, S., & BUCHNER, J. (2004). Hsp42 is the general small heat shock protein in the cytosol of *Saccharomyces cerevisiae*. *EMBO J.* **23**, 638–649.

26 ROBINSON, C. V., GROSS, M., EYLES, S. J., EWBANK, J. J., MAYHEW, M., HARTL, F. U., DOBSON, C. M., & RADFORD, S. E. (1994). Conformation of GroEL-bound alpha-lactalbumin probed by mass spectrometry. *Nature* **372**, 646–651.

27 ZAHN, R., PERRETT, S., & FERSHT, A. R. (1996). Conformational states bound by the molecular chaperones GroEL and secB: a hidden unfolding (annealing) activity. *J. Mol. Biol.* **261**, 43–61.

28 GOLDBERG, M. S., ZHANG, J., SONDEK, S., MATTHEWS, C. R., FOX, R. O., & HORWICH, A. L. (1997). Native-like structure of a protein-folding intermediate bound to the chaperonin GroEL. *Proc. Natl. Acad. Sci. U.S.A* **94**, 1080–1085.

29 CHEN, J., WALTER, S., HORWICH, A. L., & SMITH, D. L. (2001). Folding of malate dehydrogenase inside the GroEL-GroES cavity. *Nat. Struct. Biol.* **8**, 721–728.

30 FLYNN, G. C., POHL, J., FLOCCO, M. T., & ROTHMAN, J. E. (1991). Peptide-binding specificity of the molecular chaperone BiP. *Nature* **353**, 726–730.

31 LANDRY, S. J., JORDAN, R., MCMACKEN, R., & GIERASCH, L. M. (1992). Different conformations for the same polypeptide bound to chaperones DnaK and GroEL. *Nature* **355**, 455–457.

32 FENTON, W. A., KASHI, Y., FURTAK, K., & HORWICH, A. L. (1994). Residues in chaperonin GroEL required for polypeptide binding and release. *Nature* **371**, 614–619.

33 ZHU, X., ZHAO, X., BURKHOLDER, W. F., GRAGEROV, A., OGATA, C. M., GOTTESMAN, M. E., & HENDRICKSON, W. A. (1996). Structural analysis of substrate binding by the molecular chaperone DnaK. *Science* **272**, 1606–1614.

34 KATSUMATA, K., OKAZAKI, A., TSURUPA, G. P., & KUWAJIMA, K. (1996). Dominant forces in the recognition of a transient folding intermediate of alpha-lactalbumin by GroEL. *J. Mol. Biol.* **264**, 643–649.

35 GRAGEROV, A., ZENG, L., ZHAO, X., BURKHOLDER, W., & GOTTESMAN, M. E. (1994). Specificity of DnaK-peptide binding. *J. Mol. Biol.* **235**, 848–854.

36 CROUY-CHANEL, A., KOHIYAMA, M., & RICHARME, G. (1996). Specificity of DnaK for arginine/lysine and effect of DnaJ on the amino acid specificity of DnaK. *J. Biol. Chem.* **271**, 15486–15490.

37 WANG, Z., FENG, H., LANDRY, S. J., MAXWELL, J., & GIERASCH, L. M. (1999). Basis of substrate binding by the chaperonin GroEL. *Biochemistry* **38**, 12537–12546.

38 GERVASONI, P., STAUDENMANN, W., JAMES, P., GEHRIG, P., & PLUCKTHUN, A. (1996). beta-Lactamase binds to GroEL in a conformation highly protected against hydrogen/deuterium exchange. *Proc. Natl. Acad. Sci. U.S.A* **93**, 12189–12194.

39 LILIE, H. & BUCHNER, J. (1995). Interaction of GroEL with a highly structured folding intermediate: iterative binding cycles do not involve unfolding. *Proc. Natl. Acad. Sci. U.S.A* **92**, 8100–8104.

40 BRAIG, K., OTWINOWSKI, Z., HEGDE, R., BOISVERT, D. C., JOACHIMIAK, A., HORWICH, A. L., & SIGLER, P. B. (1994). The crystal structure of the

bacterial chaperonin GroEL at 2.8 Å. *Nature* **371**, 578–586.

41 BUCKLE, A. M., ZAHN, R., & FERSHT, A. R. (1997). A structural model for GroEL-polypeptide recognition. *Proc. Natl. Acad. Sci. U.S.A* **94**, 3571–3575.

42 CHEN, L. & SIGLER, P. B. (1999). The crystal structure of a GroEL/peptide complex: plasticity as a basis for substrate diversity. *Cell* **99**, 757–768.

43 CHIN, D. T., KUEHL, L., & RECHSTEINER, M. (1982). Conjugation of ubiquitin to denatured hemoglobin is proportional to the rate of hemoglobin degradation in HeLa cells. *Proc. Natl. Acad. Sci. U.S.A* **79**, 5857–5861.

44 PARAG, H. A., RABOY, B., & KULKA, R. G. (1987). Effect of heat shock on protein degradation in mammalian cells: involvement of the ubiquitin system. *EMBO J.* **6**, 55–61.

45 KARZAI, A. W., ROCHE, E. D., & SAUER, R. T. (2000). The SsrA-SmpB system for protein tagging, directed degradation and ribosome rescue. *Nat. Struct. Biol.* **7**, 449–455.

46 VAN NOCKER, S., DEVERAUX, Q., RECHSTEINER, M., & VIERSTRA, R. D. (1996). *Arabidopsis* MBP1 gene encodes a conserved ubiquitin recognition component of the 26S proteasome. *Proc. Natl. Acad. Sci. U.S.A* **93**, 856–860.

47 WEBER-BAN, E. U., REID, B. G., MIRANKER, A. D., & HORWICH, A. L. (1999). Global unfolding of a substrate protein by the Hsp100 chaperone ClpA. *Nature* **401**, 90–93.

48 PRATT, W. B. & TOFT, D. O. (1997). Steroid receptor interactions with heat shock protein and immunophilin chaperones. *Endocr. Rev.* **18**, 306–360.

49 PRATT, W. B. & TOFT, D. O. (2003). Regulation of signalling protein function and trafficking by the hsp90/hsp70-based chaperone machinery. *Exp. Biol. Med.* **228**, 111–133.

50 RICHTER, K., MUSCHLER, P., HAINZL, O., REINSTEIN, J., & BUCHNER, J. (2003). Sti1 is a non-competitive inhibitor of the Hsp90 ATPase. Binding prevents the N-terminal dimerization reaction during the atpase cycle. *J. Biol. Chem.* **278**, 10328–10333.

51 LIBEREK, K., SKOWYRA, D., ZYLICZ, M., JOHNSON, C., & GEORGOPOULOS, C. (1991). The Escherichia coli DnaK chaperone, the 70-kDa heat shock protein eukaryotic equivalent, changes conformation upon ATP hydrolysis, thus triggering its dissociation from a bound target protein. *J. Biol. Chem.* **266**, 14491–14496.

52 XU, Z., HORWICH, A. L., & SIGLER, P. B. (1997). The crystal structure of the asymmetric GroEL-GroES-(ADP)$_7$ chaperonin complex. *Nature* **388**, 741–750.

53 RANSON, N. A., FARR, G. W., ROSEMAN, A. M., GOWEN, B., FENTON, W. A., HORWICH, A. L., & SAIBIL, H. R. (2001). ATP-bound states of GroEL captured by cryo-electron microscopy. *Cell* **107**, 869–879.

54 RICHTER, K., MUSCHLER, P., HAINZL, O., & BUCHNER, J. (2001). Coordinated ATP hydrolysis by the Hsp90 dimer. *J. Biol. Chem.* **276**, 33689–33696.

55 HARDY, S. J. & RANDALL, L. L. (1991). A kinetic partitioning model of selective binding of nonnative proteins by the bacterial chaperone SecB. *Science* **251**, 439–443.

56 WEISSMAN, J. S., KASHI, Y., FENTON, W. A., & HORWICH, A. L. (1994). GroEL-mediated protein folding proceeds by multiple rounds of binding and release of nonnative forms. *Cell* **78**, 693–702.

57 FRIEDEN, C. & CLARK, A. C. (1997). Protein folding: how the mechanism of GroEL action is defined by kinetics. *Proc. Natl. Acad. Sci. U.S.A* **94**, 5535–5538.

58 SHTILERMAN, M., LORIMER, G. H., & ENGLANDER, S. W. (1999). Chaperonin function: folding by forced unfolding. *Science* **284**, 822–825.

59 JORDAN, R. & MCMACKEN, R. (1995). Modulation of the ATPase activity of the molecular chaperone DnaK by peptides and the DnaJ and GrpE heat shock proteins. *J. Biol. Chem.* **270**, 4563–4569.

60 MCLAUGHLIN, S. H., SMITH, H. W., & JACKSON, S. E. (2002). Stimulation of

the weak ATPase activity of human hsp90 by a client protein. *J. Mol. Biol.* **315**, 787–798.

61 CASHIKAR, A. G., SCHIRMER, E. C., HATTENDORF, D. A., GLOVER, J. R., RAMAKRISHNAN, M. S., WARE, D. M., & LINDQUIST, S. L. (2002). Defining a pathway of communication from the C-terminal peptide binding domain to the N-terminal ATPase domain in a AAA protein. *Mol. Cell* **9**, 751–760.

62 GAMER, J., MULTHAUP, G., TOMOYASU, T., MCCARTY, J. S., RUDIGER, S., SCHONFELD, H. J., SCHIRRA, C., BUJARD, H., & BUKAU, B. (1996). A cycle of binding and release of the DnaK, DnaJ and GrpE chaperones regulates activity of the *Escherichia coli* heat shock transcription factor sigma32. *EMBO J.* **15**, 607–617.

63 LIBEREK, K., MARSZALEK, J., ANG, D., GEORGOPOULOS, C., & ZYLICZ, M. (1991). *Escherichia coli* DnaJ and GrpE heat shock proteins jointly stimulate ATPase activity of DnaK. *Proc. Natl. Acad. Sci. U.S.A* **88**, 2874–2878.

64 PACKSCHIES, L., THEYSSEN, H., BUCHBERGER, A., BUKAU, B., GOODY, R. S., & REINSTEIN, J. (1997). GrpE accelerates nucleotide exchange of the molecular chaperone DnaK with an associative displacement mechanism. *Biochemistry* **36**, 3417–3422.

65 SULLIVAN, W. P., OWEN, B. A., & TOFT, D. O. (2002). The influence of ATP and p23 on the conformation of hsp90. *J. Biol. Chem.* **277**, 45942–45948.

66 WEGELE, H., HASLBECK, M., REINSTEIN, J., & BUCHNER, J. (2003). Sti1 is a novel activator of the Ssa proteins. *J. Biol. Chem.* **278**, 25970–25976.

67 GRAY, T. E. & FERSHT, A. R. (1991). Cooperativity in ATP hydrolysis by GroEL is increased by GroES. *FEBS Lett.* **292**, 254–258.

68 TODD, M. J., VIITANEN, P. V., & LORIMER, G. H. (1994). Dynamics of the chaperonin ATPase cycle: implications for facilitated protein folding. *Science* **265**, 659–666.

69 HASLBECK, M., WALKE, S., STROMER, T., EHRNSPERGER, M., WHITE, H. E., CHEN, S., SAIBIL, H. R., & BUCHNER, J. (1999). Hsp26: a temperature-regulated chaperone. *EMBO J.* **18**, 6744–6751.

70 SPIESS, C., BEIL, A., & EHRMANN, M. (1999). A temperature-dependent switch from chaperone to protease in a widely conserved heat shock protein. *Cell* **97**, 339–347.

71 KROJER, T., GARRIDO-FRANCO, M., HUBER, R., EHRMANN, M., & CLAUSEN, T. (2002). Crystal structure of DegP (HtrA). reveals a new protease-chaperone machine. *Nature* **416**, 455–459.

72 JAKOB, U., MUSE, W., ESER, M., & BARDWELL, J. C. (1999). Chaperone activity with a redox switch. *Cell* **96**, 341–352.

73 KOBAYASHI, T. & INOUYE, M. (1992). Functional analysis of the intramolecular chaperone. Mutational hot spots in the subtilisin pro-peptide and a second-site suppressor mutation within the subtilisin molecule. *J. Mol. Biol.* **226**, 931–933.

74 JASWAL, S. S., SOHL, J. L., DAVIS, J. H., & AGARD, D. A. (2002). Energetic landscape of alpha-lytic protease optimizes longevity through kinetic stability. *Nature* **415**, 343–346.

75 MENDOZA, J. A., ROGERS, E., LORIMER, G. H., & HOROWITZ, P. M. (1991). Unassisted refolding of urea unfolded rhodanese. *J. Biol. Chem.* **266**, 13587–13591.

76 ZHI, W., LANDRY, S. J., GIERASCH, L. M., & SRERE, P. A. (1992). Renaturation of citrate synthase: influence of denaturant and folding assistants. *Protein Sci.* **1**, 522–529.

77 AGHAJANIAN, S., WALSH, T. P., & ENGEL, P. C. (1999). Specificity of coenzyme analogues and fragments in promoting or impeding the refolding of clostridial glutamate dehydrogenase. *Protein Sci.* **8**, 866–872.

78 SZABO, A., KORSZUN, R., HARTL, F. U., & FLANAGAN, J. (1996). A zinc finger-like domain of the molecular chaperone DnaJ is involved in binding to denatured protein substrates. *EMBO J.* **15**, 408–417.

79 BOSE, S., WEIKL, T., BUGL, H., &

Buchner, J. (1996). Chaperone function of Hsp90-associated proteins. *Science* **274**, 1715–1717.

80 Freeman, B. C., Toft, D. O., & Morimoto, R. I. (1996). Molecular chaperone machines: chaperone activities of the cyclophilin Cyp-40 and the steroid aporeceptor-associated protein p23. *Science* **274**, 1718–1720.

81 Freeman, B. C. & Yamamoto, K. R. (2002). Disassembly of transcriptional regulatory complexes by molecular chaperones. *Science* **296**, 2232–2235.

82 Kimura, Y., Rutherford, S. L., Miyata, Y., Yahara, I., Freeman, B. C., Yue, L., Morimoto, R. I., & Lindquist, S. (1997). Cdc37 is a molecular chaperone with specific functions in signal transduction. *Genes Dev.* **11**, 1775–1785.

83 Nagata, K. (1996). Hsp47: a collagen-specific molecular chaperone. *Trends Biochem. Sci.* **21**, 22–26.

84 Saga, S., Nagata, K., Chen, W. T., & Yamada, K. M. (1987). pH-dependent function, purification, & intracellular location of a major collagen-binding glycoprotein. *J. Cell Biol.* **105**, 517–527.

85 Natsume, T., Koide, T., Yokota, S., Hirayoshi, K., & Nagata, K. (1994). Interactions between collagen-binding stress protein HSP47 and collagen. Analysis of kinetic parameters by surface plasmon resonance biosensor. *J. Biol. Chem.* **269**, 31224–31228.

86 Goloubinoff, P., Christeller, J. T., Gatenby, A. A., & Lorimer, G. H. (1989). Reconstitution of active dimeric ribulose bisphosphate carboxylase from an unfolded state depends on two chaperonin proteins and Mg-ATP. *Nature* **342**, 884–889.

87 Buchner, J., Schmidt, M., Fuchs, M., Jaenicke, R., Rudolph, R., Schmid, F. X., & Kiefhaber, T. (1991). GroE facilitates refolding of citrate synthase by suppressing aggregation. *Biochemistry* **30**, 1586–1591.

88 Sigler, P. B., Xu, Z., Rye, H. S., Burston, S. G., Fenton, W. A., & Horwich, A. L. (1998). Structure and function in GroEL-mediated protein folding. *Annu. Rev. Biochem.* **67**, 581–608.

89 Ehrnsperger, M., Hergersberg, C., Wienhues, U., Nichtl, A., & Buchner, J. (1998). Stabilization of proteins and peptides in diagnostic immunological assays by the molecular chaperone Hsp25. *Anal. Biochem.* **259**, 218–225.

90 Langer, T., Pfeifer, G., Martin, J., Baumeister, W., & Hartl, F. U. (1992). Chaperonin-mediated protein folding: GroES binds to one end of the GroEL cylinder, which accommodates the protein substrate within its central cavity. *EMBO J.* **11**, 4757–4765.

91 Horovitz, A., Fridmann, Y., Kafri, G., & Yifrach, O. (2001). Review: allostery in chaperonins. *J. Struct. Biol.* **135**, 104–114.

92 Stark, G. R. (1965). Reactions of cyanate with functional groups of proteins. 3. Reactions with amino and carboxyl groups. *Biochemistry* **4**, 1030–1036.

93 Grimminger, V., Richter, K., Imhof, A., Buchner, J., & Walter, S. (2004). The prion curing agent guanidinium chloride specifically inhibits ATP hydrolysis by Hsp104. *J. Biol. Chem.* **279**, 7378–7383.

94 Todd, M. J. & Lorimer, G. H. (1995). Stability of the asymmetric Escherichia coli chaperonin complex. Guanidine chloride causes rapid dissociation. *J. Biol. Chem.* **270**, 5388–5394.

95 Sörbo, B. H. (1953). Crystalline Rhodanese: The Enzyme Catalyzed Reaction. *Acta Chem. Scand.* **7**, 1137–1145.

96 Ploegman, J. H., Drent, G., Kalk, K. H., & Hol, W. G. (1978). Structure of bovine liver rhodanese. I. Structure determination at 2.5 Å resolution and a comparison of the conformation and sequence of its two domains. *J. Mol. Biol.* **123**, 557–594.

97 Mendoza, J. A., Rogers, E., Lorimer, G. H., & Horowitz, P. M. (1991). Chaperonins facilitate the in vitro folding of monomeric mitochondrial rhodanese. *J. Biol. Chem.* **266**, 13044–13049.

98 Jaenicke, R., Rudolph, R., & Heider, I. (1979). Quaternary structure, subunit activity, & in vitro association of porcine mitochondrial malic dehydrogenase. *Biochemistry* **18**, 1217–1223.

99 Staniforth, R. A., Cortes, A., Burston, S. G., Atkinson, T., Holbrook, J. J., & Clarke, A. R. (1994). The stability and hydrophobicity of cytosolic and mitochondrial malate dehydrogenases and their relation to chaperonin-assisted folding. *FEBS Lett.* **344**, 129–135.

100 Roderick, S. L. & Banaszak, L. J. (1986). The three-dimensional structure of porcine heart mitochondrial malate dehydrogenase at 3.0-Å resolution. *J. Biol. Chem.* **261**, 9461–9464.

101 Ranson, N. A., Dunster, N. J., Burston, S. G., & Clarke, A. R. (1995). Chaperonins can catalyse the reversal of early aggregation steps when a protein misfolds. *J. Mol. Biol.* **250**, 581–586.

102 Lee, G. J., Roseman, A. M., Saibil, H. R., & Vierling, E. (1997). A small heat shock protein stably binds heat-denatured model substrates and can maintain a substrate in a folding-competent state. *EMBO J.* **16**, 659–671.

103 Goloubinoff, P., Mogk, A., Zvi, A. P., Tomoyasu, T., & Bukau, B. (1999). Sequential mechanism of solubilization and refolding of stable protein aggregates by a bichaperone network. *Proc. Natl. Acad. Sci. U.S.A* **96**, 13732–13737.

104 Remington, S., Wiegand, G., & Huber, R. (1982). Crystallographic refinement and atomic models of two different forms of citrate synthase at 2.7 and 1.7 Å resolution. *J. Mol. Biol.* **158**, 111–152.

105 Haslbeck, M., Schuster, I., & Grallert, H. (2003). GroE-dependent expression and purification of pig heart mitochondrial citrate synthase in Escherichia coli. *J. Chromatogr. B Analyt. Technol. Biomed. Life Sci.* **786**, 127–136.

106 Srere, P. A., Brazil, H., & Gonen, L. (1963). Citrate synthase assay in tissue homogenates. *Acta Chem. Scand.* **17**, 129–134.

107 Buchner, J., Grallert, H., & Jakob, U. (1998). Analysis of chaperone function using citrate synthase as nonnative substrate protein. *Methods Enzymol.* **290**, 323–338.

108 Wiech, H., Buchner, J., Zimmermann, R., & Jakob, U. (1992). Hsp90 chaperones protein folding in vitro. *Nature* **358**, 169–170.

109 Shao, F., Bader, M. W., Jakob, U., & Bardwell, J. C. (2000). DsbG, a protein disulfide isomerase with chaperone activity. *J. Biol. Chem.* **275**, 13349–13352.

110 Cyr, D. M., Lu, X., & Douglas, M. G. (1992). Regulation of Hsp70 function by a eukaryotic DnaJ homolog. *J. Biol. Chem.* **267**, 20927–20931.

111 Panse, V. G., Swaminathan, C. P., Surolia, A., & Varadarajan, R. (2000). Thermodynamics of substrate binding to the chaperone SecB. *Biochemistry* **39**, 2420–2427.

112 Scholz, C., Stoller, G., Zarnt, T., Fischer, G., & Schmid, F. X. (1997). Cooperation of enzymatic and chaperone functions of trigger factor in the catalysis of protein folding. *EMBO J.* **16**, 54–58.

113 Sanger, F. (1949). Fractionation of oxidized insulin. *Biochem. J.* **44**, 126–128.

114 Horwitz, J., Huang, Q. L., Ding, L., & Bova, M. P. (1998). Lens alpha-crystallin: chaperone-like properties. *Methods Enzymol.* **290**, 365–383.

115 Stromer, T., Ehrnsperger, M., Gaestel, M., & Buchner, J. (2003). Analysis of the interaction of small heat shock proteins with unfolding proteins. *J. Biol. Chem.* **278**, 18015–18021.

116 Schonfeld, H. J. & Behlke, J. (1998). Molecular chaperones and their interactions investigated by analytical ultracentrifugation and other methodologies. *Methods Enzymol.* **290**, 269–296.

117 Blond-Elguindi, S., Cwirla, S. E., Dower, W. J., Lipshutz, R. J., Sprang, S. R., Sambrook, J. F., & Gething, M. J. (1993). Affinity

panning of a library of peptides displayed on bacteriophages reveals the binding specificity of BiP. *Cell* **75**, 717–728.

118 FOURIE, A. M., SAMBROOK, J. F., & GETHING, M. J. (1994). Common and divergent peptide binding specificities of hsp70 molecular chaperones. *J. Biol. Chem.* **269**, 30470–30478.

119 KNARR, G., GETHING, M. J., MODROW, S., & BUCHNER, J. (1995). BiP binding sequences in antibodies. *J. Biol. Chem.* **270**, 27589–27594.

120 RUDIGER, S., GERMEROTH, L., SCHNEIDER-MERGENER, J., & BUKAU, B. (1997). Substrate specificity of the DnaK chaperone determined by screening cellulose-bound peptide libraries. *EMBO J.* **16**, 1501–1507.

121 RUDIGER, S., SCHNEIDER-MERGENER, J., & BUKAU, B. (2001). Its substrate specificity characterizes the DnaJ co-chaperone as a scanning factor for the DnaK chaperone. *EMBO J.* **20**, 1042–1050.

122 KORNBERG, A., SCOTT, J. F., & BERTSCH, L. L. (1978). ATP utilization by rep protein in the catalytic separation of DNA strands at a replicating fork. *J. Biol. Chem.* **253**, 3298–3304.

123 LANZETTA, P. A., ALVAREZ, L. J., REINACH, P. S., & CANDIA, O. A. (1979). An improved assay for nanomole amounts of inorganic phosphate. *Anal. Biochem.* **100**, 95–97.

124 NORBY, J. G. (1988). Coupled assay of Na^+,K^+-ATPase activity. *Methods Enzymol.* **156**, 116–119.

125 DITZEL, L., LOWE, J., STOCK, D., STETTER, K. O., HUBER, H., HUBER, R., & STEINBACHER, S. (1998). Crystal structure of the thermosome, the archaeal chaperonin and homolog of CCT. *Cell* **93**, 125–138.

126 PRODROMOU, C., ROE, S. M., O'BRIEN, R., LADBURY, J. E., PIPER, P. W., & PEARL, L. H. (1997). Identification and structural characterization of the ATP/ADP-binding site in the Hsp90 molecular chaperone. *Cell* **90**, 65–75.

127 MEYER, P., PRODROMOU, C., HU, B., VAUGHAN, C., ROE, S. M., PANARETOU, B., PIPER, P. W., & PEARL, L. H. (2003). Structural and functional analysis of the middle segment of hsp90: implications for ATP hydrolysis and client protein and cochaperone interactions. *Mol. Cell* **11**, 647–658.

128 LEE, S., SOWA, M. E., WATANABE, Y. H., SIGLER, P. B., CHIU, W., YOSHIDA, M., & TSAI, F. T. (2003). The structure of ClpB: a molecular chaperone that rescues proteins from an aggregated state. *Cell* **115**, 229–240.

129 KIM, K. K., KIM, R., & KIM, S. H. (1998). Crystal structure of a small heat-shock protein. *Nature* **394**, 595–599.

130 MELKI, R., ROMMELAERE, H., LEGUY, R., VANDEKERCKHOVE, J., & AMPE, C. (1996). Cofactor A is a molecular chaperone required for beta-tubulin folding: functional and structural characterization. *Biochemistry* **35**, 10422–10435.

131 RUSSELL, R., JORDAN, R., & MCMACKEN, R. (1998). Kinetic characterization of the ATPase cycle of the DnaK molecular chaperone. *Biochemistry* **37**, 596–607.

132 LOPEZ-BUESA, P., PFUND, C., & CRAIG, E. A. (1998). The biochemical properties of the ATPase activity of a 70-kDa heat shock protein (Hsp70) are governed by the C-terminal domains. *Proc. Natl. Acad. Sci. U.S.A* **95**, 15253–15258.

133 BARNETT, M. E., ZOLKIEWSKA, A., & ZOLKIEWSKI, M. (2000). Structure and activity of ClpB from *Escherichia coli*. Role of the amino-and-carboxyl-terminal domains. *J. Biol. Chem.* **275**, 37565–37571.

134 SURIN, A. K., KOTOVA, N. V., KASHPAROV, I. A., MARCHENKOV, V. V., MARCHENKOVA, S. Y., & SEMISOTNOV, G. V. (1997). Ligands regulate GroEL thermostability. *FEBS Lett.* **405**, 260–262.

135 STROMER, T., FISCHER, E., RICHTER, K., HASLBECK, M., & BUCHNER, J. (2004). Analysis of the regulation of the molecular chaperone Hsp26 by temperature-induced dissociation: the N-terminal domain is important for oligomer assembly and the binding of unfolding proteins. *J. Biol. Chem.* **279**, 11222–11228.

136 HOLL-NEUGEBAUER, B., RUDOLPH, R., SCHMIDT, M., & BUCHNER, J. (1991). Reconstitution of a heat shock effect in vitro: influence of GroE on the thermal aggregation of alpha-glucosidase from yeast. *Biochemistry* **30**, 11609–11614.

137 HAYER-HARTL, M. K., EWBANK, J. J., CREIGHTON, T. E., & HARTL, F. U. (1994). Conformational specificity of the chaperonin GroEL for the compact folding intermediates of alpha-lactalbumin. *EMBO J.* **13**, 3192–3202.

138 LINDNER, R. A., KAPUR, A., & CARVER, J. A. (1997). The interaction of the molecular chaperone, alpha-crystallin, with molten globule states of bovine alpha-lactalbumin. *J. Biol. Chem.* **272**, 27722–27729.

139 SCHMIDT, M., BUCHNER, J., TODD, M. J., LORIMER, G. H., & VIITANEN, P. V. (1994). On the role of groES in the chaperonin-assisted folding reaction. Three case studies. *J. Biol. Chem.* **269**, 10304–10311.

140 SCHRODER, H., LANGER, T., HARTL, F. U., & BUKAU, B. (1993). DnaK, DnaJ and GrpE form a cellular chaperone machinery capable of repairing heat-induced protein damage. *EMBO J.* **12**, 4137–4144.

141 BUCHBERGER, A., SCHRODER, H., HESTERKAMP, T., SCHONFELD, H. J., & BUKAU, B. (1996). Substrate shuttling between the DnaK and GroEL systems indicates a chaperone network promoting protein folding. *J. Mol. Biol.* **261**, 328–333.

142 ZOLKIEWSKI, M. (1999). ClpB cooperates with DnaK, DnaJ, & GrpE in suppressing protein aggregation. A novel multi-chaperone system from *Escherichia coli. J. Biol. Chem.* **274**, 28083–28086.

143 DUMITRU, G. L., GROEMPING, Y., KLOSTERMEIER, D., RESTLE, T., DEUERLING, E., & REINSTEIN, J. (2004). DafA Cycles Between the DnaK Chaperone System and Translational Machinery. *J. Mol. Biol.* **339**, 1179–1189.

144 FARAHBAKHSH, Z. T., HUANG, Q. L., DING, L. L., ALTENBACH, C., STEINHOFF, H. J., HORWITZ, J., & HUBBELL, W. L. (1995). Interaction of alpha-crystallin with spin-labeled peptides. *Biochemistry* **34**, 509–516.

145 MILLER, A. D., MAGHLAOUI, K., ALBANESE, G., KLEINJAN, D. A., & SMITH, C. (1993). *Escherichia coli* chaperonins cpn60 (groEL) and cpn10 (groES) do not catalyse the refolding of mitochondrial malate dehydrogenase. *Biochem. J.* **291**, 139–144.

146 SPARRER, H., LILIE, H., & BUCHNER, J. (1996). Dynamics of the GroEL-protein complex: effects of nucleotides and folding mutants. *J. Mol. Biol.* **258**, 74–87.

147 HAYER-HARTL, M. K., WEBER, F., & HARTL, F. U. (1996). Mechanism of chaperonin action: GroES binding and release can drive GroEL-mediated protein folding in the absence of ATP hydrolysis. *EMBO J.* **15**, 6111–6121.

148 VIITANEN, P. V., LUBBEN, T. H., REED, J., GOLOUBINOFF, P., O'KEEFE, D. P., & LORIMER, G. H. (1990). Chaperonin-facilitated refolding of ribulosebisphosphate carboxylase and ATP hydrolysis by chaperonin 60 (groEL) are K^+ dependent. *Biochemistry* **29**, 5665–5671.

149 RYE, H. S., BURSTON, S. G., FENTON, W. A., BEECHEM, J. M., XU, Z., SIGLER, P. B., & HORWICH, A. L. (1997). Distinct actions of *cis* and *trans* ATP within the double ring of the chaperonin GroEL. *Nature* **388**, 792–798.

150 SRIRAM, M., OSIPIUK, J., FREEMAN, B., MORIMOTO, R., & JOACHIMIAK, A. (1997). Human Hsp70 molecular chaperone binds two calcium ions within the ATPase domain. *Structure* **5**, 403–414.

151 VAN MONTFORT, R. L., BASHA, E., FRIEDRICH, K. L., SLINGSBY, C., & VIERLING, E. (2001). Crystal structure and assembly of a eukaryotic small heat shock protein. *Nat. Struct. Biol.* **8**, 1025–1030.

6
Physical Methods for Studies of Fiber Formation and Structure

Thomas Scheibel and Louise Serpell

6.1
Introduction

Elucidating protein fiber assembly and analyzing the acquired regular fibrous structures have engaged the interest of scientists from different disciplines for a long time. The complexity of the problem, together with the paucity of interpretable data, has generated a large volume of literature concerned with speculations about possible assembly mechanisms and structural conformations.

Fibrous proteins are essential building blocks of life, providing a scaffold for our cells, both intracellularly and extracellularly. However, despite nearly a century of research of the assembly mechanisms and structures of fibrous protein, we have only recently begun to get insights. In the last 10 years, atomic models for some of the most important fibrous proteins have been proposed. We are increasingly knowledgeable about how fibrous proteins form and how they function (for example, the interaction of actin and myosin to produce muscular movement). Development of methods specific to the study of fibrous proteins has enabled these investigations, since fibrous proteins are generally insoluble and often heterogeneous in length, and therefore methods used for examining the structure of globular proteins cannot be used.

In this chapter, an overview of some commonly used methods to examine the assembly and structure of fibrous proteins is given. However, additional methods can be suitable to analyze fibrous proteins. Among the methods not discussed are solid-state NMR (ssNMR), near-field scanning optical microscopy (NSOM), scanning transmission electron microscopy (STEM), and light microscopy, for all of which reviews can be found elsewhere. Other complementary methods, specific for the particular protein, include cross-linking studies and antibody-binding epitope studies. These are valuable techniques that can help us to interpret structural data and contribute to the understanding of the assembly pathway, but due to space limitations they can not be discussed in this chapter.

Protein Folding Handbook. Part II. Edited by J. Buchner and T. Kiefhaber
Copyright © 2005 WILEY-VCH Verlag GmbH & Co. KGaA, Weinheim
ISBN: 3-527-30784-2

6.2
Overview: Protein Fibers Formed in Vivo

6.2.1
Amyloid Fibers

The term "amyloid" describes extracellular, fibrillar protein deposits associated with disease in humans. The deposits accumulate in various tissue types and are defined by their tinctorial and morphological properties [1]. Although the involved proteins are distinct with respect to amino acid sequence and function, the fibrils formed by different amyloidogenic proteins are structured similarly. This observation is remarkable since the soluble native forms of these proteins vary considerably: some proteins are large, some are small, and some are largely α-helical (see chapter 9 in Part I) while some are largely β-sheet (see chapter 10 in Part I). Some are intact in the fibrous form, while others are at least partially degraded. Some are cross-linked with disulfide bonds and some are not. Amyloid fibers are unusually stable and unbranched, range from 6 to 12 nm in diameter, and specifically bind the dye Congo red (CR), showing an apple-green birefringence in polarized light [2, 3]. Amyloid fibers are also rich in β-sheet structure and produce a characteristic "cross-β" X-ray diffraction pattern, consistent with a model in which stacked β-sheets form parallel to the fiber axis with individual β-strands perpendicular to the fiber axis [1, 4]. Diseases characterized by amyloids are referred to as amyloidoses (see chapters 33, 36, 37); these diseases are slow in onset and degenerative and can be found not only in humans but also in most mammals. Some of the amyloidoses are familial, while some are associated with medical treatment (e.g., hemodialysis) or infection (prion diseases), and some are sporadic (e.g., most forms of Alzheimer's disease). Certain diseases, such as the amyloidoses associated with the protein transthyretin, can be found in both sporadic and familial forms. In addition to diseases with extracellular deposits, there are others, notably Parkinson's and Huntington's disease, that appear to involve very similar aggregates, which are intracellular, and these diseases are therefore not included in the strict definition of amyloidoses [5, 6].

Amyloidogenic proteins are not found only in mammals. Studies of *Saccharomyces cerevisiae* have resulted in the discovery of proteins that have properties related to those of mammalian prions [7, 8] (see chapters 34, 35). These yeast prions have been found to convert in vitro into ordered aggregates with all the characteristics of amyloid fibers [9]. Also, amyloidogenic peptides are found in egg envelopes (chorions) of invertebrates. More than 95% of the chorion's dry mass consists of proteins that have remarkable mechanical and chemical properties that protect the oocyte and the developing embryo from a wide range of environmental hazards. About 200 proteins have been detected and classified into two major classes, A and B. Interestingly, within the chorion, protein fibrils can be detected with characteristics of amyloid fibrils. These features include a high proportion of intermolecular β-sheet, positive staining and green birefringence with CR, and detection of long, unbranched fibrils with a diameter of 4–6 nm [10, 11].

6.2.2
Silks

Arthropods produce a great variety of silks that are used in the fabrication of structures outside of the body of the animal. Silks are typically fibrous filaments or threads with diameters ranging from a few hundred nanometers to 50 µm. Insects and spiders use their silks for lifelines, nests, traps, and cocoons. Characteristically, the silk is produced in specialized glands and stored in a fluid state in the lumen of the gland. As the fluid passes through the spinning duct, a rapid transformation to the solid state takes place and the silk becomes water-insoluble [12–15]. However, some of the materials that are called silks are not produced by silk glands in the traditional sense but by extrusions of the guts, e.g., in antlions, lacewings, and many beetles [16].

The major components of traditional silks are fibrous proteins, which in the final insoluble state have a high content of regular secondary structure. Wide variations in composition and structure have been found among different silks. Therefore, the classification of silks is often based on the predominant form of regular secondary structure in the final product as determined by X-ray diffraction studies. Silks are classified as beta silks (which are rich in β-sheets; see Section 6.3.1.3), alpha silks (which resemble the X-ray diffraction pattern obtained from α-keratin and which have a much lower content of Gly and a much higher content of charged residues than any other silk; see Section 6.3.2.4), Gly-rich silks (which form a polyglycine II-like structure), and collagen-like silks (which resemble the X-ray diffraction pattern given by collagen; see Section 6.3.2.1). While silks from the class of Arachnida (spiders) are generally β-silks, in the class of Insecta all four classes of silk can be found, e.g., β-silks in the family of Bombycidae (moths, includes *Bombyx mori*, whose caterpillar is the silkworm), α-silks in the family of Apidae (bees), Gly-rich silks in the family of Blennocampinae (subspecies of sawflies), and collagen-like silks in the family of Nematinae (subspecies of sawflies) [15, 17–20]. In contrast, silks from extrusions of the guts do not form traditional silk filaments or threads; rather, they are hardened fibrous feces.

6.2.3
Collagens

Collagens are a family of highly characteristic fibrous proteins found in all multicellular animals (see chapter 30 in Part I). In combination with elastin (see Section 6.2.8), mucopolysaccharides, and mineral salts, collagens form the connective tissue responsible for the structural integrity of the animal body. The characteristic feature of a typical collagen molecule is its long, stiff, triple-stranded helical structure, in which three collagen polypeptide chains, called α-chains, are wound around each other in a rope-like superhelix (see Section 6.3.2.1). The polypeptide chains in collagens have a distinctive conformation that is intimately related to their structural function [21–25]. All collagen polypeptides have a similar primary structure, in which every third amino acid is Gly (Gly-Xaa-Yaa)$_n$ with a predominance of Pro residues at position Xaa and Yaa. Many of the Pro residues, as well

Mg^{2+}-dependent [49]. High-molecular-weight tropomyosins consist most often of 284 amino acids and bind seven actin monomers, while the low-molecular-weight tropomyosins consist of about 245–250 amino acids and span six actins.

The thick filaments of invertebrate muscles contain, in addition to myosin, the protein paramyosin, a rod-like, coiled-coil molecule similar to myosin [53]. Paramyosin filaments play an important part in certain molluscan and annelid muscles because they allow these muscles to maintain tension for prolonged periods without a large expenditure of energy.

6.2.5
Intermediate Filaments/Nuclear Lamina

Intermediate filaments are tough, durable, 10-nm wide protein fibers found in the cytoplasm of most, but not all, animal cells, particularly in cells that are subject to mechanical stress [54]. Unlike actin and tubulin, which are globular proteins, monomeric intermediate filament proteins are highly elongated, with an amino-terminal head, a carboxy-terminal tail, and a central rod domain. The central rod domain consists of an extended α-helical region containing long tandem repeats of a distinct amino acid sequence motif called heptad repeats. These repeats promote the formation of coiled-coil dimers between two parallel α-helices. Next, two of the coiled-coil dimers associate in an antiparallel manner to form a tetrameric subunit. The antiparallel arrangement of dimers implies that the tetramer and the intermediate filaments formed thereof are non-polarized structures. This distinguishes intermediate filaments from microtubules (see Section 6.2.7) and actin filaments (see Section 6.2.4), which are polarized and whose functions depend on this polarity [55–58].

The cytoplasmic intermediate filaments in vertebrate cells can be grouped into three classes: keratin filaments, vimentin/vimentin-related filaments, and neurofilaments. By far the most diverse family of these proteins is the family of α-keratins, with more than 20 distinct proteins in epithelia and over eight more so-called hard keratins in nails and hair. The α-keratins are evolutionarily distinct from β-keratins found in bird feathers, which have an entirely different structure and will not be further discussed in this context. Based on their amino acid sequence, α-keratins can be subdivided into type I (acidic) and type II (neutral/basic) keratins, which always form heterodimers of type I and type II but never homodimers [59]. Unlike keratins, the class of vimentin and vimentin-related proteins (e.g., desmin, peripherin) can form homopolymers [60]. In contrast to the first two classes of intermediate filaments, the third one, neurofilaments, is found uniquely in nerve cells, wherein they extend along the length of an axon and form its primary cytoskeletal component [61].

The nuclear lamina is a meshwork of intermediate filaments that lines the inside surface of the inner nuclear membrane in eukaryotic cells. In mammalian cells the nuclear lamina is composed of lamins, which are homologous to other intermediate filament proteins but differ from them in several ways. Lamins have a longer central rod domain, contain a nuclear localization signal, and assemble into a two-dimensional, sheet-like lattice, which is unusually dynamic. This lattice rapidly disassembles at the start of mitosis and reassembles at the end of mitosis [61, 62].

6.2.6
Fibrinogen/Fibrin

Fibrinogen is a soluble blood plasma protein, which during the process of clotting is modified to give a derived product termed fibrin. Fibrinogen is a multifunctional adhesive protein involved in a number of important physiological and pathological processes. Its multifunctional character is connected with its complex multidomain structure. Fibrinogen consists of two identical subunits, each of which is formed by three nonidentical polypeptide chains, Aα, Bβ, and γ, all of which are linked by disulfide bonds and assemble into a highly complex structure [63–65]. Conversion of fibrinogen into fibrin takes place under the influence of the enzyme thrombin. The proteolysis results in spontaneous polymerization of fibrin and formation of an insoluble clot that prevents the loss of blood upon vascular injury. This clot serves as a provisional matrix in subsequent tissue repair. The polymerization of fibrin is a multi-step process. First, two stranded protofibrils are generated, which associate with each other laterally to produce thicker fibrils. The fibrils branch to form the fibrin clot, which is subsequently stabilized by the formation of intermolecular cross-linkages [66]. Fibrinogen is a rather inert protein, but after conversion fibrin interacts with different proteins, such as tissue-type plasminogen activator, fibronectin, and some cell receptors, and participates in different physiologically important processes, including fibrinolysis, inflammation, angiogenesis, and wound healing [67–69].

6.2.7
Microtubules

Microtubules are stiff polymers formed from molecules of tubulin. Microtubules extend throughout the cytoplasm and govern the location of membrane-bounded organelles and other cell components. Particles can use the microtubules as tracks, and motor proteins, such as kinesin and dynein, can move, e.g., organelles along the microtubules [70, 71]. The protein building blocks of microtubules, tubulins, are heterodimers consisting of two closely related and tightly linked globular polypeptides, α-tubulin and β-tubulin. In mammals there are at least six forms of α-tubulin and a similar number of forms of β-tubulin. The different tubulins are very similar and copolymerize into mixed microtubules in vitro, although they can have distinct cellular locations. A microtubule can be regarded as a cylindrical structure with a 25-nm diameter, which is built from 13 linear protofilaments, each composed of alternating α- and β-tubulin subunits and bundled in parallel to form the cylinder. Since the 13 protofilaments are aligned in parallel with the same polarity, the microtubule itself is a polar structure and it is possible to distinguish both ends [71–74]. Microtubules reflect dynamic protein polymers that can polymerize and depolymerize depending on their biological function. Tubulin will polymerize into microtubules in vitro, as long as Mg^{2+} and GTP are present, by addition of free tubulin to the free end of the microtubule. Hydrolysis of the bound GTP takes place after assembly and weakens the bonds that hold the microtubule together. Without stabilization by cellular factors capping the free ends, microtubules become extremely unstable and disassemble [73, 75].

6.2.8
Elastic Fibers

The proteins elastin, fibrillin, and resilin occur in the highly elastic structures found in vertebrates and insects, which are at least five times more extensible than a rubber band of the same cross-sectional area. Elastin is a highly hydrophobic protein that, like collagen, is unusually rich in Pro and Gly residues, but unlike collagen it is not glycosylated and contains little hydroxyproline and no hydroxylysine. Elastin molecules are secreted into the extracellular space and assemble into elastic fibers close to the plasma membrane. After secretion, the elastin molecules become highly cross-linked between Lys residues to generate an extensive network of fibers and sheets. In its native form, elastin is devoid of regular secondary structure [76–79].

The elastic fibers are not composed solely of elastin, but the elastin core is covered with microfibrils that are composed of several glycoproteins including the large glycoprotein fibrillin [80, 81]. Microfibrils play an important part in the assembly of elastic fibers. They appear before elastin in developing tissues and seem to form a scaffold on which the secreted elastin molecules are deposited. As the elastin is deposited, the microfibrils become displaced to the periphery of the growing fiber [81].

Elastin and resilin are not known to be evolutionarily related, but their functions are similar. Resilin forms the major component in several structures in insects possessing rubber-like properties. Similar to elastin, the polypeptide chains are three-dimensionally cross-linked and are devoid of regular secondary structure. The amino acid composition of resilin resembles that of elastin insofar as the Gly content is high, but otherwise there is little sequence homology between these two proteins [19, 79, 82].

6.2.9
Flagella and Pili

Prokaryotes and archaea use a wide variety of structures to facilitate motility, most of which include fibrillar proteins. The best studied of all prokaryotic motility structures is the bacterial flagellum, which is composed of over 20 different proteins. The bacterial flagellum is a rotary structure driven from a motor at the base and a protein fiber acting as a propeller. The propeller consists of three major substructures: a filament, a hook, and a basal body. The filament is a tubular structure composed of 11 protofilaments about 20 nm in diameter, and the protofilaments are typically built by a single protein called flagellin. Less commonly, the filament is composed of several different flagellins [83–85]. Strikingly, monomeric flagellin does not polymerize under physiological conditions. The amino- and carboxy-terminal regions of flagellin monomers are known to be intrinsically unfolded, which prevents flagellin from assembling into a nucleus for polymerization [86, 87]. Flagellin assembly is regulated through a combination of transcriptional, translational, and posttranslational regulatory mechanisms too complex to be explained in detail in this chapter. Only the unusual filament growth should be mentioned, since individual flagellin monomers are added to the growing polymer not at the base closest to the cell surface, but at the distal tip furthest from the cell, probably by passing through the hollow interior of the tubule [85, 88].

Motility is also conferred by flagella in the domain archaea, yet these structures and the sequences of the involved proteins bear little similarity to their bacterial counterparts. Rather, archaeal flagella demonstrate sequence similarity to another bacterial motility apparatus named pili. Both types of proteins have extremely hydrophobic amino acids over the first 50 amino acids and, in contrast to bacterial flagellin, are made as preproteins with short, positively charged leader peptides that are cleaved by leader peptidases. In archaea multiple flagellins are found that form a filament with a diameter of approximately 12 nm [89, 90].

Pili are responsible for various types of flagellar-independent motility, such as twitching motility in prokaryotes, and they are formed of pilins (see chapter 29). The outer diameter of a pilus fiber is about 6 nm, and, unlike in bacterial flagellum, there is no large channel in the center of the fiber. Therefore, pili grow by adding monomers at the base closest to the cell surface. The amino terminus of pilin forms α-helices, which compose the core of the pilus fiber. The outside of the pilus fiber is made of β-sheets packed against the core and an extended carboxy-terminal tail [90–92].

6.2.10
Filamentary Structures in Rod-like Viruses

Filamentous phages comprise a family of simple viruses that have only about 10 genes and grow in gram-negative bacteria. Different biological strains have been isolated, all of which have a similar virion structure and life cycle. The most prominent filamentous phage is bacteriophage fd (M13), which infects *E. coli*. The wild-type virion is a flexible rod about 6 nm in diameter and 800–2000 nm in length. Several thousand identical α-helical proteins form a helical array around a single-stranded circular DNA molecule at the core. It is assumed that the protein subunits forming the helical array pre-assemble within the membrane of the host into ribbons with the same inter-subunit interactions as in the virion. A ribbon grows by addition of newly synthesized protein subunits at one end, and at the same time the other end of a ribbon twists out of the membrane and merges into the assembling virion [93–95].

Some proteins of plant viruses, such as the tobacco mosaic virus (TMV), also form self-assembled macromolecules. The virion of TMV is a rigid rod (18 $\times$ 300 nm) consisting of 2130 identical virion coat protein (CP) subunits stacked in a helix around a single strand of plus-sense RNA. Depending upon solution conditions, TMV CP forms three general classes of aggregates, the 4S or A protein composed of a mixture of low-ordered monomers, dimers, and trimers; the 20S disk or helix composed of approximately 38 subunits; and an extended virion-like rod. Under cellular conditions, the 20S helix makes up the predominant CP aggregate [96–99]. In the absence of RNA, the structure within the central hole of the 20S filament is disordered, but upon RNA binding structural ordering occurs, which locks the nucleoprotein complex into a virion-like helix and allows assembly to proceed [100]. Virion elongation occurs in both the 5′ and 3′ directions of the RNA. Assembly in the 5′ direction occurs rapidly and is consistent with the addition of 20S aggregates to the growing nucleoprotein rod. Assembly in the 3′ direction occurs considerably slower and is thought to involve the addition of 4S particles [97, 99].

6.2.11
Protein Fibers Used by Viruses and Bacteriophages to Bind to Their Hosts

Certain viruses and bacteriophages use fibrous proteins to bind to their host receptors. Examples are adenovirus, reovirus, and the T-even bacteriophages, such as bacteriophage T4. The fibers of adenovirus, reovirus, and T4 share common structural features: they are slim and elongated trimers with asymmetric morphologies, and the proteins consist of an amino-terminal part that binds to the viral capsid, a thin shaft, and a globular carboxy-terminal domain that attaches to a cell receptor. Strikingly, the fibers formed of these proteins are resistant to proteases, heat, and, under certain conditions, sodium dodecyl sulfate (SDS) [101]. The assembly of the tail spike from *Salmonella* phage P22 is the most extensively studied model system and has allowed detailed study of in vitro and in vivo folding pathways in light of structural information [102].

6.3
Overview: Fiber Structures

Structural elucidation of fibrous proteins is nontrivial and often involves the use of several structural techniques combined to derive a model structure. Structural studies have often involved X-ray crystallography from single crystals of monomeric or complexed proteins to elucidate the structure of fibrous protein subunits. Electron microscopy with helical image reconstruction and X-ray fiber diffraction have then been used to examine the structure of the fiber, and, where possible, modeling of a crystal structure has been matched with the electron microscopy maps or X-ray diffraction data. Protein fiber structures that have been extensively studied include natural proteins such as actin, tubulin, collagen, intermediate filaments, silk, myosin, and tropomyosin as well as misfolded protein fibers such as amyloid, paired helical filaments, and serpin polymers. In this section, a few selected examples will be discussed.

6.3.1
Study of the Structure of β-sheet-containing Proteins

6.3.1.1 Amyloid

Amyloid fibrils extracted from tissue are extremely difficult to study due to contaminants and the degree of order within the fibrils themselves. For this reason, structural studies of amyloid fibrils have concentrated on examining the structure of amyloid formed in vitro from various peptides with homology to known amyloid-forming proteins. It was clear from the early work of Eanes and Glenner [2] that amyloid fibrils resembled cross-β silk (see Section 6.3.1.3), in that the fiber diffraction pattern gave the cross-β diffraction signals. This means that the diffraction pattern consists of two major diffraction signals at 4.7 Å and around 10 Å, on perpendicular axes, that correspond to the spacing between hydrogen-bonded β-strands and the intersheet distance, respectively. Models of amyloid fibrils have been built using X-ray diffraction data [1, 103–106], solid-state NMR data [107–109], and electron microscopy [110, 111]. Many of these models have a common core structure in which the β-strands run perpendicular to the fiber axis and are hydrogen-bonded along the length of the fiber. Whether the amyloid fibrils are composed of parallel or antiparallel β-sheets may vary depending on the composite protein. X-ray diffraction and FTIR data from various amyloid fibrils have suggested an antiparallel β-sheet arrangement for some amyloid fibrils formed from short peptides [105]. Solid-state NMR data from full-length Aβ peptides appear to be consistent with a parallel arrangement [108]. Alternative models of amyloid fibril core structure involving β-helical structure have been suggested [112, 113]. The mature amyloid fibril is thought to be composed of several protofilaments; evidence for this comes from electron microscopy [111, 114] and atomic force microscopy data [115].

6.3.1.2 Paired Helical Filaments

Paired helical filaments (PHFs) are found deposited in disease and are composed of the microtubule-associated protein Tau [116]. PHFs are found in neurofibrillary tangles (NFT) in Alzheimer's disease brains as well as in a group of neurodegenerative diseases collectively known as the tauopathies. Electron microscopy structural work on PHFs has revealed that the filaments are composed of two strands twisted around one another, giving the characteristic appearance [117]. Straight filaments (SF) have also been isolated and are thought to be composed of the same building blocks. Calculated cross-sectional views showed that C-shaped subunits make up each strand for both PHFs and SFs [117]. More recently, X-ray and electron diffraction from synthetic and isolated PHFs have revealed a cross-β structural arrangement, similar to that shown for amyloid [118], indicating that the core structure of PHFs may share the generic amyloid-like structure.

6.3.1.3 β-Silks

All silk fibroins appear to contain crystalline regions or regions of long-range order, interspersed with short-range-ordered, non-crystalline regions [14]. Silks spun by different organisms vary in the arrangement, proportions, and length of the crystalline and non-crystalline regions. Generally, β-pleated sheets are packed into tight crystalline arrays with spacer regions of more loosely associated β-sheets, α-helices, and/or β-spirals [14]. X-ray fiber diffraction has been used extensively to examine the structure of silks made by different spiders and insects [14]. The crystalline domains of *Bombyx mori* fibroin silk are composed of repeating glycine-alanine residues and form an antiparallel arrangement of β-pleated strands that run parallel to the fiber axis. For this structural arrangement, a model was proposed by Marsh, Corey, and Pauling [119], in which the intersheet distance is about 5.3 Å, which accommodates the small side chains. A recent refinement of the model [120] showed that the methyl groups of the alanine residues alternately point to either side of the sheet structure along the hydrogen-bonding direction, so that the structure consists of two antipolar, antiparallel β-sheet structures. Major ampullate (dragline) silk from the spider *Nephila clavipes* has been examined by X-ray diffraction and solid-state NMR, and, again, the silk structure is dominated by β-strands aligned parallel to the fiber axis [121]. Tussah silk, which is the silk of the caterpillar of the moth *Antherea mylitta*, also has an antiparallel arrangement of β-strands parallel to the fiber axis and gives an X-ray diffraction pattern similar to that of *Bombyx mori* silk, although some significant differences are observed. These differences are thought to arise from different modes of sheet packing due to the high polyalanine content of Tussah silk [19].

Cross-β silk was identified in the egg stalk of the green lacewing fly, *Chrysopa flava* [122]. The X-ray diffraction pattern was examined in detail and found to arise from an arrangement where the β-strands run orthogonal to the fiber axis and hydrogen bonding extends between the strands forming a β-sheet that runs the length of the fiber. A model was proposed by Geddes et al. [4] in which the β-pleated strand folded back on to itself to form antiparallel β-sheet ribbons. The length of the β-strands was about 25 Å, giving the ribbon a width of 25 Å. Several ribbons then stacked together in the crystallite. This model has been used as the basis of the amyloid core structure (see Section 6.3.1.1).

6.3.1.4 β-Sheet-containing Viral Fibers

Adenovirus fiber shaft, P22 tail spike, and bacteriophage T4 short and long fibers are known to be predominantly β-sheet-structured [101]. The shaft regions of the adenovirus and bacteriophage T4 fibers contain repeating sequences, indicating that they might fold into regular, repeating structures. The adenovirus type-2 fiber shaft has been shown to fold into a novel triple β-spiral. The crystal structure of a four-repeat fragment was solved and showed that each repeat folded into an extended β-strand that runs parallel to the fiber axis [123]. Each β-strand is followed by a sharp β-turn enabled by the positioning of a proline or glycine residue, and the turn is then followed by another β-strand, which runs backward at an angle of about 45° to the fiber shaft axis. This fold results in an elongated molecule with β-strands spiraling around the fiber axis. A model of the fiber shaft suggests a diameter of around 15–22 Å and a length of around 300 Å. This fold is also thought to be included in other viral fiber structures (e.g., reovirus structure).

Several proteins have been shown to fold into a parallel β-helix, including pectate lyase [124] and the bacteriophage P22 tail spike [125]. These have right-handed, single-chain helices that have a bean-shaped cross-section, and the β-strands run perpendicular to the long axis. Left-handed β-helices (e.g., UDP N-acetylglucosamine acyltransferase) have a triangular cross-section, as each turn of the helix is made up of three β-strands, connected by short linkers [126]. A triple β-helix fold has been identified in part of the short-tail fiber of bacteriophage T4, gp12. It has a triangular cross-section and is made up of three intertwined chains [127]. The β-strands are again perpendicular to the long axis. This arrangement has also been identified in a region of P22 tail spike, and a similar fold has been found in bacteriophage T4 protein gp5. The arrangement forms the needle used by T4 to puncture host cell membranes [128]. Very recently, the structure of the T4 baseplate-tail complex was determined to 12 Å by cryo-electron microscopy, and this has enabled the known crystal structures to be fitted into the cryo-electron microscopy map to build up a full structure of the T4 tail tube and the arrangement of proteins at the baseplate [129].

6.3.2
α-Helix-containing Protein Fibers

6.3.2.1 Collagen

The structure of the collagen superfamily has been studied for over 50 years. Collagen has a complex supramolecular structure, involving the interaction of different molecules giving different functions. It has a characteristic periodic structure, forming many levels of association [130]. The primary structure consists of a repeating motif, Gly-Xaa-Yaa, interspersed by more variable regions. The secondary structure involves the central domain of collagen α-chain folding into a tight, right-handed helix with an axial residue-to-residue distance determined by the spacing of the Pro residues (Xaa) and hydroxyprolines (Yaa). This arrangement results in a Gly residue row following a slightly left-handed helix winding around the coil, and this is responsible for the packing to form the tertiary structure, a triple helix, where the Glys are buried inside the supercoiled molecules [130]. Depending on the collagen type, the triple helices may be comprised of differing collagen chains or of three identical chains [130]. The triple helix is a long, rod-like molecule around 1.5 nm wide. Recently, the crystal structure of the collagen triple helix formed synthetically from [(Pro-Pro-Gly)$_{10}$]$_3$ peptide was solved to a resolution of 1.3 Å [25] (Figure 6.1). The triple helix undergoes fibrillogenesis to form supramolecular structures. Non-helical ends that do not contain the Gly-Xaa-Yaa sequence are known as the telopeptides. Many models of the molecular packing have been presented (summarized in Ref. [130]). The most recent model of type I collagen appears to fit data from X-ray diffraction, electron microscopy, and cross-linking studies [130]. The proposed model is a one-dimensional, staggered, left-handed microfibril. The packing allows for interconnections between the microfibrils via the telopeptides [131]. This model was deduced by careful testing of several models to compare calculated and observed X-ray diffraction data [132].

tals of zinc-induced tubulin sheets in ice, and the structural interpretation was supported by the X-ray crystal structure of FtsZ [137]. Each monomer is a compact ellipsoid made up of three domains, an amino terminal nucleotide-binding domain, a smaller second domain, and a helical carboxy terminal domain [140]. The tubulin heterodimers polymerize in a polar fashion to make longitudinal protofilaments that in turn bundle to form the microtubule. The dimer structure was docked into 20-Å maps from cryo-electron microscopy of microtubules, creating a near-atomic model of the microtubule [140]. This model revealed that the carboxy terminal helices form a crest on the outside of the protofilaments and that the long loops define the microtubule lumen [140]. The microtubule is polar, with plus and minus ends, and the structure showed that the exchangeable nucleotide bound to β-tubulin was exposed at the plus end of the microtubule, whereas the proposed catalytic residue in α-tubulin was exposed at the minus end.

Cryo-electron microscopy and helical image reconstruction techniques have proved particularly powerful in examining the arrangement and interaction of motor proteins bound to microtubules [141]. Three-dimensional maps of microtubules decorated with motor domains or kinesin in different nucleotide states could be compared to naturally occurring microtubules (Figure 6.2, see p. 211) [141].

6.3.3.2 Actin and Myosin Filaments

The structure of actin filaments has been extensively studied using X-ray diffraction and electron microscopy [142]. Early work by Hanson and Lowry [143] revealed the basic geometry of F-actin by negative stain electron microscopy. Major advances were made when the crystal structure of actin was solved in complex with DNase I [144]. Since then, a more detailed picture of the structure of the actin filament has been built up [145, 146] by positioning the atomic resolution crystal structure against X-ray diffraction data. The filament is composed of actin monomers arranged in a left-handed helix with 13 monomers per six turns. The pitch is about 59 Å and the separation between monomers is about 27.5 Å. Two strands of actin monomers slowly twist around one another, with a cross-over repeat of 360–380 Å. Tropomyosin chains bind along the actin filament in the grooves between the two actin strands. Each tropomyosin has a troponin complex bound, and the positioning of the troponin marks a 385-Å repeat along the axis of the actin filaments.

X-ray fiber diffraction has been utilized to examine changes in the structure of muscle fibers in order to try to understand the mechanism of muscle contraction [147]. Muscular force is produced when myosin filaments interact with actin filaments by an ATP-dependent process. Time-resolved, low-angle diffraction patterns from muscle in various states can be taken and the positions and intensities of signal can be compared [147] to allow conclusions to be made about the movement of various components of the muscle fibers.

Myosin filaments make up the thick filaments of striated muscle and the structure has also been examined using X-ray diffraction and electron microscopy. Modeling of the arrangement of the myosin heads from X-ray and electron microscopy data have been performed, but the arrangement of the heads has been inconclusive. Four arrangements of the heads are possible, whereby the heads may be parallel or antiparallel. In each case the two heads may lie on the same helical track in a compact arrangement or on neighboring tracks in a splayed arrangement [148]. Recent work on electron microscopy data from filaments of the spider *Brachypelma smithi* (tarantula) [149] tested the four possible arrangements by fitting the crystal structure of myosin heads [150] into the map generated from electron microscopy of negatively stained thick filaments. A good fit was obtained when the myosin heads were arranged antiparallel and packed on the same helical ridge.

6.4
Methods to Study Fiber Assembly

6.4.1
Circular Dichroism Measurements for Monitoring Structural Changes Upon Fiber Assembly

6.4.1.1 Theory of CD

Circular dichroism (CD) is a phenomenon that gives information about the unequal absorption of left- and right-handed circularly polarized light by optically active molecules in solution [151–154] (see chapter 2 in Part I). CD signals are observed in the same spectral regions as the absorption bands of a particular compound, if the respective chromophores or the molecular environment of the compound are asymmetric. CD signals of proteins occur mainly in two spectral regions. The far-UV or amide region (170–250 nm) is dominated by contributions of the peptide bonds, whereas CD signals in the near-UV region (250–320 nm) originate from aromatic amino acids. In addition, disulfide bonds give rise to minor CD signals around 250 nm. The two main spectral regions give different kinds of information about protein structure.

CD signals in the amide region contain information about the peptide bonds and the secondary structure of a protein and are frequently employed to monitor changes in secondary structure in the course of structural transitions [155–157]. Since many fiber-forming proteins undergo certain conformational changes upon fiber assembly (e.g., from an intrinsically unfolded state to a β-sheet-rich fiber as seen in some amyloid-forming proteins; see Figure 6.3), far-UV CD measurements can be used to monitor the kinetics of fiber assembly.

CD signals in the near-UV region are observed when aromatic side chains are immobilized in a folded protein and thus transferred to an asymmetric environment. Since Tyr and Phe residues yield in very small near-UV CD signals, only Trp residues lead to utilizable near-UV CD spectra. The CD signal of aromatic residues is very small in the absence of ordered structure (e.g., in short peptides) and therefore serves as a good tool for monitoring fiber assembly of peptides (in case Trp residues are present). The magnitude as well as the wavelengths of the aromatic CD signals cannot be predicted, since they depend on the structural and electronic environment of the immobilized chromophores. Therefore, the individ-

ANS-protein interaction [174, 175]. Importantly, the dependence of ANS binding on electrostatic interaction between the sulfonate group and protein cationic groups does not require preexisting hydrophobic sites on or in the protein to start the binding reaction. Moreover, ANS fluorescence is quenched by water, so fluorescence is profoundly affected if a change in protein hydration or ANS accessibility to water occurs during binding [176]. Therefore, ANS can be used not only to monitor protein unfolding but also to analyze structural changes upon fiber assembly.

6.4.4.2 Experimental Guide to Using ANS for Monitoring Protein Fiber Assembly

ANS has been used to study the partially folded states of globular proteins as well as the binding pockets of a number of carrier proteins and enzymes. ANS has also been used to characterize fibrillar forms of proteins, such as fibronectin, islet amyloid polypeptide, or infectious isoforms of the prion protein PrP [177–180]. After binding to fibers, ANS can show an increased fluorescence and a blue shift of its fluorescence maxima. Protein concentrations should be in the range of 0.5–20 µM and ANS should be added in a 10- to 100-fold excess. The concentration of ANS can be determined by using a molar extinction coefficient of 6.8×10^3 M^{-1} cm^{-1} at 370 nm in methanol [181]. In a 1-cm cell the fluorescence inner-filter effect (see Section 6.4.2.2) becomes significant at ~ 30 µM ANS. Therefore, correction factors have to be used for the fluorescence of ANS when its total concentration exceeds 10 µM. The magnitude of the inner-filter effect must be experimentally determined for each instrument and whenever the instrumental configuration is altered. The magnitude of the correction depends on the wavelength range and the path length, but not on slit-width or sample turbidity for most of the instruments [182].

To determine whether ANS binds to the fibrous protein, but not to the soluble counterpart, fluorescence emission spectra should be recorded from 400–650 nm with an excitation wavelength of 380 nm. After measuring ANS fluorescence spectra before and after fiber assembly, it can be determined whether ANS serves as a tool for investigating fiber assembly of the chosen protein. Extreme care has to be taken in determining whether ANS fluorescence changes are based solely on fiber formation or whether ANS is also able to bind non-fibrous assembly intermediates. If ANS binding is specific to the fibrous form of the protein, the wavelength with maximal fluorescence differences before and after assembly can be used to monitor fiber assembly kinetics.

6.4.5
Light Scattering to Monitor Particle Growth

Hydrodynamic studies can be carried out to obtain qualitative descriptions of the shape and dimensions of macromolecules and to gain precise information on molecular geometry [183] (see chapter 19 in Part I). The interpretation of hydrodynamic measurements is always based on an assumption that the real molecule must be presented by a particle of simpler shape, characterized by a small number of geometric parameters that can be determined experimentally. As a result, the choice of an inadequate model may result in incorrect interpretation.

6.4.5.1 Theory of Classical Light Scattering

Classical light scattering is an absolute method for the determination of the molar mass of intact macromolecules, and it is particularly suited for studying large macromolecular assemblies, up to a maximum of 50×10^6 M_r. Beyond this maximum, the theory of classical light scattering, known as the "Rayleigh-Gans-Debye" approximation, is not valuable. Classical light scattering is the total, or time-integrated, intensity of light scattered by a macromolecular solution compared with the incident intensity for a range of concentrations and/or angles. The intensity of light scattered by a protein solution is measured as a function of angle with a light-scattering photometer. If the macromolecule solution is heterogeneous, as it is usually the case for fiber assembly reactions, M_r will be a weight-average molar mass M_w, which makes it difficult to determine precise fiber assembly kinetics [184, 185]. However, in combination with field-flow fractionation (FFF) (see Section 6.4.6), classical light scattering serves as a powerful tool to investigate the growth of protein fibers. The combination of classical light scattering with FFF is realized by using multi-angle laser light-scattering photometers (MALS). In addition, it is necessary to have a concentration detector, which is usually a highly sensitive differential refractometer equipped with a flow cell. Taken together, FFF-MALS is most valuable for the analysis of heterogeneous and polydispersed protein mixtures (see Section 6.4.6) [186–188].

6.4.5.2 Theory of Dynamic Light Scattering

The principle of dynamic light scattering (DLS) experiments is based on the high intensity, monochromaticity, collimation, and coherence of laser light. The primary parameter that comes from DLS measurements is the translational diffusion coefficient D. In DLS measurements, laser light is directed into a thermostated protein solution, and the intensity is recorded at either a single angle (90°) or multiple angles using a photomultiplier/photodetector. The intensities recorded will fluctuate with time, caused by Brownian diffusive motions of the macromolecules. This movement causes a "Doppler" type of wavelength broadening of the otherwise monochromatic light incident on the protein molecules. Interference with light at these wavelengths causes a fluctuation in intensity, which depends on the mobility of the protein molecules. An autocorrelator evaluates this fluctuation by a normalized intensity autocorrelation function as a function of delay time. The decay of the correlation, averaged over longer time intervals, can then be used to obtain the value of D [185]. To obtain molar mass information from the value of D, a calibration of log D versus log M_r is produced, based on globular protein standards. The logarithmic plot of the normalized autocorrelation decay is a straight line for homodispersed solutions, but will tend to curve for solutions with assembling protein fibers. The spread of diffusion coefficients is indicated by a parameter known as the polydispersity factor, which can be evaluated by computer programs. DLS is particularly valuable for the investigation of changes in macromolecular systems, when the time-scale of changes is minutes or hours and not seconds or shorter, which makes it valuable for the investigation of fiber assembly [184, 189–192].

6.4.5.3 Experimental Guide to Analyzing Fiber Assembly Using DLS

Solutions have to be as free as possible from dust and supramolecular aggregates. This requirement is met by filtering all solutions with filters of appropriate size (0.1–0.45 µm). An optimal protein concentration is 2 mg mL^{-1} for a 30-kDa protein, with proportionally lower concentrations for larger proteins. The diffusion coefficient is a sensitive function of temperature and the viscosity of the solvent in DLS measurements. Therefore, measurement and calibration have to be made at the same temperature and solvent viscosity conditions. The diffusion coefficient measured at a single concentration is an apparent one (D_{app}) because of non-ideality effects (finite volume and charge). These effects become vanishingly small as the concentration approaches zero. Importantly, for non-globular proteins (such as growing fibers) a multi-angle instrument has to be used [185].

6.4.6 Field-flow Fractionation to Monitor Particle Growth

6.4.6.1 Theory of FFF

Field-flow fractionation (FFF) is a one-phase chromatography in which high-resolution separation is achieved within a very thin flow against which a perpendicular force field is applied. This technique will be described in more detail, but it has to be mentioned that other types of fields are also used. In the cross-field technique, the flow and sample are confined within a channel consisting of two plates that are separated by a spacer foil with a typical thickness of 100–500 µm. The upper channel plate is impermeable, while the bottom channel plate is permeable and is made of a porous frit. An ultrafiltration membrane covers the bottom plate to prevent the sample from penetrating the channel. Within the flow channel, a parabolic flow profile is created due to the laminar flow of the liquid: the stream moves slower closer to the boundary edges than it does at the center of the channel flow. When the perpendicular force field is applied to the flowing, laminar stream, the analytes are driven towards the boundary layer of the channel, the so-called accumulation wall. Diffusion associated with Brownian motion in turn creates a counteracting motion. Smaller particles, which have higher diffusion rates, tend to reach an equilibrium position farther away from the accumulation wall. Thus, the velocity gradient flowing inside the channel separates different sizes of particles. Smaller particles move much more rapidly than larger particles due to their higher diffusion coefficients, which results in smaller particles eluting before larger ones. This is exactly the opposite of a size-exclusion chromatography separation, in which large molecules elute first.

With FFF separation there is no column media to interact with the samples. Even for very high-molar-mass polymers, such as protein fibers, no shearing forces are applied. The entire separation is gentle, rapid, and non-destructive, without a stationary phase that may interact, degrade, or alter the sample [193–196]. The knowledge of the actual molecular shape is of great importance for analyzing proteins using FFF. Regular and simple geometrical shapes with well-defined boundary surfaces are formed by just a few macromolecules. In contrast, most dissolved macromolecules, especially protein fibers, have to be described in terms of an asymmetrical, expanded and inhomogeneous coil. Considering the complex nature of protein fiber conformations, two parameters are important: the root-mean-square radius and the hydrodynamic radius of the macromolecule. Although a theory predicts the hydrodynamic size of eluting species as a function of elution time, absolute determination of size without reference to standards, assumptions about the conformation of the particles, etc., can be obtained only by combination of FFF with multi-angle light-scattering (MALS) detectors (see Section 6.4.3) [186–188]. FFF in combination with MALS allows separation of protein fibers of different sizes and therefore serves as a tool to monitor assembly and elongation of protein fibers.

6.4.6.2 Experimental Guide to Using FFF for Monitoring Fiber Assembly

FFF reflects a unique tool for analyzing fiber assembly, since it is independent of sample impurities and sample composition. The amount of sample injected can vary widely (0.5–500 µg) because sample size has a negligible effect on the separation efficiency, as long as it is sufficiently large to yield a good detector response and sufficiently small to avoid overloading (a few micrograms is usually a suitable amount for injection). Unlike for other techniques described in this chapter, no general experimental guidelines can be provided, since the flexible operation of FFF has to be adapted for specific needs. For the two most important parameters, cross-flow and channel flow, the optimum flow rates have to be adjusted specifically for each protein investigated. At fixed flow rates, retention time significantly increases with increasing molar mass, accompanied by a growth in bandwidth. At low flow rates, separation power is sacrificed unnecessarily and rather narrow peaks are obtained. At high flow rates, detectability decreases due to increasing dilution [196–198].

6.4.7
Fiber Growth-rate Analysis Using Surface Plasmon Resonance

6.4.7.1 Theory of SPR

Surface plasmon resonance (SPR) is an optical technique that is a valuable tool for investigating biological interactions. SPR offers real-time in situ analysis of dynamic surface events and thus is, for instance, capable of defining rates of fiber growth. When light traveling through an optically dense medium (e.g., glass) reaches an interface between this medium and a medium of a lower optical density (e.g., air), a phenomenon of total internal reflection back into the dense medium occurs [199]. Although the incident light is totally internally reflected, a component of this light, the evanescent wave, penetrates the interface into the less dense medium to a distance of one wavelength [200]. In SPR measurements a monochromatic, polarized light source is used, and the interface between the two optically dense media is coated with a thin metal film (less than one wavelength of light thick). The choice of metal used is critical, since the metal must exhibit free electron behavior. Suitable metals include silver, gold, copper, and aluminum. Silver and gold are the most commonly used, since silver provides a sharp SPR resonance peak and gold is highly stable [201]. The evanescent wave of the incoming light is able to couple with the free oscillating electrons (plasmons) in the metal film at a specific angle of incidence, and thus the plasmon resonance is resonantly excited. This causes energy from the incident light to be lost to the metal film, resulting in a reduction in the intensity of reflected light, which can be detected by a two-dimensional array of photodiodes or charge-coupled detectors. Therefore, if the refractive index directly above the metal surface changes by the adsorption of a protein layer, a change in the angle of incidence required to excite a surface plasmon will occur [202]. By monitoring the angle at which resonance occurs during an adsorption process with respect to time, an SPR adsorption profile can be obtained. The difference between the initial and the final SPR angle gives an indication of the extent of adsorption, and the positive gradient of the SPR adsorption curve determines the rate of adsorption, allowing analysis of, e.g., assembly kinetics of a growing protein fiber.

6.4.7.2 Experimental Guide to Using SPR for Fiber-growth Analysis

One of the most important parameters for SPR measurements is the surface where the molecules interact. The commercial availability of sensor chips with robust and reproducible surfaces makes SPR convenient to use. The most commonly used surface is carboxymethyl dextran (CMD) bound to a gold substrate. CMD is covalently bound to the sensor chip surface via carboxyl moieties of dextran. Functional groups on the ligand that can be used for coupling proteins include NH_2, SH, CHO, and COOH. CMD sensor chips can be regenerated by selective dissociation of the bound protein from the covalently immobilized ligand. Other commercially available sensor chips have: surfaces made of dextran matrices with a low degree of carboxylation (providing fewer negative charges), shorter carboxymethylated dextran matrices (valuable for large protein assemblies), chips with coupled nitrilotriacetic acid (designed to bind histidine-tagged molecules), chips with pre-immobilized streptavidin (for capturing biotinylated ligands), and plain gold surfaces (no dextran or hydrophobic coating), which provides the freedom to design customized surface chemistry such as self-assembled monolayers (SAM) [203]. The specific coupling chemistry has to be selected for each experiment individually. In the case of fiber growth kinetics, it is necessary to couple initiating molecules, which could be seeds, nuclei, or monomeric protein, in a manner that polymerization is not sterically hindered. This implies some fundamental knowledge on the polymerization process of the protein to be investigated. Experimental design and data processing dramatically influence the quality of SPR-derived data. However, when properly utilized, SPR can be a powerful tool for determining kinetic and equilibrium constants of molecular interactions [204]. After kinetic analysis of fiber assembly, the SPR chips can be microscopically investigated by atomic force microscopy (AFM) to obtain additional morphological insights into the assembled protein fibers (see Sections 6.4.8 and 6.5.5) [205].

6.4.8
Single-fiber Growth Imaging Using Atomic Force Microscopy

6.4.8.1 Theory of Atomic Force Microscopy

Atomic force microscopy (AFM) measures the height of a solid probe over a specimen surface. AFM has been developed as a tool for imaging biological specimens and dynamic measurements of biological interactions. AFMs operate by measuring attractive or repulsive forces between a tip and a sample, with the force between the sample and the tip causing the cantilever to bend. The magnitude of this force depends on the distance between the tip and the sample and on the chemical and mechanical properties (i.e., stiffness) of the sample. Deflection of a laser beam focused onto the end of the cantilever detects the bending of the cantilever. Since the sample is mounted on top of a ceramic piezoelectric crystal (piezo), a feedback loop between the laser detector and the piezo controls the vertical movement of the sample. Depending on the AFM design, scanners are used to translate either the sample under the cantilever or the cantilever over the sample. By scanning in either way, the local height of the sample is measured. Three-dimensional topographical maps of the surface are then constructed by plotting the local sample height versus horizontal probe tip position [206]. AFM can achieve a resolution of 10 pm and, unlike electron microscopes, can image samples in air and under liquids, thereby allowing the imaging of hydrated specimens [207, 208]. This feature can be used to monitor fiber growth as previously shown for amyloid fibril formation [115, 209–211].

In its repulsive "contact" mode, the instrument lightly touches a tip at the end of a leaf spring or "cantilever" to the sample. As a raster scan drags the tip over the sample, a detection apparatus measures the vertical deflection of the cantilever, which indicates the local sample height. Thus, in contact mode the AFM measures hard-sphere repulsion forces between the tip and sample. In tapping mode, the cantilever oscillates at its resonant frequency (often hundreds of kilohertz) and is positioned above the surface, so that it taps the surface for only a very small fraction of its oscillation period. The very short time over which this contact occurs means that lateral forces are dramatically reduced as the tip scans over the surface. Tapping mode is commonly used for poorly immobilized or soft samples. Thus, in both modes the surface topography of the biological specimen is recorded and its elevations and depressions can be accurately monitored [207, 208, 212].

6.4.8.2 Experimental Guide for Using AFM to Investigate Fiber Growth

Experiments should be carried out in tapping mode in liquid using a fluid cell, since this mode allows for higher-resolution liquid imaging than does the contact mode. The proteins should be dissolved in low-salt buffers (it is best to avoid any salts) such as 10 mM Tris/HCl or 5 mM potassium phosphate. Protein concentrations should be around 50 µg mL^{-1} to obtain high-quality images. Also, atomically flat surfaces should be used to immobilize the assembling proteins. Mica is a commonly used surface material for protein attachment and fibril growth. Since slight drifts during repeated scans can occur in experiments that take several hours, com-plate is approximately 3 min. Protein fiber adsorption to the plates has to be investigated by incubating protein fiber solutions in the plate for 10 min at 37 °C with stirring and by determining the protein concentration before and after incubation by measuring the UV absorbance.

Alternatively, fibril formation can be performed in glass vials. Ten-microliter aliquots are withdrawn from the glass vials and added directly to a fluorescence cuvette (1-cm path length semi-micro quartz cuvette) containing 1 mL of a ThT mixture (5 µM ThT, 50 mM Tris buffer, and 100 mM NaCl pH 7.5). Emission spectra are recorded immediately after addition of the aliquots to the ThT mixture from 470 nm to 560 nm (excitation at 450 nm). For each sample, the signal can be obtained as the ThT intensity at 482 nm, from which a blank measurement recorded prior to addition of protein to the ThT solution is subtracted.

6.5
Methods to Study Fiber Morphology and Structure

6.5.1
Scanning Electron Microscopy for Examining the Low-resolution Morphology of a Fiber Specimen

6.5.1.1 Theory of SEM

Scanning electron microscopy (SEM) involves the electron beam being passed over a conductive surface in a series of parallel lines, in a manner similar to that used in television imaging. Electrons leave the source and are focused onto the surface of a specimen as a very fine point. The probing electron beam penetrates only less than a few microns into the specimen and the electrons interact with the specimen surface, inducing a variety of radiations. Each leaves the specimen in a variety of directions and can be counted by a detector and displayed. The direction is not important; only the number of photons or electrons leaving the spot on the sample needs to be measured. The probing beam is moved to an adjacent spot on the specimen and the information is output to the display. This continues rapidly across the specimen, each point being amplified and magnified to form the image [220]. The type of information collected depends on the particular radiation being used as the information signal. The types of radiation utilized include visible light, X-rays, backscattered-electron current, and induced-specimen current [220]. Most commonly, secondary electrons emitted by the interaction of the beam are collected by a detector and the three-dimensional topographic image is built up. This technique, known as emissive mode, is commonly used for large specimens to examine morphological details. Coating the specimen in a conductive layer, usually gold coating, improves the specimen image, since the lower the atomic number of the specimen, the better the emission of secondary electrons [221]. The technique has been used extensively for the observation of silk fibers' overall morphology. In general, however, the range of resolution (typically 100 nm to 1 mm) is lower than is required for examination of most protein fibers but is ideally suited for topographic imaging. SEM allows a large depth of field compared to the optical microscope, permitting the visualization of rough surfaces in focus across the whole specimen. In addition, the contrast within scanning micrographs provides a three-dimensional effect when the specimen is tilted with respect to the electron beam.

6.5.1.2 Experimental Guide to Examining Fibers by SEM

SEM gives information only about the external surface of the specimen. Usually, the specimen requires fixation and dehydration. Fixation is commonly performed using osmium tetroxide or formaldehyde-glutaraldehyde as a fixative, depending on the nature of the material being examined. Dehydration is usually carried out by passing the specimen though a graded series of alcohol. Freeze-drying may be used to remove water from a specimen, and this is best achieved by freeze-drying from a nonpolar solvent rather than from water (e.g., amyl acetate can be substituted for water). After freeze-drying, the specimen is warmed to room temperature before air is re-admitted to the vacuum chamber to prevent rehydration. The specimen is often coated with a thin layer of carbon to maintain the dried state and finally coated with a layer of heavy metal. Possibly, dehydration and coating may be performed within the same vacuum chamber. Critical point drying is also a method used for dehydration of some specimens. This involves replacing the water within a specimen first by a dehydrating agent, such as ethyl alcohol, followed by an intermediate liquid (e.g., amyl acetate) and a transitional liquid (e.g., carbon dioxide). This transitional liquid will be removed from the specimen at a "critical point" of temperature and pressure, when it becomes converted to a gas. Air drying is used less often due to the surface tension effects causing distortion in the specimen, but it is occasionally used for certain specimens.

In order to improve the electrical potential of the specimen surface, specimens are coated with a thin layer of conducting materials, since the dehydrated biological specimen usually conducts poorly and may also build up beam-induced charge, which results in artifacts. Usually, a specimen will be coated with a thin layer of carbon, followed by an outer layer of a heavy metal, commonly gold. The thin layer of gold is applied by vacuum evaporation from a tungsten filament.

The method of data collection will depend considerably upon the type of specimen [220]. In general, for biological specimens, it is thought that the acceleration should be kept low (10 kV) to reduce damage. The current should also be kept low, while enabling the collection of a relatively noise-free image. Astigmatism (i.e., a different direction focus at different points) is a very common limitation to resolution and therefore should be prevented as much as possible by maintaining the microscope to a high standard and corrected during data collection. Selection of the final aperture depends on the magnification and the depth of focus required. A large aperture (200 micron) is appropriate for high-resolution work, whereas at low magnification, depth of focus may be the prime requirement and therefore a small final aperture may be used (50 micron). It may prove useful to image the specimen at different tilts to examine different angles of the material. Data recording can be on film or, more commonly, on CCD camera. The images may then be examined in detail (Figure 6.5).

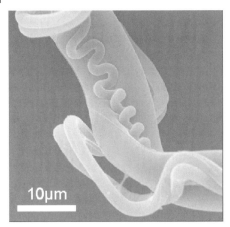

Fig. 6.5. SEM of major and minor ampullate silks collected from the garden cross spider *Araneus diadematus*.

6.5.2
Transmission Electron Microscopy for Examining Fiber Morphology and Structure

6.5.2.1 Theory of TEM

Electron micrographs represent a projection of the density distribution of the specimen onto a plane [222]. Therefore, an electron microscope image may contain sufficient information to generate the three-dimensional structure of an object. Transmission electron microscopy (TEM) is commonly used for examination of macromolecular protein assemblies, from single particles, such as viruses, to fibers, such as microtubules. Materials for TEM must be specially prepared to a thickness that allows electrons to transmit through the sample in a manner similar to how light is transmitted through materials in conventional optical microscopy. Since the wavelength of electrons is much smaller than that of light, the optimal resolution attainable for TEM images is many orders of magnitude better than that of a light microscope. This means that TEMs can reveal the finest details of internal structure – in some cases as small as individual atoms. The energy of the electrons in the TEM determines the relative degree of penetration of electrons in a specific sample or, alternatively, influences the thickness of material from which useful information may be obtained. Conventional electron microscopes tend to be run at 80–200 kV. The electron dose must be kept low to reduce radiation damage to the biological specimen. Thus, the contrast from a micrograph of biological specimens can be very poor, and therefore negative staining is required to visualize the sample. The negative stains are composed of heavy metals (such as uranyl acetate) that scatter the electrons and are electron-opaque, which results in a dark image of the stain deposited around a protein molecule or fiber.

The image visualized within the electron microscope is two-dimensional. Image reconstruction techniques can be applied to produce three-dimensional images by single-particle averaging for globular molecules or by helical image reconstruction for fibers. Further analysis of electron micrographs is useful to exceed what is possible by visual interpretation to overcome problems such as signal detection in the presence of noise and interpretation of phase contrast images. For further discussion on analysis of images, see Section 6.5.3 on cryo-electron microscopy.

6.5.2.2 Experimental Guide to Examining Fiber Samples by TEM

The way in which the sample is prepared can vary depending on the stability of the specimen. The sample must be dehydrated since it is viewed in a vacuum. In some cases the sample is fixed prior to preparing microscope grids. The grids used for TEM can consist of a layer of plastic, such as Formvar or Pioloform, overlaid with carbon or simply a thin carbon film. The grids may be purchased ready-made or can be made to the user's specifications. Detailed procedures for making electron microscope grids can be found in Ref. [223]. Improved contrast and resolution can be obtained by using a thin-layer support, and this is best achieved by making carbon support films on mica that are then transferred to the grid. Alternatively, the carbon can be evaporated onto grids covered with a plastic film. The plastic film can then either be dissolved, leaving the thin carbon film, or left in place, resulting in more robust grids but a higher background upon imaging of a specimen.

Negative Staining for Imaging Fibrous Proteins As previously mentioned, it is usually necessary to provide contrast for visualization of a biological specimen by negative staining. These are usually heavy metal salts such as uranyl acetate, potassium phosphotungstate and ammonium molybdate. Detailed considerations of the advantages of different stains and choice of concentration are summarized in Ref. [223]. Uranyl salts give the greatest amplitude contrast but have slightly larger microcrystallinity/granularity after drying than some other negative stains, whereas potassium phosphotungstate gives fine granularity but lower image contrast [223]. Spreading agents may be a useful part of preparation of some samples. Surfactants such as n-octyl-β-glucopyranoside can be used to evenly disperse the molecules. The spreading agent can be added to the biological sample or to the negative stain. However, glow-discharge treatment of the microscope grid may be sufficient to allow even spread of the sample.

The staining procedure varies considerably among different researchers. However, the simplest technique involves the placement of droplets (~ 20 µL) of the specimen, several of filtered water and stain in a row, onto parafilm. The grid can then be placed facedown on each droplet in turn and for a specified amount of time. Alternatively, droplets (~ 4 µL) may be placed onto an upturned grid and blotted away using filter paper before the next droplet is placed on the grid.

Metal Shadowing for Enhanced Contrast of Fibrous Proteins The contrast provided by negative-stain techniques may be inadequate for visualization of rod-shaped particles, and metal shadowing may increase the contrast [224], allowing better imaging of the morphological features. Metal shadowing may be unidirectional or rotary. The protein to be examined must be in a solution of glycerol (50% v/v) and

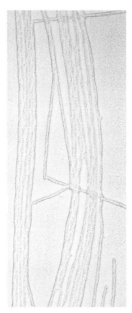

Fig. 6.6. Platinum carbon shadowing micrographs of wild-type alpha-synuclein fibrils (assembled as previously described [250]).

then be sprayed onto a piece of freshly cleaved mica and dried. The protein-covered mica is coated with platinum or another heavy metal (such as tantalum/tungsten) by evaporation and then coated with a layer of carbon. The metal/carbon replicas are then floated off onto cleaned grids and finally examined under the electron microscope. Contact prints of the micrographs provide the best images for further analysis (Figure 6.6). Details of this procedure can be found in Ref. [224].

6.5.3
Cryo-electron Microscopy for Examination of the Structure of Fibrous Proteins

6.5.3.1 Theory of Cryo-electron Microscopy
Cryo-electron microscopy allows the imaging of samples in a hydrated state encased in a thin layer of ice. No stain is required and contrast is produced by the density difference between the molecule and the vitreous (non-crystalline) ice. The samples benefit from the fact that they are not flattened by gravity, are unstained, and can be viewed at near-physiological conditions (i.e., hydrated). High-resolution data can be collected from samples at liquid nitrogen temperatures using low-dose techniques, which minimize radiation damage to the sample. Cryo-electron micrographs usually show a low signal-to-noise ratio due to the low dose of electrons, the lack of stain, and the use of small defocuses. Therefore, cryo-electron micrographs benefit from additional analysis such as Fourier filtering and

image-reconstruction techniques. High-resolution structures (to 6 Å) have been elucidated by single-particle analysis from cryo-electron micrographs of large protein molecules. Helical image reconstruction from electron micrographs can allow the elucidation of fibrous structures.

6.5.3.2 Experimental Guide to Preparing Proteins for Cryo-electron Microscopy

Cryo-electron microscopy grid preparation involves a certain amount of development to find the ideal conditions for examination of a particular protein sample. Grids are made from a holey carbon film, and the size of the holes and thickness of the carbon should be optimized for the protein molecule's size, as the thickness of the ice layer in which the protein is encased should be ideal for the size of the protein. The thickness of the ice within a carbon hole will depend on the diameter of the holes, the thickness of the carbon, and the treatment of the grid during freezing. Often, grids are glow-discharged to spread out the water layer, resulting in thinner ice, or amylamine is used to increase the thickness. Finally, the technique used for grid blotting before the sample is flash-frozen will affect the ice thickness. Holey carbon grids can be purchased or made by a procedure similar to that for making carbon films for negative-stain electron microscope grids [223]. The concentration of the fiber sample should be optimized for cryo-electron microscopy first by negative-stain EM. Ideally, the concentration should be such that several fibers are found in the scope of a micrograph but that they rarely cross over. A droplet of the fiber sample is then placed onto the optimally prepared holey carbon grids, blotted, and plunged into liquid ethane cooled by liquid nitrogen. If the procedure is carried out correctly, the grid will be covered in a thin layer of vitreous ice. The grid should be stored in liquid nitrogen until required.

Transfer to the cryo-stage specimen holder for the electron microscope must be performed at low humidity and at a temperature lower than $-160\,^\circ$C to prevent ice accumulation on the grid and to preserve the vitreous water. Specialized equipment to transfer the grid to the electron microscope specimen holder is available (Oxford Instruments, Gatan). The cryo-TEM must also have low vapor content; this is achieved by a liquid nitrogen pre-cooled anti-contaminator system to maintain the vitreous ice. The temperature of the specimen should then be maintained at less than $-125\,^\circ$C to prevent transformation of the vitreous water to crystalline cubic ice. The appearance of crystalline ice may not be visible in the microscope but can be observed as diffraction spots upon electron diffraction. Usually, cryo-electron micrographs will be collected at various defocuses in order to collect a full dataset. This is because at each defocus the contrast transfer function will pass through zero, giving rings of uncollected data in the power spectrum.

6.5.3.3 Structural Analysis from Electron Micrographs

Data collection for revealing images of protein fibers may include systematically tilting the specimen to produce a series of images with different projections of the material. However, it may be possible to reconstruct the structure of a helical molecule using only a single image, since the high symmetry of a helical fiber allows observation of many views simultaneously [225]. In order to analyze the micrograph images, they must be digitized (some electron microscopes have fitted CCD cameras, thus avoiding this step). Ideally, a fiber image will be scanned such that the fiber is vertical in order to avoid interpolation resulting in loss of data at the later stages. Examination of images can be carried out using a suitable program such as Ximdisp [226]. Regions of fiber images may be selected and must be boxed and floated. This involves masking off all density not belonging to the particle and surrounding the particle box to reduce the density step at the edge of the box [227]. Fiber images may benefit from Fourier filtering and/or background subtraction. The fibers may be straightened by a number of established techniques [228]. If necessary, the image may be interpolated. Interactive Fourier transforms are calculated to look for repeating features with the fiber. If a diffraction pattern is observed, then further analysis may be carried out on the fiber image using helical image reconstruction [227, 229, 230]. The determination of the position and tilt of the helix axis must first be performed, followed by indexing of the layer line spot positions. The indexing is tested and phases are measured, allowing a reconstruction to be carried out [227, 229, 230]. This results in a three-dimensional density function for the object. This methodology can be followed in detail [222, 229].

6.5.4
Atomic Force Microscopy for Examining the Structure and Morphology of Fibrous Proteins

6.5.4.1 Experimental Guide for Using AFM to Monitor Fiber Morphology

Minimal sample preparation is necessary for visualizing amyloid fibers by atomic force microscopy (AFM) [231]. A clean, uncontaminated solution of the sample can be placed onto freshly cleaved mica, glass, or other suitable smooth support. The substrate should be thoroughly cleaned with an excess of buffer (salt-free is best) or water. In order to get good contrast and to reduce mechanical damage of the soft biological materials, the samples can be stabilized by adding covalent cross-linking agents or certain cations that are able to link the constituents of the sample to each other or to the substrate. Cooling can also stiffen the sample to reduce sample disruption and damage.

Tapping mode is most commonly used for imaging relatively soft biomolecules. The force generated between the sample and tip depends on the amplitude of free oscillation and the set point amplitude [232]. The observed amplitude of oscillation is maintained by adjusting the vertical position of the sample. The strength of tapping must be optimized for a particular sample, since tapping too hard will reduce image resolution and quality due to damage to the tip or protein. Tapping too softly might cause adhesive forces between the sample and cantilever tip, resulting in damping of oscillation and causing spurious height data for the sample [232]. Once these conditions are optimized, image data may be collected. Visual representations of the data are often generated by assigning a color or brightness to the relative height in the image. This will enable clear images of the fiber in question to be observed above the background smooth surface (Figure 6.7). Height measurements may be taken from the data, but width cannot be measured accurately due

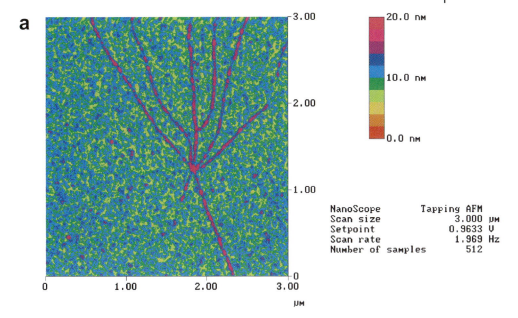

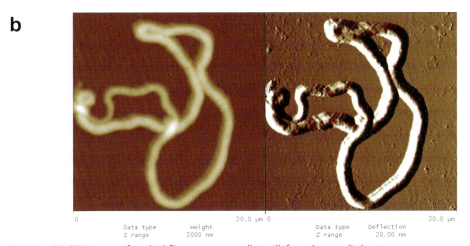

Fig. 6.7. (a) AFM image of amyloid fibers assembled with biotinylated yeast Sup35p-NM on streptavidin scanned in contact mode [251]. The scale bar gives the color code for the height information. (b) AFM images of major ampullate silk from *Araneus diadematus* on mica scanned in tapping mode. The left image reflects the sample height above the surface, and the right image depicts the tip deflection.

to the shape and size of the tip. Of course, height measurements taken from specimens in air may be affected by dehydration of the specimen, by the effect of gravity, or by the force of the cantilever compressing soft protein. However, morphological features may be clearly observed.

6.5.5
Use of X-ray Diffraction for Examining the Structure of Fibrous Proteins

6.5.5.1 Theory of X-Ray Fiber Diffraction

X-ray diffraction involves the use of proteins in an ordered arrangement. The amount of information that can be obtained is directly related to the degree of order within the diffracting object. The most ordered arrangement will be in a crystal where each identical protein molecule is arranged periodically on a three-dimensional lattice. Diffraction from crystals results in a pattern comprised of sharp spots. However, fibrous proteins usually do not crystallize due to their size, insolubility, or heterogeneity. Fibrous proteins form partly ordered structures, and standard methods of crystal structure analysis are not applicable to fibers due to the degree of disorder and number of overlapping reflections. However, X-ray fiber diffraction can be performed. Details concerning the theory of X-ray fiber diffraction may be found in Ref. [19]. For a single, isolated chain molecule, the crystalline lattice extends in only one dimension, parallel to the axis of the fiber (z-direction). The scattering appears on layer lines, and the layer planes have a separation of the reciprocal of the periodic repeat along the fiber axis. Since a single-chain molecule has no periodicity in the x- and y-directions, the scatter is a continuous function. In practice, however, a fiber specimen is composed of several chain molecules with rotational averaging. The molecules may be axial in register but arranged at all possible azimuthal orientations. These molecules are related to one another, but not by a crystallographic lattice. Therefore, a collection of chain molecules gives the scattering effect for the isolated molecules plus the effects caused by the orientation of the molecules and the packing, size, and mutual disposition of ordered regions.

Fibers are cylindrically symmetrical, so each reflection is spread out into a disk and intercepts the Ewald sphere at two symmetrical points. This means that the diffraction pattern is symmetrical. On the fiber diffraction diagram, the direction parallel to the fiber axis is known as the meridian and the perpendicular direction is the equator. Fibers are crystalline along the fiber axis and this produces strong, sharp reflections on the meridian. The degree of disorientation of the fibers relative to one another will produce more or less arcing of reflections. The appearance of diffraction arcs depends on the arrangement of crystallites (microstructure) within the material, known as the texture. Individual fibers may be aligned relative to one another, thus reducing disorientation and reducing the spread of the arcs. A sample in which the fibers have not been aligned at all will produce a diffraction pattern where the reflections appear as rings, much like a powder diffraction pattern. X-ray diffraction from a partially aligned specimen will allow distinction between the meridional and equatorial reflections. Even further orientation of the fiber within the specimen may produce a series of lines of intensity known as layer lines, running perpendicular to the meridian of the pattern. The intensities on closely spaced layer lines may overlap. In fact, it is rare to find non-overlapping reflections. If the fiber is mounted with the fiber axis normal to the X-ray beam, the diffraction pattern will be symmetrical above and below the equator (Figure 6.8).

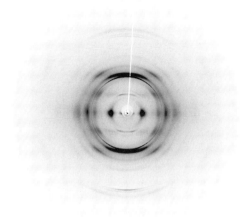

Fig. 6.8. X-ray fiber diffraction from amyloid fibrils from a short peptide homologous to the central region of Aβ peptide showing the cross-β diffraction pattern.

However, there will be regions of reciprocal space for which no data is recorded. By tilting the specimen, additional data can be collected.

6.5.5.2 Experimental Guide to X-Ray Fiber Diffraction

Sample Preparation Fibers in general are weakly diffracting, and therefore specimen preparation is particularly important. The specimen should be of an optimum size and thickness to fill the beam. Otherwise background scatter may obscure weak reflections. The amount of information that can be obtained from an X-ray fiber diffractogram is closely associated with the degree of alignment. Some fibrous proteins occur naturally in a highly oriented form (e.g., keratin and collagen). Often regions of orientation are quite limited, and therefore it can be useful to use a microfocus X-ray beam for data collection. Orientation may be improved by physical manipulation or by chemical treatment. For example, the orientation of *Chrysopa* egg stalk silk fiber orientation was increased by stretching in water or urea [4]. Preparation of oriented films and fibers from soluble fibrous proteins has involved a number of procedures that include stretching, rolling, casting films, drawing fibers from precipitate, shearing, magnetic alignment, centrifugation, and capillary flow. Long-chain molecules can be drawn from the solution in a manner similar to that used for making diffraction samples of DNA. Samples that cannot be pulled into long fibers may be aligned by hanging a drop between two wax-plugged capillary tubes arranged 1–2 mm apart [1]. Casting flat sheets is often used as a method for aligning polymers [233–235], and this can be achieved by placing a high concentration of a fiber solution on a suitable surface and allowing the solution to dry to form a thin film that can be removed from the surface (such as Teflon). The film will show alignment of the fibers in the direction parallel to the film surface but no alignment in the perpendicular direction. Further improvement to alignment of fiber bundles or films may be achieved by annealing in high humidity. Some fibrous protein specimens are diamagnetically anisotropic [236] and may benefit from magnetic alignment [105, 237, 238]. The effect of the magnetic field on a single molecule is small, and therefore it is necessary to embed the molecules in a liquid crystal so that the molecules act together and to use a high-energy magnet [239]. The alignment can be fixed in the specimen by slow drying while the specimen is still in the magnetic field.

Data Collection Since fibrous proteins are often weak scatterers, long exposure times may be needed. Cameras specially designed for the particular qualities of the specimen may be helpful. The arrangement of the apparatus will depend on whether a high- or low-angle diffraction is being collected. The use of a helium chamber and an evacuated camera can be helpful to reduce air scatter as well as to allow the humidity around the specimen to be controlled if required. In the past, the recording of diffraction data was performed using photographic film. This allowed the placement of several films in the detector to record a range of intensities. However, it is now more common to use detectors used for crystallography, such as a MarResearch image plate or CCD cameras. If film has been used, then the diffraction image must be digitized and converted to an appropriate format for data analysis.

Data Analysis There are a number of fiber diffraction processing programs available to facilitate processing of fiber diffraction data. CCP13 [240] is a collection of programs designed to analyze high-quality diffraction patterns. FIT2D (www.esrf.fr/computing/expg/subgroups/data_analysis/FIT2D) can be used to analyze one- and two-dimensional datasets. The CCP4 suite [241] is intended for protein crystallography, but many of its components such as ipdisp and Mosflm can be utilized for examining the diffraction pattern. The data from the detector must be converted into a format compatible with the other programs in the chain. Examples of such programs are XCONV (CCP13) [147], marcvt (www.marresearch.com/software.htm) and Denzo [242]. The diffraction image must be carefully centered to avoid a systematic error while measuring spot locations. Mosflm allows this to be done by manually moving the image with respect to a series of overlaid concentric circles. XFIX (CCP13) allows the user to find the center of the pattern and to calibrate the specimen to film distance. One may also wish to correct for the shape of the image plate and determine the sample's tilt and rotation [243]. Background subtraction is often beneficial to correct for detector fog, white radiation, and X-ray scattering from air, camera components, the sample holder, and amorphous material in the specimen such as solvent and disordered polymer [244, 245]. The positions of the reflection can then be assessed. An initial survey can be carried out by noting spot locations and resolutions. This is straightforward in Mosflm, since the resolutions are calculated by clicking on spot maxima using a zoomed image. Resolution calculation can also be done by hand using an approximation to Bragg's law. This is the result of the product of the wavelength of the X-rays and the distance between the sample and the image plate, divided by the distance of the diffraction spot from the center of the diffraction pattern. For high-quality diffraction patterns containing diffraction peaks on layer lines, the CCP13 suite of programs allows the user to quadrant-average the diffraction pattern given the calculated tilt/rotation of the image. This image can then be processed further to include background subtraction and modeling of the position, size, shape, and intensity of the reflections.

Model Building An X-ray diffraction pattern from an unknown structure cannot directly lead to a structure since only amplitudes, but not phases, can be measured from the diffractogram. One method of solving this phase problem is by trial and error, so that from consideration of the diffraction pattern, symmetry, and chemical composition, a model is built and refined until a calculated diffraction can be generated that is similar to the observed [239]. A final model will predict the observed diffraction pattern so well that the possibility of the existence of another model that also predicts the diffraction pattern is virtually eliminated. Fibers by their nature are highly symmetrical and consist of repeating units, which reduces the number of possibilities for arrangement. Structural information about the subunits may already be available, either from crystal structures of individual monomers or from other biophysical techniques that indicate a secondary or tertiary structure. A molecular structure may be modeled using software such as LALS [246] or Cerius2 (http://www.accelrys.com/cerius2/cerius248/), which then allows the calculation of a simulated diffractogram to compare with the empirical data. Fxplor (www.molbio.vanderbilt.edu/fiber/software.html) is an extension of X-Plor, a program useful for atomic model refinement.

6.5.6
Fourier Transformed Infrared Spectroscopy

6.5.6.1 Theory of FTIR

Fourier transformed infrared spectroscopy (FTIR) has been used to gain information about the conformation of fibrous proteins. Proteins yield three major infrared bands of interest that arise from the peptide backbone. These are the amide I absorption at 1600–1700 cm^{-1}, arising from the C=O stretch; the amide II absorption at 1500–1600 cm^{-1}, arising from the N–H deformation; and the amide III absorption at 1200–1350 cm^{-1}. A N–H stretch also absorbs at about 3300 cm^{-1} [247]. The presence of hydrogen bonding within the sample shifts the energies of the peptide variations, therefore allowing the determination of the presence of α-helical and β-sheet conformations in theory. However, this is complicated in practice due to coupling between individual peptide vibrations, which leads to splitting of the bands into a series of bands. The symmetry of α-helical fibers leads to a distinctive set of bands upon aggregation from the monomer [247, 248], whereas the spectra observed from the formation of all-β-sheet protein fibers are more subtle [248]. Observed and calculated infrared spectra from α-helix and antiparallel and parallel β-structures can be found in Ref. [247].

6.5.6.2 Experimental Guide to Determining Protein Conformation by FTIR

FTIR data collection is complicated by the fact that the bending vibrations H–O–H from water are near 1640 cm^{-1}, which can obscure the amide I band. This problem can be overcome by examination of the protein in the solid state, when little water is present, or by exchanging the hydrogen for deuterium. The D–O–D bending vibration is shifted to 1220 cm^{-1}, but care must be taken to fully exchange the H for D, since partial exchange will result in bands in the amide I region [248]. However, for examining the structure of *ex vivo* samples of fibrous proteins, exchange of deuterium may not be possible, and in order to overcome this, the signal-to-noise resolution of the data must be high. For solution samples, the use of a short path length (4–10 μm) and high protein concentration (1–50 mg mL^{-1}) should reduce the water absorbance relative to the protein signal at 1640 cm^{-1} [248]. To minimize scattering, samples may be prepared in dry state in KBr pellets by combining oven-dried KBr with the protein. The sample is then kept under vacuum and pressed (5–10 kbar) using a Carver press to form a thin, translucent pellet. This pellet can then be transferred to the transmission holder and a spectrum can be collected [248]. The spectrum is then compared with the spectrum collected from a KBr-only pellet. However, this harsh method may cause denaturation of the protein.

Alternative methods include attenuated total reflectance (ATR) FTIR, diffuse reflectance FTIR, and transmission-mode FTIR from a thin film. For ATR-FTIR, the sample is placed on a surface composed of a material of high refractive index, which is known as the internal reflection element (IRE) and is often composed of germanium or zinc selenide. The beam penetrates to the boundary between the IRE and the sample and is not transmitted through the sample. This makes this method ideal for examining fibrous proteins, and the method may be used for samples prepared as a thin film or suspension [249]. For ATR experiments, the maximum amount of light throughput is required and the amount of water vapor is reduced (by purging the system with dry nitrogen or dry air) to increase the sensitivity. Oberg et al. [249] recommend the use of out-of-compartment, horizontal, trapezoid-shaped IREs. To prepare a sample for ATR-FTIR, thin films can be prepared of the fibrous protein using a solution (50–100 μL of 0.5–1.0 mg mL^{-1}) or suspension placed onto the IRE and dried using nitrogen gas. The thin films are partly hydrated, maintaining the protein structure. Further, the limited-depth penetration of the ATR method reduces the contribution of water to the spectra [248]. Lyophilized proteins may also be spread onto the IRE for collection of spectra from powdered proteins. Alternatively, the diffuse reflectance (DRIFT) method can be used for dry samples [248]. In contrast to the ATR method, in transmission-mode FTIR, the beam passes directly through the sample [248]. High protein concentrations (2–60 mg mL^{-1}) and small path lengths (less than 10 μm) are required, although for a D$_2$O sample, lower protein concentrations and longer path lengths

(25–50 µm) may be used. Thin films can be prepared by drying the protein suspension or solution onto thin silver chloride disks.

Data analysis of FTIR spectra involves some expertise. Water or buffer subtraction must be performed before analysis of the spectrum. The conformation of the polypeptide chain has a signature in the amide I band. However, the bands may overlap, making interpretation difficult. This can be tackled in two ways. One approach is resolution enhancement methods or deconvolution to decrease the widths of the infrared bands and to allow determination of the overlapping components. A second approach is to use pattern-recognition methods for which software packages are available (e.g., GRAMS, Galactic Corp.).

6.6 Concluding Remarks

In the present chapter, fiber-forming proteins and methods to determine their fiber assembly and fiber morphology have been described. There is a wide variation among fibrous proteins with regard to the content of regular secondary structure. Fibers of proteins such as collagen and tropomyosin with highly elongated molecules tend to have a high content of regular secondary structure (generally greater than 90%). In contrast, fibers of globular proteins, such as actin, feather keratin, and flagellin, generally have much lower contents of regular structure (often less than 50%). In another group of proteins, comprising elastin, resilin, and the matrix proteins of mammalian keratins, the content of regular secondary structure is very small. In these cases the protein fibers are cross-linked and have rubber-like elastic properties.

So far, any attempt to correlate sequence with structure and function in fibrous proteins is highly speculative. Therefore, from a given sequence it cannot be concluded whether a protein forms fibers or not. However, the structure-function relation based on sequence data is the aim of many biologists, chemists, physicists, and material scientists. One main conformation-determining factor in fibrous proteins appears to be regions with repeating sequences. The simplest examples of repeating sequences are found among the silks, which have, for instance, runs of Gly_n. The corresponding sections of the chains adopt conformations that appear to be identical to those observed in synthetic homopolypeptides. Four other conformation-determining factors in fibrous proteins are interactions involving side chains, interactions between apolar side chains, hydrogen bond formation, and the content and distribution of Gly and Pro residues.

Aggregation studies on amyloids, collagens, myofibrils, and silks illustrate the capacity of fibrous proteins for self-assembly into filaments and other organized aggregates. The mode of aggregation is generally found to be sensitive to changes in pH, ionic strength, and the presence of other solutes, indicating that the assembly process depends upon a fine balance among various types of weak secondary bonding. Packing considerations are also clearly involved, but in the case of assembly in aqueous environment, these will be less important than in a solvent-free crystalline phase. Several features of the self-assembly process in vivo are still not well understood, such as nucleation and termination of fiber formation. For termination of fiber formation, two possibilities can be envisaged: the assembly process could be intrinsically self-limiting, with the information residing in the individual protein molecules, or the size of the aggregate is potentially unlimited, but regulation of size is achieved through some external agency.

The future aim in investigating fibrillar proteins is to understand the kinetics of fiber assembly (which will be a basis for controlling fiber assembly) and to gain more insights into fibrous structure to learn about sequence-structure relationships. Such knowledge will be a basis for elucidating the relevance of assembly processes in cellular development, cellular structure, and cellular stability; for assessing fundamental influences leading to protein-folding diseases; and for developing products through material science.

Acknowledgements

T.S. is supported by the Deutsche Forschungsgemeinschaft (SFB 563 A9); SCHF 603/4-1, the Fonds der Chemischen Industrie, and the Leonhard-Lorenz-Stiftung. L.S. is funded by a Welcome Trust Research Career Development fellowship. The authors would like to thank Linda Amos for the donation of Figure 6.2, Bettina Richter for technical assistance, and Daniel Hümmerich, Sumner Makin, and Thusnelda Stromer for critical reading of the manuscript.

References

1 SUNDE, M. & BLAKE, C. (1997). The structure of amyloid fibrils by electron microscopy and X-ray diffraction. *Adv. Protein Chem.* **50**, 123–159.
2 EANES, E. D. & GLENNER, G. G. (1968). X-ray diffraction studies on amyloid filaments. *J. Histochem. Cytochem.* **16**, 673–677.
3 TEPLOW, D. B. (1998). Structural and kinetic features of amyloid beta-protein fibrillogenesis. *Amyloid* **5**, 121–142.
4 GEDDES, A. J., PARKER, K. D., ATKINS, E. D. & BEIGHTON, E. (1968). "Cross-beta" conformation in proteins. *J. Mol. Biol.* **32**, 343–358.
5 ROCHET, J. C. & LANSBURY, P. T. (2000). Amyloid fibrillogenesis: themes and variations. *Curr. Opin. Struct. Biol.* **10**, 60–68.
6 DOBSON, C. M. (2001). The structural basis of protein folding and its links with human disease. *Philos. Trans. R. Soc. Lond. B Biol. Sci.* **356**, 133–145.
7 LINDQUIST, S. (1997). Mad cows meet psi-chotic yeast: the expansion of the prion hypothesis. *Cell* **89**, 495–498.
8 WICKNER, R. B., EDSKES, H. K., MADDELEIN, M. L., TAYLOR, K. L. & MORIYAMA, H. (1999). Prions of yeast and fungi. Proteins as genetic material. *J. Biol. Chem.* **274**, 555–558.
9 SCHEIBEL, T. (2002). [PSI]-chotic yeasts: protein-only inheritance of a yeast prion. in *Recent Research in Molecular Microbiology* **1** (ed. PANDALAI, S. G.), Hindustan Publ. Corp., Delhi, 71–89.
10 ICONOMIDOU, V. A., VRIEND, G. & HAMODRAKAS, S. J. (2000). Amyloids protect the silkmoth oocyte and embryo. *FEBS Lett.* **479**, 141–145.

11 PODRABSKY, J. E., CARPENTER, J. F. & HAND, S. C. (2001). Survival of water stress in annual fish embryos: dehydration avoidance and egg envelope amyloid fibers. *Am. J. Physiol. Regul. Integr. Comp. Physiol.* **280**, R123–131.

12 GUHRS, K. H., WEISSHART, K. & GROSSE, F. (2000). Lessons from nature – protein fibers. *J. Biotechnol.* **74**, 121–134.

13 KNIGHT, D. P. & VOLLRATH, F. (2001). Comparison of the spinning of selachian egg case ply sheets and orb web spider dragline filaments. *Biomacromolecules* **2**, 323–334.

14 CRAIG, C. L. & RIEKEL, C. (2002). Comparative architecture of silks, fibrous proteins and their encoding genes in insects and spiders. *Comp. Biochem. Physiol. B Biochem. Mol. Biol.* **133**, 493–507.

15 VALLUZZI, R., WINKLER, S., WILSON, D. & KAPLAN, D. L. (2002). Silk: molecular organization and control of assembly. *Philos. Trans. R. Soc. Lond. B Biol. Sci.* **357**, 165–167.

16 VOLLRATH, F. (1999). Biology of spider silk. *Int. J. Biol. Macromol.* **24**, 81–88.

17 WARWICKER, J. O. (1960). Comparative studies of fibroins II. The crystal structures of various fibroins. *J. Mol. Biol.* **2**, 350–362.

18 RUDALL, K. M. & KENCHINGTON, W. (1971). Arthropod silks: The problem of fibrous proteins in animal tissues. *Annu. Rev. Entomol.* **16**, 73–96.

19 FRASER, R. D. & MACRAE, T. P. (1973) in *Conformation in fibrous proteins and related synthetic polypeptides* (eds. HORECKER, B., KAPLAN, N. O., MARMUR, J. & SCHERAGA, H. A.) Academic Press London, 293–343.

20 VOLLRATH, F. (2000). Strength and structure of spiders' silks. *J. Biotechnol.* **74**, 67–83.

21 BURGESON, R. E. (1988). New collagens, new concepts. *Annu. Rev. Cell Biol.* **4**, 551–577.

22 VAN DER REST, M. & GARRONE, R. (1991). Collagen family of proteins. *FASEB J.* **5**, 2814–2823.

23 MAYNE, R. & BREWTON, R. G. (1993). New members of the collagen superfamily. *Curr. Opin. Cell Biol.* **5**, 883–890.

24 FRATZL, P., MISOF, K., ZIZAK, I., RAPP, G., AMENITSCH, H. & BERNSTORFF, S. (1998). Fibrillar structure and mechanical properties of collagen. *J. Struct. Biol.* **122**, 119–122.

25 BERISIO, R., VITAGLIANO, L., MAZZARELLA, L. & ZAGARI, A. (2002). Recent progress on collagen triple helix structure, stability and assembly. *Protein Pept. Lett.* **9**, 107–116.

26 CREIGHTON, T. E. (1993) in *Proteins – structures and molecular properties* W. H. Freeman New York, 193–198.

27 BRODSKY, B. & RAMSHAW, J. A. (1997). The collagen triple-helix structure. *Matrix Biol.* **15**, 545–554.

28 GROSS, J. (1963). Comparative biochemistry of collagen. *Comp. Biochem.* **5**, 307–345.

29 ADAMS, E. (1978). Invertebrate collagens. *Science* **202**, 591–598.

30 SPIRO, R. G., LUCAS, F. & RUDALL, K. M. (1971). Glycosylation of hydroxylysine in collagens. *Nat. New Biol.* **231**, 54–55.

31 KNIGHT, D. P., FENG, D., STEWART, M. & KING, E. (1993). Changes in the macromolecular organization in collagen assemblies during secretion in the nidamental gland and formation of the egg capsule wall in the dogfish *Scyliorhinus canicula*. *Phil. Trans. R. Soc. Lond. B* **341**, 419–436.

32 KRAMER, J. M., COX, G. N. & HIRSCH, D. (1992). Comparisons of the complete sequences of two collagen genes from *Caenorhabditis elegans*. *Cell* **30**, 599–606.

33 MANN, K., GAILL, F. & TIMPL, R. (1992). Amino acid sequence and cell adhesion activity of fibril-forming collagen from the tube worm *Riftia pachyptila* living at deep sea hydrothermal vents. *Eur. J. Biochem.* **210**, 839–847.

34 GARRONE, R. (1978). Phylogenesis of connective tissue. In *Frontiers of matrix biology Vol. 5* (ed. ROBERT, L.) Karger Basel, 1–250.

35 MERCER, E. H. (1952). Observations on the molecular structure of byssus

36 RUDALL, K. M. (1955). The distribution of collagen and chitin. *Symp. Soc. Exp. Biol.* **9**, 49–71.

37 MELNICK, S. C. (1958). Occurrence of collagen in the phylum Mollusca. *Nature* **181**, 1483.

38 DEVORE, D. P., ENGEBRETSON, G. H., SCHACTELE, C. F. & SAUK, J. J. (1984). Identification of collagen from byssus threads produced by the sea mussel *Mytilus edulis*. *Comp. Biochem. Physiol.* **77B**, 529–531.

39 BENEDICT, C. V. & WAITE, J. H. (1986). Location and analysis of byssal structural proteins. *J. Morphol.* **189**, 261–270.

40 PIKKARAINEN, J., RANTANEN, J., VASTAMAKI, M., LAPIAHO, K., KARI, A. & KULONEN, E. (1968). On collagens of invertebrates with special reference to *Mytilus edulis*. *Eur. J. Biochem.* **4**, 555–560.

41 QIN, X. X. & WAITE, J. H. (1995). Exotic collagen gradients in the byssus of the mussel Mytilus edulis. *J. Exp. Biol.* **198**, 633–644.

42 COYNE, K. J., QIN, X. X. & WAITE, J. H. (1997). Extensible collagen in mussel byssus: a natural block copolymer. *Science* **277**, 1830–1832.

43 CARLIER, M. F. (1991). Actin: protein structure and filament dynamics. *J. Biol. Chem.* **266**, 1–4; KABSCH, W. & VANDEKERCKHOVE, J. (1992). Structure and function of actin. *Annu. Rev. Biophys. Biomol. Struct.* **21**, 49–76.

44 JANMEY, P. A., SHAH, J. V., TANG, J. X. & STOSSEL, T. P. (2001). Actin filament networks. *Results Probl. Cell. Differ.* **32**, 181–199.

45 CHENEY, R. E., RILEY, M. A. & MOOSEKER, M. S. (1993). Phylogenetic analysis of the myosin superfamily. *Cell Motil. Cytoskeleton* **24**, 215–223.

46 TITUS, M. A. (1993). Myosins. *Curr. Opin. Cell. Biol.* **5**, 77–81.

47 BERG, J. S., POWELL, B. C. & CHENEY, R. E. (2001). A millennial myosin census. *Mol. Biol. Cell.* **12**, 780–794.

48 PERRY, S. V. (2001). Vertebrate tropomyosin: distribution, properties and function. *J. Muscle Res. Cell Motil.* **22**, 5–49.

49 WEGNER, A. (1979). Equilibrium of the actin-tropomyosin interaction. *J. Mol. Biol.* **131**, 839–853.

50 YANG, Y.-Z., KORN, E. D. & EISENBERG, E. (1979). Cooperative binding of tropomyosin to muscle and *Acanthamoeba* actin. *J. Biol. Chem.* **254**, 7137–7140.

51 ZOT, A. S. & POTTER, J. D. (1987) Structural aspects of troponin-tropomyosin regulation of skeletal muscle contraction. *Ann. Rev. Biophys. Biophys. Chem.* **16**, 535–539.

52 STOKES, D. L. & DEROSIER D. J. (1987). The variable twist of actin and its modulation by actin-binding proteins. *J. Cell Biol.* **104**, 1005–1017.

53 PANTE, N. (1994). Paramyosin polarity in the thick filament of molluscan smooth muscles. *J. Struct. Biol.* **113**, 148–163.

54 ALBERS, K. & FUCHS, E. (1992). The molecular biology of intermediate filament proteins. *Int. Rev. Cytol.* **134**, 243–279.

55 CHOU, Y. H., BISCHOFF, J. R., BEACH, D. & GOLDMAN, R. D. (1990). Intermediate filament reorganization during mitosis is mediated by p34cdc2 phosphorylation of vimentin. *Cell* **62**, 1063–1071.

56 STEWART, M. (1993). Intermediate filament structure and assembly. *Curr. Opin. Cell Biol.* **5**, 3–11.

57 HERRMANN, H. & AEBI, U. (2000). Intermediate filaments and their associates: multi-talented structural elements specifying cytoarchitecture and cytodynamics. *Curr. Opin. Cell Biol.* **12**, 79–90.

58 STRELKOV, S. V., HERRMANN, H. & AEBI, U. (2003). Molecular architecture of intermediate filaments. *Bioessays* **25**, 243–251.

59 COULOMBE, P. A. (1993). The cellular and molecular biology of keratins: beginning a new era. *Curr. Opin. Cell Biol.* **5**, 17–29.

60 GARROD, D. R. (1993). Desmosomes and hemidesmosomes. *Curr. Opin. Cell Biol.* **5**, 30–40.

61 NIGG, E. A. (1992). Assembly and cell

62. Rzepecki, R. (2002) The nuclear lamins and the nuclear envelope. *Cell. Mol. Biol. Lett.* **7**, 1019–1035.
63. Privalov, P. L. & Medved, L. V. (1982). Domains in the fibrinogen molecule. *J. Mol. Biol.* **159**, 665–683.
64. Doolittle, R. F. (1984). Fibrinogen and fibrin. *Annu. Rev. Biochem.* **53**, 195–229.
65. Medved, L. V., Litvinovich, S., Ugarova, T., Matsuka, Y. & Ingham, K. (1997). Domain structure and functional activity of the recombinant human fibrinogen gamma-module (gamma148–411). *Biochemistry* **36**, 4685–4693.
66. Budzynski, A. Z. (1986). Fibrinogen and fibrin: biochemistry and pathophysiology. *Crit. Rev. Oncol. Hematol.* **6**, 97–146.
67. Mosher, D. F. & Johnson, R. B. (1983). Specificity of fibronectin–fibrin cross-linking. *Ann. N.Y. Acad. Sci.* **408**, 583–594.
68. Varadi, A. & Patthy, L. (1983). Location of plasminogen-binding sites in human fibrin(ogen). *Biochemistry* **22**, 2440–2446.
69. Bach, T. L., Barsigian, C., Yaen, C. H. & Martinez, J. (1998). Endothelial cell VE-cadherin functions as a receptor for the beta15–42 sequence of fibrin. *J. Biol. Chem.* **273**, 30719–30728.
70. Skoufias, D. A. & Scholey, J. M. (1993). Cytoplasmic microtubule-based motor proteins. *Curr. Opin. Cell. Biol.* **5**, 95–104.
71. Mandelkow, E. & Mandelkow, E. M. (1995). Microtubules and microtubule-associated proteins. *Curr. Opin. Cell. Biol.* **7**, 72–81.
72. Amos, L. A. & Baker, T. S. (1979). The three-dimensional structure of tubulin protofilaments. *Nature* **279**, 607–612.
73. Nogales, E. (2001). Structural insight into microtubule function. *Annu. Rev. Biophys. Biomol. Struct.* **30**, 397–420.
74. Job, D., Valiron, O. & Oakley, B. (2003). Microtubule nucleation. *Curr. Opin. Cell. Biol.* **15**, 111–117.
75. Mandelkow, E. M. & Mandelkow, E. (1992). Microtubule oscillations. *Cell. Motil. Cytoskeleton* **22**, 235–244.
76. Gosline, J. M. (1976). The physical properties of elastic tissue. *Int. Rev. Connect Tissue Res.* **7**, 211–249.
77. Ellis, G. E. & Packer, K. J. (1976). Nuclear spin-relaxation studies of hydrated elastin. *Biopolymers* **15**, 813–832.
78. Urry, D. W. (1988). Entropic elastic processes in protein mechanisms. II. Simple (passive) and coupled (active) development of elastic forces. *J. Protein Chem.* **7**, 1–34.
79. Tatham, A. S. & Shewry, P. R. (2002). Comparative structures and properties of elastic proteins. *Philos. Trans. R. Soc. Lond. B Biol. Sci.* **357**, 229–234.
80. Cleary, E. G., & Gibson, M. A. (1983). Elastin-associated microfibrils and microfibrillar proteins. *Int. Rev. Connect Tissue Res.* **10**, 97–209.
81. Vaughan, L., Mendler, M., Huber, S., Bruckner, P., Winterhalter, K. H., Irwin, M. I. & Mayne, R. (1988). D-periodic distribution of collagen type IX along cartilage fibrils. *J. Cell Biol.* **106**, 991–997.
82. Ardell, D. H. & Andersen, S. O. (2001). Tentative identification of a resilin gene in Drosophila melanogaster. *Insect. Biochem. Mol. Biol.* **31**, 965–970.
83. Morgan, D. G., Owen, C., Melanson, L. A. & DeRosier, D. J. (1995). Structure of bacterial flagellar filaments at 11 A resolution: packing of the alpha-helices. *J. Mol. Biol.* **249**, 88–110.
84. Mimori, Y., Yamashita, I., Murata, K., Fujiyoshi, Y., Yonekura, K., Toyoshima, C. & Namba, K. (1995). The structure of the R-type straight flagellar filament of Salmonella at 9 A resolution by electron cryomicroscopy. *J. Mol. Biol.* **249**, 69–87.
85. Aldridge, P. & Hughes, K. T. (2002). Regulation of flagellar assembly. *Curr. Opin. Microbiol.* **5**, 160–165.
86. Vonderviszt, F., Kanto, S., Aizawa, S. & Namba, K. (1989). Terminal

regions of flagellin are disordered in solution. *J. Mol. Biol.* **209**, 127–133.

87 NAMBA, K., YAMASHITA, I. & VONDERVISZT, F. (1989). Structure of the core and central channel of bacterial flagella. *Nature* **342**, 648–654.

88 YONEKURA, K., MAKI-YONEKURA, S. & NAMBA, K. (2002). Growth mechanism of the bacterial flagellar filament. *Res. Microbiol.* **153**, 191–197.

89 THOMAS, N. A., BARDY, S. L. & JARRELL, K. F. (2001). The archaeal flagellum: a different kind of prokaryotic motility structure. *FEMS Microbiol. Rev.* **25**, 147–174.

90 BARDY, S. L., NG, S. Y. & JARRELL, K. F. (2003). Prokaryotic motility structures. *Microbiology* **149**, 295–304.

91 FOREST, K. T. & TAINER, J. A. (1997). Type-4 pilus-structure: outside to inside and top to bottom – a mini-review. *Gene* **192**, 165–169.

92 WALL, D. & KAISER, D. (1999). Type IV pili and cell motility. *Mol. Microbiol.* **32**, 1–10.

93 MARVIN, D. A. & WACHTEL, E. J. (1976). Structure and assembly of filamentous bacterial viruses. *Philos. Trans. R. Soc. Lond. B Biol. Sci.* **276**, 81–98.

94 MARVIN, D. A. (1998). Filamentous phage structure, infection and assembly. *Curr. Opin. Struct. Biol.* **8**, 150–158.

95 PEDERSON, D. M., WELSH, L. C., MARVIN, D. A., SAMPSON, M., PERHAM, R. N., YU, M. & SLATER, M. R. (2001). The protein capsid of filamentous bacteriophage PH75 from Thermus thermophilus. *J. Mol. Biol.* **309**, 401–421.

96 DURHAM, A. C., FINCH, J. T. & KLUG, A. (1971). States of aggregation of tobacco mosaic virus protein. *Nat. New Biol.* **1971**, 229, 37–42.

97 SCHUSTER, T. M., SCHEELE, R. B., ADAMS, M. L., SHIRE, S. J., STECKERT, J. J. & POTSCHKA, M. (1980). Studies on the mechanism of assembly of tobacco mosaic virus. *Biophys. J.* **32**, 313–329.

98 DIAZ-AVALOS, R. & CASPAR, D. L. (1998). Structure of the stacked disk aggregate of tobacco mosaic virus protein. *Biophys. J.* **74**, 595–603.

99 CULVER, J. N. (2002). Tobacco mosaic virus assembly and disassembly: determinants in pathogenicity and resistance. *Annu. Rev. Phytopathol.* **40**, 287–308.

100 BUTLER, P. J. (1984). The current picture of the structure and assembly of tobacco mosaic virus. *J. Gen. Virol.* **65**, 253–279.

101 MITRAKI, A., MILLER, S. & VAN RAAIJ, M. J. (2002). Review: conformation and folding of novel beta-structural elements in viral fiber proteins: the triple beta-spiral and triple beta-helix. *J. Struct. Biol.* **137**, 236–247.

102 BETTS, S. & KING, J. (1999). There's a right way and a wrong way: in vivo and in vitro folding, misfolding and subunit assembly of the P22 tailspike. *Structure Fold. Des.* **7**, R131–139.

103 INOUYE, H., FRASER, P. & KIRSCHNER, D. (1993). Structure of β-Crystallite Assemblies by Alzheimer β Amyloid Protein Analogues: Analysis by X-Ray Diffraction. *Biophys. J.* **64**, 502–519.

104 BLAKE, C. & SERPELL, L. (1996). Synchrotron X-ray studies suggest that the core of the transthyretin amyloid fibril is a continuous b-helix. *Structure* **4**, 989–998.

105 SIKORSKI, P., ATKINS, E. D. T. & SERPELL, L. C. (2003). Structure and textures of fibrous crystals of the Abeta(11–25) fragment from Alzheimer's Abeta amyloid protein. *Structure* **11**, 1–20.

106 MALINCHIK, S., INOUYE, H., SZUMOWSKI, K. & KIRSCHNER, D. (1998). Structural analysis of Alzheimer's b(1–40) amyloid: proto-filament assembly of tubular fibrils. *Biophys. J.* **74**, 537–545.

107 GRIFFITHS, J. M., ASHBURN, T. T., AUGER, M., COSTA, P. R., GRIFFIN, R. G. & LANSBURY, P. T. (1995). Rotational resonance solid state NMR elucidates a structural model of pancreatic amyloid. *J. Am. Chem. Soc.* **117**, 3539–3546.

108 BALBACH, J. J., PETKOVA, A. T., OYLER, N. A., ANTZUKIN, O. N., GORDON, D. J., MEREDITH, S. C. & TYCKO, R.

109 Benzinger, T. L. S., Gregory, D. M., Burkoth, T. S., Miller-Auer, H., Lynn, D. G., Botto, R. E. & Meredith, S. C. (2000). Two-dimensional structure of b-amyloid (10–35) fibrils. *Biochemistry* **39**, 3491–3499.

108 (2002). Supramolecular structure in full-length Alzheimer's b-amyloid fibrils: evidence for parallel b-sheet organisation from solid state nuclear magnetic resonance. *Biophys. J.* **83**, 1205–1216.

110 Serpell, L. & Smith, J. (2000). Direct visualisation of the beta-sheet structure of synthetic Alzheimer's amyloid. *J. Mol. Biol.* **299**, 225–231.

111 Jimenez, J. L., Guijarro, J. I., Orlova, E., Zurdo, J., Dobson, C., Sunde, M. & Saibil, H. (1999). Cryo-electron microscopy structure of an SH3 amyloid fibril and model of the molecular packing. *EMBO J.* **18**, 815–821.

112 Lazo, N. & Downing, D. (1998). Amyloid fibrils may be assembled from b-helical protofibrils. *Biochemistry* **37**, 1731–1735.

113 Perutz, M. F., Finch, J. T., Berriman, J. & Lesk, A. (2002). Amyloid fibers are water-filled nanotubes. *Proc Natl Acad Sci USA* **99**, 5591–5595.

114 Serpell, L., Sunde, M., Benson, M., Tennent, G., Pepys, M. & Fraser, P. (2000). The protofilament substructure of amyloid fibrils. *J Mol Biol.* **300**, 1033–1039.

115 Goldsbury, C., Kistler, J., Aebi, U., Arvinte, T. & Cooper, G. J. (1999) Watching amyloid fibrils grow by time-lapse atomic force microscopy. *J. Mol. Biol.* **285**, 33–39.

116 Crowther, R. (1993). TAU Protein and Paired Helical Filaments of Alzheimers Disease. *Curr. Opin. Struct. Biol.* **3**, 202–206.

117 Crowther, R. (1991). Structural Aspects of Pathology in Alzheimers Disease. *Biochim. Biophys. Acta* **1096**, 1–9.

118 Berriman, J., Serpell, L. C., Oberg, K. A., Fink, A. L., Goedert, M. & Crowther, R. A. (2003). Tau filaments from human brain and from in vitro assembly of recombinant protein show cross-beta structure. *Proc. Natl. Acad. Sci. USA* **100**, 9034–9038.

119 Marsh, R. E., Corey, R. B. & Pauling, L. (1955). An investigation of the structure of silk fibroin. *Biochim. Biophys. Acta* **16**, 1–34.

120 Takahashi, Y., Gehoh, M. & Yuzuriha, K. (1999). Structure refinement and diffuse streak scattering of silk (Bombyx mori). *Int. J. Biol. Macromol.* **24**, 127–138.

121 Parkhe, A. D., Seeley, S. K., Gardner, K., Thompson, L. & Lewis, R. V. (1997). Structural studies of spider silk proteins in the fiber. *J. Mol. Recognit.* **10**, 1–6.

122 Parker, K. & Rudall, K. (1957). The silk of the egg-stalk of the green lacewing fly. *Nature* **179**, 905–907.

123 van Raaij, M. J., Mitraki, A., Lavigne, G. & Cusack, S. (1999). A triple b-spiral in the adenovirus fiber shaft reveals a new structural motif for a fibrous protein. *Nature* **401**, 935–938.

124 Yoder, M., Keen, N. & Jurnak, F. (1993). New domain motif – the structure of pectate lyase-C, a secreted plant virulence factor. *Science* **260**, 1503–1506.

125 Steinbacher, S., Seckler, R., Miller, D., Steipe, B., Huber, R. & Reinemer, P. (1994). Crystal-structure of P22 tailspike protein – interdigitated subunits in a thermostable trimer. *Science* **265**, 383–386.

126 Raetz, C. & Roderick, S. (1995). A left-handed parallel b-helix in the structure of UDP-N-Acetylglucosamine acyltransferase. *Science* **270**, 997–1000.

127 van Raaij, M. J., Schoehn, G., Jaquinod, M., Ashman, K., Burda, M. R. & Miller, S. (2001). Crystal structure of a heat- and protease-stable part of the bacteriophage t4 short tail fiber. *J. Mol. Biol.* **314**, 1137–1146.

128 Kanamaru, S., Leiman, P. G., Kostyuchenko, V. A., Chipman, P. R., Mesyanzhinov, V. V., Arisaka, F. & Rossman, M. G. (2002). The structure of the bacteriophage T4 cell-puncturing device. *Nature* **415**, 553–557.

129 KOSTYUCHENKO, V. A., LEIMAN, P. G., CHIPMAN, P. R., KANAMARU, S., VAN RAAIJ, M. J., ARISAKA, F., MESYANZHINOV, V. V. & ROSSMAN, M. G. (2003). Three-dimensional structure of the bacteriophage T4 baseplate. *Nat. Struct. Biol.* **10**, 688–693.

130 OTTANI, V., MARTINI, D., FRANCHI, M., RUGGERI, A. & RASPANTI, M. (2002). Hierarchical structures in fibrillar collagens. *Micron* **33**, 587–596.

131 WESS, T. J., HAMMERSLEY, A. P., WESS, L. & MILLER, A. (1998). A consensus model for molecular packing of type I collagen. *J. Struct. Biol.* **122**, 92–100.

132 WESS, T. J., HAMMERSLEY, A. P., WESS, L. & MILLER, A. (1995). Type I Collagen packing conformation of the triclinic unit cell. *J. Mol. Biol.* **248**, 487–493.

133 PHILLIPS, G. N., FILLERS, J. P. & COHEN, C. (1986). Tropomyosin crystal structure and muscle regulation. *J. Mol. Biol.* **192**, 111–131.

134 WHITBY, F. G., KENT, H., STEWART, F., STEWART, M., XIE, X., HATCH, V., COHEN, C. & PHILLIPS, G. N. (1992). Structure of tropomyosin at 9 Angstroms resolution. *J. Mol. Biol.* **227**, 441–452.

135 CRICK, F. H. C. (1952). Is alpha-keratin a coiled coil. *Nature* **170**, 882–883.

136 STRELKOV, S. V., HERRMAN, H., GEISLER, N., WEDIG, T., ZIMBELMANN, R., AEBI, U. & BURKHARD, P. (2002). Conserved segments 1A and 2B of the intermediate filament dimer: their atomic structures and role in filament assembly. *EMBO J.* **21**, 1255–1266.

137 LOWE, J. & AMOS, L. A. (1998). Crystal structure of the bacterial cell-division protein FtsZ. *Nature* **391**, 203–206.

138 LOWE, J. & AMOS, L. A. (1999). Tubulin-like protofilaments in Ca2+ induced FtsZ sheets. *EMBO J.* **18**, 2364–2371.

139 NOGALES, E., WOLF, S. & DOWNING, K. H. (1998). Structure of the alpha-beta tubulin dimer by electron crystallography. *Nature* **391**, 199–203.

140 NOGALES, E., WHITTAKER, M., MILLIGAN, R. A. & DOWNING, K. H. (1999). High-resolution model of the microtubule. *Cell* **96**, 79–88.

141 HIROSE, K., AMOS, W. B., LOCKHART, A., CROSS, R. A. & AMOS, L. A. (1997). Three-dimensional cryo-electron microscopy of 16-protofilament microtubules: structure, polarity and interaction with motor proteins. *J. Struct. Biol.* **118**, 140–148.

142 SQUIRE, J. & MORRIS, E. (1998). A new look at thin filament regulation in vertebrate skeletal muscle. *FASEB J.* **12**, 761–771.

143 HANSON, E. J. & LOWRY, J. (1963). The structure of F-actin and of actin filaments isolated from muscle. *J. Mol. Biol.* **6**, 48–60.

144 KABSCH, W., MANNHERZ, H. G., SUCK, D., PAI, E. F. & HOLMES, K. C. (1990). Atomic structure of the actin filament. *Nature* **347**, 44–49.

145 HOLMES, K., POPP, D., GERHARD, W. & KABSCH, W. (1990). Atomic Model of the Actin Filament. *Nature* **347**, 44–49.

146 LORENZ, M., POOLE, K., POPP, D., ROSENBAUM, G. & HOLMES, K. (1995). An Atomic Model of the Unregulated Thin Filament Obtained by X-ray Fiber Diffraction on Oriented Actin-Tropomyosin Gels. *J. Mol. Biol.* **246**, 108–119.

147 SQUIRE, J. M., KNUPP, C., AL-KHAYAT, H. A. & HARFORD, J. J. (2003). Milli-second time-resolved low-angle X-ray fiber diffraction: a powerful high-sensitivity technique for modelling real-time movements in biological macromolecular assemblies. *Fib. Diff. Rev.* **11**, 28–35.

148 MORRIS, E., SQUIRE, J. & FULLER, G. (1991). The 4-Stranded Helical Arrangement of Myosin Heads on Insect (Lethocerus) Flight Muscle Thick Filaments. *J. Struct. Biol.* **107**, 237–249.

149 PADRON, R., ALAMO, L., MURGICH, J. & CRAIG, R. (1998). Towards an atomic model of the thick filaments of muscle. *J. Mol. Biol.* **275**, 35–41.

150 RAYMENT, I., RYPNIEWSKI, W., SCHMIDT-BASE, K., SMITH, R.,

Tomcheck, D., Benning, M., Winkelmann, D., Wesenberg, G. & Holden, H. (1993). 3D-Structure of Myosin Subfragment-1: A Molecular Motor. *Science* **261**, 50–57.

151 Adler, A. J., Greenfield, N. J. & Fasman, G.D. (1973). Circular dichroism and optical rotatory dispersion of proteins and polypeptides. *Methods Enzymol.* **27**, 675–735.

152 Johnson, W. C. (1990). Protein secondary structure and circular dichroism: a practical guide. *Proteins* **7**, 205–214.

153 Kuwajima, K. (1995). Circular dichroism. *Methods Mol. Biol.* **40**, 115–135.

154 Woody, R. W. (1995). Circular dichroism. *Methods Enzymol.* **246**, 34–71.

155 Bloemendal, M. & Johnson, W. C. (1995). Structural information on proteins from circular dichroism spectroscopy possibilities and limitations. *Pharm. Biotechnol.* **7**, 65–100.

156 Greenfield, N. J. (1996). Methods to estimate the conformation of proteins and polypeptides from circular dichroism data. *Anal. Biochem.* **235**, 1–10.

157 Kelly, S. M. & Price, N. C. (2000). The use of circular dichroism in the investigation of protein structure and function. *Curr. Protein Pept. Sci.* **1**, 349–384.

158 Penzer, G. R. (1980) in *An introduction to spectroscopy for biochemists* (ed. Brown, S. B.) Academic Press London, 70–114.

159 Royer, C. A. (1995). Fluorescence spectroscopy. *Methods Mol. Biol.* **40**, 65–89M.

160 Roy, S. & Bhattacharyya, B. (1995). Fluorescence spectroscopic studies of proteins. *Subcell. Biochem.* **24**, 101–114.

161 Schmid, F. X. (1997) in *Protein structure – a practical approach* (ed. Creighton, T. E.) Oxford University Press, 261–298.

162 McLaughlin, M. L. & Barkley, M. D. (1997). Time-resolved fluorescence of constrained tryptophan derivatives: implications for protein fluorescence. *Methods. Enzymol.* **278**, 190–202.

163 Beechem, J. M. & Brand, L. (1985). Time-resolved fluorescence of proteins. *Annu. Rev. Biochem.* **54**, 43–71.

164 Papp, S. & Vanderkooi, J. M. (1989). Tryptophan phosphorescence at room temperature as a tool to study protein structure and dynamics. *Photochem. Photobiol.* **49**, 775–784.

165 Mathies, R. A., Peck, K. & Stryer, L. (1990). Optimization of high-sensitivity fluorescence detection. *Anal. Chem.* **62**, 1786–1791.

166 Haugland, R. P. (1996) in *Handbook of fluorescent probes and research chemicals* (ed. Spence, M. T. Z.) Molecular Probes Inc.

167 Gorman, P. M., Yip, C. M., Fraser, P. E. & Chakrabartty, A. (2003). Alternate aggregation pathways of the Alzheimer beta-amyloid peptide: Abeta association kinetics at endosomal pH. *J. Mol. Biol.* **325**, 743–757.

168 Leissring, M. A., Lu, A., Condron, M. M., Teplow, D. B., Stein, R. L., Farris, W. & Selkoe, D. J. (2003). Kinetics of amyloid beta-protein degradation determined by novel fluorescence- and fluorescence polarization-based assays. *J. Biol. Chem.* **278**, 37314–37320.

169 Scheibel, T., Bloom, J. & Lindquist, S. L. The elongation of yeast prion fibers involves separable steps of association and conversion. *Proc. Natl. Acad. Sci. USA*, **101**, 2287–2292.

170 Hermanson, G. (1996) in *Bioconjugate techniques* Academic Press London.

171 Udenfriend, S. (1962) in *Fluorescence assay in biology and medicine* (eds. Horecker, B., Kaplan, N. O., Marmur, J. & Scheraga, H. A.) Academic Press London, 223–229.

172 Slavik, I. (1982). Anilinonaphthalene sulfonate as a probe of membrane composition and function. *Biochim. Biophys. Acta* **694**, 1–25.

173 Förster, T. (1951) in *Fluoreszenz organischer Verbindungen*, Vandenhoeck and Ruprecht Göttingen.

174 Matulis, D. & Lovrien, R. (1998). 1-Anilino-8-naphthalene sulfonate anion-protein binding depends primarily on ion pair formation. *Biophys. J.* **72**, 422–429.

175 MATULIS, D., BAUMANN, C. G., BLOOMFIELD, V. A. & LOVRIEN, R. (1999). 1-anilino-8-naphthalene sulfonate as a protein conformational tightening agent. *Biopolymers* **49**, 451–458.

176 KIRK, W., KURIAN, E. & PRENDERGAST, F. (1996). Affinity of fatty acid for (r)rat intestinal fatty acid binding protein: further examination. *Biophys. J.* **70**, 69–83.

177 SAFAR, J., ROLLER, P. P., GAJDUSEK, D. C. & GIBBS, C. J. (1984). Scrapie amyloid (prion) protein has the conformational characteristics of an aggregated molten globule folding intermediate. *Biochemistry* **33**, 8375–8383.

178 LITVINOVICH, S. V., BREW, S. A., AOTA, S., AKIYAMA, S. K., HAUDENSCHILD, C. & INGHAM, K. C. (1998). Formation of amyloid-like fibrils by self-association of a partially unfolded fibronectin type III module. *J. Mol. Biol.* **280**, 245–258.

179 KAYED, R., BERNHAGEN, J., GREENFIELD, N., SWEIMEH, K., BRUNNER, H., VOELTER, W. & KAPURNIOTU, A. (1999). Conformational transitions of islet amyloid polypeptide (IAPP) in amyloid formation in vitro. *J. Mol. Biol.* **287**, 781–796.

180 BASKAKOV, I. V., LEGNAME, G., BALDWIN, M. A., PRUSINER, S. B. & COHEN, F. E. (2002). Pathway complexity of prion protein assembly into amyloid. *J. Biol. Chem.* **277**, 21140–21148.

181 MANN, C. J. & MATTHEWS, C. R. (1993). Structure and stability of an early folding intermediate of Escherichia coli trp aporepressor measured by far-UV stopped-flow circular dichroism and 8-anilino-1-naphthalene sulfonate binding. *Biochemistry* **32**, 5282–5290.

182 SUBBARAO, N. K. & MACDONALD, R. C. (1993). Experimental method to correct fluorescence intensities for the inner filter effect. *Analyst* **118**, 913–916.

183 BENOIT, H., FREUND, L. & SPACH, G. (1967) in: *Poly-α-amino acids* (ed. FASMAN, G. D.) Marcel Dekker Inc. New York, 105–155.

184 SHEN, C. L., SCOTT, G. L., MERCHANT, F. & MURPHY, R. M. (1993). Light scattering analysis of fibril growth from the amino-terminal fragment beta(1–28) of beta-amyloid peptide. *Biophys. J.* **65**, 2383–2395.

185 HARDING, S. E. (1997) in *Protein structure – a practical approach* (ed. CREIGHTON, T. E.) Oxford University Press, 219–251.

186 WYATT, P. J. (1991). Combined differential light scattering with various liquid chromatography separation techniques. *Biochem. Soc. Trans.* **19**, 485.

187 ROESSNER, D. & KULICKE, W. M. (1994). On-line coupling of flow field-flow fractionation and multi-angle laser light scattering *J. Chromatogr.* **687**, 249–258.

188 WYATT, P. J. & VILLALPANDO, D. (1997). High-Precision Measurement of Submicrometer Particle Size Distributions. *Langmuir* **13**, 3913–3914.

189 HARDING, S. E. (1986). Applications of light scattering in microbiology. *Biotechnol. Appl. Biochem.* **8**, 489–509.

190 WALSH, D. M., LOMAKIN, A., BENEDEK, G. B., CONDRON, M. M. & TEPLOW, D. B. (1997). Amyloid beta-protein fibrillogenesis. Detection of a protofibrillar intermediate. *J. Biol. Chem.* **272**, 22364–22372.

191 TSENG, Y., FEDOROV, E., MCCAFFERY, J. M., ALMO, S. C. & WIRTZ, D. (2001). Micromechanics and ultrastructure of actin filament networks crosslinked by human fascin: a comparison with alpha-actinin. *J. Mol. Biol.* **310**, 351–366.

192 KITA, R., TAKAHASHI, A., KAIBARA, M. & KUBOTA, K. (2002). Formation of fibrin gel in fibrinogen-thrombin system: static and dynamic light scattering study. *Biomacromolecules* **3**, 1013–1020.

193 GIDDINGS, J. C., YANG, F. J. & MYERS, M. N. (1977). Flow field-flow fractionation as a methodology for protein separation and characterization. *Anal. Biochem.* **81**, 394–407.

194 WAHLUND, K. G. & GIDDINGS, J. C. (1987). Properties of an asymmetrical

flow field-flow fractionation channel having one permeable wall. *Anal. Chem.* **59**, 1332–1339.

195 GIDDINGS, J. C. (1989). Field-flow fractionation of macromolecules. *J. Chromatogr.* **470**, 327–335.

196 LIU, M. K., LI, P. & GIDDINGS, J. C. (1993). Rapid protein separation and diffusion coefficient measurement by frit inlet flow field-flow fractionation. *Protein Sci.* **2**, 1520–1531.

197 NILSSON, M., BIRNBAUM, S. & WAHLUND, K. G. (1996). Determination of relative amounts of ribosome and subunits in Escherichia coli using asymmetrical flow field-flow fractionation. *J. Biochem. Biophys. Methods* **33**, 9–23.

198 ADOLPHI, U. & KULICKE, W. M. (1997). Coil dimensions and conformation of macromolecules in aqueous media from flow field-flow fractionation/multi-angle laser light scattering illustrated by studies on pullulan. *Polymer* **38**, 1513–1519.

199 KRETCHMANN, E. (1971). The determination of the optical constants of metals by excitation of surface plasmons. *Z. Phys.* **241**, 313–324.

200 FAGERSTAM, L. G., FROSTELL-KARLSSON, A., KARLSSON, R., PERSSON, B. & RONNBERG, I. (1992). Biospecific interaction analysis using surface plasmon resonance detection applied to kinetic, binding site and concentration analysis. *J. Chromatogr.* **597**, 397–410.

201 DE BRUIJN, H. E., KOOYMAN, R. P. & GREVE, J. (1992). Choice of metal and wavelength for SPR sensors: some considerations. *Appl. Opt.* **31**, 440–442.

202 GREEN, R. J., FRAZIER, R. A., SHAKESHEFF, K. M., DAVIES, M. C., ROBERTS, C. J. & TENDLER, S. J. (2000). Surface plasmon resonance analysis of dynamic biological interactions with biomaterials. *Biomaterials* **21**, 1823–1835.

203 RICH, R. L. & MYSZKA, D. G. (2000). Advances in surface plasmon resonance biosensor analysis. *Curr. Opin. Biotechnol.* **11**, 54–61.

204 MYSZKA, D. G., WOOD, S. J., BIERE, A. L. (1999). Analysis of fibril elongation using surface plasmon resonance biosensors. *Methods Enzymol.* **309**, 386–402.

205 VIKINGE, T. P., HANSSON, K. M., BENESCH, J., JOHANSEN, K., RANBY, M., LINDAHL, T. L., LIEDBERG, B., LUNDSTOM, I. & TENGVALL, P. (2000). Blood plasma coagulation studied by surface plasmon resonance. *J. Biomed. Opt.* **5**, 51–55.

206 BINNIG, G., QUATE, C. F. & GERBER, C. (1986). Atomic force microscope. *Phys. Rev. Lett.* **56**, 930–933.

207 STOLZ, M., STOFFLER, D., AEBI, U. & GOLDSBURY, C. (2000). Monitoring biomolecular interactions by time-lapse atomic force microscopy. *J. Struct. Biol.* **131**, 171–180.

208 YANG, Y., WANG, H. & ERIE, D. A. (2003). Quantitative characterization of biomolecular assemblies and interactions using atomic force microscopy. *Methods* **29**, 175–187.

209 HARPER, J. D., LIEBER, C. M. & LANSBURY, P. T. (1997). Atomic force microscopic imaging of seeded fibril formation and fibril branching by the Alzheimer's disease amyloid-beta protein. *Chem. Biol.* **4**, 951–959.

210 BLACKLEY, H. K., SANDERS, G. H., DAVIES, M. C., ROBERTS, C. J., TENDLER, S. J. & WILKINSON, M. J. (2000). In-situ atomic force microscopy study of beta-amyloid fibrillization. *J. Mol. Biol.* **298**, 833–840.

211 DEPACE, A. H., WEISSMAN, J. S. (2002). Origins and kinetic consequences of diversity in Sup35 yeast prion fibers. *Nat. Struct. Biol.* **9**, 389–396.

212 DA SILVA, L. P. (2002). Atomic force microscopy and proteins. *Protein Pept. Lett.* **9**, 117–126.

213 PUCHTLER, H. & SWEAT, F. (1965). Congo red as a stain for fluorescence microscopy of amyloid. *J. Histochem. Cytochem.* **13**, 693–694.

214 KLUNK, W. E., JACOB, R. F. & MASON, R. P. (1999). Quantifying amyloid by congo red spectral shift assay. *Methods Enzymol.* **309**, 285–305.

215 DELELLIS, R. A., GLENNER, G. G. & RAM, J. S. (1968). Histochemical observations on amyloid with

reference to polarization microscopy. *J. Histochem. Cytochem.* **16**, 663–665.

216 KHURANA, R., UVERSKY, V. N., NIELSEN, L. & FINK, A. L. (2001). Is Congo red an amyloid-specific dye? *J. Biol. Chem.* **276**, 22715–22721.

217 NAIKI, H., HIGUCHI, K., HOSOKAWA, M., TAKEDA, T. (1989). Fluorometric determination of amyloid fibrils in vitro using the fluorescent dye, thioflavin T1. *Anal. Biochem.* **177**, 244–249.

218 LEVINE, H. (1997). Stopped-flow kinetics reveal multiple phases of thioflavin T binding to Alzheimer beta (1–40) amyloid fibrils. *Biochem. Biophys.* **342**, 306–316.

219 BAN, T., HAMADA, D., HASEGAWA, K., NAIKI, H. & GOTO, Y. (2003). Direct observation of amyloid fibril growth monitored by thioflavin T fluorescence. *J. Biol. Chem.* **278**, 16462–16465.

220 HAYES, T. L. (1973). Scanning electron microscope techniques in biology. In *Advanced Techniques in Biological Electron Microscopy* (ed. KOEHLER, J. K.). Springer-Verlag, Berlin, Heidelberg, New York.

221 ECHLIN, P. (1971). The application of SEM to biological research. *Phil. Trans. Soc. Lond. B* **261**, 51–59.

222 DEROSIER, D. J. & MOORE, P. B. (1970). Reconstruction of three-dimensional images from electron micrographs of structures with helical symmetry. *J. Mol. Biol.* **52**, 355–369.

223 HARRIS, J. R. (1997). *Negative staining and cryoelectron microscopy*. Royal Microscopical Society, Microscopy handbook series, Bios scientific publishers limited, Oxford.

224 GLENNEY, J. R. (1987). Rotary metal shadowing for visualizing rod-shaped proteins. In *Electron microscopy in molecular biology* (eds. SOMMERVILLE, J. & SCHEER, U.), IRL press, Oxford, Washington DC, 167–178.

225 DEROSIER, D. J. & KLUG, A. (1968). Reconstruction of three-dimensional structures from electron micrographs. *Nature* **217**, 130–134.

226 SMITH, J. M. (1999). XIMDISP – A visualization tool to aid in structure determination from electron micrograph images. *J. Struct. Biol.* **125**, 223–228.

227 CROWTHER, R. A., HENDERSON, R. & SMITH, J. (1996). MRC Image processing programs. *J. Struct. Biol.* **116**, 9–16.

228 EGELMAN, E. H. (1986). An algorithm for straightening images of curved filamentous structures. *Ultramicroscopy* **19**, 367–373.

229 STEWART, M. (1988). Computer image processing of electron micrographs of biological structures with helical symmetry. *J. Electron microscopy technique* **9**, 325–358.

230 FRANK, J. (1973). Computer Processing of Electron Micrographs. In *Advanced Techniques in Biological Electron Microscopy* (ed. KOEHLER, J. K.). Springer-Verlag, Berlin, Heidelberg, New York.

231 STINE, W. B., SNYDER, S. W., LADROR, U. S., WADE, W. S., MILLER, M. F., PERUN, T. J., HOLZMAN, T. F. & KRAFFT, G. A. (1996) The nanometer-scale structure of amyloid-beta visualized by atomic force microscopy. *J. Protein Chem.* **15**, 193–203.

232 DING, T. T. & HARPER, J. D. (1999). Analysis of amyloid-beta assemblies using tapping mode atomic force microscopy under ambient conditions. In *Methods in Enzymology: Amyloid, prions and other protein aggregates* (ed. WETZEL, R.) Academic press New York, **309**, 510–525.

233 SIKORSKI, P. & ATKINS, E. D. T. (2001). The three-dimensional structure of monodisperse 5-amide nylon 6 crystals in the lambda phase. *Macromolecules* **34**, 4788–4794.

234 KREJCHI, M. T., COOPER, S. J., DEGUCHI, Y., ATKINS, E. D. T., FOURNIER, M. J., MASON, T. L. & TIRRELL, D. A. (1997). Crystal structures of chain-folded antiparallel beta-sheet assemblies from sequence-designed periodic polypeptides. *Macromolecules* **17**, 5012–5024.

235 FANDRICH, M. & DOBSON, C. (2002). The behaviour of polyamino acids reveals an inverse side-chain effect in amyloid structure formation. *EMBO Journal* **21**, 5682–5690.

236 WORCESTER, D. (1978). Structural Origins of Diamagnetic Anisotropy in Proteins. *Proc. Natl. Acad. Sci. USA* **75**, 5475–5477.

237 TORBET, J., FREYSSINET, J.-M. & HUDRY-CLERGRON, G. (1981). Oriented Fibrin Gels Formed by Polymerization in Strong Magnetic Fields. *Nature* **289**, 91–93.

238 TORBET, J. & DICKENS, M. J. (1984). Orientation of skeletal muscle actin in strong magnetic field. *FEBS Lett.* **173**, 403–406.

239 MARVIN, D. A. & NAVE, C. (1982). X-ray fiber diffraction. In *Structural molecular biology* (eds. DAVIES, D. B., SRENGER, W. & DANYLUK, S. S.). Plenum Publishing corporation.

240 SQUIRE, J., AL-KHAYAT, H., ARNOTT, S., CRAWSHAW, J., DENNY, R., DIAKUN, G., DOVER, D., FORSYTH, T., HE, A., KNUPP, C., MANT, G., RAJKUMAR, G., RODMAN, M., SHOTTON, M. & WINDLE, A. (2003). New CCP13 software and the strategy behind further developments: stripping and modelling of fiber diffraction data. *Fiber diffraction review* **11**, 13–19.

241 CCP4. (1994). The CCP4 Suite Programs for Crystallography. *Acta Cryst.* **D50**, 760–763.

242 OTWINOWSKI, Z. & MINOR, W. (1997). Processing of X-ray diffraction data collected in oscillation mode. In *Methods in Enzymology* (eds. CARTER, C. W. & SWEET, R. M.) Academic press New York, **276**, 307–326.

243 FRASER, R. D. B., MACRAE, T. P., MILLER, A. & ROWLANDS, R. J. (1976). Digital Processing of Fiber Diffraction Patterns. *J. Appl. Cryst.* **9**, 81–94.

244 MILLANE, R. P. & ARNOTT, S. (1985). Background subtraction in X-ray fiber diffraction patterns. *J. Appl. Cryst.* **18**, 419–423.

245 IVANOVA, M. I. & MAKOWSKI, L. (1998). Iterative low-pass filtering for estimation of the background in fiber diffraction patterns. *Acta Cryst. A* **54**, 626–631.

246 OKADA, K., NOGUCHI, K., OKUYAMA, K. & ARNOTT, S. (2003). WinLALS for a linked-atom least squares refinement program for helical polymers on Windows PCs. *Computational Biology and Chemistry* **3**, 265–285.

247 CANTOR, C. R. & SCHIMMEL, P. R. (1980). In *Biophysical Chemistry. Part II: Techniques for the study of biological structure and function.* W. H. Freeman and company New York.

248 SESHADRI, S., KHURANA, R. & FINK, A. L. (1999). Fourier transform infrared spectroscopy in analysis of protein deposits. In *Methods in Enzymology: Amyloid, prions and other protein aggregates* (ed. WETZEL, R.) Academic press New York, **309**, 559–576.

249 OBERG, K. A. & FINK, A. L. (1998). A new attenuated total reflectance Fourier transform infrared spectroscopy method for the study of proteins in solution. *Anal. Biochem.* **256**, 92–106.

250 SERPELL, L. C., BERRIMAN, J., JAKES, R., GOEDERT, M. & CROWTHER, R. A. (2000). Fiber diffraction of synthetic a-synuclein filaments shows amyloid-like cross-b conformation. *Proc. Natl. Acad. Sci. USA* **97**, 4897–4902.

251 SCHEIBEL, T., KOWAL, A., BLOOM, J. & LINDQUIST, S. (2001). Bi-directional amyloid fiber growth for a yeast prion determinant. *Curr. Biol.* **11**, 366–369.

7
Protein Unfolding in the Cell

Prakash Koodathingal, Neil E. Jaffe, and Andreas Matouschek

7.1
Introduction

Regulated protein unfolding is a key step in biological processes such as translocation across membranes and degradation by ATP-dependent proteases (Figure 7.1). Most cellular proteins are encoded within the nucleus and subsequently imported into different organelles such as mitochondria, chloroplasts, microsomes, and peroxisomes. The dimensions of protein import channels in many of these compartments are such that native proteins simply do not fit through them. In the case of ATP-dependent proteases, the unfolding requirement is imposed by the sequestration of the proteolytic active sites deep inside the protease structure. Access to this site is controlled by the narrow opening at the entrance of the degradation channel. Unfolding is catalyzed during translocation across membranes and degradation by ATP-dependent proteases. The mechanisms of catalyzed unfolding in these processes resemble each other and differ from global unfolding processes induced by chaotropic agents. In this chapter, we will discuss findings that led to the current understanding of unfolding observed in cell.

7.2
Protein Translocation Across Membranes

7.2.1
Compartmentalization and Unfolding

The eukaryotic cytoplasm is divided into several functionally distinct compartments or organelles. Most proteins are synthesized in the cytosol and then have to be imported into compartments. In order for proteins to fit through import channels, unfolding is required. The best-understood example of a compartment that requires unfolding for protein transport is the mitochondria [1–3]. Mitochondria are surrounded by two well-defined membranes. The protein import channel across the outer mitochondrial membrane is rigid and has a diameter of approxi-

Protein Folding Handbook. Part II. Edited by J. Buchner and T. Kiefhaber
Copyright © 2005 WILEY-VCH Verlag GmbH & Co. KGaA, Weinheim
ISBN: 3-527-30784-2

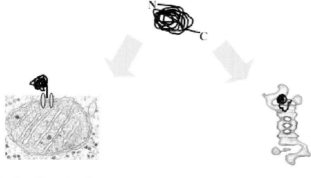

Fig. 7.1. Protein unfolding occurs during translocation across some membranes (left) and during degradation by ATP-dependent proteases (right).

mately 24 Å [4–6]. The channel across the inner membrane is flexible, but its maximum diameter is smaller than that of the outer-membrane channel [6–8]. The majority of mitochondrial proteins are synthesized in the cytosol as preproteins with positively charged N-terminal targeting signals [9]. After synthesis, preproteins are localized to the mitochondria.

Several lines of evidence indicate that most precursor proteins are not in their native conformation during translocation across the mitochondrial matrix. First, import of mouse dihydrofolate reductase (DHFR) is blocked when its unfolding is inhibited by the tightly binding ligand methotrexate [10]. Second, studies using *Neurospora crassa* mitochondria at a low temperature identified a translocation intermediate during the import of F_1-ATPase β-subunit and cytochrome c_1 precursors [11]. This intermediate appeared to span both mitochondrial membranes, which is a conformation that is possible only when the precursor protein is at least partially unfolded. Third, attaching oxidized BPTI (containing three intramolecular disulfide bridges) to the C-terminus of mouse DHFR prevented the import of the DHFR moiety completely inside the mitochondria [12]. Finally, a study using barnase mutants showed that a single disulfide bridge in a precursor protein slowed import [7]. It is now widely accepted that proteins are in a fully unfolded conformation during import.

The next question is when and where preproteins are unfolded. One possibility could be that proteins never fold before import. However, it is thought that multidomain proteins fold co-translationally in eukaryotes [13]. The N-terminal domain of nascent polypeptide chains folds before the synthesis of the C-terminal domain is complete [13, 14]. In vitro studies using three model proteins demonstrated that pre-sequences of differing lengths do not affect the stability or folding and unfolding kinetics of the mature proteins [15–17]. Two in vivo experiments show that precursor proteins can be in the native conformation prior to import. Studies in *S. cerevisiae* showed that a hybrid protein consisting of the amino-terminal third of cytochrome b_2 followed by the DHFR domain accumulates outside the mitochondrial matrix as a translocation intermediate in the presence of the substrate analogue aminopterin [18]. This shows that the DHFR domain was completely translated and folded before translocation across the membrane. Similarly, precursors containing the tightly folded heme-binding domain could be completely imported into mitochondria only when the pre-sequence could actively engage the mitochondrial unfolding machinery [19], suggesting that an authentic mitochondrial protein folds in the cytosol.

A subset of precursors, such as subunits of larger complexes or integral membrane proteins, is unable to fold in the cytosol and interact with chaperones [20, 21]. These precursors require ATP outside the mitochondrial matrix for import. The requirement for cytosolic ATP can be negated by denaturing the precursors with urea [21]. Thus, cytosolic chaperones can facilitate import of precursors that are unable to fold in the cytosol (precursors prone to aggregation). Although cytosolic chaperones affect import of some proteins, it is well documented that purified mitochondria can import chemically pure folded precursor proteins [10, 22].

7.2.2
Mitochondria Actively Unfold Precursor Proteins

Mitochondria can import folded precursor proteins much faster than the spontaneous unfolding measured in free solution [22, 23]. In other words, mitochondria actively unfold proteins during import. Two models can explain the nature of the unfolding activity during translocation into mitochondria. One possibility is that the mitochondrial membrane surface destabilizes importing proteins. For example, it has been shown that lipid vesicles induce the partially unfolded molten-globule state observed in diphtheria toxin [24–28]. However, the mitochondrial membrane surface does not appear to affect the stability of importing proteins. This question has been studied extensively using artificial precursor proteins composed of barnase. Barnase is a small ribonuclease that is bound by its stabilizing ligand barstar. Mutations had the same effect on the stability of barnase at the import site on the mitochondria and in free solution [29]. In addition, the dissociation constant of the barnase-barstar complex at the mitochondrial surface coincides with that in free solution [29].

It appears that mitochondria can accelerate unfolding of at least some precursor proteins by changing their unfolding pathway (Figure 7.2, see p. 258). Experiments that measured the effect of mutations on the unfolding rate and stability of barnase showed that the spontaneous unfolding pathway of barnase begins with a specific sub-domain (formed by the second and third α-helices and some loops packed against the edge of a β-sheet) and follows with the rest of the protein (the N-terminal α-helix and the five-stranded β-sheet located at the C-terminus). The catalyzed unfolding pathway during mitochondrial import is different: the mitochondrial import machinery begins unfolding by unraveling the N-terminal α-helix and then processively unravels the protein to its C-terminus [15]. Thus, mitochondria unfold this precursor protein by unraveling it from its targeting signal [15]. Additional evidence for this model comes from the observation that some model substrates that are stabilized against spontaneous unfolding are not stabilized during unfolding by the mitochondrial translocase [15]. However, stabilizing other proteins can completely prevent their import, and, therefore, mitochondrial unfolding machinery relies on the structure of the precursor protein near the targeting signal.

7.2.3
The Protein Import Machinery of Mitochondria

Import of precursor proteins into the mitochondrial matrix is catalyzed by the translocation machinery of the outer and inner mitochondrial membranes (Figure 7.3). In the outer membrane, components of the TOM complex (translocase of the outer mitochondrial membrane) recognize and translocate precursors. The TOM complex of N. crassa and yeast is composed of at least eight proteins [1, 30]. In the inner membrane, two separate TIM complexes (translocase of the inner mitochondrial membrane) are required for import, one for proteins destined for the inner membrane and one for proteins destined for the matrix and inter-membrane space. The structures and roles of the TOM and TIM complexes are discussed in chapter 30.

Import into the mitochondrial matrix requires both an electrochemical gradient across the inner membrane and the action of the Hsp70 homologue found in the mitochondrial matrix (mHsp70). mHsp70 is required for the matrix import of all mitochondrial proteins [31–33] and interacts directly with translocating precursors. The TIM complex recruits mHsp70 to the import site and facilitates the stepwise movement of the polypeptide across the translocation pore [34, 35]. The electrical potential across the inner membrane was found to act on pre-sequences of precursor proteins before they interact with the mHsp70 physically [36], and the potential is required for import of all proteins into the matrix [37, 38]. Respiring yeast mitochondria maintain an electrical potential of approximately 150 mV across the inner membrane, which is positive at the outer surface and negative at the inner surface. The electrical potential performs two main functions in precursor unfolding and import. First, it enhances the dimerization of Tim23, a component of the import channel, presumably to augment the interaction between the targeting signal and the import channel. Second, it unfolds precursor proteins. Reducing the electrical potential with an uncoupler of respiration reduces the import rate of precursors whose targeting sequence reaches mitochondrial inner membranes [39]. The slower import rate is due to a reduced unfolding activity. Experiments in which the charges of the mitochondrial targeting sequences are changed by mutating the sequence suggest that the potential acts directly on the charged targeting sequence [39]. Precursors with targeting sequences of approximately 50 amino acids or more can interact with mHsp70 while the mature domain is still folded at the mitochondrial surface. For these precursors the membrane potential is required

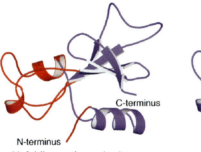 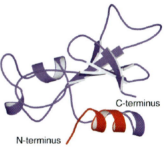

Fig. 7.2. Unfolding pathways of barnase. Structure of barnase, color-coded according to the order in which structure is lost (left) during spontaneous global unfolding in vitro and (right) during import into mitochondria. The parts of the structure shown in red unfold early, whereas those shown in blue unfold late. Figure reproduced from Ref. [15] with permission.

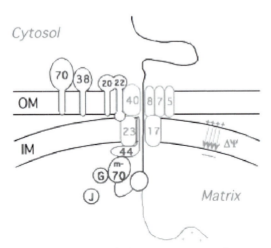

Fig. 7.3. The mitochondrial protein import machinery. Proteins in the outer/inner membrane are called Tom/Tim, followed by the number indicated in the figure. The number reflects their approximate molecular weight. During import, precursor proteins first interact with the Tom20 and Tom22 receptors through their targeting sequence. The Tom70 receptor binds precursors associated with cytosolic chaperones. Targeting sequences insert into the Tom40 channel and pass through the Tom23 complex into the matrix. Import into the matrix always requires an electrical potential across the inner membrane and the ATP-dependent action of mHsp70. mHsp70 is found bound to the import machinery through Tim44 and free in the matrix. Precursors begin to interact with mHsp70 while they are still associated with the import channels. G, J: mGrpE, Mdj1, two co-chaperones of Hsp70; IM/OM: inner/outer membrane; m-70: mHsp70. Figure reproduced from Ref. [40] with permission.

only for the insertion of the targeting sequence in to the import channel. Unfolding can be induced by mHsp70 in the absence of an electrical potential. The actual length of natural targeting sequences is uncertain. However, databases indicate that the mean cleavage site is at amino acid position 31. This suggests that most targeting signals are too short to interact with mHsp70 before they unfold [40].

7.2.4
Specificity of Unfolding

The susceptibility to unfolding by the mitochondrial import machinery depends on the structure of the preprotein near the targeting signal. When a collection of random peptides were attached to the N-terminus of subunit IV of cytochrome oxidase (COX IV), approximately a quarter of these constructs were imported into mitochondria [41]. In addition, there have been some reports that the mature domain affects import efficiency [42, 43]. Indeed, when proteins of different stabilities and folds are attached to the same targeting sequence, the import efficiencies of these constructs vary considerably, with precursors that are more difficult to unfold importing less efficiently [15, 23, 44, 45]. Together, these results suggest that the susceptibility of the mature domain of precursor proteins to unfolding contributes to the specificity of protein import into mitochondria.

7.2.5
Protein Import into Other Cellular Compartments

Unfolding may also play an important role in protein import into ER and chloroplasts. Chloroplasts are surrounded by two membranes and are divided into two compartments, the stroma and the thylakoids. Most chloroplast proteins are imported posttranslationally from the cytosol and, presumably, some precursors will fold before translocation. However, in the case of chloroplast protein import, it is not clear whether unfolding is always required. For example, one study indicates that chloroplast membranes can import small folded proteins [46]. On the other hand, import is blocked when large proteins or protein complexes are stabilized against unfolding [47–50]. Within the chloroplast, translocation of proteins from stroma into thylakoids occurs through two different machineries that seem to impose different steric requirements on the transporting proteins. One subset of proteins uses the ATP-dependent Sec system [51], while other proteins are transported by a mechanism that does not require ATP but is instead dependent on the pH gradient across the thylakoidal membrane [51]. The Sec-related pathway requires protein unfolding, whereas the ΔpH-dependent pathway tolerates folded proteins.

Protein translocation into the endoplasmic reticulum generally does not require unfolding because it occurs mostly co-translationally. However, at least in *S. cerevisiae*, there are examples of posttranslational translocation, and it is possible that some of these precursor proteins fold in the cytosol. The internal diameter of the translocon complex is 20–40 Å at its narrowest point, and experiments suggest that several proteins have to be in an unfolded conformation to fit through the translocation channel [9, 52]. Together these findings suggest that protein unfolding may play a role in the import of a subset of ER-associated proteins. The ER import machinery resembles those of mitochondria and chloroplasts in that an Hsp70 homologue is located at the exit of the protein import channel [53].

Protein unfolding may also play a key role in ER retro-translocation of misfolded or unassembled polypeptides back into the cytoplasm for degradation [54–56]. Misfolded or unassembled polypeptides inside the ER are directed to proteasome-mediated degradation in the cytoplasm in a process called ER-associated degradation (ERAD) [54, 56]. During this process, the polypeptide chain may be translocated backwards into the cytoplasm by a pulling force generated by cytoplasmic chaperones such as CDC48/p97/VCP, by proteins involved in the ubiquitination process, or by the proteasome itself [55, 57–59].

7.3
Protein Unfolding and Degradation by ATP-dependent Proteases

Protein unfolding and degradation by ATP-dependent proteases is a key step in many biological processes, including signal transduction, cell cycle control, DNA transcription, DNA repair, angiogenesis, and apoptosis. In eukaryotic cells, the majority of regulated protein degradation is mediated by the ubiquitin-proteasome pathway [60]. The proteasome also removes proteins that are misfolded as a result of mutations and various stresses. In prokaryotes, similar functions are performed by functional analogues of the proteasome such as ClpAP, ClpXP, Lon, and FtsH (on the membrane) and HslUV (ClpYQ) proteases [60–63]. Unfolding processes driven by ATP-dependent proteases and by the mitochondrial import machinery exhibit several similarities. Both processes involve the hydrolysis of ATP, and unfolding is coupled to the movement of the extended polypeptide chain through a narrow channel. We will now examine protein unfolding by ATP-dependent proteases and draw generalizations about protein unfolding in the cell.

7.3.1
Structural Considerations of Unfoldases Associated With Degradation

ATP-dependent proteases share similar overall structures [60, 61, 63–67]. The 26S eukaryotic proteasome is a 2-MDa structure composed of two structurally and functionally separated subunits: a catalytic subunit, called the 20S core particle, and the ATPase subunit, called the 19S regulatory particle [60, 64]. The core particle is composed of four stacked heptameric rings, two of which consist of α-subunits and two of β-subunits. The two β-rings are stacked together at the core of the proteasome and contain the active sites for proteolysis. One ring of α-subunits flanks the β-rings on each side. Together, the α- and β-rings form a cylindrical structure that is flanked on each side by the regulatory subunit. The 19S cap, which has a molecular weight of 700 kDa, contains 18 different subunits that recognize ubiquitinated proteins and present the substrate proteins to the proteolytic core in an ATP-dependent manner [60, 61].

Substrate proteins are targeted to the proteasome by the covalent attachment of multiple ubiquitin chains to their surface-exposed lysine residues [68]. A cascade of reactions, catalyzed by ubiquitin-activating enzyme (E1), ubiquitin-conjugating enzymes (E2), and ubiquitin ligases (E3), forms isopeptide linkages between the C-terminus of the ubiquitin moieties and the ε-amino group of lysine residues on the acceptor protein. Successive rounds of ubiquitination result in the formation of polyubiquitin chains, and at least four ubiquitins are required for efficient targeting [69]. The polyubiquitin chains are recognized by the S5a subunit of the 19S cap of the 26S proteasome [70]. However, at least one protein (ornithine decarboxylase [71]) has been identified that is targeted to the proteasome without ubiquitin modification. In addition, the 19S regulatory cap can interact with misfolded or natively unfolded intermediates in an ubiquitin-independent manner [72].

Other ATP-dependent proteases, such as ClpAP and ClpXP, share structural similarities to the 26S proteasome (Figure 7.4). ClpAP and ClpXP are elongated cylindrical complexes composed of hexameric ATPase single rings (ClpA and ClpX), juxtaposed to a proteolytic component of two stacked heptameric rings (ClpP) [60, 61, 63, 65, 66]. The active sites of proteolysis in the Clp proteases are located on the inside of the ClpP ring. Lon and the HflB proteases, unlike other ATP-dependent proteases, are homo-oligomeric complexes. However, they share the overall cylindrical structure of other proteases, with the active sites buried deep inside the central cavity [61, 62, 67].

Substrates of prokaryotic ATP-dependent proteases are recognized through their targeting tags, which are mostly N- or C-terminal extensions of varying length. The sequences of these targeting signals often show some homology, but definitive consensus sequences for these targeting signals are still being determined and may be specific to the various proteases. Interestingly, there are cases where specific targeting signals can specify more than one ATP-dependent protease. For example, substrates tagged with the ssrA peptide are targeted to ClpAP, ClpXP, and HflB [73–75]. Yet ClpAP, ClpXP, and HflB can also possess specific substrate preferences [62, 76, 77]. Specificity is often conferred by adaptor proteins, which bind substrates and deliver them in a *trans*-targeting mechanism to the protease [78].

7.3.2
Unfolding Is Required for Degradation by ATP-dependent Proteases

A requirement for unfolding for degradation is imposed by the sequestration of the proteolytic active sites deep inside the protease structure. Access to these active sites is possible only through a narrow channel that runs along the long axis of the cylindrical particle [63, 79]. The entrance to the degradation channel of the proteasome is also blocked by the N-termini of the α-subunits at the small end of the cylindrical particle [79]. The requirement for unfolding during proteasome degradation is demonstrated experimentally by biochemical studies where stabilizing the folded state of the substrate protein, DHFR, prevented its degradation [80].

Eukaryotic Proteasome

Bacterial ClpAP Protease

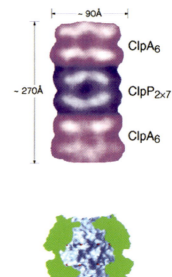

Fig. 7.4. Structures of the proteasome and ClpAP. Top left: Structure of the eukaryotic proteasome holo-enzyme from *Xenopus laevis* as determined by electron microscopy; the core particle is shown in blue, and the ATPase caps are in pink. Bottom left: Medial sections of the proteasome core particles from yeast as determined by crystallography; active sites of proteolysis are indicated by red dots, and the slice surface is shown in green. Top right: Structure of the ClpAP holo-enzyme as determined by electron microscopy. Bottom right: Structure of the ClpP particle, the proteolytic core of ClpAP, as determined by crystallography. Figure reproduced from Ref. [62] with permission.

7.3.3
The Role of ATP and Models of Protein Unfolding

Degradation by ATP-dependent proteases requires ATP, which serves to power the unfolding and translocation of the polypeptide chain from the outer surface of the protease to the proteolytic sites deep inside the cylindrical structure [61, 63, 65, 81, 82] (Figure 7.5). Additionally, ATP is required for the cooperative association of the cap structures at both ends of the catalytic cylinder [81, 82]. Not only ATP binding but also ATP hydrolysis is required for proteolysis to occur [81, 82], and there is a direct relationship between the amount of ATP hydrolyzed by ClpXP and the length and stability of the substrate [83]. The exact mechanism by which ATP binding and ATP hydrolysis induce unfolding has not yet been elucidated. Two general models have emerged for the mechanism of ATP-driven unfolding and translocation of substrate proteins into the proteolytic chamber. The first model proposes that ATP drives the translocation of the substrate's polypeptide chain from the

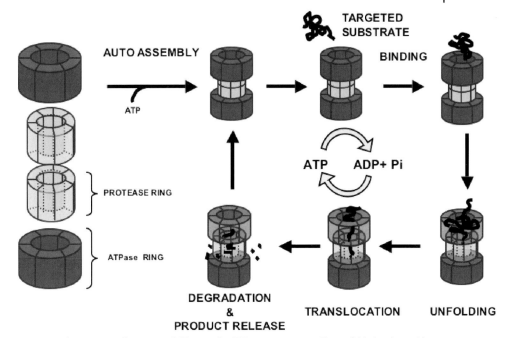

Fig. 7.5. The process of protein unfolding and degradation by ATP-dependent proteases. ATP binding and hydrolysis are required for the assembly of a functional ATP-dependent protease. Targeted substrates for degradation bind noncovalently to the ATPase particle. Repeated cycles of ATP binding and hydrolysis drive unfolding of the substrate within the ATPase component. The unfolded polypeptide chain is fed into the proteolytic chamber in a process called translocation. Subsequently, the polypeptide reaches the active sites of proteolysis, where it is rapidly degraded and released. The assembled ATP-dependent protease can unfold and degrade several substrates before disassembly.

substrate-binding site at the ends of the cylindrical protease particle to the proteolytic chamber where degradation occurs. The translocation would result in a pulling force on the native polypeptide, which could help to collapse the structure of the substrate [83, 84, 85]. An alternative model proposes that unfolding is not coupled to translocation into the proteolytic compartment [86]. Instead, unfolding occurs on the surface of the ATPase ring, where ATP binding and hydrolysis cause the substrate-binding domains to undergo concerted conformational changes that exert mechanical strain on the bound protein and result in unfolding. The proposed role of ATP in this model draws parallels to the role of ATP in the action of the molecular chaperone GroEL [87–89].

7.3.4
Proteins Are Unfolded Sequentially and Processively

ATP-dependent proteases degrade their substrates by sequentially unraveling them from the degradation signal [84, 85]. Sequential degradation was demonstrated using multi-domain proteins consisting of barnase and DHFR. Experiments showed that stabilizing the domain closer to the degradation tag by tightly binding ligands protected the downstream domain from proteolysis. The sequential degradation of substrates by ClpAP was also demonstrated by FRET experiments [85]. In these experiments the energy donor was attached to the protease subunits of ClpAP and a fluorescence acceptor was attached to either the N-terminus or the C-terminus of substrate proteins [85]. Kinetic analysis of ATP-triggered reactions showed that the probe near the degradation signal interacted with the protease probe before the probe farther away from the degradation signal did, reflecting a directional translocation of substrates from ClpA to ClpP. In addition, biochemical and electron microscopy experiments demonstrated the sequential interaction of substrate proteins first with the ATPase subunits and then with the protease subunits of ClpAP and ClpXP [90–92].

7.3.5
The Influence of Substrate Structure on the Degradation Process

The local structure of the protein near the degradation tag determines the ability of the protein to be degraded [83, 84]. This is best exemplified by studies with circular permutants of DHFR where the original N- and C-termini are connected by a short series of glycine residues and the structure is disrupted to create new termini [93]. The resulting proteins have almost identical structure and enzymatic activities [84, 93] and differ primarily in the location of the signals. However, these circular permutants showed substantially different susceptibilities towards unfolding and degradation by ATP-dependent proteases [84]. In these studies, susceptibility to unfolding correlated with the local structure adjacent to the targeting signal, not with the stability against spontaneous global unfolding. When the degradation signal leads into a stretch of polypeptide chain that forms an α-helix or a surface loop, the proteins were unfolded and degraded efficiently. Substrates were more difficult to unfold and degrade when the degradation signal led into an internal β-strand [84].

7.3.6
Unfolding by Pulling

The susceptibility to unfolding correlates with neither the global thermodynamic stability nor the kinetic stability determined by in vitro experiments [94]. In a study using titin, it was found that only mutations near the degradation signal influence unfolding by ClpXP [83]. Mutations elsewhere have no effect. These studies, together with experiments demonstrating the sequential unfolding and degradation of proteins, suggest that ATP-dependent proteases catalyze unfolding by altering the unfolding pathway of their substrates, just as mitochondria unfold proteins [83, 84].

Although it is clear that unfolding occurs as the polypeptide chain is translocated into the proteolytic chamber, the mechanics of force exertion are still under investigation. In unfolding during both translocation and the degradation process, the unraveling mechanism is more reminiscent of unfolding by atomic force microscopy than of unfolding by chemical denaturants or heat. The spontaneous unfolding pathway of proteins by chaotropic agents follows a global mechanism. When unfolded by AFM, substrate proteins are unfolded by pulling mechanically on the polypeptide chain [95–99]. However, the processes of protein unfolding during AFM experiments and during protein degradation and translocation are not identical [79, 100]. During AFM experiments, a mechanical force is applied continuously from both ends of the polypeptide chain until the protein unfolds. This resembles a stretching force. In unfolding mediated by ATP-dependent proteases and mitochondrial import machinery, a force would be applied by pulling the substrate against the sterically restrictive entrance of the translocation channel. In addition, studies on ClpXP-mediated protein unfolding and degradation suggest that the unfolding force is applied iteratively on the substrates, possibly by tugging at them repeatedly [83].

7.3.7
Specificity of Degradation

The fact that the susceptibility to degradation depends on the local structure near the degradation tag may contribute to the specificity of degradation. This particular mechanism may have several physiological consequences. First, it allows the proteases to degrade specific subunits of a multi-protein complex without affecting other components. For example, the proteasome specifically degrades the cell-cycle inhibitor Sic1 while it is associated with the yeast cyclin-cyclin-dependent kinase (CDK) complex to release active CDK [101]. Second, the observation that β-structures are difficult to unfold is important in the context of diseases that are characterized by the accumulation of large intracellular protein aggregates, such as Parkinson's and Huntington's diseases [102, 103]. Protein aggregates are found associated with ubiquitin and components of the proteasome, suggesting that the cell tries to degrade the aggregates but is unable to do so. The aggregates associated with amyloid diseases are characterized by the accumulation of long fibers with extensive β-sheet character [104–106]. Third, in multi-domain proteins, differences in susceptibility to unfolding of individual domains will influence the end product of the degradation reaction and may provide a mechanism for processing proteins by partial degradation. Experimental evidence suggests that this mechanism explains the activation of NFκB, which plays a central role in the regulation of immune and inflammatory responses in mammals [84, 107]. The p105 precursor of the p50 subunit of NF-kappa B is processed by proteasome-mediated degradation. The proteasome degrades the domain at the C-terminus of p105 and spares the N-terminal p50 domain. Processing by a partial degradation mechanism may occur elsewhere in the cell. For example, the transcription factors cubitus interruptus in *Drosophila* [108] and Spt23 and Mga2 in yeast [109] are activated by partial proteasome processing.

7.4
Conclusions

Regulated protein unfolding is a critical step in several processes in the cell, including translocation across membranes and degradation by ATP-dependent proteases.

Translocases and ATP-dependent proteases catalyze unfolding using machineries driven by ATP. These machineries denature proteins and translocate their polypeptide chains across the barrier. In both of these processes, the proteins unravel sequentially and processively as the machinery pulls the polypeptide chain at one end, resulting in a cooperative collapse of the protein. As a consequence, the susceptibility of a protein to unfolding depends on its local structure as well as on its overall stability.

7.5
Experimental Protocols

7.5.1
Size of Import Channels in the Outer and Inner Membranes of Mitochondria

The diameters of the import channels in the outer and inner membranes of purified yeast mitochondria are measured by using precursor proteins in which the C-termini are cross-linked to rigid compounds of specific dimensions [6]. The effects of these modifications on translocation across the outer and inner mitochondrial membranes are then determined.

Radioactive precursor proteins are synthesized by in vitro transcription and translation in a rabbit reticulocyte lysate supplemented with [^{35}S] methionine (Promega). Ribosomes and their associated incompletely translated polypeptide chains are removed by centrifugation at 150 000 g for 15 min. Precursor proteins are then partially purified by precipitation with 50% (v/v) saturated ammonium sulfate for at least 30 min on ice, pelleted by centrifugation at 20 800 g for 15 min, and resuspended in import buffer (50 mM HEPES-KOH, pH 7.4, 50 mM KCl, 10 mM MgCl$_2$, 2 mM KH$_2$PO$_4$, 5 mM unlabeled methionine, 1 mg mL^{-1} fatty acid-free BSA). Size probes (Monomaleimido-Nanogold and Monomaleimido-Undecagold) are then attached to barnase precursors containing a single cysteine residue at their C-termini. After incubation for 2 h at room temperature, modified precursors are used directly in the import experiments. To assess the diameter of the inner-membrane protein import pore, the outer membrane of mitochondria is ruptured by hypo-osmotic shock before the import experiments.

7.5.2
Structure of Precursor Proteins During Import into Mitochondria

The structure of precursor proteins during translocation into mitochondria is determined by looking into the import kinetics of a series of precursors. These precursors have a β-sheet where two, three, four, or five of the strands are chemically cross-linked by disulfide bridges. Because mitochondria have no mechanism for reducing disulfides at their surface, this modification makes it impossible for the import machinery to separate the cross-linked strands.

Radioactive precursor proteins are synthesized by in vitro transcription and translation in rabbit reticulocyte lysate. To induce disulfide bridge formation between cysteine residues, precursors are oxidized with 10 mM $K_3Fe(CN)_6$ for 2 min at room temperature before precipitation with ammonium sulfate. To test for disulfide bridge formation, unrelated cysteine residues are modified with 4-acetamido-4'-maleididylstilbene-2,2' disulfonic acid (stilbene disulfonate maleimide [SDSM]) and detected through a change in mobility of the modified proteins in SDS-PAGE.

7.5.3
Import of Barnase Mutants

Determining the import characteristics of a series of barnase mutants allows understanding of the pathway of unfolding of this substrate during import. Unfolding of barnase in free solution follows a defined pathway, with the sub-domain formed by the second and third α-helices and several loops packing against the edges of the β-sheet unfolding before the remaining portion of the protein. In contrast, import-driven unfolding begins at the N-terminus of the protein; therefore, mutations that destabilize this part of the protein accelerate the unfolding. Additionally, stabilizing the N-terminal portion by cross-linking to the rest of the protein slows down the import rate. To lock the N-terminus, residues 5 and 78 are mutated to cysteines and a disulfide bridge is induced by oxidation with 10 mM $K_3Fe(CN)_6$. Import experiments are then performed. The stability of cross-linked precursors under import conditions is assessed by proteinase K digestion. After incubation of the substrate with proteinase K at 35°C, the protease is inhibited with 1 mM phenylmethylsulfonyl fluoride and the sample is analyzed by SDS-PAGE and electronic autoradiography.

The role of mitochondrial membrane potential in precursor unfolding and import is tested by measuring the effect of CCCP (cyanide m-chlorophenylhydrazone), an uncoupler of respiration, on the import rates of precursor proteins. Reducing the membrane potential with CCCP leads to a progressive reduction in the initial import rates of a barnase precursor whose rate of import is limited by the unfolding step. For titrating the protonophore CCCP, 0.4 mM NADH and different amounts of CCCP are added to 0.4 mg mL^{-1} mitochondria in import mix. Oligomycin (5 g mL^{-1}) is added to prevent the reverse action of F_oF_1-ATPase to generate membrane potential. The import mix is incubated at 25°C for 10 min. Import experiments are then performed.

7.5.4
Protein Degradation by ATP-dependent Proteases

Proteins to be degraded by proteasome are targeted by the covalent attachment of ubiquitin; one mechanism by which ubiquitination can occur is the N-end rule pathway. Ubiquitination occurs on two lysine residues in a 40-amino-acid extension at the N-terminus of the substrate proteins. This system can be used to target model proteins to proteasome in reticulocyte lysate [80, 84]. Substrate translated in vitro is added to ATP-depleted reticulocyte lysate supplemented with 1 mM DTT and is incubated for 20 min at 25°C to allow cleavage of the substrate protein. Ubiquitination and degradation are initiated by addition of ATP and an ATP-regenerating system (0.5 mM ATP, 10 mM creatine phosphate, 0.1 mg mL^{-1} creatine phosphokinase; final concentrations) and incubation is continued at 25 °C. At designated time points, aliquots are transferred to ice-cold 5% trichloroacetic acid (TCA), and the TCA-insoluble fractions are analyzed by SDS-PAGE and quantified by electronic autoradiography.

Proteins are targeted to prokaryotic proteases by the attachment of specific degradation signals to the N- or C-terminus of the polypeptide chain. A typical degradation experiment is performed in a total of 120 μL degradation buffer (50 mM Tris-HCl pH 8.0, 100 mM KCl, 0.02% Triton X-100, 20 mM MgCl$_2$). Substrate protein is added to 100 μL prewarmed degradation buffer containing 4 mM ATP, 1 mM DTT, 20 mM creatine phosphate, 0.1 mg mL^{-1} creatine kinase, and appropriate concentration of the protease.

7.5.5
Use of Multi-domain Substrates

The clever use of multi-domain substrates in degradation experiments helps to analyze the mechanism of action of different unfoldases associated with various cellular proteases. Stabilizing individual domains against unfolding by ligand binding helps to prevent the unfolding and degradation of domains away from the degradation signal. A fusion protein consisting of the degradation tag followed first by DHFR and then by barnase is fully degraded by different proteases in both the presence and absence of the barnase ligand barstar. However, when DHFR is stabilized by a ligand, neither DHFR nor barnase is degraded. Thus, the DHFR domain protects the downstream barnase from degradation. Similarly, in a construct that consists of the degradation tag followed by barnase and then DHFR, the fusion protein is fully degraded in the presence or absence of the barnase ligand. However, in this construct, stabilizing DHFR no longer protects barnase from degradation. This mechanism implies that in multi-domain proteins, the protease would unravel domains sequentially from the degradation tag.

7.5.6
Studies Using Circular Permutants

The observation that different ATP-dependent proteases degrade multi-domain proteins sequentially suggests that the structure adjacent to the degradation tag influences its susceptibility to unfold. The use of circular permutants helps to determine which aspect of the substrate structure is easily unraveled by the protease. In circular permutants, the original N- and C-termini are connected by a linker, and the structure is interrupted to produce new N- and C-termini. The resulting proteins have almost identical structures and similar enzymatic activities, differing only in the location of the N- and C-termini within their structure [93]. In the easily digested proteins, the degradation signals lead directly into α-helices or surface loops, whereas in the structures that can be too stable to be degraded, the degradation tags are attached to β-strands. These findings implicate that the local structure near the degradation signal strongly influences the ability of a protein to be degraded by an ATP-dependent protease.

References

1 NEUPERT, W. (1997). Protein import into mitochondria. *Annu. Rev. Biochem.* **66**, 863–917.

2 PFANNER, N., CRAIG, E. A. and HONLINGER, A. (1997). Mitochondrial preproteins translocase. *Annu Rev Cell Dev Biol.* **13**, 25–51.

3 NEUPERT, W., HARTL, F. U., CRAIG, E. A. and PFANNER, N. (1990). How do polypeptides cross the mitochondrial membranes? *Cell*, **63**, 447–450.

4 KUNKELE, K. P., HEINS, S., DEMBOWSKI, M., NARGANG, F. E., BENZ, R., THIEFFRY, M., WALZ, J., LILL, R., NUSSBERGER, S. and NEUPERT, W. (1998). The preprotein translocation channel of the outer membrane of mitochondria. *Cell*, **93**, 1009–1019.

5 HILL, K., MODEL, K., RYAN, M. T., DIETMEIER, K., MARTIN, F., WAGNER, R. and PFANNER, N. (1998). Tom40 forms the hydrophilic channel of the mitochondrial import pore for preproteins. *Nature*, **395**, 516–521.

6 SCHWARTZ, M. P. and MATOUSCHEK, A. (1999). The dimensions of the protein import channels in the outer and inner mitochondrial membranes. *Proc. Natl. Acad. Sci. USA*, **96**, 13086.

7 SCHWARTZ, M. P., HUANG, S. and MATOUSCHEK, A. (1999). The structure of precursor proteins during import into mitochondria. *J. Biol. Chem.* **274**, 12759–12764.

8 TRUSCOTT, K. N., KOVERMANN, P., GEISSLER, A., MERLIN, A., MEIJER, M., DRIESSEN, A. J., RASSOW, J., PFANNER, N. and WAGNER, R. (2001). A presequence- and voltage-sensitive channel of the mitochondrial preprotein translocase formed by Tim23. *Nat. Struct. Biol.* **8**, 1074–1082.

9 ZHENG, N. and GIERASCH, L. M. (1996). Signal sequences: the same yet different. *Cell*, **86**, 849–852.

10 EILERS, M. and SCHATZ, G. (1986). Binding of a specific ligand inhibits import of a purified precursor protein into mitochondria. *Nature*, **322**, 228–232.

11 SCHLEYER, M. and NEUPERT, W. (1985). Transport of proteins into mitochondria: Translocational intermediates spanning contact sites between outer and inner membranes. *Cell*, **43**, 339–350.

12 VESTWEBER, D. and SCHATZ, G. (1988). A chimeric mitochondrial precursor protein with internal disulfide bridges blocks import of authentic precursors into mitochondria and allows quantification of import sites. *J. Cell Biol.* **107**, 2037–2043.

13 NETZER, W. and HARTL, F. (1997). Recombination of protein domains facilitated by co-translational folding in eukaryotes. *Nature*, **388**, 343–349.

14 LIN, L., DEMARTINO, G. N. and GREENE, W. C. (1998). Cotranslational biogenesis of NF-κB p50 by the 26S proteasome. *Cell*, **92**, 819–828.

15 HUANG, S., RATLIFF, K. S., SCHWARTZ, M. P., SPENNER, J. M. and MATOUSCHEK, A. (1999). Mitochondria unfold precursor proteins by unraveling them from their N-termini. *Nat. Struct. Biol.* **6**, 1132–1138.

16 ENDO, T. and SCHATZ, G. (1988). Latent membrane perturbation activity of a mitochondrial precursor protein is exposed by unfolding. *EMBO. J.* **7**, 1153–1158.

17 MATTINGLY, JR., J. R., IRIARTE, A. and MARTINEZ-CARRION, M. (1993).

Structural features which control folding of homologous proteins in cell-free translation systems. The effect of a mitochondrial-targeting presequence on aspartate aminotransferase. *J. Biol. Chem.* **268**, 26320–26327.

18 WIENHUES, U., BECKER, K., SCHLEYER, M., GUIARD, B., TROPSCHUG, M., HORWICH, A. L., PFANNER, N. and NEUPERT, W. (1991). Protein folding causes an arrest of preprotein translocation into mitochondria in vivo. *J. Cell Biol.* **115**, 1601–1609.

19 BOMER, U., MEIJER, M., GUIARD, B., DIETMEIER, K., PFANNER, N. and RASSOW, J. (1997). The sorting route of cytochrome b_2 branches from the general mitochondrial import pathway at the preprotein translocase of the inner membrane. *J. Biol. Chem.* **272**, 30439–30446.

20 MIHARA, K. and OMURA, T. (1996). Cytoplasmic chaperones in precursor targeting to mitochondria: the role of MSF and hsp70. *Trends. Cell. Biol.* **6**, 104–108.

21 WACHTER, C., SCHATZ, G. and GLICK, B. S. (1994). Protein import into mitochondria: the requirement for external ATP is precursor-specific whereas intramitochondrial ATP is universally needed for translocation into the matrix. *Mol. Biol. Cell*, **5**, 465–474.

22 LIM, J. H., MARTIN, F., GUIARD, B., PFANNER, N. and VOOS, W. (2001). The mitochondrial Hsp70-dependent import system actively unfolds preproteins and shortens the lag phase of translocation. *EMBO J.* **20**, 941–950.

23 MATOUSCHEK, A., AZEM, A., RATLIFF, K., GLICK, B. S., SCHMID, K. and SCHATZ, G. (2000). Active unfolding of precursor proteins during mitochondrial protein import. *EMBO J.* **16**, 6727–6736.

24 MUGA, A., MANTSCH, H. H. and SUREWICZ, W. K. (1991). Membrane binding induces destabilization of cytochrome *c* structure. *Biochemistry* **30**, 7219–7224.

25 VAN DER GOOT, F. G., GONZALEZ-MANAS, J. M., LAKEY, J. H. and PATTUS, F. (1991). A 'molten-globule' membrane insertion intermediate of the pore-forming domain of colicin A. *Nature (London)*, **354**, 408–410.

26 DE JONGH, H. H., KILLIAN, J. A. and DE KRUIJFF, B. (1992). A water-lipid interface induces a highly dynamic folded state in apocytochrome *c* and cytochrome *c*, which may represent a common folding intermediate. *Biochemistry*. **31**, 1636–1643.

27 BYCHKOVA, V. E., DUJSEKINA, A. E., KLENIN, S. I., TIKTOPULO, E. I., UVERSKY, V. N. and PTITSYN, O. B. (1996). Molten globule-like state of cytochrome c under conditions simulating those near the membrane surface. *Biochemistry*, **35**, 6058–6063.

28 REN, J., KACHEL, K., KIM, H., MALENBAUM, S. E., COLLIER, R. J. and LONDON, E. (1999). Interaction of diphtheria toxin T domain with molten globule-like proteins and its implications for translocation. *Science*, **284**, 955–957.

29 HUANG, S., MURPHY, S. and MATOUSCHEK, A. (2000). Effect of the protein import machinery at the mitochondrial surface on precursor stability. *Proc. Natl. Acad. Sci. USA*, **97**, 12991–12996.

30 SCHNEIDER, H. C., BERTHOLD, J., BAUER, M. F., DIETMEIER, K. and GUIARD, B. (1994). Mitochondrial Hsp70/MIM44 complex facilitates protein import. *Nature*, **371**, 768–774.

31 KANG, P. L., OSTERMANN, J., SHILLING, J., NEUPERT, W., CRAIG, E. A. and PFANNER, N. (1990). Requirement for hsp70 in the mitochondrial matrix for translocation and folding of precursor proteins. *Nature*, **348**, 137–143.

32 GAMBILL, B. D., VOOS, W., KANG, P. J., MIAO, B., LANGER, T., CRAIG, E. A. and PFANNER, N. (1993). A dual role for mitochondrial heat shock protein 70 in membrane translocation of preproteins. *J. Cell Biol.* **123**, 109–117.

33 KRONIDOU, N. G., OPPLIGER, W., BOLLIGER, L., HANNAVY, K. and GLICK, B. S. (1994). Dynamic interaction between Isp45 and mitochondrial

hsp70 in the protein import system of the yeast mitochondrial inner membrane. *Proc. Natl. Acad. Sci. USA*, **91**, 12818–12822.
34. HARTL, F. U. (1996). Molecular chaperones in cellular protein folding. *Nature*, **381**, 571–579.
35. CYR, D. M., LANGER, T. and DOUGLAS, M. G. (1994). A matrix ATP requirement for presequence translocation across the inner membrane of mitochondria. *Trends Biochem. Sci.* **19**, 176–181.
36. MARTIN, J., MAHLKE, K. and PFANNER, N. (1991) Role of an energized inner membrane in mitochondrial protein import: $\Delta\varphi$ drives the movement of pre-sequences. *J. Biol. Chem.* **266**, 18051–18057.
37. SCHLEYER, M., SCHMIDT, B. and NEUPERT, W. (1982). Requirement of a membrane potential for the post-translational transfer of proteins into mitochondria. *Eur. J. Biochem.* **125**, 109–116.
38. GASSER, S. M., DAUM, G. and SCHATZ, G. (1982). Import of proteins into mitochondria. Energy-dependent uptake of precursors by isolated mitochondria. *J. Biol. Chem.* **257**, 13034–13041.
39. HUANG, S., RATLIFF, K. S. and MATOUSCHEK, A. (2002). Protein unfolding by the mitochondrial membrane potential. *Nature Struct. Biol.* **9(4)**, 301–307.
40. MATOUSCHEK, A. (2003). Protein unfolding – an important process in vivo? *Curr. Opin. Struct. Biol.* **13(1)**, 98–109.
41. LEMIRE, B. D., FANKHAUSER, C., BAKER, A. and SCHATZ, G. (1989). The mitochondrial targeting function of randomly generated peptide sequences correlates with predicted helical amphiphilicity. *J. Biol. Chem.* **264**, 20206–20215.
42. VAN STEEG, H., OUDSHOORN, P., VAN HELL, B., POLMAN, J. E. and GRIVELL, L. A. (1986). Targeting efficiency of a mitochondrial pre-sequence is dependent on the passenger protein. *EMBO J.* **5**, 3643–3650.
43. VERNER, K. and LEMIRE, B. D. (1989). Tight folding of a passenger protein can interfere with the targeting function of a mitochondrial pre-sequence. *EMBO J.* **8**, 1491–1495.
44. VESTWEBER, D. and SCHATZ, G. (1988). Point mutations destabilizing a precursor protein enhance its post-translational import into mitochondria. *EMBO J.* **7**, 1147–1151.
45. EILERS, M., HWANG, S. and SCHATZ, G. (1988). Unfolding and refolding of a purified precursor protein during import into isolated mitochondria. *EMBO J.* **7**, 1139–1145.
46. CLARK, S. A. and THEG, S. M. (1997). A folded protein can be transported across the chloroplast envelope and thylakoid membranes. *Mol. Biol. Cell*, **8**, 923–934.
47. GUERA, A., AMERICA, T., VAN WAAS, M. and WEISBEEK, P. J. (1993). A strong protein unfolding activity is associated with the binding of precursor chloroplast proteins to chloroplast envelopes. *Plant Mol. Biol.* **23**, 309–324.
48. AMERICA, T., HAGEMAN, J., GUERA, A., ROOK, F., ARCHER, K., KEEGSTRA, K. and WEISBEEK, P. (1994). Methotrexate does not block import of a DHFR fusion protein into chloroplasts. *Plant Mol. Biol.* **24**, 283–294.
49. WU, C., SEIBERT, F. S. and KO, K. (1994). Identification of chloroplast proteins in close physical proximity to a partially translocated chimeric precursor protein. *J. Biol. Chem.* **269**, 32264–32271.
50. DELLA-CIOPPA, G. and KISHORE, G. (1988). Import of precursor protein is inhibited by the herbicide Glyphosate. *EMBO. J.* **7**, 1299–1305.
51. ENDO, T., KAWAKAMI, M., GOTO, A., AMERICA, T., WEISBEEK, P. and NAKAI, M. (1994). Chloroplast protein import. Chloroplast envelopes and thylakoids have different abilities to unfold proteins. *Eur. J. Biochem.* **225**, 403–409.
52. MULLER, G. and ZIMMERMANN, R. (1988). Import of honeybee pre-promelittin into the endoplasmic reticulum: energy requirements for

membrane insertion. *EMBO J.* **7**, 639–648.
53 Schatz, G. and Dobberstein, B. (1996). Common principles of protein translocation across membranes. *Science*, **271**, 1519–1526.
54 Johnson, A. E. and Haigh, N. G. (2000). The ER translocon and retrotranslocon: is the shift into reverse manual or automatic? *Cell*, **102**, 709–712.
55 Brodsky, J. L., Hamamoto, S., Feldheim, D., Schekman, R. (1993). Reconstitution of protein translocation from solubilized yeast membranes reveals topologically distinct roles for BiP and cytosolic Hsc70. *J. Cell Biol.* **120**, 95–102.
56 Plemper, R. K. and Wolf, D. H. (1999). Retrograde protein translocation: ERADication of secretory proteins in health and disease. *Trends Biochem. Sci.* **24**, 266–270.
57 Haigh, N. G. and Johnson, A. E. (2002). A new role for BiP: closing the aqueous translocon pore during protein integration into the ER membrane. *J Cell Biol.* **156**, 261–270.
58 Hamman, B. D., Hendershot, L. M. and Johnson, A. E. (1998). BiP maintains the permeability barrier of the ER membrane by sealing the lumenal end of the translocation pore before and early in translocation. *Cell*, **92**, 747–758.
59 Ooi, C. E. and Weiss, J. (1992). Bidirectional movement of a nascent polypeptide across microsomal membranes reveals requirements for vectorial translocation of proteins. *Cell*, **71**, 87–96.
60 Wickner, S., Maurizi, M. R. and Gottesman, S. (1999). Posttranslational quality control: folding, refolding, and degrading proteins. *Science*, **286**, 1888–1893.
61 Lupas, A., Flanagan, J. M., Tamura, T. and Baumeister, W. (1997). Self-compartmentalizing proteases. *Trends Biochem. Sci.* **22**, 399–404.
62 Larsen, C. N. and Finley, D. (1997). Protein translocation channels in the proteasome and other proteases. *Cell*, **91**, 431–434.

63 Wang, J., Hartling, J. A. and Flanagan, J. M. (1997). The structure of ClpP at 2.3 A resolution suggests a model for ATP-dependent proteolysis. *Cell*, **91(4)**, 447–456.
64 Lowe, J., Stock, D., Jap, B., Zwickl, P., Baumeister, W. and Huber, R. (1995). Crystal structure of the 20S proteasome from the archaeon T. acidophilum at 3.4 A resolution. *Science*, **268(5210)**, 533–9.
65 Grimaud, R., Kessel, M., Beuron, F., Steven, A. C. and Maurizi, M. R. (1998). Enzymatic and structural similarities between the Escherichia coli ATP-dependent proteases, ClpXP and ClpAP. *J Biol Chem*, **273(20)**, 12476–81.
66 Wang, J., Hartling, J. A. and Flanagan, J. M. (1998). Crystal structure determination of Escherichia coli ClpP starting from an EM-derived mask. *J. Struct Biol.* **124(2–3)**, 151–63.
67 Sousa, M. C., Trame, C. B., Tsuruta, H., Wilbanks, S. M., Reddy, V. S., McKay, D. B. (2000). Crystal and solution structures of an HslUV protease-chaperone complex. *Cell*, **103(4)**, 633–43.
68 Ciechanover, A. (1998). The ubiquitin-proteasome pathway: on protein death and cell life. *EMBO J.* **17**, 7151–7160.
69 Thrower, J. S., Hoffman, L., Rechsteiner, M., Pickart, C. M. (2000). Recognition of the polyubiquitin proteolytic signal. *EMBO. J.* **19(1)**, 94–102.
70 Lam, Y. A., Lawson, T. G., Velayutham, M., Zweier, J. L. and Pickart, C. M. (2002). A proteasomal ATPase subunit recognizes the polyubiquitin degradation signal. *Nature*, **416**, pp. 763–766.
71 Zhang, M., Pickart, C. and Coffino, P. (2003). Determinants of proteasome recognition of ornithine decarboxylase, a ubiquitin-independent substrate. *EMBO. J.* **22(7)**, 1488–1496.
72 Braun, B. C. et al. (1999). The base of the proteasome regulatory particle exhibits chaperone-like activity. *Nature Cell Biol.* **1**, 221–226.
73 Keiler, K. C., Waller, P. R. and

SAUER, R. T. (1996). Role of a peptide tagging system in degradation of proteins synthesized from damaged messenger RNA. *Science*, 16;271, (5251):990–3.

74 SMITH, C. K., BAKER, T. A. and SAUER, R. T. (1999). Lon and Clp family proteases and chaperones share homologous substrate-recognition domains. *Proc. Natl. Acad. Sci. USA*, 8;96(12), 6678–82.

75 HERMAN, C., THEVENET, D., BOULOC, P., WALKER, G. C. and D'ARI, R. (1998). Degradation of carboxy-terminal-tagged cytoplasmic proteins by the Escherichia coli protease HflB (FtsH). *Genes Dev.* 1;12(9):1348–55.

76 HOSKINS, J. R., SINGH, S. K., MAURIZI, M. R. and WICKNER, S. (2000). Protein binding and unfolding by the chaperone ClpA and degradation by the protease ClpAP. *Proc. Natl. Acad. Sci. USA*, 97, 8892–8897.

77 HERMAN, C., PRAKASH, S., LU, C. Z., MATOUSCHEK, A. and GROSS, C. A. (2003). Lack of a Robust Unfoldase Activity Confers a Unique Level of Substrate Specificity to the Universal AAA Protease FtsH. *Mol. Cell*, 11(3), 659–669.

78 DOUGAN, D. A., MOGK, A., ZETH, K., TURGAY, K. and BUKAU, B. (2002). AAA+ proteins and substrate recognition, it all depends on their partner in crime. *FEBS Lett.* 529, 6–10.

79 GROLL, M., DITZEL, L., LOWE, J., STOCK, D., BOCHTLER, M., BARTUNIK, H. D., HUBER, R. (1997). Structure of the 20S proteasome from yeast at 2.4 Å resolution. *Nature*, 386, 463–471.

80 JOHNSTON, J. A., JOHNSON, E. S., WALLER, P. R. H. and VARSHAVSKY, A. (1995). Methotrexate inhibits proteolysis of dihydrofolate reductase by the N-end rule pathway. *J. Biol. Chem.* 270, 8172–8178.

81 MAURIZI, M. R., SINGH, S. K., THOMPSON, M. W., KESSEL, M. and GINSBURG, A. (1998). Molecular properties of ClpAP protease of Escherichia coli: ATP-dependent association of ClpA and ClpP. *Biochemistry*, 37(21), 7778–7786.

82 SINGH, S. K., GUO, F. and MAURIZI, M. R. (1999). ClpA and ClpP Remain Associated during Multiple Rounds of ATP-Dependent Protein Degradation by ClpAP Protease. *Biochemistry*, 38, 14906–14915.

83 KENNISTON, J. A., BAKER, T. A., FERNANDEZ, J. M., SAUER, R. T. (2003). Linkage between ATP Consumption and Mechanical Unfolding during Protein Processing Reactions of an AAA+ Degradation Machine. *Cell*, 114, 511–520.

84 LEE, C., SCHWARTZ, M. P., PRAKASH, S., IWAKURA, M. and MATOUSCHEK, A. (2001). ATP-dependent proteases degrade their substrates by processively unraveling them from the degradation signal. *Mol. Cell*, 7, 627–637.

85 REID, B., G. FENTON, W. A., HORWICH, A. L., WEBER-BAN, E. U. (2001). ClpA mediates directional translocation of the substrate proteins into the ClpP protease. *Proc. Natl. Acad. Sci. USA*, 98(7), 3768–3772.

86 NAVON, A. and GOLDBERG, A. L. (2001). Proteins are unfolded on the surface of the ATPase ring before transport into the proteasome. *Mol. Cell*, 8, 1339–1349.

87 SAIBIL, H. R., HORWICH, A. L. and FENTON, W. A. (2001). Allostery and protein substrate conformational change during GroEL/GroES-mediated protein folding. *Adv Protein Chem*, 59, 45–72.

88 ROSEMAN, A. M., CHEN, S., WHITE, H., BRAIG, K. and SAIBIL, H. R. (1996). The chaperonin ATPase cycle: mechanism of allosteric switching and movements of substrate-binding domains in GroEL. *Cell*, 87(2): p. 241–51.

89 RANSON, N. A., FARR, G. W., ROSEMAN, A. M., GOWEN, B., FENTON, W. A., HORWICH, A. L. and SAIBIL, H. R. (2001). ATP-bound states of GroEL captured by cryo-electron microscopy. *Cell*, 107(7), 869–79.

90 KIM, Y. I., BURTON, R. E., BURTON, B. M., SAUER, R. T. and BAKER, T. A. (2000). Dynamics of substrate denaturation and translocation by the

ClpXP degradation machine. *Mol. Cell*, **5**, 639–648.

91 HOSKINS, J. R., SINGH, S. K., MAURIZI, M. R. and WICKNER, S. (2000). Protein binding and unfolding by the chaperone ClpA and degradation by the protease ClpAP. *Proc. Natl. Acad. Sci. USA*, **97**, 8892–8897.

92 SINGH, S. K., GRIMAUD, R., HOSKINS, J. R., WICKNER, S. and MAURIZI, M. R. (2000). Unfolding and internalization of proteins by the ATP-dependent proteases ClpXP and ClpAP. *Proc. Natl. Acad. Sci. USA*, **97**, 8898–8903.

93 IWAKURA, M., NAKAMURA, T., YAMANE, C., MAKI, K. (2000). Systematic circular permutation of an entire protein reveals essential folding elements. *Nature Struct. Biol.* **7(7)**, 680–585.

94 BURTON, R. E., SIDDIQUI, S. M., KIM, Y. I., BAKER, T. A. and SAUER, R. T. (2001). Effects of protein stability and structure on substrate processing by the ClpXP unfolding and degradation machine. *EMBO J.* **20(12)**, 3092–3100.

95 KELLERMAYER, M. S. Z., SMITH, S. B., GRANZIER, H. L. & BUSTAMANTE, C. (1997). Folding-Unfolding Transitions in Single Titin Molecules Characterized with Laser Tweezers. *Science*, **276**, 1112–1116.

96 RIEF, M., GAUTEL, M., OESTERHELT, F., FERNANDEZ, J. M. & GAUB, H. E. (1997). Reversible Unfolding of Individual Titin Immunoglobulin Domains by AFM. *Science*, **276**, 1109–1112.

97 KELLERMAYER, M. S., BUSTAMANTE, C. & GRANZIER, H. L. (2003). Mechanics and structure of titin oligomers explored with atomic force microscopy. *Biochim. Biophys. Acta.* **1604**, 105–114.

98 BROCKWELL, D. J., PACI, E., ZINOBER, R. C., BEDDARD, G. S., OLMSTED, P. D., SMITH, D. A., PERHAM, R. N. & RADFORD, S. E. (2003). Pulling geometry defines the mechanical resistance of a β-sheet protein. *Nat. Struct. Biol.* **10(9)**, 731–737.

99 CARRION-VAZQUEZ, M., LI, H., LU, H., MARSZALEK, P. E., OBERHAUSER, A. F. & FERNANDEZ, J. M. (2003). The mechanical stability of ubiquitin is linkage dependent. *Nat. Struct. Biol.* **10(9)**, 738–743.

100 MATOUSCHEK, A. and BUSTAMANTE, C. (2003). Finding a proteins Achilles heel. *Nat. Struct. Biol.* 2003, **13(1)**, 98–109.

101 VERMA, R., MCDONALD, H., YATES, III. J. R. and DESHAIES, R. J. (2001). Selective degradation of ubiquitinated Sic1 by purified 26S proteasome yields active S phase cyclin-Cdk. *Mol. Cell*, **8**, 439–448.

102 SHERMAN, M. Y. and GOLDBERG, A. L. (2001). Cellular defenses against unfolded proteins: a cell biologist thinks about neurodegenerative diseases. *Neuron*. **29**, 15–32.

103 ROCHET, J. C. and LANSBURY, JR., P. T. (2000). Amyloid fibrillogenesis: themes and variations. *Curr. Opin. Struct. Biol.* **10**, 60–68.

104 SUNDE, M. and BLAKE, C. (1997). The structure of amyloid fibrils by electron microscopy and X-ray diffraction. *Adv. Protein Chem.* **50**, 123–159.

105 JIMENEZ, J. L., NETTLETON, E. J., BOUCHARD, M., ROBINSON, C. V., DOBSON, C. M. and SAIBIL, H. R. (2002). The protofilament structure of insulin amyloid fibrils. *Proc. Natl. Acad. Sci. USA*, **99**, 9196–9201.

106 JIMENEZ, J. L., GUIJARRO, J. I., ORLOVA, E., ZURDO, J., DOBSON, C. M., SUNDE, M. and SAIBIL, H. R. (1999). Cryo-electron microscopy structure of an SH3 amyloid fibril and model of the molecular packing. *EMBO J.* **18**, 815–821.

107 PERKINS, N. D. (2000). The Rel/NF-κB family: friend and foe. *Trends Biochem. Sci.* **25**, 434–440.

108 JIANG, J. and STRUHL, G. (1998). Regulation of the Hedgehog and Wingless signaling pathways by the F-box/WD40-repeat protein Slimb. *Nature*, **391**, 493–496.

109 HOPPE, T., MATUSCHEWSKI, K., RAPE, M., SCHLENKER, S., ULRICH, H. D. and JENTSCH, S. (2000). Activation of a membrane-bound transcription factor by regulated ubiquitin/proteasome-dependent processing. *Cell*, **102**, 577–586.

8
Natively Disordered Proteins

*Gary W. Daughdrill, Gary J. Pielak, Vladimir N. Uversky,
Marc S. Cortese, and A. Keith Dunker*

8.1
Introduction

To understand natively disordered proteins, it is first important to introduce the structure-function paradigm, which dominates modern protein science. We then discuss the terminology used to describe natively disordered proteins and present a well-documented example of a functionally disordered protein. Finally, we compare the standard structure-function paradigm with structure-function relationships for natively disordered proteins and, from this comparison, suggest an alternative for relating sequence, structure, and function.

8.1.1
The Protein Structure-Function Paradigm

The structure-function paradigm states that the amino acid sequence of a protein determines its 3-D structure and that the function requires the prior formation of this 3-D structure. This view was deeply engrained in protein science long before the 3-D structure of a protein was first glimpsed almost 50 years ago. In 1893 Emil Fischer developed the "lock-and-key" hypothesis from his studies on different types of similar enzymes, one of which could hydrolyze α- but not β-glycosidic bonds and another of which could hydrolyze β- but not α-glycosidic bonds [1]. By 1930 it had become clear that globular proteins lose their native biological activity (i.e., they become denatured) when solution conditions are altered by adding heat or solutes. Anson and Mirsky showed in 1925 that denatured hemoglobin could be coaxed back to its native state by changing solution conditions [2]. This reversibility is key because it means that the native protein and the denatured protein can be treated as separate thermodynamic states. Most importantly, this treatment leads directly to the idea that a protein's function is determined by the *definite structure* of the native state. In short, the structure is known to exist because it is destroyed by denaturation. Wu stated as much in 1931, but his work was probably unknown outside China, even though his papers were published in English [3]. The West

Protein Folding Handbook. Part II. Edited by J. Buchner and T. Kiefhaber
Copyright © 2005 WILEY-VCH Verlag GmbH & Co. KGaA, Weinheim
ISBN: 3-527-30784-2

had to wait until 1936, when Mirsky and Pauling published their review of protein denaturation [4].

Anfinsen and colleagues used the enzyme ribonuclease to solidify these ideas. Their work led in 1957 to the "thermodynamic hypothesis." Anfinsen wrote [5], "This hypothesis states that the three-dimensional structure of a native protein in its normal physiological milieu is the one in which the Gibbs free energy of the whole system is lowest: that is, the native conformation is determined by the totality of interatomic interactions and hence by the amino acid sequence, in a given environment." Merrifield and colleagues then performed an elegant experiment that drove home the idea; they synthesized ribonuclease in the test tube from the amino acid sequence [6]. Their experiments provided direct evidence that the amino acid sequence determines all other higher-order structure, function, and stability.

As mentioned above, the observation that denaturation is reversible allows the application of equilibrium thermodynamics. It was soon realized that many small globular proteins exist in only two states, the native state or the denatured state. That is, each protein molecule is either completely in the native state or completely in the denatured state. Such two-state behavior leads directly to expressions for the equilibrium constant, K_D, and free energy, ΔG_D, of denaturation:

$$K_D = [D]/[N] \tag{1}$$

$$\Delta G_D = -RT \ln(K_D) \tag{2}$$

where R is the gas constant, T is the absolute temperature, and $[N]$ and $[D]$ represent the concentrations of the native and denatured states, respectively (see chapters 2 and 3 in Part I).

The definition of K_D, Eq. (1), is straightforward, but quantifying K_D is more difficult than defining it. The difficulty arises because K_D usually cannot be quantified at the conditions of interest, i.e., at room temperature in buffered solutions near physiological pH. Specifically, most biophysical techniques can only sensitively measure K_D between values of about 10 and 0.1, but the overwhelming majority ($> 99.9\%$) of two-state globular protein molecules are in the native state at the conditions of interest. The difficulty can be overcome by extrapolation. Increasing the temperature or adding solutes such as urea or guanidinium chloride pushes K_D into the quantifiable region. Plots of $-RT \ln(K_D)$ versus temperature or solute concentration are then extrapolated back to the unperturbed condition to give ΔG_D at the conditions of interest. Many such studies indicate that small globular proteins have a stability of between about 1 and 10 kcal mol^{-1} at room temperature near neutral pH. We introduced this formal definition of stability so that later we can discuss the idea that the higher-order structure within some intrinsically disordered proteins is simply unstable.

What exactly is higher-order structure? Pauling showed that the protein chain was organized in definite local structures, helices and sheets [7], but it was unclear how these structures interact to form the native state. Because the conceptual accessibility of physical representations is greater than that of thermodynamics, the advent of X-ray crystallography strongly reinforced the sequence-structure-function paradigm. In 1960, Kendrew and Perutz used X-ray crystallography to reveal the intricate, atomic-level structures of myoglobin [8] and hemoglobin [9], effectively locking in the sequence-to-structure paradigm. The paradigm took on the aura of revealed truth when Phillips solved the first structure of an enzyme, lysozyme, in 1965 [10]. The position of the bound inhibitor revealed the structure of the active site, making it clear that the precise location of the amino acid side chains is what facilitates catalysis.

Given all this evidence, it appeared that the case was closed: the native state of every protein possesses a definite and stable three-dimensional structure, and this structure is required for biological function. But even early on there were worrying observations. Sometimes loops were missing from high-resolution structures, and these loops were known to be required for function [11, 12]. Nuclear magnetic resonance spectroscopy also showed that some proteins with known biological functions did not possess stable, defined structure in solution [13].

Perhaps the most important difference to bear in mind when relating the sequence-structure-function paradigm to intrinsically disordered proteins is the difference between a structural state and a thermodynamic state. The native state of a globular protein is *both* a structural state and thermodynamic state, but the disordered (and denatured) state is *only* a thermodynamic state. That is, all the molecules in a sample of the native state of a globular protein have nearly the same structure, and this structure is what is lost upon denaturation. On the other hand, the denatured state consists of a broad ensemble of molecules – each having a different conformation. Therefore, averaged quantities have different meanings for native and disordered states. For a native globular protein, an averaged quantity, such as the CD signal (see Section 8.2.4), gives information about each molecule in the sample because nearly all the molecules are in the same structural state. For a denatured or disordered protein, an averaged quantity contains information about the ensemble, and this information may or may not be applicable to individual molecules in the sample.

8.1.2
Natively Disordered Proteins

Many proteins carry out function by means of regions that lack specific 3-D structure, existing instead as ensembles of flexible, unorganized molecules. In some cases the proteins are flexible ensembles along their entire lengths, while in other cases only localized regions lack organized structure. Still other proteins contain regions of disorder without ascribed functions, but functions might be associated with these regions at a later date. Whether the lack of specific 3-D structure occurs wholly or in part, such proteins do not fit the standard paradigm that 3-D structure is a prerequisite to function.

Various terms have been used to describe proteins or their regions that fail to form specific 3-D structure, including: flexible [14], mobile [15], partially folded [16], rheomorphic [400], natively denatured [17], natively unfolded [18], intrinsically unstructured [19], and intrinsically disordered [20].

None of these terms or combinations is completely appropriate. "Flexible," "mobile," and "partially folded" have the longest histories and most extensive use; however, all three of these terms are used in a variety of ways, including many that are not associated with proteins that exist as structural ensembles under apparently native conditions. For example, ordered regions with high B factors are often called flexible or mobile. "Partially folded" is often used to describe transient intermediates involved in protein folding. With regard to the term "natively," it is difficult to know whether a protein is in its native state. Even when under apparently physiological conditions, a protein might fail to acquire a specific 3-D structure due to the absence of a critical ligand or because the crowded conditions inside the cell are needed to promote folding. Because of such uncertainties, "intrinsically" is often chosen over "natively." Unfolded and denatured are often used interchangeably, so the oxymoron "natively denatured" has a certain appeal but has not gained significant usage. "Unfolded" and "unstructured" both imply lack of backbone organization, but natively disordered proteins often have regions of secondary structure, sometimes transient and sometimes persistent. There are even examples of apparently native proteins with functional regions that resemble molten globules [21]; the molten globule contains persistent secondary structure but lacks specific tertiary structures, having instead regions of non-rigid side chain packing that leads to mobile secondary structure units [22, 23, 32] (see chapters 23 and 24 in Part I). Rheomorphic, which means "flowing structure" [400], provides an interesting alternative to random coil, but again would not indicate native molten globules. Since "disordered" encompasses both extended and molten globular forms, this descriptor has some advantage, but the widespread use of "disorder" in association with human diseases complicates computer searches and gives a negative impression.

Here we use "natively disordered," which has been infrequently used if at all, mainly to distinguish this work from previous manuscripts on this topic. Developing a standardized vocabulary in this field would be of great benefit. We propose "collapsed disorder" for proteins and domains that exist under physiological conditions primarily as molten globules and "extended disorder" for proteins and regions that exist under physiological conditions primarily as random coil.

A significant body of work suggests that the unfolded state is not a true random coil but instead possesses substantial amounts of an extended form that resembles the polyproline II helix [24–26] as well as other local conformations that resemble the native state (see chapter 20 in Part I). For this reason, extended disorder may be a preferable term to random coil, but the latter term continues to have widespread usage; therefore, for convenience, we will continue to use this term here – with the understanding that by the term random coil we do not mean the true random coil defined by the polymer chemist.

It is useful to introduce the topic of natively disordered proteins with a specific, very clear example. Calcineurin (Figure 8.1A) makes a persuasive case for the existence and importance of native disorder [27–29]. This protein contains a catalytic A subunit and a B subunit with 35% sequence identity to calmodulin. The A subunit is a serine-threonine phosphatase that becomes activated upon association with the Ca^{2+}-calmodulin complex. Thus, calcineurin, which is widespread among the eukaryotes, connects the very important signaling systems based on Ca^{2+} levels and

8.1 Introduction | 279

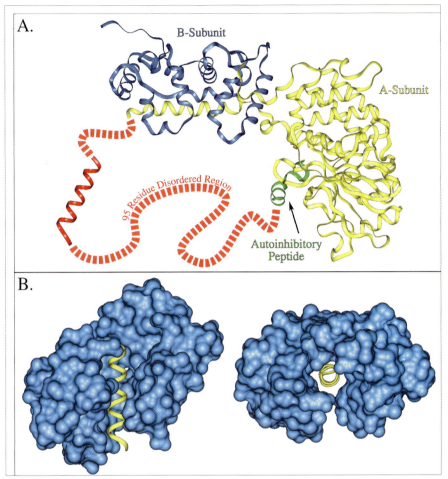

Fig. 8.1. (A) 3-D structure of calcineurin, showing the A subunit (yellow), the B subunit (blue), the autoinhibitory peptide (green), and the location of a 95-residue disordered region (red). The calmodulin-binding site (red helix) is located within the disordered region. (B) Side and top views of calmodulin (blue) binding a target helix (yellow). Note that the calmodulin molecule surrounds the target helix when bound.

phosphorylation/dephosphorylation. When Ca^{2+}-calmodulin binds to its target helix within calcineurin, the autoinhibitory peptide becomes displaced from its association with the active site, and by this means the phosphatase activity is turned on. Since Ca^{2+}-calmodulin wraps around its target helix (Figure 8.1B), this helix must lack tertiary contacts and therefore lies in a disordered region. The disordered character of the region surrounding the target helix in calcineurin has been shown by its sensitivity to trypsin digestion [27, 28] and by the missing electron density of over 95 consecutive residues in the crystal structure [29]. In this example, these lines of evidence combine to support the importance of intrinsic disorder for biological function.

8.1.3
A New Protein Structure-Function Paradigm

The standard view is that amino acid sequence codes for 3-D structure and that 3-D structure is a necessary prerequisite for protein function. In contrast, not only calcineurin but also many additional proteins, as we have shown elsewhere [30, 31], are natively disordered or contain natively disordered regions. For such proteins or protein regions, lack of specific 3-D structures contributes importantly to their functions. Here we explore the distinctions between the standard and our alternative view of protein structure-function relationships.

The standard structure-function paradigm as discussed above arose initially from the study of enzymes. The original lock-and-key proposal [1] was based on distinctive substrate recognition by a pair of enzymes. In the approximately 110 years since this initial proposal, studies of enzyme catalysis have continuously reinforced this view.

While more than 20 proposals have been made to explain enzyme catalysis [33], it has become accepted that the most profitable way to think about the problem is in terms of transition-state stabilization [34]. That is, as first suggested by Polanyi [35] and later by Pauling [36], for an enzyme to carry out catalysis, it must bind more tightly to the transition state than to the ground state. This tighter binding lowers the transition-state energy and accelerates the reaction rate [34–36]. Transition states are derived from ground states by very slight movements of atoms (typically half the length or so of a chemical bond). Attempts to understand the mechanistic details of how an enzyme can bind more tightly to the transition state have occupied a number of protein chemists and theoreticians, and this is not completely clear in every aspect even today [37], with some researchers arguing that electrostatic contributions provide the dominant effects [38] (see chapter 7 in Part I) and others emphasizing the importance of entropic effects [39] or other contributions [40]. Despite these uncertainties, there seems to be universal agreement that tighter binding to the transition state depends on an accurate prior positioning of the key residues in the enzyme. This prior positioning requires a well-organized protein 3-D structure. Thus, in short, the evolution of ordered structure in proteins was likely reinforced by or resulted directly from the importance of enzyme catalysis.

For not only the calcineurin example given above but also for most of the natively disordered proteins we have studied [30, 31], the function of the disorder is for signaling, regulation, or control. Compared to order, disorder has several clear advantages for such functions. When disordered regions bind to signaling partners, the free energy required to bring about the disorder-to-order transition takes away from the interfacial, contact free energy, with the net result that a highly specific interaction can be combined with a low net free energy of association [20, 41, 77]. High specificity coupled with low affinity seems to be a useful pair of properties for a signaling interaction so that the signaling interaction is reversible. It would appear to be more difficult to evolve a highly specific yet weak interaction between two ordered structures. In addition, a disordered protein can readily bind to multiple partners by changing shape to associate with different targets [20, 42,

43]. Multiple interactions are now being commonly documented, and proteins having 20 or more partners are being described. In protein interaction or signaling networks, proteins with multiple partners are often called hubs. We previously suggested that the ability to interact with multiple partners may depend on regions of native disorder [44], but so far we have investigated only a limited number of examples. Whether hub proteins utilize regions of native disorder to enable their binding diversity is an important question for proteins of this class.

Given the above information, we propose a new protein structure-function paradigm. Simply put, we propose a two-pathway paradigm, with sequence → 3-D structure → function for catalysis and sequence → native disorder → function for signaling and regulation.

8.2
Methods Used to Characterize Natively Disordered Proteins

8.2.1
NMR Spectroscopy

Intrinsically disordered proteins have dynamic structures that interconvert on a number of timescales. Nuclear magnetic resonance spectroscopy can detect this molecular motion as well as any transient secondary or tertiary structure that is present. Several reviews have focused on the use of NMR to characterize the structure and dynamics of intrinsically disordered proteins [45–49]. There is also a rich body of literature reviewing the use of NMR to characterize the structure and dynamics of nonnative states of globular proteins [50–55] (see chapters 18 and 21 in Part I). Due to the high activation barrier for studying the structure and dynamics of intrinsically disordered proteins, much of the work reviewed here was performed on nonnative states of globular proteins. Fortunately, these nonnative states have many similarities to intrinsically disordered proteins. The barrier to studying intrinsically disordered proteins seems to prevail regardless of their functional relevance and the abundance of NMR techniques available to characterize their structure and dynamics. This barrier is based on an entrenched attitude of most protein chemists regarding the relationship between protein structure and function. For example, when graduate students present an HSQC spectrum of an intrinsically disordered protein displaying limited ^{1}H chemical shift dispersion and narrow resonance line shape to their advisor, they are likely to be ridiculed for making an error during the purification of the protein. This attitude prevails despite more than a decade of overwhelming evidence that intrinsic protein disorder exists and can be essential for function. It is time to move beyond the limited view that defined 3-D structure is a requirement for protein function and to acknowledge the growing body of evidence that natively disordered proteins exist and have functions.

In this report, a systematic approach for using NMR to investigate the structure, dynamics, and function of intrinsically disordered proteins will be suggested based on a subset of NMR experiments. Our objective is to develop a comprehensive picture of the structure and dynamics of intrinsically disordered proteins. This objective is most easily accomplished by unifying the different types of NMR data collected on intrinsically disordered proteins while avoiding the pitfall described in the following quotation from a classic paper on NMR and protein disorder [56]: "[U]ntil the amount of structural information provided by NMR methods increases by several orders of magnitude, descriptions of non-native structure will probably consist of simple lists of estimates of fractional population of secondary-structure segments and of side-chain interactions."

It is interesting that the author of this quote has provided one of the most complete descriptions of the structure and dynamics of an unfolded protein, a fragment of staphylococcal nuclease referred to as $\Delta 131 \Delta$ [56–64].

The most complete description of the structure and dynamics of intrinsically disordered proteins would include the population and structure of the different members of the rapidly interconverting ensemble along with the rate of interconversion between unique structures. Some of these parameters can be estimated from NMR chemical shifts and resonance line shape measurements [48, 50, 51, 65–67]. Because liquid-state NMR detects the ensemble average, chemical shifts can be treated as macroscopic equilibrium constants for secondary structure formation [50, 51]. This information can be combined with measurements of hydrodynamic radii, using pulsed field gradient methods, to place limits on the conformational space that is being explored, which ultimately facilitates structure calculation of ensemble members [68–70]. Of course, it remains unclear how many discrete states are represented in the resonance line shape measurements, which presents the fundamental limitation of current NMR practice and theory to provide a more detailed description of the ensemble members. Many consider this problem insoluble, given the large number of possible conformations, even for a small polypeptide. However, there is ample evidence that intrinsically disordered proteins do not explore all of these conformations. Further, several systems have been characterized where resonance line shape measurements were deconvoluted to provide information about the number of ensemble members or to characterize the interconversion rate between conformations [71–73, 397].

8.2.1.1 Chemical Shifts Measure the Presence of Transient Secondary Structure

Resonance assignments are the first step in characterizing the structure and dynamics of any protein using NMR. Resonance assignments specify the atomic identity of unique frequencies observed in the spectrum. These frequencies are generally normalized to some standard and are reported in parts per million as the chemical shift. Chemical shift measurements for $^{1}H_{\alpha}$, $^{13}C_{\alpha}$, $^{13}C_{\beta}$, and $^{13}C(O)$ nuclei are sensitive to $\phi\Psi$ dihedral angles and deviate systematically from random coil values for helical and beta conformations [65, 67, 74, 75]. The deviations are diagnostic for the presence of secondary structure, regardless of stability, as long as the interconversion rate is fast and the deviation from the random coil chemical shift value is greater than the spectral resolution. For stably folded proteins, the chemical shift measurement of $^{1}H_{\alpha}$, $^{13}C_{\alpha}$, $^{13}C_{\beta}$, and $^{13}C(O)$ nuclei provide a picture of secondary structure that represents a lower limit on the equilibrium constant for folding from the unfolded state [50]. The extension of this relationship to intrinsically disordered proteins makes the same two-state assumption. In this case, the fraction of helix or strand present can be determined by comparing the chemical shift value observed in an intrinsically disordered protein to the value expected in a stably folded protein [50, 51]. These data combined with knowledge of the Stokes radius can be used to restrict the available conformations in a structure calculation. This assumption is in agreement with a recent analysis of the effective hydrodynamic radius of protein molecules in a variety of conformationnal states [76]. In fact, based on the analysis of 180 proteins in different conformational states, it has been shown that the overall protein dimension could be predicted based on the chain length, i.e., the protein molecular weight, with an accuracy of 10%. Furthermore, it has been emphasized that the incorporation of biophysical constraints, which can be rationalized based on conventional biophysical measurements, might lead to considerable improvement in structure simulation procedures. Clearly, the size and shape of the bounding volume used for structure simulations play a crucial role in determining the efficiency and accuracy of any algorithm. The incorporation of a size/shape constraint derived from experimental data might lead to considerable improvement of simulation procedures [76].

In a study of an intrinsically disordered negative regulator of flagellar synthesis, FlgM, chemical shift deviations from random coil values were observed for $^{13}C_\alpha$, $^{13}C_\beta$, and $^{13}C(O)$ nuclei [77]. Similar deviations were not observed for $^1H_\alpha$ nuclei, and this may be a general property of intrinsically disordered proteins. For FlgM, the chemical shift deviations indicated the presence of helical structure in the C-terminal half of the protein and a more extended flexible structure in the N-terminal half. Two regions with significant helical structure were identified, containing residues 60–73 and 83–90. Additional NMR and genetic studies demonstrated that these two helical regions were necessary for interactions with the sigma factor. The chemical shift deviations observed for FlgM exhibited a characteristic variation where the central portion of the helix had a larger helical chemical shift difference than the edges. To test whether the helical structure based on chemical shifts represented an ensemble of nonrandom conformations, the $^{13}C_\alpha$ chemical shifts were measured in the presence of increasing concentrations of the chemical denaturant urea. As the concentration of urea was increased, the $^{13}C_\alpha$ chemical shifts moved toward the random coil value in a manner characteristic of a non-cooperative unfolding transition.

In a study of the basic leucine zipper transcription factor GCN4, transient helical structure was observed in the basic region based on chemical shift deviations from random coil values for $^{13}C_\alpha$, $^{13}C_\beta$, and $^{13}C(O)$ nuclei [78]. The temperature dependence of the three chemical shifts was also monitored. For the folded leucine zipper region, a small dependence between chemical shift and temperature was observed. Conversely, the unfolded basic region showed large changes in $^{13}C_\alpha$, $^{13}C_\beta$, and $^{13}C(O)$ chemical shifts when the temperature was changed [78]. The dynamic behavior of the basic region and rapidly interconverting structures are responsible for the temperature dependence of the chemical shifts [45]. In contrast, the temperature dependence of the chemical shifts in the folded leucine zipper was small, and this behavior is generally observed for both folded and random coil regions.

8.2.1.2 Pulsed Field Gradient Methods to Measure Translational Diffusion

Another easily applied NMR method is measurement of the translational diffusion coefficient, D, using pulsed field gradients (PFG) [68–70]. This approach relies on the fact that a protein molecule undergoing translational diffusion will differentially sense a gradient of signal designed to destroy any acquired magnetization. An array of gradient strength or time can be used to calculate D [69]. Knowledge of D can be used to calculate the hydrodynamic radius. The hydrodynamic radius can then be compared to the expected value based on molecular weight to determine whether the protein is compact and globular or extended and flexible. Empirical relationships were recently established between the hydrodynamic radius determined using PFG translational diffusion measurements and the number of residues in the polypeptide chain for native folded proteins and highly denatured states [68]. This study provided evidence for significant coupling between local and global features of the conformational ensembles adopted by disordered polypeptides. As expected, the hydrodynamic radius of the polypeptide was dependent on the level of persistent secondary structure or the presence of hydrophobic clusters.

8.2.1.3 NMR Relaxation and Protein Flexibility

NMR relaxation is the premier method to investigate protein flexibility [45, 47, 56, 57, 79]. It is the first-order decay of an inductive signal back to equilibrium with the applied field. In particular, "The relaxation mechanism for the NH spin system arises from the local time-varying magnetic fields generated at the ^{1}H and ^{15}N nuclei due to global tumbling and internal mobility of the various N–H bond vectors" [80:2676].

This means relaxation of the NH spin system is sensitive to both local and global motion. Measuring the longitudinal relaxation rate, R_1, the transverse relaxation rate, R_2, and the heteronuclear Overhauser effect between the amide proton and its attached nitrogen, NH-NOE, is useful for describing the internal dynamics and molecular motions associated with proteins and typically measurable for every amide N–H pair in proteins less than 40 kDa [79–83].

In particular, the NH-NOE experiment provides a fast, powerful, and easy-to-interpret diagnostic tool for the presence of intrinsic protein disorder. The sign of the NH-NOE resonance is sensitive to the rotational correlation time and is positive for N–H bond vectors with a long rotational correlation time (> 1–10 ns) and negative for N–H bond vectors with a short rotational correlation time (< 0.1–1 ns). The NH-NOE experiment can be performed on any uniformly ^{15}N-labeled protein sample that can be concentrated without aggregation to between 0.1 and 1.0 mM. The relatively short rotational correlation time observed for intrinsically disordered proteins results in sharp, narrow lines in the NMR spectrum. Because of this property, the NH-NOE experiment can be used to detect intrinsic disorder in proteins up to 200 kDa [84].

We argue that the NH-NOE experiment is such a valuable diagnostic tool for the presence of intrinsic protein disorder that it should be incorporated into all screening protocols developed for structural genomics. It is generally a waste of time and resources to directly pursue crystallization of proteins that are intrinsically disordered. Crystallization meets with limited success for proteins containing intrinsically disordered regions. This is because the presence of structural interconversions for intrinsically disordered proteins prohibits the formation of an isomorphous lattice, unless there are one or a few stable, low-energy conformations that can be populated. During screening, proteins determined to have intrinsically disordered regions can be marked for further NMR analysis. On a technical note, a range of relaxation and saturation delays should be used when measuring the NH-NOE values in intrinsically disordered proteins [85].

Some of the most valuable contributions to understanding the structure and dynamics of intrinsically disordered proteins have come from using R_1, R_2, and NH-NOE data to model molecular motion or to solve the spectral density function. The two most successful models for relaxation data analysis are the so called "model-free" approach of Lipari and Szabo [86, 87] and reduced spectral density mapping introduced by Peng and Wagner [80, 88].

8.2.1.4 Using the Model-free Analysis of Relaxation Data to Estimate Internal Mobility and Rotational Correlation Time

In the interpretation of ^{15}N relaxation data, it is assumed that the relaxation properties are governed solely by the ^{1}H-^{15}N dipolar coupling and the chemical shift anisotropy [89]. Using these assumptions, a spherical molecule with an overall correlation time, τ_m, and an effective correlation time for fast internal motions, τ_e, will have a spectral density function of the following form [86, 87]:

$$J(\omega) = \frac{2}{5}\left\{\frac{S^2 \tau_m}{(1+(\omega\tau_m)^2)} + \frac{(1+S^2)\tau}{(1+(\omega\tau)^2)}\right\} \quad (3)$$

where $1/\tau = 1/\tau_m + 1/\tau_e$ and S^2 is the square of the generalized order parameter describing the amplitude of internal motions. The overall correlation time, τ_m, is determined based on an assumption of isotropic Brownian motion. For a folded protein of known structure, the diffusion tensor can be incorporated into the model to compensate for anisotropic motions [90–92]. It is unclear how valid the assumption of isotropic rotation is for intrinsically disordered proteins. It is assumed that anisotropic structures will populate the conformational ensemble of an intrinsically disordered protein. This is in the absence of little direct evidence on the subject. It has been suggested that the effects of rapid interconversion between more isotropic structures will tend to smooth out static anisotropy and result in an isotropic average conformation [57].

Regardless of the analytical limitations, the model-free analysis of several natively disordered proteins has provided to be a useful qualitative picture of the heterogeneity in rotational diffusion that is observed for these systems [58, 77, 93–95]. Some general trends are observed: (1) correlation times are greater than those calculated based on polymer theory, and (2) transient secondary structure tends to induce rotational correlations.

Changes in S^2 have been used to estimate changes in conformational entropy due to changes in ns-ps bond vector motions during protein folding and for intrinsically disordered protein folding that is coupled to binding [45, 63, 77, 96, 97]. In aqueous buffer and near neutral pH, the N-terminal SH3 domain of the *Drosophila* signal transduction protein Drk is in equilibrium between a folded, ordered structure and an unfolded, disordered ensemble. The unfolded ensemble is stabilized by 2 M guanidine hydrochloride and the folded structure is stabilized by 0.4 M sodium sulfate. Order parameters were determined for both the unfolded and folded Drk SH3 domains. The unfolded ensemble had an average S^2 value of 0.41 ± 0.10 and the folded structure had an average S^2 value of 0.84 ± 0.05. Based on the difference in S^2 between the unfolded and folded structures, the average conformational entropy change per residue was estimated to be 12 J $(\text{molK})^{-1}$. This approach does not address additional entropy contributions from slower motional processes or the release of solvent. Interestingly, the estimate of 12 J $(\text{molK})^{-1}$ is similar to the average total conformational entropy change per residue estimated from other techniques (~ 14 J $(\text{molK})^{-1}$ [96]. Another study has also suggested that changes in order parameters provide a reliable estimate of the total conformational entropy changes that occur during protein folding [63].

8.2.1.5 Using Reduced Spectral Density Mapping to Assess the Amplitude and Frequencies of Intramolecular Motion

The Stokes-Einstein equation defines a correlation time for rotational diffusion of a spherical particle:

$$\tau_c = \frac{\eta V_{sph}}{Tk_B} \tag{4}$$

In Eq. (4), this rotational correlation time, τ_c, depends directly on the volume of the sphere (V_{sph}) and the solution viscosity (η) and inversely on the temperature (T). The spectral density function, $J(\omega)$, describes the frequency spectrum of rotational motions of the N–H bond vector relative to the external magnetic field and is derived from the Fourier transform of the spherical harmonics describing the rotational motions [80]. For isotropic Brownian motion, $J(\omega)$ is related to τ_c in a frequency-dependent manner:

$$J(\omega) = \frac{2}{5} \frac{\tau_c}{1 + \omega^2 \tau_c^2} \tag{5}$$

Reduced spectral density mapping uses the conveniently measured ^{15}N R_1, R_2, and NH-NOE to estimate the magnitude of the spectral density function at 0, ^1H, and ^{15}N angular frequencies. According to Eq. (5), $J(\omega)$ at 0 frequency is equal to $2/5$ τ_c. This relationship represents an upper limit on $J(0)$, which is usually reduced by fast internal motions such as local anisotropic N–H bond vector fluctuations. It is also important to note that chemical exchange that is in the microsecond to millisecond range contributes positively to $J(0)$ but that this effect can be attenuated by measuring R_2 under spin-lock conditions [80].

Reduced spectral density mapping is a robust approach for analyzing intrinsic disorder because it does not depend on having a model of the molecular motions under investigation. Several groups have characterized intrinsic protein disorder using reduced spectral density mapping, providing new insights into protein dynamics [77, 78, 80, 94, 98–101]. In one of these studies, reduced spectral density mapping was used to help demonstrate that the intrinsically disordered anti-sigma factor, FlgM, contained two disordered domains [77]. One domain, representing the N-terminal half of the protein, was characterized by fast internal motions and small $J(0)$ values. The second domain, representing the C-terminal half of the protein, was characterized by larger $J(0)$ values, representing correlated rotational motions induced by transient helical structure. NMR was also used to help demonstrate that the C-terminal half of FlgM contained the sigma factor–binding domain, and it was proposed that the transient helical structure was stabilized by binding to the sigma factor. However, more recent NMR studies of FlgM showed that this structure appears to be stabilized in the cell due to molecular crowding, which would reduce the role of conformational entropy on the thermodynamics of the FlgM/sigma factor interaction [102].

In another study, a temperature-dependent analysis of the reduced spectral density map was performed on the GCN4 bZip DNA-binding domain [78]. In the absence of DNA, GCN4 exists as a dimer formed through a coiled-coil C-terminal domain and a disordered N-terminal DNA-binding domain. This disordered DNA-binding domain becomes structured when DNA is added [54]. In the GCN4 study, the reduced spectral density map was evaluated at three temperatures; 290 K, 300 K, and 310 K. $J(0)$ values increased in a manner consistent with changes in solvent viscosity induced by increasing temperature. When $J(0)$ was normalized for changes in solvent viscosity, values for the C-terminal dimerization domain and the intrinsically disordered N-terminal DNA-binding domain were identical within experimental error. This behavior suggested that the correlated rotational motion of the folded leucine zipper had the dominant influence on the solution-state backbone dynamics of the basic leucine zipper of GCN4. A recently completed study monitored the temperature dependence of the reduced spectral density map for a partially folded fragment of thioredoxin [401]. Unlike GCN4, the thioredoxin fragment does not contain a stably folded domain. In this study, normalizing $J(0)$ for changes in solvent viscosity induced a trend in the data, with $J(0)$ increasing with increasing temperature. In the absence of chemical exchange, these data suggest that we are monitoring an increase in the hydrodynamic volume of the fragment (G. W. Daughdrill, unpublished data).

8.2.1.6 Characterization of the Dynamic Structures of Natively Disordered Proteins Using NMR

In 1994 Alexandrescu and Shortle described the first complete NMR relaxation analysis of a partially folded protein under non-denaturing conditions [57, 58]. It was a fragment of staphylococcal nuclease referred to as Δ131Δ. For the majority of Δ131Δ residues, their experimental data was best described by a modified "model-free" formalism that included contributions from internal motions on intermediate and fast timescales and slow overall tumbling. They observed that the generalized order parameter S^2 correlated with sequence hydrophobicity and the fractional populations of three alpha helices in the protein. In a more recent study, residual dipolar couplings were measured for native staphylococcal nuclease and Δ131Δ [60–63]. Native-like dipolar couplings were observed in Δ131Δ, even in high concentrations of denaturant.

An extensive NMR dataset was used to identify an ensemble of three-dimensional structures for the N-terminal SH3 domain of the *Drosophila* signal transduction protein Drk, with properly assigned population weights [103, 104]. This was accomplished by calculating multiple unfolding trajectories of the protein using the solution structure of the folded state as a starting point. Of course this approach is limited to proteins that have a compact rigid conformation. However, the ability to integrate multiple types of data describing the structural and dynamic properties of disordered proteins is pertinent to this discussion. Population weights of the structures calculated from the unfolding trajectories were assigned by optimizing their fit to experimental data based on minimizing pseudo energy terms defined for each type of experimental constraint. This work marks the first time that NOE, J-coupling, chemical shifts, translational diffusion coefficients, and tryptophan solvent-accessible surface area data were used in combination to estimate ensemble members. As seen with many other studies of the structure and dynamics of intrinsically disordered proteins, the unfolded ensemble for this domain was significantly more compact than a theoretical random coil (see chapter 20 in Part I).

8.2.2
X-ray Crystallography

Wholly disordered proteins would not be expected to crystallize and thus could not be studied by X-ray crystallography. On the other hand, proteins consisting of both ordered and disordered regions can form crystals, with the ordered parts forming the crystal and the disordered regions occupying spaces between the ordered parts. Regions of disorder vary in location from one molecule to the next and therefore fail to scatter X-rays coherently. The lack of coherent scattering leads to missing electron density.

A given protein crystal is made up of identical repeating unit cells, where each unit cell contains one to several protein molecules depending on the symmetry of the crystal. Each atom, j, in the protein occupies a particular position, xj, yj, and zj. It is convenient to express the positions as dimensionless coordinates that are fractions of the lengths of the unit cell, yielding Xj, Yj, and Zj as the indicators of the position of atom j. The result of an X-ray diffraction experiment is a 3-D grid of spots that are indexed by three integers, h, k, and l. The positions of the spots are determined by the crystal lattice. The intensities of the spots are determined by both the symmetries in the crystal and the positions of the atoms in the molecule. Specifically, each spot has an intensity that is the square of the magnitude of the structure factor, $F(h, k, l)$, which is given by the following equation:

$$F_{(h,k,l)} = \sum_{j=1}^{atoms} f_{(j)} \exp[2\pi \cdot i(hX_j + kY_j + lZ_j)] \qquad (6)$$

where h, k, and l are the indices of the spots in the diffraction pattern, $f_{(j)}$ is the scattering power of atom j (dependent on the square of the number of electrons in the atom), i is the square root of -1, and X_j, Y_j, and Z_j are the coordinates of atom j given as fractions of the unit cell dimensions as mentioned above.

Each $F_{(h,k,l)}$ has both a magnitude and a phase. The magnitude of $F_{(h,k,l)}$ is determined by the square root of the intensity of each spot in the diffraction pattern. For wavelets from two different spots, the phase is the shift in the peak values and is 0 degrees for two wavelets that are exactly in phase (constructive interference) and 180 degrees for two wavelets that are totally out of phase (which leads to destructive interference). The phase for two arbitrary wavelets can be found to be any value between 0 and 360 degrees. Thus, given a structure, both the magnitude and phase of each $F_{(h,k,l)}$ can be calculated by carrying out the summation in Eq. (6) over all of the atoms in the structure. Taking the Fourier transform of Eq. (6) gives:

$$\rho_{(x,y,z)} = \frac{1}{V} \sum_h \sum_k \sum_l F_{(h,k,l)} \exp[-2\pi \cdot i(hx + ky + lz)] \qquad (7)$$

where $\rho_{(x,y,z)}$ is the electron density of the protein at x, y, and z and V is the volume of the unit cell (the other terms have been defined above).

Thus, to determine a structural model of the given protein, one merely has to carry out the summation of Eq. (7) and then fit the resulting electron density map with a set of atoms that correspond to the connected set of residues in the structure. To carry out this triple-vector sum, it is necessary to know both the magnitude and the phase of each $F_{(h,k,l)}$. While the magnitude is obtainable simply as the square root of the intensity of the spot, the phase information is lost during data collection because the time of arrival of the peak of each wavelet relative to those of the others cannot be recorded with current technologies. The loss of this critical information is commonly called the phase problem.

Three main methods have been developed to find the missing phase information for each structure factor, $F_{(h,k,l)}$. Each of these methods has advantages and disadvantages.

The earliest successful approach for proteins was to add heavy atoms. Since scattering power depends on the number of electrons squared, even a small number of heavy atoms can perturb the diffraction values enough to allow the determination of the phase values. The mathematical details are fairly involved, but for this approach to work, the positions of the heavy atoms must be determined by comparing the phase intensities with and without the heavy atoms. In addition, the addition of the heavy atoms cannot significantly perturb the structure of the protein. With regard to this second point, the protein structure with and without the heavy atom must be isomorphous. Thus, this approach is called isomorphous replacement, and at least two such heavy atom replacements must be made to determine each phase value.

A second approach is to use multiple X-ray wavelengths that traverse the absorption edge of a selected heavy atom. The change in scattering over the absorption edge becomes substantial, which enables the phases to be determined. This approach is called multi-wavelength anomalous dispersion, or MAD. For protein structure determination by this approach, it is common to introduce selenium in the form of selenomethionine. This is a convenient atom that has an absorption edge at an appropriate wavelength value, and often (but not always) substitution of methionine with selenomethionine does not significantly affect the structure or activity of the protein of interest.

Finally, if the structure of a closely related protein is already known, it is often possible to use the phases from the related protein with the intensities from the protein of interest to generate the initial model structure. This approach is called molecular replacement. Often the homologous structure is not similar enough, so the phases are too inaccurate to give a reasonable starting structure. A second major difficulty is determining the correct orientation of the known structure relative to that of the unknown so that the phases can be correctly associated with the intensities. Since a very large number of possible orientations exist, a complex search of the possibilities needs to be developed. Evolutionary algorithms have recently been found to be useful for this task [105].

Structure determination by X-ray crystallography typically involves not only straightforward calculation and equation solving (as indicated above) but also significant modeling and simulation. Once the phase problem has been solved by experiment, an electron density map is generated by the calculations described in Eq. (7). However, there are important technical difficulties in solving the phases, so the phase values usually contain large errors and the resulting electron density map contains many mistakes. To give an example of the difficulty, it is unclear whether the heavy atom derivatives are truly isomorphous; even small protein movements upon heavy atom binding can lead to significant errors in the estimation of the phases. Modeling is then used to fit the amino acid sequence into the error-containing electron density map. The structural model that emerges is typically adjusted by dynamics and simulation to improve bond lengths, bond angles, and inter-atom contacts. The model is then used to calculate improved phases and the whole process is reiterated until the structure no longer changes in successive cycles.

In the overall process of protein crystallography, there are no clear-cut rules for dealing with regions of low intensity or missing electron density. Some crystallographers are more aggressive than others in attempting to fit (or model) such low-density or missing regions. This variability means that missing coordinates in the Protein Data Bank (PDB) have not been assigned by uniform standards, which in turn leads to variability in the assignment of disorder. All of this is compounded by the imperfections (packing defects) in real crystals, which can additionally contribute to the absence of electron density.

Crystallographers have classified regions of missing electron density as being either static or dynamic [106, 107]. Static disorder is trapped into different conformations by the crystallization process, while dynamic disorder is mobile. A dynamically disordered region could potentially freeze into a single preferred position upon cooling, while static disorder would be fixed regardless of temperature. By collecting data and determining structures at a variety of temperatures, dynamic disorder sometimes becomes frozen and thus distinguished from static disorder [108].

From our point of view, more important than static or dynamic is whether the region of missing electron density has a single set of coordinates along the backbone or whether the region exists as an ensemble of structures. A missing region with one set of coordinates could be trapped in different positions by the crystal lattice (statically disordered) or could be moving as a rigid body (dynamically disorder). In either case, the wobbly domain [20] is not an ensemble of structures, i.e., is not natively disordered, but rather is an ordered region that adopts different positions due to a flexible hinge.

Given the difficulties in ascribing disorder to regions of missing electron density or to missing coordinates in the PDB, it is useful to use a second method, such as NMR or protease sensitivity, to confirm that a missing region is due to intrinsic disorder rather than some other cause. While such confirmation has been carried out for a substantial number of proteins [20], the importance of disorder has not been generally recognized, so such confirmation is not routine.

Despite the uncertainties described above, missing electron density provides a useful sampling of native disorder in proteins. To estimate the frequency of such regions, a representative set of proteins was studied for residues with missing coordinates [109]. The representative set, called PDB_Select_25, was constructed by first grouping the PDB into subsets of proteins with greater than 25% sequence identity and then choosing the highest-quality structure from each subset [110]. Out of 1223 chains with 239 527 residues, only 391 chains, or 32%, displayed no residues with missing backbone atoms (Table 8.1). Thus, 68% of the

Tab. 8.1. Disorder in PDB_Select_25*.

Parameter	Number	Percentage
Total chains	1223	
Chains with no disorder	391	32
Chains with disorder	832	68
Disordered regions	1168	
Disordered regions/disordered chains	1.4	
Disordered regions > 30 residues in length	68	5.8
Total residues	239 527	
Disordered residues	12 138	5.1
Residues in disordered regions > 30	3710	1.5

*Data extracted from PDB as of 10/1/2001.

non-redundant proteins contained some residues that lacked electron density. The 832 chains with missing electron density contained 1168 distinct regions of disorder, corresponding to ~1.4 such regions/chain. These 1168 disordered regions contained 12138 disordered residues, or ~10 residues/disordered region on average. While most of the disordered regions are short, a few are quite long: 68 of the disordered regions are greater than 30 residues in length. Overall, the residues with missing coordinates are about 5% of the total.

A substantial fraction of the disorder-containing proteins in PDB are fragments rather than whole proteins. Since disordered regions tend to inhibit crystallization, they are often separated from ordered domains by genetic engineering or protease digestion prior to crystallization attempts. In either case, longer regions of disorder become shortened, presumably leading to improved probability of obtaining protein crystals. Given that wholly disordered proteins do not crystallize, and given that the proteins in PDB often contain truncated disordered regions or are ordered fragments that have been separated from flanking regions of disorder, it is clear that the PDB substantially underrepresents the amount of protein disorder in nature.

To further understand the distribution of order and disorder in crystallized proteins, we carried out a comparison between the PDB and the Swiss-Prot databases. Swiss-Prot provides an easy mechanism to extract information about various proteins [111], so this comparison allows a convenient means to study disorder across the different kingdoms. The overall results of this comparison are given in Table 8.2. Of 4175 exact sequence matches between proteins in PDB_Select_25 and Swiss-Prot, 2258 were from eukaryotes, 1490 were from bacteria, 170 were from archaea, and 257 were from viruses. Proteins with no disorder or only short disorder were estimated by considering proteins for which at least 95% of the primary structure was represented as observed coordinates in the PDB structures. The number and percentage of proteins in each set for these mostly ordered proteins were 428 (19%) for eukaryotes, 594 (40%) for bacteria, 82 (48%) for archaea, and 42 (16%) for viruses. Proteins with substantial regions of disorder were estimated by considering proteins for which the crystal structure contained less than half of the entire sequence. The number and percentage of proteins in each set for these proteins likely to contain substantial amounts of disorder were 713 (32%) for eu-

Tab. 8.2. Comparison of disorder between the PDB_Select_25 and Swiss-Prot databases.

	Number listed in both databases	Proteins with ≥90% assigned coordinates		Proteins with ≤50% assigned coordinates	
		Number	Percentage	Number	Percentage
Eukaryotes	2258	428	19	713	32
Bacteria	1490	594	40	97	6.5
Archaea	170	82	48	4	2.3
Viruses	257	42	16	123	48

karyotes, 97 (6.5%) for bacteria, 4 (2.3%) for archaea, and 123 (48%) for viruses. These data suggest that both eukaryotes and viruses are likely to have proteins with large regions of disorder flanking fragmentary regions of ordered, crystallizable domains.

8.2.3
Small Angle X-ray Diffraction and Hydrodynamic Measurements

Both small-angle X-ray scattering (SAXS) and hydrodynamic methods such as gel-exclusion chromatography or dynamic light scattering (see chapter 19 in Part I) have been used to estimate the sizes of protein molecules in solution. Analytical ultracentrifugation is another method for determining the molecular weight and the hydrodynamic and thermodynamic properties of a protein [112]. The observed size of a given protein can then be compared with the size of a globular protein of the same mass. As expected, by these methods molten globules have overall sizes similar to those of the globular proteins of the same mass, whereas random coil forms are considerably extended and thus have a significantly larger size than their globular counterparts. Natively disordered proteins and globular proteins can be compared using several features of these techniques.

With regard to SAXS, one approach is to plot the normalized $I(S)S^2$ versus S, where I is the scattering intensity at a given scattering angle and S is the given scattering angle. This method emphasizes changes in the signal at higher scattering angles, which in turn strongly depends on the dimensions of the scattering molecule. The resulting graph, called a Kratky plot [113–115], readily distinguishes random coil proteins from globular structures. That is, globular proteins give parabolas, with scattering intensity increasing and then dropping sharply due to the reduced scattering intensity at large angles. On the other hand, random coil forms give monotonically increasing curves. Importantly, the natively disordered proteins with extended disorder are characterized by low (coil-like) intramolecular packing density, reflected by the absence of a maximum on their Kratky plots [116–122]. This statement is illustrated by Figure 8.2, which compares the Kratky plots of five natively disordered proteins with those of two rigid globular proteins. One can see that the Kratky plots of natively disordered proteins do not exhibit maxima. Maxima on Kratky plots are typical of the folded conformations of globular proteins such as staphylococcal nuclease. In other words, the absence of a maximum indicates the lack of a tightly packed core under physiological conditions in vitro [123].

A second SAXS approach is to compare the radius of gyration, R_g, with that from a globular protein of the same size. The R_g value is estimated from the Guinier approximation [124]:

$$I(S) = I(0) \exp\left(-\frac{S^2 \cdot R_g^2}{3}\right) \qquad (8)$$

where $I(S)$ is the scattering intensity at angle S. The value of R_g can then be deter-

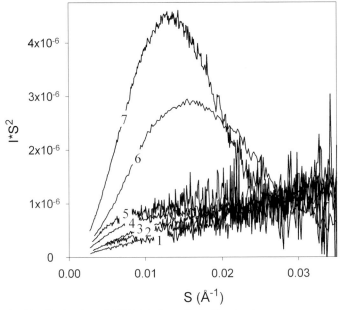

Fig. 8.2. Kratky plots of SAXS data for natively disordered α-synuclein (1), β-synuclein (2), prothymosin α (3), caldesmon 636–771 fragment (4), and core histones (5). The Kratky plots of native globular staphylococcal nuclease (6) and hexameric insulin (7) are shown for comparison.

mined from plots of $\ln[I(S)]$ versus S^2. As tabulated by Millet, Doniach, and Plaxco [125], the radii of gyration of proteins unfolded by low pH, methanol, urea, or guanidinium chloride are typically 1.5–2.5 times larger than the radii of gyration of the same proteins in their native globular states. Note, however, that denaturation does not always result in complete conversion to extended coil formation: some structure may persist [126, 127].

SAXS can also be used to estimate the maximum dimension of the protein. The distance distribution function $P(r)$ can be estimated by Fourier inversion of the scattering intensity, $I(S)$ [128, 129], where $P(r)$ is the probability of finding a dimension of length r. A plot of $P(r)$ versus r yields the maximum dimension in the limit as $P(r)$ goes to zero. The observed maximal dimension can then be compared with those observed for globular and random coil forms of various proteins [130, 131].

A fourth method is to compare hydrodynamic volumes. Globular proteins differ significantly from fully or partly unstructured proteins in their hydrodynamic properties. For both native, globular proteins and fully denatured proteins, empirical relationships have been determined between the Stoke's radius, R_s, and the number of residues in the chain [68, 132, 133]. These relationships in turn provide estimates of the overall hydrodynamic sizes compared to the respective controls.

Two common methods for estimating the Stoke's radius are gel-exclusion chromatography and dynamic light scattering. In the former, the mobility of the protein of interest is compared to the mobilities of a collection of protein standards [132, 133]. In the latter, the translational diffusion coefficient is estimated and the Stoke's radius is calculated from the Stoke's-Einstein equation.

For a given Stoke's radius, Rs, the hydrodynamic volume is given by:

$$Vh = \left(\frac{4}{3}\right)\pi(Rs)^3 \quad (9)$$

where the Stoke's radius of a given protein can be estimated by comparison to known standards using gel-exclusion chromatography. Alternatively, the Stoke's radius can also be estimated from diffusion values estimated by dynamic light-scattering measurements as described previously [68].

Plots of the logarithm of the hydrodynamic volume versus number of residues yield distinct straight lines for three different classes of reference proteins: (1) native, globular proteins, (2) molten globules, and (3) proteins unfolded by guanidinium chloride [123, 134]. As expected, the molten globule forms are just slightly larger than their ordered, globular counterparts, while the unfolded proteins exhibit significantly larger hydrodynamic volumes.

This conclusion is illustrated by Figure 8.3, which compares the $\log(R_S)$ versus $\log(M)$ curves for natively disordered proteins with those of native, molten globule, pre-molten globule, and guanidinium chloride-unfolded globular proteins [123]. Additionally, Figure 8.3 shows that the $\log(R_S)$ versus $\log(M)$ dependencies for different conformations of globular proteins can be described by the following set of straight lines:

$$\log(R_S^N) = -(0.204 \pm 0.023) + (0.357 \pm 0.005) \cdot \log(M) \quad (10)$$

$$\log(R_S^{MG}) = -(0.053 \pm 0.094) + (0.334 \pm 0.021) \cdot \log(M) \quad (11)$$

$$\log(R_S^{PMG}) = -(0.210 \pm 0.180) + (0.392 \pm 0.041) \cdot \log(M) \quad (12)$$

$$\log(R_S^{U\text{-}GdmCl}) = -(0.723 \pm 0.033) + (0.543 \pm 0.007) \cdot \log(M) \quad (13)$$

where N, MG, PMG, and U-GdmCl correspond to the native, molten globule, pre-molten globule, and guanidinium chloride-unfolded globular proteins, respectively.

For the non-molten globular natively disordered proteins, their $\log(R_S)$ versus $\log(M)$ dependence may be divided in two groups: natively disordered proteins behaving as random coils in poor solvent (denoted as NU-coil in Eq. (14)) and essentially more compact proteins, which are similar to pre-molten globules with respect to their hydrodynamic characteristics (denoted as NU-PMG in Eq. (15)) [123, 135]:

$$\log(R_S^{NU\text{-}coil}) = -(0.551 \pm 0.032) + (0.493 \pm 0.008) \cdot \log(M) \quad (14)$$

$$\log(R_S^{NU\text{-}PMG}) = -(0.239 \pm 0.055) + (0.403 \pm 0.012) \cdot \log(M) \quad (15)$$

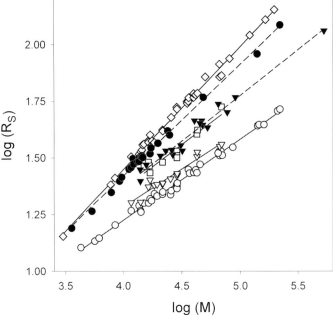

Fig. 8.3. Dependencies of the hydrodynamic dimensions, R_S, on protein molecular mass, M, for native globaular (open circles), molten globule (open triangles), pre-molten globule (open squares), 6 M guanidinium chloride-unfolded conformational states of globular proteins (open diamonds), natively disordered proteins with coil-like (black circles) and pre-molten globule-like (black inverse triangles) properties [123].

Non-molten globular, natively disordered proteins might be expected to contain subregions of various lengths, each having differing degrees of partial folding. If this were indeed the case, plots of the logarithm of the hydrodynamic volumes versus residue numbers for a set of such natively disordered proteins would give a scatter plot having values randomly located between the lines specified by the extensively unfolded and the molten globular reference forms. Instead of this expectation, two distinct types of natively disordered proteins were observed [123]. One type closely resembles the reference proteins that were extensively unfolded by guanidinium chloride. The other type gives a log volume versus residue line between that of the fully extended protein molecules and that of the molten globules. The line for this second type is nearly superimposable [123] with a previously described denaturation intermediate called the pre-molten globule [136].

The finding of two distinct classes of extended disorder – random coil-like and pre-molten globule-like – is difficult to understand. Polymers in good solvents tend to be collapsed but still unstructured. Uversky (2002) suggested that sequence differences could be responsible for whether a natively disordered protein behaves more like a random coil or more like a pre-molten globule; however, a clear sequence distinction between the two classes has not been found [123, 135]. But even if a sequence distinction is found, it remains difficult to understand the origin of this partition into two distinct classes. For example, if a hybrid protein were created, with one half being random coil-like and one-half being pre-molten globule-like, would its hydrodynamic properties lie between those of the two classes? Why haven't such chimeras been found in nature? Perhaps the simple explanation is that not enough natively disordered proteins have been characterized. But if a wide variety of natively disordered proteins have not been found in the current sampling, even though such proteins do exist, what is leading to the partition observed so far? This question deserves further study.

8.2.4
Circular Dichroism Spectropolarimetry

Circular dichroism (CD) measures the difference in the absorbance of left versus right circularly polarized light and is therefore sensitive to the chirality of the environment [137]. There are two types of UV-absorbing chromophores in proteins: side chains of aromatic amino acid residues and peptide bonds [137, 138]. CD spectra in the near-ultraviolet region (250–350 nm), also known as the aromatic region, reflect the symmetry of the aromatic amino acid environment and, consequently, characterize the protein tertiary structure. Proteins with rigid tertiary structure are typically characterized by intense near-UV CD spectra, with unique fine structure, which reflect the unique asymmetric environment of individual aromatic residues. Thus, natively disordered proteins are easily detected since they are characterized by low-intensity, near-UV CD spectra of low complexity. The far-UV region of a protein's absorbance spectrum (190–240 nm) is dominated by the electronic absorbance from peptide bonds. The far-UV CD spectrum provides quantifiable information about the secondary structure of a protein because each category of secondary structure (e.g., α-helix, β-sheet) has a different effect on the chiral environment of the peptide bond. Because the timescale for electronic absorbance ($\sim 10^{-12}$ seconds) is so much shorter than that of folding or unfolding reactions (at least $\sim 10^{-6}$ seconds), CD provides information about the weighted average structure of all the peptide bonds in the beam.

CD provides different information about native globular versus denatured or disordered proteins. In a native globular protein, the weighted average can be applied to each molecule because nearly all the molecules are in the native state and each molecule of the native state has a similar structure. For instance, if the CD data indicate that 30% of the peptide bonds in a sample of a 100-residue native globular protein are in the α-helical conformation and 70% are in the β-sheet conformation, then 30 peptide bonds in each protein molecule are helical and the other 70 are sheet. Furthermore, it is the same 30 and 70 in each molecule. For a denatured or disordered protein, the CD spectrum provides information about the ensemble of conformations, and the information will not be generally applicable to any single molecule. For instance, changing the above example to a disordered protein, we can say that 30% of the peptide bonds in the sample are helical and 70% are sheet, but with only these data in hand, nothing can be inferred about the structure of an individual molecule – neither the helix and sheet percentage nor the location of those structures along the chain.

A good correlation exists between the relative decrease in hydrodynamic volume and the increase in secondary structure content. This correlation was shown for a set of 41 proteins that had been evaluated by both far-UV CD and hydrodynamic methods [139]. Study of the equilibrium unfolding of these globular proteins revealed that the Stokes radii (R_S) and secondary structure of native and partially folded intermediates were closely correlated. Results of this analysis are presented in Figure 8.4 as the dependence of $(R_S^U/R_S)^3$ versus $[\theta]_{222}/[\theta]_{222}^U$, which represent relative compactness and relative content of ordered secondary structure, respectively. Significantly, Figure 8.4 illustrates that data for both classes of conformations (native globular and partially folded intermediates) are accurately described by a single dependence (correlation coefficient, $r^2 = 0.97$) [139]:

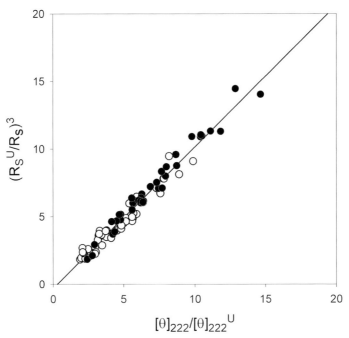

Fig. 8.4. Correlation between the degree of compactness and the amount of ordered secondary structure for native globular proteins (filled circles) and their partially folded intermediates (open circles). The degree of compactness, $(R_S^U/R_S)^3$, was calculated for different conformational states as the decrease in hydrodynamic volume relative to the volume of the unfolded conformation, while the amount of ordered secondary structure, $[\theta]_{222}/[\theta]_{222}^U$, was calculated from far-UV CD spectra as the increase in negative ellipticity at 222 nm relative to the unfolded conformation. The data used to plot these dependencies are taken from Uversky [139].

$$\left(\frac{R_S^U}{R_S}\right)^3 = (1.047 \pm 0.010) \cdot \left(\frac{[\theta]_{222}}{[\theta]_{222}^U}\right) - (0.31 \pm 0.12) \qquad (16)$$

This correlation means that the degree of compactness and the amount of ordered secondary structure are conjugate parameters. In other words, there is no compact equilibrium intermediate lacking secondary structure or any highly ordered but non-compact species among the proteins analyzed. Therefore, hydrophobic collapse and secondary structure formation occur simultaneously rather than as two subsequent independent processes. This conclusion generalizes earlier observations made for several individual proteins including DnaK [140], apomyoglobin [141], and staphylococcal nuclease [142].

Tiffany and Krimm noted over 35 years ago that CD spectra of reversibly denatured proteins resemble that of a left-handed polyproline II helix [143]. It was only recently, however, that Creamer revisited these data and put them on a more sound footing with studies of non-proline model peptides [144, 145]. Polyproline helices made from non-proline residues have a negative CD band at 195 nm and positive band at 218 nm. These wavelengths are shorter than those of true polyproline helices, (205 nm and 228 nm, respectively) because of the different absorption maxima for primary and secondary amides. Inspection of the literature shows that many intrinsically disordered proteins exhibit a negative feature near 195 nm, with near-zero ellipticity near 218 nm [17, 18, 146–149], and some exhibit both the negative and the positive features [150, 151]. The interpretation of CD spectra in terms of polyproline II helix content remains controversial for several reasons. First, it is not yet clear what the difference is between the CD spectrum of a random coil and the CD spectrum of a polyproline II helix. Second, there is not yet a way to quantify the amount of polyproline II helix.

The conclusion is clear: many intrinsically disordered proteins resemble reversibly denatured proteins and may exist, on average, in a polyproline II helix. However, some of these proteins exhibit small amounts of other secondary structures [77]. Sometimes [102, 152–157], but not always [148, 158], secondary structure can be induced by molecular crowding and "structure-inducing" co-solutes.

8.2.5
Infrared and Raman Spectroscopy

Infrared (IR) spectra are the result of intramolecular movements (bond stretching, change of angles between bonds along with other complicated types of motion) of various functional groups (e.g., methyl, carbonyl, amide, etc.) [159]. The merits of spectral analysis in the IR region originate from the fact that the modes of vibration for each group are sensitive to changes in chemical structure, molecular conformation, and environment. In the case of proteins and polypeptides, two infrared bands that are connected with vibrational transitions in the peptide backbone and reflect the normal oscillations of simple atom groups are of the most interest. These bands correspond to the stretching of N–H and C=O bonds (amide I band) and the deformation of N–H bonds (amide II band). The amide I and amide

II bands are characterized by frequencies within the ranges of 1600–1700 cm^{-1} and 1500–1600 cm^{-1}, respectively [159–161]. The position of the bands changes due to the formation of hydrogen bonds. Thus, analysis of IR spectra allows determination of the relative content of α-helical, β-, and irregular structure in proteins by monitoring the intensity of the bands within the amide I and amide II regions.

The main advantage of using IR spectroscopy to determine the secondary structure of protein is that this method is based on a simple physical phenomenon, the change of vibrational frequency of atoms upon the formation of hydrogen bonds. Thus, it is possible to calculate the parameters of normal vibrations of the main asymmetrical units forming secondary structure and to compare these calculated values with experimental data. Calculations of normal vibrational parameters for α-helices and β-sheets were done by Miyazawa [162]. Further, Chirgadze et al. have shown that there is a good correlation between the calculated and experimental data for the amide I and amide II bands [159, 161]. Thus, such calculations may serve as a rational basis for the interpretation of polypeptide and protein IR spectra (Figure 8.5). The application of IR spectroscopy to evaluate protein secondary structure is based on the following assumptions about protein structure: (1) protein consists of a limited set of different types of secondary structure (α-helix, parallel and antiparallel β-structures, β-turns, and irregular structure), (2) the IR spectra of protein is a simple sum of the spectra of these structures taken with the

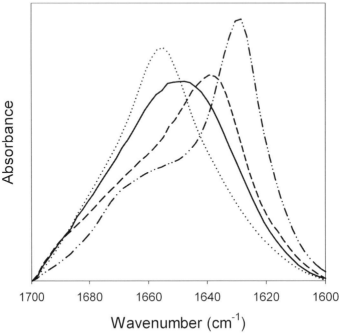

Fig. 8.5. FTIR spectra in the amide I region measured for the natively disordered α-synuclein (solid line), α-helical human α-fetoprotein (dotted line), β-structural *Yersinia pestis* capsular protein Caf1 (dashed line), and α-synuclein amyloid fibril with cross-β-structure (dash-dot-dotted line). Spectra are normalized to have same area.

weights corresponding to their content in the protein, and (3) the spectral characteristics of secondary structure are the same for all proteins and for all the structural elements of one type in the protein. The finding that Fourier transform infrared spectroscopy (FTIR) exhibited a high sensitivity to the conformational state of macromolecules resulted in numerous studies where this approach was used to analyze protein molecular structure (including a number of natively disordered proteins) and to investigate the processes of protein denaturation and renaturation.

Raman is complementary to IR, depending on inelastic light scattering rather than absorption to obtain vibrational information, and is further complementary to IR in emphasizing vibrations lacking changes in dipole moments. Raman spectroscopy has long been used for the analysis of protein structure and protein conformational changes [398]. Recently, specialized Raman methods, namely surface enhanced resonance Raman [399] and Raman optical activity [400], have demonstrated significant usefulness for the characterization of natively disordered proteins.

FTIR also provides a means of keeping track of conformational changes in proteins. These are followed by monitoring the changes in the frequencies of the IR bands that result from deuteration (substitution of hydrogen atoms by deuterium) of the molecule [159]. Since it is usually known which band belongs to which functional group (carbonyl, oxy-, or amino group), one can identify the exchangeable groups by observing the changes in band position as a result of deuteration. The rate of deuteration depends on the accessibility of a given group to the solvent that, in turn, varies according to changes in conformation. Thus, by tracking the changes in the deuterium-exchange rates for different solvent compositions and other environmental parameters, one can obtain information about the resultant conformational changes within a given protein.

8.2.6
Fluorescence Methods

8.2.6.1 Intrinsic Fluorescence of Proteins

Proteins contain only three residues with intrinsic fluorescence. These chromophores form the following series: tryptophan > tyrosine > phenylalanine according to their quantum yield. The fluorescence of tryptophan is most commonly used for analysis of proteins since the quantum yield of phenylalanine fluorescence is extremely low and tyrosine fluorescence is strongly quenched in the majority of cases. Quenching of tyrosine fluorescence can be due to ionization, proximity to amide or carboxyl groups, the energy transfer to tryptophan [163]. Application of intrinsic fluorescence to the study of protein conformational analysis relies on the fact that the parameters of tryptophan emission (intensity and wavelength of maximal fluorescence) depend essentially on environmental factors, including solvent polarity, pH, and the presence or absence of quenchers [163]. For example, a completely solvated tryptophan residue (e.g., free tryptophan in water or tryptophan in an unfolded polypeptide chain) has a maximum fluorescence in the vicinity of 350 nm, whereas embedding this chromophore into the nonpolar interior of a compact globular protein results in a characteristic blue shift of its fluorescence maximum (Stokes shift) by as much as 30–40 nm [163–165]. This means that λ_{max} of tryptophan fluorescence contains some basic information about whether the protein is compact under the experimental conditions. For this reason, the analysis of intrinsic protein fluorescence is frequently used to study protein structure and conformational change.

8.2.6.2 Dynamic Quenching of Fluorescence

Additional information about the accessibility of protein chromophores to solvent (and thus on relative compactness of a protein molecule) can be obtained from the analysis of dynamic quenching of intrinsic fluorescence by small molecules. Fluorescence quenching data are frequently analyzed using the general form of the Stern-Volmer equation [166]:

$$\frac{I_0}{I} = (1 + K_{SV}[Q])e^{V[Q]} \tag{17}$$

where I_0 and I are the fluorescence intensities in the absence and presence of quencher, K_{SV} is the dynamic quenching constant, V is a static quenching constant, and $[Q]$ is the quencher concentration.

To some extent, the information obtained from dynamic quenching of intrinsic fluorescence is similar to that obtained from studies of deuterium exchange (see Section 8.2.8 and chapter 18 in Part I), since it reflects the accessibility of defined protein groups to the solvent. However, in distinction from the deuterium exchange, this method can be used to evaluate the amplitude and timescale of dynamic processes by using quenchers of different size, polarity, and charge. In one example, it has been shown that the rate of diffusion of oxygen (which is one of the smallest and most efficient quenchers of intrinsic protein fluorescence) within a protein molecule is only two to four times slower than in an aqueous solution [167, 168]. Furthermore, oxygen was shown to affect even those tryptophan residues that, according to X-ray structural analysis, should not be accessible to the solvent. These observations clearly demonstrated the presence of substantial structural fluctuations in proteins on the nanosecond timescale [167, 168].

Acrylamide is one of the most widely used quenchers of intrinsic protein fluorescence [169, 170]. Acrylamide, like oxygen, is a neutral quencher but has a much larger molecular size. This size difference results in a dramatic decrease in the rate of protein fluorescence quenching over that of oxygen [170, 171]. This decrease is due to the inaccessibility of the globular protein interior to the acrylamide molecule. Thus, acrylamide actively quenches only the intrinsic fluorescence of solvent-exposed residues. As applied to conformational analysis, acrylamide quenching was shown to decrease by two orders of magnitude as unstructured polypeptide chains transitioned to globular structure [170, 171]. Importantly, the degree of shielding of tryptophan residues by the intramolecular environment of the molten globule state was shown to be close to that determined for the native globular proteins, whereas the accessibility of tryptophans to acrylamide in the pre-molten globule state was closer to that in the unfolded polypeptide chain [142]. The fluorescence quenching by acrylamide of the single tryptophan residue in the beta 2 subunit of tryptophan synthase was used to verify the presence of a conformational transition induced by interaction with the cofactor pyridoxal 5′-phosphate [172].

Simultaneous application of quenchers of different size, polarity, and charge (oxygen, nitrite, methyl vinyl ketone, nitrate, acrylate, acrylamide, acetone, methyl ethyl ketone, succinimide, etc.) could be more informative since it may yield information not only about protein dynamics but also about peculiarities of the local environment of chromophores. Information on the local environment of chromophores can also be retrieved from simple quenching experiments. In an example of simultaneous application of multiple quenchers, the heterogeneous fluorescence of yeast 3-phosphoglycerate kinase was resolved into two approximately equal components, one that was accessible and one that was inaccessible to the quencher succinimide [173]. The fluorescence of the inaccessible component was shown to be blue-shifted, and it exhibited a heterogeneous fluorescence decay that had steady-state acrylamide quenching properties and a temperature dependence typical of a single tryptophan in a buried environment. This component was assigned to the buried tryptophan W333. The presence of succinimide greatly simplified the fluorescence, allowing the buried tryptophan to be studied with little interference from the exposed tryptophan [173].

8.2.6.3 Fluorescence Polarization and Anisotropy

Useful information about the mobility and aggregation state of macromolecules in solution can be obtained from analysis of fluorescence polarization or anisotropy. If the excitation beam is polarized and passed through a protein solution, fluorescence will be depolarized or remain partially polarized. The degree of fluorescence depolarization results from the following factors that characterize the structural state of the protein molecules: (1) mobility of the chromophores (strongly dependent on the viscosity of their environment) and (2) energy transfer between similar chromophores [163, 165, 174–178]. Furthermore, the relaxation times of tryptophan residues determined from polarized luminescence data are a reliable indicator of the compactness of the polypeptide chain. For example, it has been noted that the retention of intact disulfide bonds in the unfolded state often results in a nonessential decrease of intrinsic fluorescence polarization, whereas reduction of the disulfide bonds leads to a dramatic decrease in the luminescence polarization to values that are approximately equal for all proteins [177, 178]. The relaxation times of tryptophan residues determined by fluorescence polarization for α-lactalbumin [23, 179] and bovine carbonic anhydrase B [180] showed a high degree of protein compactness in both the native and the molten globule states.

8.2.6.4 Fluorescence Resonance Energy Transfer

Along with intrinsic protein fluorescence, the fluorescence of extrinsic chromophore groups is widely used in conformational studies. Extrinsic chromophores are divided into covalently attached labels and noncovalently interacting probes. Fluorescence labels are indispensable tools in studies of energy transfer between two chromophores. The essence of the phenomenon is that in the interaction of oscillators at small distances the electromagnetic field of the excited (donor) oscillator can induce oscillation in the non-excited (acceptor) oscillator [163, 165, 181] (see chapter 17 in Part I). It should be noted that the transfer of excitation energy between the donor and the acceptor originates only upon the fulfillment of several conditions: (1) the absorption (excitation) spectrum of the acceptor overlaps with the emission (luminescence) spectrum of the donor (an essential prerequisite for resonance), (2) spatial proximity of the donor and the acceptor (within a few dozen angstroms), (3) a sufficiently high quantum yield of the donor, and (4) a favorable spatial orientation of the donor and acceptor. The biggest advantage, and hence the attractiveness, of fluorescence resonance energy transfer (FRET) is that it can be used as a molecular ruler to measure distances between the donor and acceptor. According to Förster, the efficiency of energy transfer, E, from the excited donor, D, to the non-excited acceptor, A, located from the D at a distance R_{DA} is determined by the equation [181]

$$E = \frac{1}{1 + \left(\frac{R_{DA}}{R_o}\right)^6} \tag{18}$$

where R_o is the characteristic donor-acceptor distance, so-called Förster distance, which has a characteristic value for any given donor-acceptor pair given by the equation [181]

$$R_o^6 = \frac{9000 \ln 10}{128 \pi^5 N} \frac{\langle k^2 \rangle \phi_D}{n^4} \int_0^\infty F_D(\lambda) \varepsilon_A(\lambda) \lambda^4 \, d\lambda \tag{19}$$

where the parameter ϕ_D is the fluorescence quantum yield of the donor in the absence of the acceptor, n is the refractive index of the medium, N is the Avogadro's number, λ is the wavelength, $F_D(\lambda)$ is the fluorescence spectrum of the donor with the total area normalized to unity, and $\varepsilon_A(\lambda)$ is the molar extinction coefficient of the acceptor. These parameters are obtained directly from independent experiments. Finally, $\langle k^2 \rangle$ represents the effect of the relative orientations of the donor and acceptor transition dipoles on the energy transfer efficiency. For a particular donor-acceptor orientation, this parameter is given as

$$k = (\cos \alpha - 3 \cos \beta \cos \gamma) \tag{20}$$

where α is the angle between the transition moments of the donor and the acceptor and β and γ are the angles between the donor and acceptor transition moments and the donor-acceptor vector, respectively. Experimentally, the efficiency of direct energy transfer, E, is calculated as the relative loss of donor fluorescence due to the interaction with the acceptor [163, 181]:

$$E = 1 - \frac{\phi_{D,A}}{\phi_D} \tag{21}$$

where ϕ_D and $\phi_{D,A}$ are the fluorescence quantum yields of the donor in the absence and the presence of acceptor, respectively.

For FRET experiments, one typically uses the intrinsic chromophores (tyrosines or tryptophans) as donors and covalently attached chromophores (such as dansyl) as acceptors. According to Eq. (18), the efficiency of energy transfer is proportional to the inverse sixth power of the distance between donor and acceptor. Obviously, structural changes within a protein molecule might be accompanied by changes in this distance, giving rise to the considerable changes in energy transfer. FRET has been used to show that the urea-induced unfolding of proteins is accompanied by a considerable increase in hydrodynamic dimensions. This expansion resulted in a significant decrease in the efficiency of energy transfer from the aromatic amino acids within the protein (donor) to the covalently attached dansyl (acceptor).

An elegant approach based on the unique spectroscopic properties of nitrated tyrosine (which has maximal absorbance at $\sim$350 nm, does not emit light, and serves as an acceptor for tryptophan) has been recently elaborated [182–187]. For these experiments tyrosine residues are modified by reaction with tetranitromethane to convert them to a nitro form, $Tyr(NO_2)$. The extent of decrease of tryptophan fluorescence in the presence of $Tyr(NO_2)$ provides a measure of average distance (R_{DA}) between these residues. This approach has been applied to study the 3-D structure of apomyoglobin in different conformational states [187]. These conformations included the native, molten globule, and unfolded conformations. The A-helix of horse myoglobin contains Trp7 and Trp14, the G-helix contains Tyr103, and the H-helix contains Tyr146. Both tyrosine residues can be converted successively into the nitro form [186]. Comparison of tryptophan fluorescence in unmodified and modified apomyoglobin permits the evaluation of the distances from tryptophan residues to individual tyrosine residues (i.e., the distances between identified points on the protein). Employing this energy transfer method to a variety of nonnative forms of horse apomyoglobin revealed that the helical complex formed by the A-, G-, and H-helices exists in both the molten globule and premolten globule states [187].

8.2.6.5 ANS Fluorescence

Hydrophobic fluorescent probes can be used to detect the hydrophobic regions of protein molecules exposed to the solvent. Hydrophobic fluorescent probes characteristically exhibit intense fluorescence upon interaction with protein and low fluorescence intensity in aqueous solutions. One such probe that is frequently used for studying the structural properties of protein molecules is 8-anilino-1-naphthalene sulfonate (ANS) [188–190]. Interest in this probe reached its peak after it was shown that there was a predominant interaction of ANS with the equilibrium and kinetic folding intermediates in comparison with the native and completely unfolded states of globular proteins [180, 190–193]. The interaction of ANS with protein molecules in the molten globular state is accompanied by a pronounced blue shift of maximal fluorescence and a significant increase of the probe fluorescence intensity, making the latter property a useful tool for the detection of partially folded intermediates in the process of protein folding (Figure 8.6A). Furthermore, the interaction of ANS with protein is also accompanied by a change in the fluorescence lifetime [192, 193]. Fluorescence decay of free ANS is well described by the

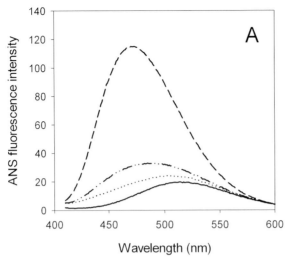

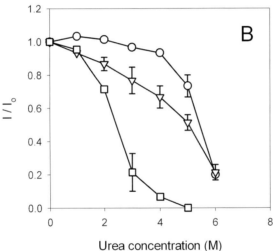

Fig. 8.6. ANS interaction with proteins. (A) ANS fluorescence spectra measured for free dye (solid line) and in the presence of natively disordered coil-like α-synuclein (dotted line), natively disordered pre-molten globule-like caldesmon 636–771 fragment (dash-dot-dotted line), and molten globule state of α-lactalbumin (dashed line). The native molten globular domain of clusterin has an ANS spectrum comparable with that of the molten globular state of α-lactalbumin (see text). (B) Dependence of the ANS fluorescence intensity on urea concentration measured for a series of rigid globular proteins capable of ANS binding: bovine serum albumin (circles), apomyoglobin, pH 7.5 (squares), and hexokinase (inverse triangles).

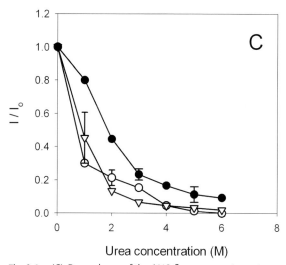

Fig. 8.6. (C) Dependence of the ANS fluorescence intensity on urea concentration measured for a series of molten globular proteins: apomyoglobin, pH 2.0 (open circles), α-lactalbumin, pH 2.0 (open inverse triangles), and clusterin, a protein with a native molten globular domain (black circles).

monoexponential law. However, in the case of formation of complexes between ANS and proteins, fluorescence decay has a more complex dependence. In fact, analysis of the ANS fluorescence lifetimes of a number of proteins revealed that there are at least two types of ANS-protein complexes. The complexes of the first type are characterized by a fluorescence lifetime ranging from 1 to 5 ns. In this type, the probe molecules are bound to the surface hydrophobic clusters of the protein and remain relatively accessible to the solvent. Complexes of the second type are characterized by a fluorescence lifetime of 10 to 17 ns, with the probe molecules embedding themselves inside the protein molecule and therefore being poorly accessible to the solvent [192]. The interaction of ANS with both molten globules and pre-molten globules of different proteins exhibits fluorescence lifetimes characteristic of the second type, with molten globules reacting more strongly than pre-molten globules [134]. Furthermore, an increased affinity for ANS was shown to be a characteristic property of several natively disordered proteins.

Some native globular proteins also possess significant affinity to ANS [188–190]. However, ordered proteins unfold cooperatively, whereas the unfolding of molten globular forms are typically much less cooperative [194]. It has been found that urea titration of ANS fluorescence could be used to distinguish between the binding of ANS to the hydrophobic pocket of an ordered protein and the binding of ANS to a molten globule. This experimental approach has been successfully applied in studies of clusterin. Results like those in Figure 8.6B indicate that this protein likely contains a molten globule-like domain in its native state [21]. This conclusion follows from the comparison of denaturation profiles of clusterin with those for rigid and molten globular forms of a globular protein (Figure 8.6C).

An alternative method is based on comparing Stern-Volmer quenching curves for a polar quencher, acrylamide, with the quenching curves for a nonpolar quencher, trichloroethanol (TCE). The essence of this method is based on the following reasoning. If the hydrophobic groups surrounding a chromophore are rigidly packed, then both acrylamide and TCE are excluded from the potential contact with chromophore and therefore exhibit little quenching. On the other hand, if the hydrophobic groups surrounding the chromophore are loosely packed and dynamic, then the hydrophilic quencher, acrylamide, is still excluded and continues to show little quenching. However, the hydrophobic quencher, TCE, partitions into the hydrophobic region surrounding the chromophore, leading to quenching that is much stronger than if the chromophores were completely exposed on the protein surface. These concepts were proved during the characterization of three different forms of fd phage. The fluorescence emission maxima and intensities for the tryptophans in all three forms are nearly identical, suggesting that all of the environments had very similar overall polarities. For the fd filament there was a very little difference in the quenching by TCE or acrylamide, suggesting that the indole rings were in tightly packed environments. On the other hand, for the two contracted forms, quenching by TCE was much stronger than quenching by acrylamide. Furthermore, the quenching of the contracted forms by TCE was shown to be even stronger than the quenching of a free indole ring in water. This result was interpreted as indicating that the residues surrounding the tryptophans in the contracted forms of fd phage were likely to be highly dynamic (similar to the inside of an SDS micelle), thus leading to accumulation of TCE and therefore increased quenching [195]. We believe that this approach has useful features that provided additional insights into the molten globular state.

8.2.7
Conformational Stability

Intrinsic disorder may be detected by the analysis of protein conformational stability. For example, the presence or absence of a cooperative transition in a calorimetric melting curve is a simple and convenient criterion that indicates the presence or absence of a rigid tertiary structure [196–198] (see chapter 4 in Part I). Furthermore, the response of a protein to denaturing conditions may be used to discriminate between collapsed (molten globule-like) and extended intrinsic disorder (coil-like and pre-molten globule-like conformations). In fact, the increase in temperature and changes in pH (as well as the increase in urea or guanidinium chloride concentration) will induce relatively cooperative loss of the residual ordered structure in molten globule-like disordered proteins, whereas temperature and pH will bring formation of residual structure in native coils and native pre-molten globules (see sections 8.2.7.1 and 8.2.7.2). Furthermore, the steepness of urea- or guanidinium chloride-induced unfolding curves depends strongly on whether a given protein has a rigid tertiary structure (i.e., is a native globular protein) or is already denatured and exists as a molten globule [194, 199]. To perform this type of analysis, the values of Δv^{eff} (the difference in the number of denaturant molecules "bound" to one protein molecule in each of the two states) should be determined. This quantity should be compared to the $\Delta v^{\text{eff}}_{\text{N}\to\text{U}}$ and $\Delta v^{\text{eff}}_{\text{MG}\to\text{U}}$ values corresponding to the native-to-coil and molten globule-to-coil transitions in globular protein of that given molecular mass [194].

8.2.7.1 Effect of Temperature on Proteins with Extended Disorder

It has been pointed out that at low temperatures, natively disordered proteins with extended disorder show far-UV CD spectra typical of an unfolded polypeptide chain (see Section 8.2.4). However, as the temperature increases, the spectrum changes, consistent with temperature-induced formation of secondary structure [123, 135]. In fact, such behavior was observed for natively disordered proteins such as α-synuclein [119], the phosphodiesterase γ-subunit [200], the caldesmon 636–771 fragment [121], the extracellular domain of the nerve growth factor receptor [201] and α_{s}-casein [202]. Thus, an increase in temperature can induce the partial folding of natively disordered proteins, rather than the unfolding typical of globular proteins. The effects of elevated temperatures may be attributed to increased strength of the hydrophobic interaction at higher temperatures, leading to a stronger hydrophobic driving force for folding.

8.2.7.2 Effect of pH on Proteins with Extended Disorder

Several natively disordered proteins with extended disorder, including α-synuclein [119], prothymosin α [116], pig calpastatin domain I [203], histidine rich protein II [204], naturally occurring human peptide LL-37 [205], and several other proteins, show intriguing dependence of their structural parameters on pH. In fact, these proteins possess low structural complexity at neutral pH but have significant residual structure under conditions of extreme pH. These observations show that a decrease (or increase) in pH induces partial folding of natively disordered proteins due to the minimization of their large net charge present at neutral pH, thereby decreasing charge/charge intramolecular repulsion and permitting hydrophobic-driven collapse to partially folded conformations.

8.2.8
Mass Spectrometry-based High-resolution Hydrogen-Deuterium Exchange

To obtain high-resolution structural information using X-ray crystallography, it is necessary to produce crystals. This can be impossible for proteins with substantial amounts of intrinsic disorder. High-resolution structural information can also be provided by NMR, but this technique is limited by protein size and the need for high concentrations of protein. Mass spectrometry-based high-resolution hydrogen-deuterium exchange (MSHDX) shows promise in becoming a third source of high-resolution structural information.

Monitoring the exchange rate between main chain amides and solvent hydrogens as a method to study the structure of proteins has seen increased usage over the past 40 years [206]. Hydrogen-deuterium exchange (HDX) rates are dependent on thermodynamics and dynamic behavior and thus yield information regarding the structural stability of the protein under study. The study of HDX as applied to proteins was initiated by Kaj Linderstrøm-Lang in the 1950s as a method to investigate Pauling's newly postulated α-helix and β-sheet secondary structures [207]. The two-step model of HDX is described as [208]

$$C^H \underset{k_{cl}}{\overset{k_{op}}{\rightleftharpoons}} O^H \underset{k_{ch}}{\overset{k_{ch}}{\rightleftharpoons}} O^D \underset{k_{op}}{\overset{k_{cl}}{\rightleftharpoons}} C^D \tag{22}$$

where H and D denote protonated and deuterated forms, respectively, of the "closed" or C (non-exchanging) and "open" or O (exchanging) conformations; k_{op} is the rate constant for the conformational change that exposes the hydrogen; k_{ch} is the rate constant for the chemical exchange reaction; and k_{cl} is the rate constant for the return to the closed conformation.

The fundamental equation for the observed exchange rate formulated by Linderstrøm-Lang still applies:

$$k_{ex} = \left(\frac{k_{op} \cdot k_{ch}}{k_{cl} + k_{op}} \right) \tag{23}$$

The EX2 HDX mechanism, where $k_{cl} \gg k_{ch}$, describes most proteins at or below neutral pH. In this case, k_{ex} can be found using the following equation:

$$k_{ex} = \left(\frac{k_{op} \cdot k_{ch}}{k_{cl}} \right) = K_{op} \cdot k_{ch} \tag{24}$$

where K_{op} is the unfolding equilibrium constant that is also equal to the ratio of the observed rate (k_{ex}) to that of a random coil amide (k_{ch}).

HDX rates for random coil amides (k_{ch}), which depend on pH, primary sequence, and temperature, can be obtained from studies of peptide behavior [209]. The ratio of the observed rate of exchange (k_{ex}) to that of a random coil (k_{ch}) relates to the free energy of unfolding by the following equation:

$$\Delta G_{op} = -RT \ln \left(\frac{k_{ex}}{k_{ch}} \right) = -RT \ln(K_{op}) \tag{25}$$

Eq. (25) shows that there is a logarithmic correlation between the exchange rate and the conformation stability, i.e., the faster the exchange rate, the less stable the folded state. Being initially developed to study protein folding on a global scale, this formalism could also be used to analyze the intrinsic dynamic behavior of proteins and, most attractively, to quantify the stability of localized regions within protein molecules.

Hydrogen bonding, accessibility to the protein surface, and flexibility of the immediate and adjacent regions affect the HDX rate. The combined contribution of these structural and dynamic factors is termed the "protection factor," which has been observed to vary by as much as 10^8-fold [210]. Based on these observations, it has been suggested that each amide hydrogen in the polypeptide can be viewed as a sensor of the thermodynamic stability of localized regions within the protein structure [211]. HDX can yield data to single-amide resolution.

In the 1960s, HDX rates were determined using tritium incorporation, size-exclusion chromatography, and liquid scintillation counting [212]. Structural resolution remained a limitation of this procedure. Improvements in resolution were facilitated by separating fragments using HPLC and analysis of tritium or deuterium incorporation using mass spectrometric methods [212–214, 405]. Recent improvements in automation, consisting of solid-phase proteolysis, automated liquid handling and sample preparation, online ESI MS, and specialized data reduction software, have enhanced throughput and sequence coverage of MSHDX [211]. As high as six-amide resolution has been achieved [211]. The use of multiple proteases to generate more overlaps in the peptide map can be used to increase the resolution even further.

Crystallization success rates can be improved by detecting and excluding sequences coding for unstructured regions when designing recombinant expression constructs. As part of a structural genomics effort, 24 proteins from *Thermotoga maritima* were analyzed using MSHDX to detect unstructured regions [215]. Prior to HDX analysis, parameters affecting fragmentation were optimized to maximize the number of fragments available for analysis. These conditions included denaturant concentration, protease type, and duration of incubation. As a prerequisite for MSHDX analysis, the ability to generate complete fragment maps using SYQUEST (Thermo Finnigan, San Jose, CA) and DXMS data reduction software was tested for each protein. Satisfactory fragment maps were generated for 21 of the 24 proteins. Next, labeling conditions were optimized for discrimination of fast-exchanging amide protons. In this procedure, the proteins were subjected to proteolysis prior to exchange and HDX analysis. An exchange duration of 10 s was determined to be sufficient to allow differentiation between deuterated freely solvated and non-deuterated inaccessible amides. MSHDX analysis detected regions of rapidly changing amides that were greater than 10% of total protein length in five of the proteins. From these five, MSHDX analysis led to the construction of deletion mutants for two recalcitrant proteins. Proteins produced from these two deletion mutants were subsequently crystallized, and their structures solved [215].

8.2.9
Protease Sensitivity

Numerous enzymes, including trypsin, pepsin and carboxypeptidase, catalyze the hydrolysis of peptide bonds, including trypsin, pepsin, carboxy peptidase, etc. Those that cleave at specific sites within protein chains have long been used to probe protein structure. In the 1920s Wu and others showed that native (ordered) proteins are typically much more resistant to protease digestion than are denatured (disordered) forms [216]. These early indications were substantiated over the next

~20 years [217]. Increasing the stability of a protein, e.g., by ligand or substrate binding, in turn often leads to reduced digestion rates [218].

In essentially all biochemistry textbooks, the structure of proteins is described as a hierarchy. The amino acid sequence is the primary structure; the local folding of the backbone into helices, sheets, turns, etc. is the secondary structure; and the overall arrangement and interactions of the secondary structural elements make up the tertiary structure. Many details of this hierarchy were determined from the first protein 3-D structure [16], but the original proposal for the primary-secondary-tertiary hierarchy was based on differential protease digestion rates over the three types of structure [217].

While acknowledging that flexible protein regions are easy targets for proteolysis, some researchers argued that the surface exposure of susceptible amino acids in appropriate conformations could provide important sites for accelerated digestion rates [219]. Indeed, proteins that inhibit proteases, such as soybean trypsin inhibitor [220], have local regions of 3-D structure that fit the active sites and bind irreversibly to their target proteases. Despite the examples provided by the inhibitors, susceptible surface digestion sites in ordered proteins are not very common [221, 222]. Despite the paucity of examples, the importance of surface exposure rather than flexibility or disorder seems to have become widely accepted as the main feature determining protease sensitivity.

In the well-studied protease trypsin, most of the lysine and arginine target residues are located on protein surfaces, but, as mentioned above, very few of these are actually sites of digestion when located in regions of organized structure. Starting with a few protein examples having specific surface-accessible digestion sites, attempts were made to use molecular simulation methods to dock the backbones of susceptible residues into the active site of trypsin. These studies suggested the need to unfold more than 10 residues to fit each target residue's backbone into trypsin's active site [223]. Indeed, recent studies show that backbone hydrogen bonds are protected from water by being surrounded or wrapped with hydrophobic side chains from nearby residues, with essentially every ordered residue being well wrapped [224]. Backbone hydrogen bond wrapping and binding the backbone deep within trypsin's active site are mutually exclusive, and so the requirement for local unfolding (unwrapping) is easily understood.

Direct support for the importance of local unfolding has been provided by comparison of the digestion rates of myoglobin and apomyoglobin [225]. Myoglobin is digested very slowly by several different proteases. Removal of the heme leads to local unfolding of the F-helix. This disordered loop becomes the site of very rapid digestion by several different proteases including trypsin. The trypsin digestion rate following the order-to-disorder transition of the F-helix is at least five orders of magnitude faster.

Many fully ordered proteins, such as the myoglobin example given above, are digested by proteases without observable intermediates of significant size. That is, if digestion progress were followed using PAGE, myoglobin would be seen to disappear and small fragments would appear, but mid-sized intermediates would not be observed. In contrast, digestion of the F-helix in apomyoglobin leads to two relatively stable mid-sized fragments. The likely cause of the lack of observable mid-sized intermediates for fully ordered proteins is that digestion proceeds at multiple sites within the ordered regions of the protein. Access to these sites is mediated by transient unfolding events within ordered regions that allow initial digestion events. These initial cuts perturb the local structure and inhibit refolding. The sum of these changes leads to the exposure of multiple protease sites, which are rapidly cleaved in a random order. Additional transient unfolding events elsewhere in the structure lead to the same outcome. Thus the regions of the protein where structure has been disturbed become rapidly digested into a multitude of different-sized intermediate fragments without the accumulation of an observable quantity of any particular mid-sized intermediate [225].

Molten globules have also been probed by protease digestion [226]. Protease digestion of molten globules leads rapidly to multiple mid-sized fragments; upon further digestion, the intermediates convert to smaller-sized fragments consistent with nearly complete digestion. The transient stability of several mid-sized intermediates suggests that molten globules have multiple regions of high flexibility that are easy to digest interspersed with stabler structured regions that are resistant to digestion.

At this time we have not found published, systematic studies of fully unfolded, natively disordered proteins. One would expect such proteins to undergo digestion in a random fashion with all of the target residues equally accessible to digestion and with only sequence-dependent effects to modulate the digestion rates at the various sites. If this general picture were true, then proteolytic digestion would lead to rapid conversion of the primary protein into small fragments, without significant amounts of transient, mid-sized intermediates. Highly disordered proteins such as p21$^{\text{Waf1/Cip1/Sdi1}}$ [402, 403] and cytochrome c [404] undergo digestion to small fragments without any observable intermediates, as suggested here. The digestion patterns of extensively unfolded proteins likely resemble those of fully folded proteins: both evidently lack the accumulation of mid-sized intermediates. However, the two types of proteins can be distinguished by their vastly different digestion rates: extensively unfolded proteins would be expected to be digested at least 10^5 times faster than fully folded forms.

8.2.10
Prediction from Sequence

Amino acid sequence codes for protein 3-D structure [5]. Since native disorder can be viewed as a type of structure, we wanted to test whether the amino acid compositions of disordered regions differed from the compositions of ordered regions and, if so, whether the types of enrichments and depletions gave insight into the disorder. Thus, we studied amino acid compositions of collections of both ordered and natively disordered protein regions longer than 30 amino acids. We chose such long regions to lessen the chance that non-local interactions would complicate our analysis.

Natively disordered sequences characterized by X-ray diffraction, NMR spectroscopy, and CD spectroscopy all contain similar amino acid compositions, and these are very different from the compositions of the ordered parts of proteins in PDB. For example, compared to the ordered proteins, natively disordered proteins contain (on average) 50% less W, 40% less F, 50% less isoleucine, 40% less Y, and 25% or so less C, V, and L, but 30% more K, 40% more E, 40% more P, and 25% more S, Q, and R [31]. Thus, natively disordered proteins and regions are substantially enriched in typically hydrophilic residues found on protein surfaces and significantly depleted of hydrophobic, especially aromatic, residues found in protein interiors. These data help to explain why natively disordered proteins fail to form persistent, well-organized 3-D structure.

Because ordered and disordered proteins contain significantly different amino acid compositions, it is possible to construct order/disorder predictors that use amino acid sequence data as inputs. By now, several researchers have published predictors of order and disorder [13, 30, 109, 227–233]. The first predictor, developed by R. J. P. Williams, separated just two natively disordered proteins from a small set of ordered proteins by noting the abnormally high charge/hydrophobic ratio for the two natively disordered proteins. This paper is significant as being the first indication that natively disordered proteins have amino acid compositions that differ substantially from those of proteins with 3-D structure. The predictor by Uversky et al. [227] also utilizes the relative abundance of charged and hydrophobic groups, but on a much larger set of proteins. Most of the other predictors [30, 109, 229–233] utilize neural networks.

Predicting order and disorder was included in the most recent cycle of the Critical Assessment of Protein Structure Prediction (CASP) [234]. While the summary publication contained papers from two sets of researchers [109, 233], several additional groups participated. All of the groups achieved similar levels of accuracy [234]. Furthermore, prediction accuracies were on the order of 90% or so for regions of order and 75–80% for regions of intrinsic disorder. These values are significantly above what could be expected by chance. The predictability of native disorder from sequence further supports the conjecture that natively disordered proteins and regions lack specific 3-D structures as a result of their amino acid sequences. That is, amino acid sequence codes for both order and disorder.

8.2.11
Advantage of Multiple Methods

Given the limitations of the various physical methods, it is useful if natively disordered proteins are characterized by multiple methods. Each approach gives a slightly different view, with better understanding arising from the synthesis of the different perspectives (see chapter 2 in Part I).

A significant difficulty with long disordered regions identified by X-ray diffraction is that the missing coordinates could indicate either native disorder or a wobbly domain that moves as a folded, ordered, rigid body. Thus, it is especially useful if X-ray-indicated disorder is confirmed by additional methods. The X-ray-indicated regions of disorder in calcineurin [29, 235], clotting factor Xa [236–238], histone H5 (C. Crane-Robinson, personal communication) [239–241], and topoisomerase

II (J. M. Berger, personal communication) [242–244] were all confirmed by limited proteolysis. The X-ray-indicated regions of disorder in Bcl-X_L [245–248], the gene 3 protein (g3p) of filamentous phage [249–251], and the negative factor of HIV1 (Nef) [252–255] were all confirmed by NMR spectroscopy. Given the importance of natively disordered regions, it would be helpful if disordered regions indicated by X-ray were routinely subjected to further study by an alternative method.

To illustrate the advantages of combining X-ray and NMR analysis, a hypothetical example is given here. Suppose an X-ray structure reveals that a protein has a short N-terminal disordered region, a central region composed of an ordered region flanked by two disordered segments, and a long C-terminal disordered region. Figure 8.7 shows the expected $^{15}N\{^1H\}$-NOE spectra for two different possibilities. Figure 8.7A shows the expected result if all three regions of missing electron density are truly disordered with very little secondary structure: in this case all three regions of missing electron density give negative values, indicating unfolded, peptide-like motions for these segments. In Figure 8.7B, the C-terminal region shows a short region of disorder followed by a large region of order; this would be consistent with an ordered but wobbly domain that would lead to missing electron density in the X-ray structure.

The nucleoprotein N of the measles virus contains more than 100 disordered residues at its C-terminus, a region called Ntail. The disordered character of Ntail was first identified by prediction and was confirmed by Ntail's hypersensitivity to protease digestion and by its NMR and CD spectra [256]. Predictions on the N proteins of related Paramyxoviridae indicate that the Ntail regions, which have hypervariable sequences, are all predicted to be natively disordered [257]. Cloning, expressing, and isolation of the Ntail region have enabled multiple studies of this region. The native disorder of this region by now has been confirmed by CD spectroscopy, SAXS, and dynamic light scattering (for the determination of R_s) [257]. The Ntail is apparently of the pre-molten globule type. Furthermore, although natively disordered and separated from the remainder of the molecule, Ntail retains its biological function of binding to a protein partner [258].

8.3
Do Natively Disordered Proteins Exist Inside Cells?

8.3.1
Evolution of Ordered and Disordered Proteins Is Fundamentally Different

8.3.1.1 The Evolution of Natively Disordered Proteins
Low-resolution structural models of novel proteins can be generated based on sequence similarity and evolutionary relationships to proteins with known structures [259–265]. This process assumes that the proteins being compared adopt compact rigid structures. Natively disordered proteins have important biological functions, and analysis of genome sequence data has revealed that proteins with intrinsically disordered segments longer than 50 amino acids are common in nature [20, 30, 266–271]. The lack of biophysical data characterizing the partially collapsed dy-

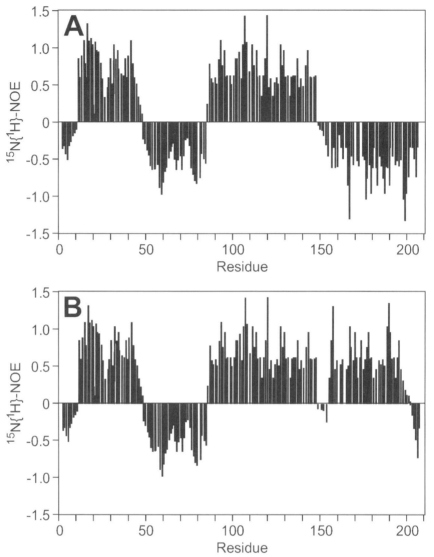

Fig. 8.7. Idealized steady-state heteronuclear $^{15}N\{^1H\}$-NOE for the backbone amides of the hypothetical protein discussed in the text. The domain profile of this protein based on the X-ray crystal structure is: short disordered N-terminal region/ordered region/disordered region/ordered region/long disordered C-terminal region. (A) $^{15}N\{^1H\}$-NOE that is in agreement with X-ray-determined structure; all regions of missing electron density in the X-ray crystal structure have negative values. (B) $^{15}N\{^1H\}$-NOE indicating that the C-terminal domain is a wobbly domain (a structured domain attached via a short, flexible linker) and is not totally disordered as determined by X-ray crystallography.

namic structures of natively disordered proteins limits our ability to predict their existence based on sequence data. It also limits our understanding of how the sequences of such regions specify function, the presence or absence of residual structure, and degree of flexibility.

The evolution of globular protein structures depends on the maintenance of a nonpolar interior and a polar exterior to promote collapse and folding through the hydrophobic effect [261]. This is achieved by the placement and distribution of nonpolar amino acids so that those making favorable nonpolar contacts in the folded structure will be far apart in the linear sequence (so-called long-range interactions). Therefore, the evolution of globular protein structure is partly dependent on selection for this property. In addition to the hydrophobic effect, desolvation of backbone hydrogen bonds appears to be of similar importance [224].

The evolution of intrinsically disordered protein structure seems to depend on selection for other properties. Natively disordered proteins with extended disorder do not form globular structures and therefore do not have a requirement to maintain long-range interactions such as those required by globular proteins. This creates a potential for these sequences to accumulate more variation and to generate more sequence divergence than globular proteins (see Section 8.3.1.2). For globular proteins, the selection for structural motifs is apparent by sequence and structural comparison of evolutionarily related protein structures. In an attempt to develop evolutionary relationships for intrinsically disordered proteins, the Daughdrill lab is currently investigating the structure, dynamics, and function of a conserved flexible linker from the 70-kDa subunit of replication protein A (RPA70). The flexible linker for human RPA70 is ~70 residues long [269, 272]. For the handful of sequenced RPA70 homologues, the similarity among linker sequences varies significantly, going from 43% sequence identity between *Homo sapiens* and *Xenopus laevis* to no significant similarity between *Homo sapiens* and *Saccharomyces cerevisiae*. It is unclear which selective processes have resulted in the observed sequence variations among RPA70 linkers. It is also unclear how the observed sequence variation affects the structure and function of the linkers. If natural selection works to preserve flexible structures, then one would expect that the linkers from different species have evolved to the same level of flexibility although adopting different sequences. By testing this hypothesis, we will begin understanding the rules governing the evolution of natively disordered protein sequences.

8.3.1.2 Adaptive Evolution and Protein Flexibility

The ability to align multiple sequences and reconstruct phylogenies based on sequence data is essential to understanding molecular evolution. The development of reliable algorithms that can align multiple sequences and reconstruct phylogenies is encumbered by the presence of highly divergent segments within otherwise obviously related sequences [259, 260, 273, 274]. This problem appears to be especially acute for totally disordered proteins and disordered protein domains. We hypothesize that intrinsically disordered regions will have a higher rate of evolution than compact rigid structures because their rate of evolution is not constrained by the requirement to maintain long-range interactions. Genetic distance measurements of the RPA70 homologues lend support to this hypothesis and suggest that the linker has evolved at a rate 1.5 to 5 times faster than the rest of RPA70 [275]. This study tested the evolutionary rate heterogeneity between intrinsically disordered regions and ordered regions of proteins by estimating the pairwise genetic distances among the ordered and disordered regions of 26 protein families that have at least one member with a region of disorder of 30 or more consecutive residues that has been characterized by X-ray crystallography or NMR. For five of the protein families, there were no significant differences in pairwise genetic distances between ordered and disordered sequences. Disordered regions evolved more rapidly than ordered regions for 19 of the 26 families. The known functions for some of these disordered regions were diverse, including flexible linkers and binding sites for protein, DNA, or RNA. The functions of other disordered regions were unknown. For the two remaining families, the disordered regions evolved significantly more slowly than the ordered regions. The functions of these more slowly evolving disordered regions included sites for DNA binding. According to the authors, much more work is needed to understand the underlying causes of the variability in the evolutionary rates of intrinsically ordered and disordered proteins. Figure 8.8 illustrates the point, showing the contrast between the protein sequence alignment for the flexible linker and a fragment of the ssDNA binding domain from eight RPA70 homologues. Because of the known functional consequences of linker deletion in RPA70 [272, 276–279], we are interested in how the level of functional selection for flexible regions is related to their evolutionary rates. As is observed for folded proteins, most likely this relationship depends on the function under selection [259–261, 273]. However, because the substitution rate for flexible regions can be uncoupled from the maintenance of long-range interactions, the exact dependence between the rate of evolution and functional selection should differ significantly from that observed for folded proteins. Experiments planned to test the functional consequences of swapping divergent flexible linkers between species will begin to address this subject.

8.3.1.3 Phylogeny Reconstruction and Protein Structure

We are interested in developing reliable algorithms that can align multiple sequences and reconstruct phylogenies for highly divergent sequences. We hypothesize that flexible linkers and other disordered protein domains will often be highly divergent. Recent efforts at phylogeny reconstruction using models that incorporate the effects of secondary structure and solvent accessibility on amino acid substitution rate and type have yielded significantly improved maximum likelihood scores [280–282]. These models depend on an empirically determined three-dimensional structure for at least one member in a homologous family and incorporate four classes of secondary structure (helix, sheet, turn, and coil) as well as two solvent accessibility classes (buried and exposed). In particular, the inclusion of a flexible secondary structure class (coil) is necessary to model the evolution of natively disordered protein domains. The coil category was based on relatively short regions (≤ 5 residues) whose ends are constrained by the compact globular part of the protein. For proteins with longer flexible regions, the current model must be adapted

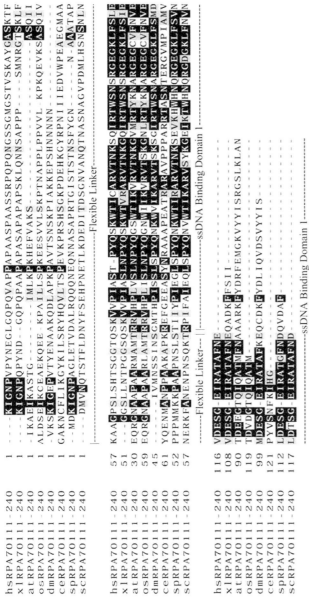

Fig. 8.8. Protein sequence alignment of the RPA70 flexible linker and a fragment of the first ssDNA-binding domain from *Homo sapiens* (hs), *Xenopus laevis* (xl), *Arabidopsis thaliana* (at), *Oriza sativa* (os), *Drosophila melanogaster* (dm), *Caenorhabditis elegans* (ce), *Saccharomyces pombe* (sp), and *Saccharomyces cerevisiae* (sc). Dark shading indicates identity, and light shading indicates conservative substitutions. The alignment was performed using Clustal 1.8 for residues 111–240 of all eight sequences. The alignment shows the stark contrast in the level of sequence similarities between the ssDNA-binding domain and the flexible linker regions.

to incorporate structural classes that are based on the presence of residual secondary structure and the degree of flexibility estimated from NMR relaxation measurements.

Another important feature of protein evolution is the pattern of amino acid substitutions observed over time. For ordered proteins, the change of an amino acid into one of a similar chemical type is commonly observed, whereas the change to a chemically dissimilar one is rare. For example, isoleucine to leucine, aspartate to glutamate, and arginine to lysine are all commonly observed in related ordered proteins, while tryptophan to glycine, lysine to aspartate, and leucine to serine are all uncommon changes, especially for buried residues.

Patterns of amino acid substitutions are readily observed in substitution matrices used to assign values to various possible sequence alignments. These scoring matrices are constructed by first assembling a set of aligned sequences, typically with low rates of change in pairs of sequences so that there is confidence in the alignments. Given such a set of high-confidence alignments, the frequencies of the various amino acid substitutions are then calculated and used to build a scoring matrix. Important and commonly used scoring matrices are the PAM [283] and the Blossum [284] series.

To improve alignments of disordered regions, a scoring matrix for disordered regions was developed [285]. First, homologous groups of disordered proteins were aligned by standard protocols using the Blossum 62 scoring matrix. From this set of aligned sequences, a new scoring matrix was calculated and then realignment was carried out with this new scoring matrix. As compared to the first alignment, the new alignment was improved as estimated by a reduction in the sizes and number of gaps in the pairwise alignments. The new set of alignments was then used to develop a new scoring matrix, and a new set of alignments was generated. These steps were repeated in an iterative manner until little or no change was observed in successive sets of alignments and in successive scoring matrices.

The resulting scoring matrix for disordered regions showed significant differences from the Blossum 62 or PAM 250 scoring matrices [286]. Glycine/tryptophan, serine/glutamate, and alanine/lysine substitutions, for example, were much more common in aligned regions of native disorder as compared to aligned regions of ordered proteins. Since natively disordered regions lack specific structure and the accompanying specific amino acid interactions, substitutions of disparate amino acids are not so strongly inhibited by the need to conserve structure. Thus the commonness of such disparate substitutions in natively disordered regions is readily understood.

The commonly observed higher rates of amino acid change [287] and the distinctive pattern of amino acid substitutions [286] both strongly support the notion that native disorder exists inside the cell.

8.3.2
Direct Measurement by NMR

The inside of cells are extremely crowded, and proteins themselves do most of the crowding since they occupy 40% of a cell's volume and achieve concentrations of greater than 500 g L^{-1} [288–290]. Despite the crowded nature of the cell's interior, almost all proteins are studied outside cells and in dilute solutions, i.e., total solute concentrations of less than 1 g L^{-1}. It is also clear that macromolecular and small-molecule crowding can increase protein stability [291–293]. Could some disordered proteins be artifacts of the way proteins are studied? Are some just unstable proteins that unfold in dilute solution? A recent study by Dedmon et al. [102] on a protein called FlgM shows that the answer to the question is yes and no. Some disordered proteins probably have structure in cells while others do not.

Bacteria use rotating flagella to move through liquids [294, 295]. The protein FlgM is part of the system controlling flagellar synthesis. It binds the transcription factor σ^{28}, arresting transcription of the genes encoding the late flagellar proteins. Transcription can resume when FlgM leaves the cell, most probably via extrusion through the partially assembled flagellum.

Free FlgM is mostly disordered in dilute solution, but NMR studies in dilute solution indicate that the C-terminal half of FlgM becomes structured on binding σ^{28}, as is shown by the disappearance of cross-peaks from residues in the C-terminal half of FlgM in the FlgM-σ^{28} complex [77, 267]. One signature of protein structure can be the absence of cross-peaks in ^{1}H-^{15}N HSQC NMR spectra because of conformational exchange [296–298]. The disappearance of cross-peaks results from chemical exchange between a disordered and more ordered form. Specifically, cross-peaks broaden until they are undetectable when the rate of chemical exchange between states is about the same as the difference in the resonance frequencies of the nuclei undergoing exchange. The bipartite behavior of FlgM (i.e., disappearance of cross-peaks from the C-terminal half with retention of cross-peaks from the N-terminal half) provides a valuable built-in control for studying the response of FlgM to different solution conditions.

How do these observations about the ability of FlgM to gain structure on binding its partner relate to what might happen in cells? Until recently, all protein NMR was performed in vitro on purified protein samples in dilute solution. Two years ago, Dötsch and colleagues showed the feasibility of obtaining the spectra of ^{15}N-labeled proteins inside living *Escherichia coli* cells [299–301]. Overexpression is key to the success of in-cell NMR. The protein of interest must contain a large proportion of the ^{15}N in the sample so that the spectrum of the overexpressed protein can be observed on top of signals arising from other ^{15}N-enriched proteins in the cell, which contribute to a uniform background.

^{15}N-enriched FlgM was found to give excellent in-cell NMR data [102]. About half the cross-peaks disappear in cells. Most importantly, the cross-peaks that disappear are the same ones that disappear on σ^{28} binding in simple buffered solution, and the cross-peaks that persist are the same ones that persist on σ^{28} binding. These data suggest that FlgM gains structure all by itself under the crowded conditions found in the cell, but it is important to rule out alternative explanations. There is a homologue of *S. typhimurium* σ^{28} in *E. coli*, but there is not enough of the homologue present in *E. coli* (i.e., FlgM is overexpressed, the σ^{28} homologue is not) for σ^{28}-FlgM binding to explain the results. Furthermore, the same behavior is observed in vitro – in the complete absence of σ^{28} – when intracellular crowding was mimicked by using 450 g L^{-1} glucose, 400 g L^{-1} bovine serum albumin, or

450 g L^{-1} ovalbumin. The lack of cross-peaks from the C-terminal half of the protein is not caused by degradation, because the FlgM can be isolated intact at the end of the in-cell experiment. The gain of structure in cells does not seem to be an artifact of FlgM overexpression because the total protein concentration is independent of FlgM expression. Two observations show that the presence or absence of cross-peaks is not simply a matter of viscosity. First, cross-peaks from the N-terminal half of FlgM are present under all conditions tested even though the relative viscosities of the solutions differ dramatically. Second, the absence of cross-peaks does not correlate with increased viscosity. Taken together, these data strongly suggest that even in the absence of its binding partner, the C-terminal portion of FlgM gains structure in cells and in crowded in vitro conditions.

Does this observation about the C-terminal part of FlgM mean that all intrinsically disordered proteins will gain structure under crowded conditions? No. First, the N-terminal part of FlgM remains disordered under crowded conditions both inside and outside the cell. Second, the same observation has been made for another protein under crowded conditions in vitro [302]. Third, as discussed below, several functions of disordered proteins require the absence of stable structure. Perhaps the N-terminal half of FlgM gains structure only upon binding some yet unknown molecule, or maybe it needs to remain disordered to ensure its exit from the cell. It is also important to note, however, that crowding can induce compaction even when it does not introduce structure [158].

In summary, some, but certainly not all, so-called natively disordered proteins will gain structure in cells. In terms of the equilibrium thermodynamics of protein stability (Eqs. (1) and (2)), these proteins are best considered as simply very unstable. This instability may be essential for function, e.g., allowing rapid degradation or facilitating exit from the cell. And, finally, it is important to consider this discussion in terms of Anfinsen's thermodynamic hypothesis. Specifically the last four words of his statement [5]: "the native conformation is determined by the totality of inter-atomic interactions and hence by the amino acid sequence, *in a given environment*."

8.4
Functional Repertoire

8.4.1
Molecular Recognition

8.4.1.1 The Coupling of Folding and Binding

Molecular recognition is an essential requirement for life. Protein/protein, protein/nucleic acid, and protein/ligand interactions initiate and regulate most cellular processes. Many natively disordered proteins can fold upon binding to other proteins or DNA. The loss of conformational entropy that occurs during folding can influence the kinetics and thermodynamics of binding [78, 267, 303–305]. The most appealing model for the coupling of folding and binding proposes an initial encounter complex that forms nonspecifically while the protein is still unfolded

[303]. The release of solvent will provide a favorable entropic contribution to the binding. The extent of this favourable effect depends on the amount of hydrophobic burial that occurs during the formation of the nonspecific encounter complex. This encounter complex will undergo a sequential selection to achieve a specific complex. Sequential selection is achieved by consecutive structural interconversions that increase the surface complementarity. A subsequent increase in surface complementarity leads to a stabler complex. Of course, this model does not address the possibility that folding to a single structure does not occur but rather the bound structure is dynamic and the overall stability is governed by a collection of competing interactions.

8.4.1.2 Structural Plasticity for the Purpose of Functional Plasticity

There is increasing evidence that multiple binding modes can be accommodated in protein/ligand, protein/protein, and protein/DNA interactions [306–311]. In some cases, different segments of a single polypeptide can be used for recognition of different substrates. In other cases a single protein surface can accommodate the coupled binding and folding of multiple polypeptide sequences into different structures. For instance, the phosphotyrosine-binding domain of the cell fate determinant Numb can recognize peptides that differ in both primary and secondary structure by engaging various amounts of the binding surface [308].

The importance of structural plasticity in molecular recognition is made more concrete by considering a specific example: the activation of calcineurin (CaN) by calmodulin (CaM). The calcium-dependent binding of CaM to CaN brings about exposure of CaN's serine-threonine phosphatase active site by displacement of the autoinhibitory peptide and thereby turns on the phosphatase activity (Figure 8.1). This interaction forms a bridge between phosphorylation-dephosphorylation-based signaling and calcium-based signaling. The CaN-CaM interaction plays important roles in a wide variety of eukaryotic cells. For example, dephosphorylation of Nfat by Ca^{2+}-CaM-activated CaN leads to killer T cell proliferation and foreign tissue rejection [312]; blockage of CaN activation by complexation with FK506-FK-binding protein leads to suppression of the rejection [313].

The intrinsic disorder (plasticity) of CaM-binding domains enables them to bind to a wide variety of target sequences [314, 315]. The four EF hands of CaM undergo disorder-to-order transitions upon Ca^{2+} binding [316]. The two domains of CaM are connected by a helix in the crystal, but NMR shows the central region to be melted in solution, thus providing a flexible hinge that enables CaM to wrap around its target helix [317]. On the CaN side, intrinsic disorder flanks the CaM target and thus provides space for CaM to wrap around its target helix [235]. Indeed, before the CaN disordered region was revealed to be missing electron density in its X-ray crystal structure, trypsin digestion had already indicated disorder in CaN's CaM-binding region [318, 319]. Similar trypsin digestion analyses show that many CaM-binding sites are within disordered regions.

8.4.1.3 Systems Where Disorder Increases Upon Binding

Two studies have shown that disorder can increase upon binding to hydrophobic ligands and proteins [320, 321]. Increased protein backbone conformational entropy was observed for the mouse major urinary protein (MUP-I) upon binding the hydrophobic mouse pheromone 2-*sec*-butyl-4,5-dihydrothiazole [320]. ^{15}N relaxation measurements of free and pheromone-bound MUP-I were fitted using the Lipari-Szabo model-free approach [86, 87]. Order parameter differences were observed between free and bound MUP-I that were consistent with an increase in the conformational entropy. Out of 162 MUP-I residues, 68 showed significant reductions in S^2 upon pheromone binding. The changes were distributed throughout the β-barrel structure of MUP-I, with a particular prevalence in a helical turn proposed to form a ligand entry gate at one end of the binding cavity. The authors called into question assumptions that the protein backbone becomes more restricted upon ligand binding.

Increased side chain entropy facilitates effector binding to the signal transduction protein Cdc42Hs [321]. Cdc42Hs is a member of the Ras superfamily of GTP-binding proteins that displays a wide range of side chain flexibility. Methyl axis order parameters ranged from 0.3 ± 0.1 (highly disordered) in regions near the effector-binding site to 0.9 ± 0.1 in some helices. Upon effector binding, the majority of methyl groups showed a significant reduction in their order parameters, indicating increased entropy. Many of the methyl groups that showed increased disorder were not part of the effector-binding interface. The authors propose that increased methyl dynamics balance entropy losses as the largely unstructured effector peptide folds into an ordered structure upon binding.

Actually, there is an entire class of proteins for which ligand binding may be accompanied by destabilization of the native state [306, 307]. For example, the introduction of a Ca^{2+}-binding amino acid sequence did not affect the structure or stability of the T4 lysozyme in the absence of calcium. However, in the presence of this cation, the stability of the mutant protein was detectably less than that of wild-type T4 lysozyme. This instability suggests that the binding of Ca^{2+} might be accompanied by considerable conformational changes in the modified loop that lead to destabilization of the protein [322]. Similar effects have been described for the calcium-binding N-domain fragments of *Paramecium* calmodulin [323], rat calmodulin [324], and isolated domains of troponin C [325].

Similarly, the tertiary structure of calreticulin, a 46.8-kDa chaperone involved in the conformational maturation of glycoproteins in the lumen and endoplasmic reticulum, was distorted by Zn^{2+} binding, which resulted in a concomitant decrease in the conformational stability of this protein [326]. The zinc-induced structural perturbations and destabilization are characteristic of several calcium-binding proteins, including α-lactalbumin, parvalbumin, and recoverin [307].

Of special note are some biomedical implications of the observed destabilization upon binding. For example, β2-microglobulin is able to bind Cu^{2+}. The binding is accompanied by a significant destabilization of the protein, suggesting that the ion has a higher affinity for the unfolded form [327]. β2-microglobulin is a 12-kDa polypeptide that is necessary for the cell surface expression of the class I major histocompatibility complex (MHC). Turnover of MHC results in release of soluble β2-microglobulin, followed by its catabolism in the kidney. In patients suffering from kidney disease treated by dialysis, β2-microglobulin forms amyloid deposits principally in the joints, resulting in a variety of arthropathies. Importantly, the zinc-induced destabilization of the β2-microglobulin native structure was implicated as the driving force of this amyloidosis [327].

8.4.2
Assembly/Disassembly

Protein crystallographers are often frustrated in attempts to ply their trade because disordered N- and C-termini prevent crystallization. The idea that such sequences have been retained by nature to frustrate crystallographers, although enticing, is probably invalid since the termini have been around longer than crystallographers. Instead, disordered termini are often conserved to facilitate assembly and disassembly of complex objects, such as viruses.

The main idea of disorder-assisted assembly is that sequences from several different proteins are required for the disordered termini to fold into a defined structure that stabilizes the assembly. In some instances, only one player is disordered, while in others disordered chains must interact to gain structure. Disorder-assisted assembly accomplishes two beneficial goals. First, it ensures assembly only when all the players are in their correct positions. Second, it prevents aggregation of individual components. For example, the N-terminal 20 residues of porcine muscle lactate dehydrogenase form a disordered tail in the monomer that is essential for tetramerization [328]. Namba has reviewed several of these systems, including examples from the assembly of tobacco mosaic virus, bacterial flagella, icosahedral viruses, and DNA/RNA complexes [329].

8.4.3
Highly Entropic Chains

These are the elite of protein disorder – proteins, whole or in part, that function only when disordered. Terms such as "entropic bristles," "entropic springs," and "entropic clocks" have been used to describe these systems, but these terms can be misleading because all matter, except for perfect crystals at 0 Kelvin, has nonzero entropy values.

Hoh set down the main concept in his contribution about entropic bristle domains [330]. Disordered regions functioning as entropic bristles within a binding site will block binding until the bristle is modified. The modification causes the disordered region to move to one side of the binding site. Members of this class can be recognized by the effects of deletion. That is, removal of the bristle should lead to permanent activation [330]. The C-terminal region of p53 has been shown to function as an entropic bristle domain of this type [330, 331]. Bristles can also act as springs when two sets of disordered proteins are brought into contact with each other. The interactions of neurofilaments may be controlled in this way [332–334].

An extended unfolded region is important for the timed inactivation of some voltage-gated potassium channels [335]. The extended disordered region functions as one component of an entropic clock. Charge migrations within the tetrameric pore proteins are associated with the majority of state changes of voltage-gated K^+ ion channels [336]. However, the timing of the inactivation step is determined by the time it takes for a mobile domain to find and block the channel. The movement of the mobile domain is restricted by a tether composed of ~60 disordered residues (Figure 8.9). The timing of channel inactivation is a function of the length of the disordered tether [337]. Since ion channels serve to modulate the excitability of nerve cells, their malfunction can have substantial impact on human health.

One further example of entropic disordered region function is length adjustment within the muscle protein [338]. Please note, however, that for each of the three

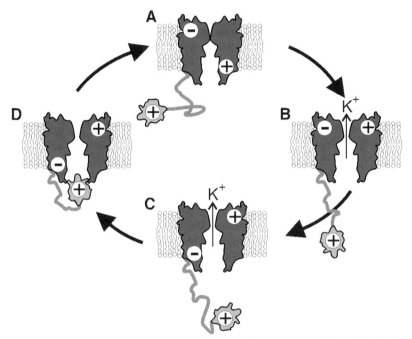

Fig. 8.9. Example of an entropic clock. Simplified model of a Shaker-type voltage-gated K^+ ion channel (dark grey) with a "ball-and-chain" timing mechanism. The ball-and-chain is comprised of an inactivation, or ball, domain (light grey) that is tethered to the pore assembly by a disordered chain of ~60 residues. For simplicity, only four of the proposed 10 states are shown [336]. The cytoplasmic side of the assembly is oriented downward. (A) Closed state prior to membrane depolarization. Note that conformational changes in the pore have sealed the channel and that a positive charge on the cytoplasmic side of the pore assembly excludes binding of the ball domain. (B) Open state following membrane depolarization. (C) After depolarization, the cytoplasmic side of the pore opening assumes a negative charge that facilitates interaction with the positively charged ball domain. (D) Inactivation of the channel occurs when the ball domain occludes the pore. The transition from (C) to (D) does not involve charge migration and can be modeled as a random walk of the ball domain towards the pore opening. (Portions of the figure are based on Antz et al. [396]).

examples described above, there are counter examples from related systems or proteins where the region either is absent or is replaced by a globular domain.

8.4.4
Protein Modification

As discussed above, protease digestion occurs preferentially in unfolded regions of proteins. The need to protect backbone hydrogen bonds in folded structure [224] and the need for extensive contacts with the backbone residues to bring about hydrolysis [222] are mutually exclusive. This can explain the observations that there is a very strong preference for protease-sensitive regions to be located in disordered regions [339, 340]. It is much less clear whether protein modification involving side chains would occur preferentially in ordered or disordered regions.

In a study of the functions associated with more than 100 long disordered regions, many were found to contain sites of protein modification [30, 31]. These modifications included phosphorylation, acetylation, fatty acylation, methylation, glycosylation, ubiquitination, and ADP ribosylation. These observations suggest the possibility that protein modifications commonly occur in regions of disorder.

Phosphorylation by the kinases and dephosphorylation by the phosphatases provide an extremely important signaling system for eukaryotic cells, with an estimate that up to one-third of eukaryotic proteins are phosphorylated [341]. As mentioned above, many sites of protein phosphorylation were found to be in regions structurally characterized as natively disordered [30, 31]. Thus, further study of the relationship between phosphorylation and disorder seems appropriate.

Several lines of evidence support the view that protein phosphorylation in eukaryotic cells occurs primarily in regions of disorder: (1) despite the very high interest in phosphorylation, very few structures in PDB exist for both the unphosphorylated and phosphorylated forms of the same protein [342, 343] (a possible explanation is that the prevalence of disorder in proteins that become phosphorylated tends to inhibit their crystallization [343, 344]); (2) nine structures of eukaryotic kinase substrates in their unphosphorylated forms show that the residues of the phosphorylation site have extended, irregular conformation that are consistent with disordered structure [343]; (3) the structures of substrate or inhibitor polypeptides indicate that the residues corresponding to the sites of phosphorylation are within segments that lack secondary structure and that are held in place not only by side chain burial but also by backbone hydrogen bonds to the surrounding kinase side chains [345–350] (just as for protease digestion, this is a strong indicator that the substrate must be locally unfolded before binding to its enzyme partner); (4) in a database of more than 1500 well-characterized sites of phosphorylation and a larger number of sites that are not phosphorylated, the residues flanking the sites of phosphorylation are substantially and systematically enriched in the same amino acids that promote protein disorder and are depleted in the amino acids that promote protein order [343]; (5) the sequence complexity distribution of the residues flanking phosphorylation sites matches almost exactly the complexity distribution obtained for a collection of experimentally characterized regions of disorder, while the complexity distribution of the residues flanking non-phosphorylation sites matches almost exactly the complexity distribution obtained for a collection of ordered proteins. In addition, there is a high correspondence between the prediction of disorder and the occurrence of phosphorylation and, conversely, the prediction of order and the lack of phosphorylation (unpublished observations). This is expected from the amino acid compositions of the residues flanking phosphorylation and non-phosphorylation sites. A new predictor of phosphorylation exhibited small but significant improvement if predictions of order and disorder were added [343]. These observations support the suggestion that sites of protein phosphorylation occur preferentially in regions of native disorder.

Data support the suggestion that protease digestion and phosphorylation both occur preferentially within regions of disorder. Also, several other types of protein modification, such as acetylation, fatty acid acylation, ubiquitination, and methylation, have also been observed to occur in regions of intrinsic disorder [30, 31]. From these findings, it is tempting to suggest that sites of protein modification in eukaryotic cells universally exhibit a preference for natively disordered regions.

What might be the basis of a preference for locating sites of modification within regions of native disorder? For all of the examples discussed above, the modifying enzyme has to bind to and modify similar sites in a wide variety of proteins. If all the regions flanking these sites are disordered before binding to the modifying enzyme, it is easy to understand how a single enzyme could bind to and modify a wide variety of protein targets. If instead all of these regions bind as ordered structures, then there is the complicating feature that the proteins being regulated by the modification must all adopt the same local structure at the site of modification. This imposes significant constraints on the site of modification. The structural simplification that arises from locating the sites of modification within regions of disorder is herein proposed to be an important principle. Elsewhere we point out that a particular advantage of disorder for regulatory and signaling regions is that changes such as protein modification lead to large-scale disorder-to-order structural transitions: such large-scale structural changes are not subtle and so could be an advantage for signaling and regulation as compared to the much smaller changes that would be expected from the decoration of an ordered protein structure (see Section 8.1.3).

8.5
Importance of Disorder for Protein Folding

Interest in the unfolded protein state has increased markedly in recent years. A major motivation has been to better understand the structural transitions that occur as a protein acquires 3-D structure, both from the point of view of the mechanism of folding and from the point of view of the energetics. A major effort has been to connect structural models of the unfolded polypeptide chain with experimental data supporting the given model. The overall sizes of guanidinium chloride-denatured proteins fit the values expected for random coils when excluded volume effects were taken into account [351]. For a true random coil, the Φ and Ψ angles of a given dipeptide are independent of the angles of the dipeptides before and after. This is often called the Flory isolated-pair hypothesis [352]. Both experiments and calculations call the isolated-pair hypothesis into question [353]. For example, using primarily repulsive terms, Pappu et al. [354] computed the effects of steric clash (or excluded volume) on blocked polyalanines of various lengths. Contrary to Flory's isolated hypothesis, they found that excluded volume effects were sufficient to lead to preferred backbone conformations, with the polyproline II helix being among the most preferred. Representing unfolded proteins as polymer chain models, which have the advantage of simplifications that result from ignoring most atomic details, is proving useful for modeling experimental data pertaining to protein stability, folding, and interactions [355].

While there has been substantial emphasis on the study of unfolded proteins as precursors to 3-D structures, as briefly described above, to our knowledge there has been no systematic discussion or studies of how natively disordered regions affect protein 3-D structure and folding kinetics. Here we will consider several possibilities, each of which is a simple hypothesis that has not yet been tested by experiment. These hypotheses form the basis for experiments into the effects of native disorder on protein structure and folding.

First, a well-known example with ample experimental support is provided. Trypsinogen folds into a stable 3-D structure, but compared to trypsin, the folding is incomplete and the protein is inactive. Trypsinogen's folding does involve the formation of the crucial catalytic triad, and yet trypsinogen remains inactive evidently because the binding pocket for the substrate lysine or arginine fails to form completely, e.g., remains natively disordered [12, 356]. Thus, trypsinogen, the precursor to trypsin, remains inactive and thus does not harm the interior of the cell. The folding into active trypsin is inhibited by a short region of native disorder at the amino terminus; this region is completely missing in the electron density map of trypsinogen [357]. Once trypsinogen has been exported from the cell, this disordered region is cleaved by trypsin. The amino terminus changes from a highly charged moiety into a hydrophobic terminus of isoleucine followed by valine (IV). In the absence of the natively disordered, charged extension, this IV terminus becomes capable of binding into a particular site elsewhere in the structure. This binding of the IV moieties brings about a disorder-to-order transition of trypsin's binding pocket, now enabling the protein to bind arginine or lysine and thereby converting inactive trypsinogen into active trypsin [358]. Even in the absence of the cleavage, high levels of an IV dipeptide can stimulate the protease activity of uncleaved trypsinogen by binding into the site used by the IV terminus [12]. We speculate that proteolytic removal of a natively disordered region may be a common mechanism for regulating protein folding and function, but we have not yet searched systematically for other examples.

While trypsinogen provides an example of a protein for which a region of native disorder inhibits protein folding, we speculate that there are proteins for which regions of native disorder promote protein folding. Suppose, for example, that a protein contained a short, highly charged region of native disorder and a folded domain with an opposite charge of significant magnitude. From various analyses suggesting that a high net charge can lead to a natively unfolded protein [123, 227, 359], the region of oppositely charged disorder might be essential for overall charge balance and, if so, would be required for protein folding. A simple experiment is proposed here: (1) identify example proteins in the PDB with short, highly charged tails and oppositely charged, folded domains; (2) determine which domains are likely not to fold without their oppositely charged tail by means of the net charge versus hydropathy plot [227]; (3) remove the charged tail by proteolysis or genetic engineering; and (4) compare the folding rate of the shorter protein with that of the full-length protein. The prediction is that, if located in the unfolded region of the net charge hydropathy plot, the shorter protein would fail to fold or would require higher ionic strength to fold as compared to the full-length protein.

Various theoretical analyses on protein folding have suggested relationships among the size, stability, and topology of a protein fold and the rate and mechanism by which the fold is achieved. The characterization of the folding of a large number of simple, single-domain proteins enabled detailed studies to test the various models and assertions regarding the mechanism of protein folding. The simple proteins in this set are characterized by having single domains; by the lack of prosthetic groups, disulfide bonds, and *cis* proline residues; and by two-state refolding kinetics. Despite the simplicity of this set, more than a million-fold variation in refolding rates was observed. A remarkable finding was that the statistically significant correlations among the folding rate, the transition state placement, and the relative contact order are observed, where the transition state placement is an estimate of the fraction of the burial of the hydrophobic surface area in the transition state as compared to the native state, and where the relative contact order is the length-normalized sequence separation between contacting residues in the native state [360]. A number of alternative empirical measures of topology were later shown to correlate about equally well with folding rates as does the relative contact order, including local secondary structure content [406], the number of sequence-distant contacts per residue [361], the fraction of contacts that are distant in the sequence [362], and the total contact distance [363]. The quality of these alternative measures supports the conjecture that contact order predicts rates, not because it is directly related to the mechanism of folding, but rather because it is related to an alternative physical parameter of importance [364].

Attempts to reconcile the observed relationship between contact order or other measures of topology and folding rates has led to the topomer search model [365], which was based substantially on similar prior models [366, 367]. Simply put, the topomer search model stipulates that the rate-limiting step in protein folding is the search for an unfolded conformation with the correct overall topology. The unfolded form with the correct topology then rapidly folds into the native state [364].

Assuming for the sake of discussion the basic correctness of the topomer search model, the expected effects of native disorder can be readily described. First let us consider natively disordered regions at the amino or carboxy terminus that do not affect the overall protein stability and that do not stabilize misfolded intermediates.

Such natively disordered regions would be expected to have very small effects on the folding rates. On the other hand, natively disordered internal loops would be expected to slow the rate of folding in a length-dependent manner. Indeed, a systematic study of the folding rates of simple, two-state proteins with natively disordered loops of varying lengths could provide a useful test of the topomer search model.

The relationship between the log of the folding rates of simple two-state proteins and the contact order value exhibits an r^2 value of about 0.8, suggesting that this measure captures in excess of 3/4 of the variance in the logarithms of the reported folding rates [364]. The topomer search model, which evidently captures the dependence of protein folding on the contact order, ignores variation in foldability along the sequence. On the other hand, predictors of order and disorder capture local sequence tendencies for order or disorder. Such a local tendency for disorder could be overridden by non-local interactions in the final native structure, and so a local tendency for disorder could be important for a folded protein even if the native structure does not exhibit actual disorder. We wonder whether there is any relationship between protein folding rates and variations in order/disorder tendencies along an amino acid sequence. That is, we wonder whether the 1/4 variability in the logarithms of the folding rate that is not captured by relative contact order could be related to differences in the amounts or in the organization of disorder tendencies along the amino acid sequences. Local regions with high tendencies for disorder could, for example, lead to very non-uniform polymer chain models by local alterations in the stiffness or persistence length; locating such anomalies at topologically critical sites could greatly speed up or slow down protein folding rates. These ideas could be tested both computationally and experimentally.

8.6
Experimental Protocols

In this section we include several protocols for NMR spectroscopy, X-ray crystallography, and circular dichroism spectropolarimetry. We focus on these methods because they are traditionally the most frequently used techniques to characterize native disorder in proteins. Protocols for small angle X-ray diffraction, hydrodynamic measurement, fluorescence methods, conformational stability, mass spectrometry-based high-resolution hydrogen-deuterium exchange, protease sensitivity, and prediction from sequence are available from the references cited in the above methods section.

8.6.1
NMR Spectroscopy

8.6.1.1 General Requirements
The following methods assume that the experimenter has access to a high field digital nuclear magnetic resonance spectrometer and a soluble, homogeneous pro-

tein sample at a concentration of ∼1 mM. These methods also assume that the incorporation of an NMR-active isotope such as ^{15}N and/or ^{13}C is possible.

8.6.1.2 Measuring Transient Secondary Structure in Secondary Chemical Shifts

Resonance assignments are the first step in the analysis of protein structure using NMR. A convenient outcome of these measurements is the determination of protein secondary structure. The C_α, C_β, and C' chemical shifts are the most sensitive to phi and psi angles. To measure these chemical shifts, sensitivity-enhanced HNCACB and HNCO experiments can be performed on uniformly ^{15}N- and ^{13}C-labeled protein samples [368–370]. An appropriate digital resolution for the HNCACB and HNCO experiments is 9.8 Hz/pt in ^{1}H with 512 complex points, 47.1 Hz/pt in ^{13}C with 128 complex points, and, 34.4 Hz/pt in ^{15}N with 32 complex points. After transformation of the data, it is essential to apply the appropriate referencing before comparing the protein chemical shifts to random coil standards [67].

8.6.1.3 Measuring the Translational Diffusion Coefficient Using Pulsed Field Gradient Diffusion Experiments

Pulsed field gradient diffusion measurements can be made using a variation of the water-sLED sequence developed by Altieri et al. 1995 [69]. Collect data at an appropriate field strength, temperature, sweep width, gradient pulse width, and delay time. The amplitude of the gradient pulses should be varied from ∼1–31 gauss cm^{-1} in increments of 2 gauss cm^{-1}. However, this range will vary depending on the protein and typically must be determined empirically. It is also possible to perform the experiment by varying the length of the gradient pulses and holding the magnitude constant [371]. The NMR data must be processed so that resonance intensity measurements can be made. An entire region of the spectrum can be integrated, or individual resolved resonances can be measured. Resonance assignments are not necessary, the experiment can be performed on an unlabeled sample, and it is probably best to integrate the resonances of non-labile nuclei. Data should be collected over at least a four-fold reduction in signal. Resonance intensity measurements should be normalized, averaged, and fit to the function relating the normalized resonance intensity, A, to the translational diffusion coefficient, D:

$$A = \exp(-Dg^2\delta^2\gamma^2(\Delta - \delta/3)) \tag{26}$$

where g and δ are, respectively, the magnitude and duration of the gradient pulses, Δ is the time between gradient pulses, and γ is the gyromagnetic ratio of the observed nucleus [372]. Nonlinear least-squares regression of a decaying exponential function onto the data can be used to extract D. It is useful to combine these measurements with another technique such as sedimentation to obtain information about the shape of the protein.

8.6.1.4 Relaxation Experiments

For ^{15}N relaxation experiments the spin-lattice relaxation rates (R_1), spin-spin relaxation rates (R_2), and ^{1}H-^{15}N NOEs can be measured by inverse-detected two-dimensional NMR methods [81]. Spin-lattice relaxation rates are typically determined by collecting 8–12 two-dimensional spectra using relaxation delays from 10–1500 ms. Spin-spin relaxation rates can determined by collecting 8–12 two-dimensional spectra using spin-echo delays of 10–300 ms. Peak heights from each series of relaxation experiments are then fitted to a single decaying exponential. To measure the ^{1}H-^{15}N NOEs, one spectrum is acquired with a 3-s mixing time for the NOE to buildup and another spectrum is acquired with a 3-s recycle delay for a reference. It is often necessary to optimize this delay, as intrinsically disordered proteins and longer delays will attenuate larger-than-predicted negative NHNOE values for highly dynamic regions of the protein [85]. For all experiments, water suppression can be achieved by using pulsed field gradients. Uncertainties in measured peak heights are usually estimated from baseline noise level and, with good tuning and shimming, are typically less than 1% of the peak heights from the first R_1 and R_2 delay points.

8.6.1.5 Relaxation Data Analysis Using Reduced Spectral Density Mapping

Relaxation data are typically analyzed using the reduced spectral density mapping approach [80, 88, 373, 374]. The ^{15}N chemical shift anisotropy and the dipolar coupling between the amide ^{15}N nucleus and the attached proton have the greatest influence on the ^{15}N nuclear relaxation [89]:

$$R_1 = \left(\frac{d^2}{4}\right)[3J(\omega_N) + 6J(\omega_H + \omega_N) + J(\omega_H - \omega_N)] + c^2 J(\omega_N) \tag{27}$$

$$R_2 = \left(\frac{d^2}{8}\right)[4J(0) + 3J(\omega_N) + 3J(\omega_H + \omega_N) + 6J(\omega_H) + J(\omega_H - \omega_N)]$$

$$+ \left(\frac{c^2}{6}\right)[J(0) + 6J(\omega_N)] + R_{ex} \tag{28}$$

$$NOE = 1 + \left(\frac{d^2}{4R_1}\right)\left(\frac{\gamma_H}{\gamma_N}\right)[6J(\omega_H + \omega_N) - J(\omega_H - \omega_N)] \tag{29}$$

where $d = \left(\frac{\mu_0 h \gamma_H \gamma_N}{8\pi^2}\right)\langle r_{NH}^{-3}\rangle$ and $c = \frac{\omega_N \Delta\sigma}{\sqrt{3}}$, μ_0 is the permeability of free space, h is Planck's constant, γ_H and γ_N are the gyromagnetic ratios of ^{1}H and ^{15}N, respectively, r_{NH} is the amide bond length (1.02 Å), $\Delta\sigma$ is the chemical shift anisotropy (−160 ppm), and R_{ex} is the chemical exchange contribution to R_2. $J(\omega)$ is the power spectral density function defining the reorientation of the N–H bond vector by stochastic (global) and intramolecular motions. Reduced spectral density mapping uses an average value of $J(\omega_H)$ for the linear combinations of $J(\omega_H + \omega_N)$, $J(\omega_H)$, and $J(\omega_H - \omega_N)$ leading to values of $J(0)$, $J(\omega_N)$, and $J(\omega_H)$ that are given by:

$$\sigma_{NH} = R_1(NOE - 1)\frac{\gamma_N}{\gamma_H} \tag{30}$$

$$J(\omega_H) = \frac{4\sigma_{NH}}{5d^2} \tag{31}$$

$$J(\omega_N) = \frac{[4R_1 - 5\sigma_{NH}]}{[3d^2 + 4c^2]} \tag{32}$$

$$J(0) = \frac{[6R_2 - 3R_1 - 2.72\sigma_{NH}]}{[3d^2 + 4c^2]} \tag{33}$$

This approach estimates the magnitude of the spectral density function at the given frequencies, making no assumptions about the form of the spectral density function or about the molecular behavior giving rise to the relaxation.

8.6.1.6 In-cell NMR

Although a potentially powerful technique, there are few published studies involving in-cell NMR [102, 299–301, 375]. Therefore, the suggestions here will have to be quite general. Most of the following suggestions are distilled from publications by Volker Dötsch's group, and a few are from the Pielak laboratory.

Most importantly, protein expression is, at present, more of an art than a science; the suitability of each protein expression system (i.e., inducer concentration, induction time, cell density at induction) must be determined empirically. Although yeast and insect cells have been tried, the technique currently works best for proteins expressed in *E. coli*. Overexpression of the protein to be studied in ^{15}N- or ^{13}C-enriched media is essential. Overexpression is operationally defined as the ability to isolate greater than 10 mg of pure protein from 1 L of saturated culture, using the same media and conditions that will be used for the in-cell NMR experiment. It is best to check expression using unenriched media first. The HSQC experiment works well for in-cell NMR. It is important to know what the "background" spectrum looks like when the experiment fails. Background spectra contain cross-peaks from metabolites that become enriched in ^{15}N. That is, a spectrum without protein overexpression (i.e., untransformed cells) should be collected. It can often take as long as 15 h to obtain a spectrum with a conventional probe. Using a cryoprobe can dramatically decrease this time. When the experiment is over, it is important to perform dilution and plating experiments to show that the cells are still alive and that the protein of interest has remained inside the cells. Gentle centrifugation of the NMR sample followed by SDS-PAGE analysis of both the supernatant and the pellet is the best way to show that the protein was overexpressed and did not leave the cells.

8.6.2
X-ray Crystallography

Determination of protein structure by X-ray crystallography has become almost routine once crystals are obtained. Many excellent books and papers provide the details of protein structure determination by X-ray crystallography. Two books that emphasize practical experimental approaches are by Stout and Jensen [376] and by McRee [377]. In this section, we will concentrate on how to obtain crystals in the face of native disorder.

The presence of large regions of disorder can block attempts to crystallize proteins. Failure to obtain crystals is the single greatest experimental problem in X-ray crystallography. Here, we present calsequestrin as a case study illustrating the successful crystallization of a natively disordered protein.

Calsequestrin is a calcium-storage protein located within the sarcoplasmic reticulum that binds 40–50 calcium ions with ∼1 mM affinity. Calsequestrin is a highly acidic protein with many of its acidic residues located in clusters. The physiochemical properties of purified calsequestrin as revealed by tryptophan fluorescence [378–381], circular dichroism [378, 380–383], Raman spectroscopy [380, 384], NMR [380], and proteolytic digestion [381, 384, 385] indicated that calsequestrin was mostly unfolded at low ionic strength. As ionic strength was increased, folding into a compact structure was observed. This structure can be induced by calcium as well as by other ions such as Na^+, Zn^{2+}, Sr^{2+}, Tb^{2+}, K^+, and H^+. The high Ca^{2+}-binding capacity of calsequestrin was believed to require the formation of aggregates, thus transitioning from a soluble disordered form to a solid crystalline form [381, 386, 387]. However, crystallization attempts in the presence of Ca^{2+} resulted in needle-like crystals. These liner polymeric forms were also observed in vivo, suggesting that this was the physiologically relevant form of the complex [388]. These narrow crystals were unsuitable for structure determination but did demonstrate that crystallization of calsequestrin was possible. The facts that calsequestrin adopted structure in the presence of mono- and divalent cations other than calcium and that these forms were not observed to aggregate and precipitate as the needle-like crystals did suggested that alternate crystal forms could be possible.

The following hypothesis was formed: the growth of needle-like crystals of calsequestrin is a two-step process [389]. The first step was the induction of structure by high ionic strength. The second was the calcium-specific cross-bridging of individual calsequestrin molecules by means of the unneutralized charges remaining after initial nonspecific binding of ions by the monomers. This cross-bridging could account for growth in a single direction, thereby producing the needles. To test this hypothesis, crystallization in the absence of Ca^{2+} was attempted [389, 390].

Calsequestrin was purified from the skeletal muscle of New Zealand white rabbits as previously described [381]. Approximately 500 initial crystallization experiments were conducted using the hanging-drop vapor-diffusion method [391] with an incomplete factorial approach [392] to cover a wide range of conditions. Each condition was tested in a volume of 1 μL containing 5–10 μg of calsequestrin. Conditions with high monovalent cation concentrations were emphasized, along with a range of 16 different precipitation reagents. Good-sized non-needle crystals were obtained when 2-methyl-2,4 pentane diol (MPD) was used as the precipitating reagent. The best crystals were grown from a solution containing 10% (v/v) MPD, 0.1 M sodium citrate, 0.05 M sodium cacodylate, pH 6.5, and 5 mg mL^{-1} calsequestrin. The nominal initial Na^+ concentration was 0.35 M. The rectangular crystals formed within one week and grew to $0.2 \times 0.2 \times 0.8$ mm by the second week. Structure determination details can be found elsewhere [389, 390].

Surprisingly, the resulting structure of calsequestrin exhibited three identical domains, each with a thioredoxin protein fold. The three domains interact to form a disk-like shape with an approximate radius of 32 Å and a thickness of 35 Å. No clues to this three-domain structure were obtained from sequence analysis [393, 394]. Rather, analysis pointed towards a globular N-terminal domain and a C-terminal disordered region [381]. Even in hindsight, no significant similarities among the three similar domains could be deduced from the sequence.

A common approach to disordered regions in proteins is to remove the coding regions for the disordered parts from the recombinant expression constructs so that these regions do not prevent the protein from crystallizing [344]. In the case of calsequestrin, the fact that structure could be induced in the disordered protein by increasing cation concentration led to attempts at crystallization under non-intuitive conditions. The result was the elucidation of an interesting and important structure.

8.6.3
Circular Dichroism Spectropolarimetry

There is an excellent book on protein circular dichroism that contains several sections on the collection and interpretation of spectra [137]. As discussed above (see Section 8.2.4), the interpretation of spectra from disordered proteins remains controversial. However, the two most important experimental concepts are well agreed upon.

First, collect data as far into the ultraviolet region as possible. The far-UV region, from 260 nm to 190 nm, contains a great deal of information about secondary structure. With a powerful UV source (which often means a new lamp) and a well-behaved sample, data can be obtained down to wavelengths below 190 nm. Second, only the electronic transitions of the protein being studied should contribute to the absorbance. Any extrinsic absorbance degrades the instrument's ability to detect the small differences between left and right circularly polarized light absorbed by the protein under study.

The two most common sources of unwanted absorbance are light scattering and buffer/co-solute absorbance. The protein must not aggregate to such an extent that the sample scatters light. It is important to pass every sample through a 0.2-micron filter prior to data acquisition. A common mistake is to use a buffer that absorbs in the ultraviolet region. Histidine is a common buffer because it is used to elute His-tagged proteins from Ni^{2+}-affinity columns, but at the concentrations used for elution, the histidine contributes to an excessive amount of ultraviolet absorbance. Tris is also a poor choice. Phosphate and acetate are more useful buffers, at least in terms of absorbance. Absorbance can also be a problem when collecting spectra at high co-solute (e.g., sugars, urea, etc.) concentrations.

The best way to proceed is to collect a spectrum of filtered water or buffer and then compare this spectrum to that of the solution. After choosing appropriate solution conditions, there is no better protocol for preparing the sample than thorough dialysis followed by filtration. The reference solution is then made from a filtered sample of the solution from outside the dialysis bag.

CD analysis is especially sensitive to error due to misestimation of protein concentration. It is important that extreme care be taken during concentration determination. Additionally, we prefer the method of Gill and von Hippel [395] rather than any of the popular colorimetric assays. This method is based on the calculation of a molar extinction coefficient based on the amino acid content of the protein under study and is typically accurate to ±5%.

Acknowledgements

A.K.D. and M.S.C. are supported by NIH grant R01 LM 007688 (Indiana University School of Medicine). A.K.D. is additionally supported by NIH grants R43 CA099053, R43 CA 097629, R43 CA 097568, and R43 GM 066412 (Molecular Kinetics, Inc.). G.J.P. thanks the NSF and NIH for support and Trevor Creamer for helpful discussions. G.W.D. is supported by NIH grant P20 RR 16448 from the COBRE Program of the National Center for Research Resources.

References

1 FISHER, E. (1894). Einfluss der configuration auf die wirkung den enzyme. *Ber. Dtsch. Chem. Ges.* **27**, 2985–2993.
2 ANSON, M. L. & MIRSKY, A. E. (1925). On some general properties of proteins. *J. Gen. Physiol.* **9**, 169–179.
3 EDSALL, J. T. (1995). Hsien Wu and the first theory of protein denaturation (1931). *Adv. Protein Chem.* **46**, 1–5.
4 MIRSKY, A. E. & PAULING, L. (1936). On the structure of native, denatured and coagulated proteins. *Proc. Natl. Acad. Sci. U.S.A.* **22**, 439–447.
5 ANFINSEN, C. B. (1973). Principles that govern the folding of protein chains. *Science* **181**, 223–230.
6 GUTTE, B. & MERRIFIELD, R. B. (1969). The total synthesis of an enzyme with ribonuclease A activity. *J. Am. Chem. Soc.* **91**, 501–2.
7 PAULING, L., COREY, R. B. & BRANSON, R. H. (1951). The structure of proteins: two hydrogen-bonded configurations of the polypeptide chain. *Proc. Natl. Acad. Sci. U.S.A.* **37**, 205–210.
8 KENDREW, J. C., DICKERSON, R. E. & STRANDBERG, B. E. (1960). Structure of myoglobin: a three-dimensional Fourier synthesis at 2 Å resolution. *Nature* **206**, 757–763.
9 PERUTZ, M. F., ROSSMANN, M. P., CULLIS, A. F., MUIRHEAD, H., WILL, G. & NORTH, A. C. (1960). Structure of haemoglobin: a three-dimensional Fourier synthesis at 5.5 Å resolution, obtained by X-ray analysis. *Nature* **185**, 416–422.
10 BLAKE, C. C., KOENIG, D. F., MAIR, G. A., NORTH, A. C., PHILLIPS, D. C. & SARMA, V. R. (1965). Structure of hen egg-white lysozyme. A three-dimensional Fourier synthesis at 2 Ångstrøm resolution. *Nature* **206**, 757–761.
11 BLOOMER, A. C., CHAMPNESS, J. N., BRICOGNE, G., STADEN, R. & KLUG, A. (1978). Protein disk of tobacco mosiac virus at 2.8 Å resolution showing the interactions within and between subunits. *Nature* **276**, 362–368.
12 BODE, W., SCHWAGER, P. & HUBER, R. (1978). The transition of bovine trypsinogen to a trypsin-like state upon strong ligand binding. The refined crystal structures of the bovine trypsinogen-pancreatic trypsin inhibitor complex and of its ternary complex with Ile-Val at 1.9 Å resolution. *J. Mol. Biol.* **118**, 99–112.
13 WILLIAMS, R. J. (1978). The conformational mobility of proteins and its

14 PULLEN, R. A., JENKINS, J. A., TICKLE, I. J., WOOD, S. P. & BLUNDELL, T. L. (1975). The relation of polypeptide hormone structure and flexibility to receptor binding: the relevance of X-ray studies on insulins, glucagon and human placental lactogen. *Mol. Cell. Biochem.* **8**, 5–20.

15 CARY, P. D., MOSS, T. & BRADBURY, E. M. (1978). High-resolution proton-magnetic-resonance studies of chromatin core particles. *Eur. J. Biochem.* **89**, 475–82.

16 LINDERSTRØM-LANG, K. U. & SCHELLMAN, J. A. (1959). Protein structure and enzyme activity. In *The Enzymes* (BOYER, P. D., LARDY, H. & MYRBACK, K., eds.), Vol. 1, pp. 443–510. Academic Press, New York.

17 SCHWEERS, O., SCHÖNBRUNN-HANEBECK, E., MARX, A. & MANDELKOW, E. (1994). Structural studies of tau protein and Alzheimer paired helical filaments show no evidence for β-structure. *J. Biol. Chem.* **269**, 24290–24297.

18 WEINREB, P. H., ZHEN, W., POON, A. W., CONWAY, K. A. & LANSBURY, P. T., JR. (1996). NACP, a protein implicated in Alzheimer's disease and learning, is natively unfolded. *Biochemistry* **35**, 13709–13715.

19 WRIGHT, P. E. & DYSON, H. J. (1999). Intrinsically unstructured proteins: reassessing the protein structure-function paradigm. *J. Mol. Biol.* **293**, 321–331.

20 DUNKER, A. K., LAWSON, J. D., BROWN, C. J., WILLIAMS, R. M., ROMERO, P., OH, J. S., OLDFIELD, C. J., CAMPEN, A. M., RATLIFF, C. M., HIPPS, K. W., AUSIO, J., NISSEN, M. S., REEVES, R., KANG, C., KISSINGER, C. R., BAILEY, R. W., GRISWOLD, M. D., CHIU, W., GARNER, E. C. & OBRADOVIC, Z. (2001). Intrinsically disordered protein. *J. Mol. Graph. Model.* **19**, 26–59.

21 BAILEY, R. W., DUNKER, A. K., BROWN, C. J., GARNER, E. C. & GRISWOLD, M. D. (2001). Clusterin: a binding protein with a molten globule-like region. *Biochemistry* **40**, 11828–11840.

22 KUWAJIMA, K. (1989). The molten globule state as a clue for understanding the folding and cooperativity of globular-protein structure. *Proteins* **6**, 87–103.

23 DOLGIKH, D. A., GILMANSHIN, R. I., BRAZHNIKOV, E. V., BYCHKOVA, V. E., SEMISOTNOV, G. V., VENYAMINOV, S. & PTITSYN, O. B. (1981). Alpha-lactalbumin: compact state with fluctuating tertiary structure? *FEBS Lett.* **136**, 311–315.

24 TIFFANY, M. L. & KRIMM, S. (1968). New chain conformations of poly(glutamic acid) and polylysine. *Biopolymers* **6**, 1379–82.

25 SHI, Z., WOODY, R. W. & KALLENBACH, N. R. (2002). Is polyproline II a major backbone conformation in unfolded proteins? *Adv Protein Chem* **62**, 163–240.

26 CREAMER, T. P. & CAMPBELL, M. N. (2002). Determinants of the polyproline II helix from modeling studies. *Adv. Protein Chem.* **62**, 263–82.

27 KLEE, C. B., CROUCH, T. H. & KRINKS, M. H. (1979). Calcineurin: a calcium- and calmodulin-binding protein of the nervous system. *Proc. Natl. Acad. Sci. U.S.A.* **76**, 6270–6273.

28 KLEE, C. B., DRAETTA, G. F. & HUBBARD, M. J. (1988). Calcineurin. *Adv. Enzymol. Relat. Areas Mol. Biol.* **61**, 149–200.

29 KISSINGER, C. R., PARGE, H. E., KNIGHTON, D. R., LEWIS, C. T., PELLETIER, L. A., TEMPCZYK, A., KALISH, V. J., TUCKER, K. D., SHOWALTER, R. E., MOOMAW, E. W., GASTINEL, L. N., HABUKA, N., CHEN, X., MALDONADO, F., BARKER, J. E., BACQUET, R. & VILLAFRANCA, J. E. (1995). Crystal structures of human calcineurin and the human FKBP12-FK506-calcineurin complex. *Nature* **378**, 641–644.

30 DUNKER, A. K., BROWN, C. J., LAWSON, J. D., IAKOUCHEVA, L. M. & OBRADOVIC, Z. (2002). Intrinsic disorder and protein function. *Biochemistry* **41**, 6573–82.

31 DUNKER, A. K., BROWN, C. J. & OBRADOVIC, Z. (2002). Identification and functions of usefully disordered proteins. *Adv. Protein Chem.* **62**, 25–49.

32 PTITSYN, O. (1995). Molten globule and protein folding. *Adv. Protein Chem.* **47**, 83–229.

33 PAGE, M. I. (1987). Enzyme Mechanisms. In *Enzyme Mechanisms* (PAGE, M. I. & WILLIAMS, A., eds.), pp. 1–13. Royal Society of Chemistry, London.

34 KRAUT, J. (1988). How do enzymes work? *Science* **242**, 533–40.

35 POLANYHI, M. (1921). Über Adsorptionskatalyse. *Zeitschrift für Electrochemie* **27**, 142–150.

36 PAULING, L. (1946). Molecular architecture and biological reactions. *Chemical Engineering News* **24**, 1375–1377.

37 BORMAN, S. (2004). Much ado about enzyme mechanisms. *Chemical and Engineering News* **82**, 35–39.

38 WARSHEL, A. (1998). Electrostatic origin of the catalytic power of enzymes and the role of preorganized active sites. *J Biol Chem* **273**, 27035–27038.

39 BRUICE, T. C. & BENKOVIC, S. J. (2000). Chemical basis for enzyme catalysis. *Biochemistry* **39**, 6267–74.

40 HUR, S. & BRUICE, T. C. (2003). Just a near attack conformer for catalysis (chorismate to prephenate rearrangements in water, antibody, enzymes, and their mutants). *J Am Chem Soc* **125**, 10540–2.

41 SCHULZ, G. E. (1979). Nucleotide Binding Proteins. In *Molecular Mechanism of Biological Recognition* (BALABAN, M., ed.), pp. 79–94. Elsevier/North-Holland Biomedical Press, New York.

42 KARUSH, F. (1950). Heterogeneity of the binding sites of bovine serum albumin. *J. Am. Chem. Soc.* **72**, 2705–2713.

43 KRIWACKI, R. W., HENGST, L., TENNANT, L., REED, S. I. & WRIGHT, P. E. (1996). Structural studies of p21$^{Waf1/Cip1/Sdi1}$ in the free and Cdk2-bound state: conformational disorder mediates binding diversity. *Proc. Natl. Acad. Sci. U.S.A.* **93**, 11504–11509.

44 DUNKER, A. K. & OBRADOVIC, Z. (2001). The protein trinity – linking function and disorder. *Nat. Biotechnol.* **19**, 805–806.

45 BRACKEN, C. (2001). NMR spin relaxation methods for characterization of disorder and folding in proteins. *J. Mol. Graph. Model.* **19**, 3–12.

46 DYSON, H. J. & WRIGHT, P. E. (1998). Equilibrium NMR studies of unfolded and partially folded proteins. *Nat. Struct. Biol.* **5 Suppl**, 499–503.

47 DYSON, H. J. & WRIGHT, P. E. (2001). Nuclear magnetic resonance methods for elucidation of structure and dynamics in disordered states. *Methods Enzymol.* **339**, 258–270.

48 DYSON, H. J. & WRIGHT, P. E. (2002). Insights into the structure and dynamics of unfolded proteins from nuclear magnetic resonance. *Adv. Protein Chem.* **62**, 311–40.

49 BARBAR, E. (1999). NMR characterization of partially folded and unfolded conformational ensembles of proteins. *Biopolymers* **51**, 191–207.

50 YAO, J., CHUNG, J., ELIEZER, D., WRIGHT, P. E. & DYSON, H. J. (2001). NMR structural and dynamic characterization of the acid-unfolded state of apomyoglobin provides insights into the early events in protein folding. *Biochemistry* **40**, 3561–71.

51 ELIEZER, D., YAO, J., DYSON, H. J. & WRIGHT, P. E. (1998). Structural and dynamic characterization of partially folded states of apomyoglobin and implications for protein folding. *Nat. Struct. Biol.* **5**, 148–155.

52 ARCUS, V. L., VUILLEUMIER, S., FREUND, S. M., BYCROFT, M. & FERSHT, A. R. (1994). Toward solving the folding pathway of barnase: the complete backbone 13C, 15N, and 1H NMR assignments of its pH-denatured state. *Proc. Natl. Acad. Sci. U.S.A.* **91**, 9412–6.

53 ARCUS, V. L., VUILLEUMIER, S., FREUND, S. M., BYCROFT, M. & FERSHT, A. R. (1995). A comparison of the pH, urea, and temperature-denatured states of barnase by heteronuclear NMR: implications for the initiation of protein folding. *J. Mol. Biol.* **254**, 305–21.

54 WEISS, M. A., ELLENBERGER, T., WOBBE, C. R., LEE, J. P., HARRISON, S. C. & STRUHL, K. (1990). Folding transition in the DNA-binding domain

of GCN4 on specific binding to DNA. *Nature* **347**, 575–578.

55 DOBSON, C. & HORE, P. (1998). Kinetic studies of protein folding using NMR spectroscopy. *Nat. Struct. Biol.* **5**, 504–507.

56 SHORTLE, D. R. (1996). Structural analysis of non-native states of proteins by NMR methods. *Curr. Opin. Struct. Biol.* **6**, 24–30.

57 ALEXANDRESCU, A. T. & SHORTLE, D. (1994). Backbone dynamics of a highly disordered 131 residue fragment of staphylococcal nuclease. *J. Mol. Biol.* **242**, 527–46.

58 ALEXANDRESCU, A. T., ABEYGUNA-WARDANA, C. & SHORTLE, D. (1994). Structure and dynamics of a denatured 131-residue fragment of staphylococcal nuclease: a heteronuclear NMR study. *Biochemistry* **33**, 1063–1072.

59 GILLESPIE, J. R. & SHORTLE, D. (1997). Characterization of long-range structure in the denatured state of staphylococcal nuclease. II. Distance restraints from paramagnetic relaxation and calculation of an ensemble of structures. *J. Mol. Biol.* **268**, 170–84.

60 GILLESPIE, J. R. & SHORTLE, D. (1997a). Characterization of long-range structure in the denatured state of *staphylococcal* nuclease. I. Paramagnetic relaxation enhancement by nitroxide spin labels. *J. Mol. Biol.* **268**, 158–169.

61 SHORTLE, D. & ACKERMAN, M. S. (2001). Persistence of native-like topology in a denatured protein in 8 M urea. *Science* **293**, 487–9.

62 SINCLAIR, J. F. & SHORTLE, D. (1999). Analysis of long-range interactions in a model denatured state of staphylococcal nuclease based on correlated changes in backbone dynamics. *Protein Sci.* **8**, 991–1000.

63 WRABL, J., SHORTLE, D. & WOOLF, T. (2000). Correlation between changes in nuclear magnetic resonance order parameters and conformational entropy: molecular dynamics simulations of native and denatured staphylococcal nuclease. *Proteins* **38**, 123–133.

64 WRABL, J. O. & SHORTLE, D. (1996). Perturbations of the denatured state ensemble: modeling their effects on protein stability and folding kinetics. *Protein Sci.* **5**, 2343–52.

65 WISHART, D. S., SYKES, B. D. & RICHARDS, F. M. (1992). The chemical shift index: a fast and simple method for the assignment of protein secondary structure through NMR spectroscopy. *Biochemistry* **31**, 1647–51.

66 WISHART, D. S. & SYKES, B. D. (1994). The 13C chemical-shift index: a simple method for the identification of protein secondary structure using 13C chemical-shift data. *J. Biomol. NMR* **4**, 171–80.

67 WISHART, D. S. & SYKES, B. D. (1994). Chemical shifts as a tool for structure determination. *Methods Enzymol.* **239**, 363–92.

68 WILKINS, D. K., GRIMSHAW, S. B., RECEVEUR, V., DOBSON, C. M., JONES, J. A. & SMITH, L. J. (1999). Hydrodynamic radii of native and denatured proteins measured by pulse field gradient NMR techniques. *Biochemistry* **38**, 16424–31.

69 ALTIERI, A. S., HINTON, D. P. & BYRD, R. A. (1995). Association of biomolecular systems via pulsed field gradient NMR self-diffusion measurements. *J. Am. Chem. Soc.* **117**, 7566–67.

70 PAN, H., BARANY, G. & WOODWARD, C. (1997). Reduced BPTI is collapsed. A pulsed field gradient NMR study of unfolded and partially folded bovine pancreatic trypsin inhibitor. *Protein Sci.* **6**, 1985–92.

71 DAVIS, D. G., PERLMAN, M. E. & LONDON, R. E. (1994). Direct measurements of the dissociation-rate constant for inhibitor-enzyme complexes via the T1 rho and T2 (CPMG) methods. *J. Magn. Reson. B* **104**, 266–75.

72 FARROW, N. A., ZHANG, O., FORMAN-KAY, J. D. & KAY, L. E. (1994). A heteronuclear correlation experiment for simultaneous determination of 15N longitudinal decay and chemical exchange rates of systems in slow equilibrium. *J. Biomol. NMR* **4**, 727–34.

73 FEENEY, J., BATCHELOR, J. G., ALBRAND, J. P. & ROBERTS, G. C. K. (1979). Effects of intermediate

exchange processes on the estimation of equilibrium-constants by NMR. *J. Magn. Reson.* **33**, 519–529.

74 SPERA, S. & BAX, A. (1991). Empirical correlation between protein backbone conformation and C-alpha and C-beta C-13 nuclear-magnetic-resonance chemical-shifts. *J. Am. Chem. Soc.* **113**, 5490–5492.

75 WISHART, D. S., SYKES, B. D. & RICHARDS, F. M. (1991). Relationship between nuclear magnetic resonance chemical shift and protein secondary structure. *J. Mol. Biol.* **222**, 311–33.

76 TCHERKASSKAYA, O., DAVIDSON, E. A. & UVERSKY, V. N. (2003). Biophysical constraints for protein structure prediction. *J. Proteome Res.* **2**, 37–42.

77 DAUGHDRILL, G. W., HANELY, L. J. & DAHLQUIST, F. W. (1998). The C-terminal half of the anti-sigma factor FlgM contains a dynamic equilibrium solution structure favoring helical conformations. *Biochemistry* **37**, 1076–1082.

78 BRACKEN, C., CARR, P. A., CAVANAGH, J. & PALMER, A. G., 3rd. (1999). Temperature dependence of intra-molecular dynamics of the basic leucine zipper of GCN4: implications for the entropy of association with DNA. *J. Mol. Biol.* **285**, 2133–2146.

79 PALMER, A. G., 3rd. (1993). Dynamic properties of proteins from NMR spectroscopy. *Curr. Opin. Biotechnol.* **4**, 385–91.

80 LEFEVRE, J. F., DAYIE, K. T., PENG, J. W. & WAGNER, G. (1996). Internal mobility in the partially folded DNA binding and dimerization domains of GAL4: NMR analysis of the N–H spectral density functions. *Biochemistry* **35**, 2674–86.

81 KAY, L. E., TORCHIA, D. A. & BAX, A. (1989). Backbone dynamics of proteins as studied by 15N inverse detected heteronuclear NMR spectroscopy: application to staphylococcal nuclease. *Biochemistry* **28**, 8972–9.

82 SKELTON, N. J., PALMER, A. G., AKKE, M., KORDEL, J., RANCE, M. & CHAZIN, W. J. (1993). Practical aspects of 2-dimensional proton-detected N-15 spin relaxation measurements. *J. Magn. Reson. B* **102**, 253–264.

83 KORDEL, J., SKELTON, N. J., AKKE, M., PALMER, A. G. & CHAZIN, W. J. (1992). Backbone dynamics of calcium-loaded calbindin D9k studied by 2-dimensional proton-detected N15 NMR spectroscopy. *Biochemistry* **31**, 4856–4866.

84 MCEVOY, M. M., DE LA CRUZ, A. F. & DAHLQUIST, F. W. (1997). Large modular proteins by NMR. *Nat. Struct. Biol.* **4**, 9.

85 RENNER, C., SCHLEICHER, M., MORODER, L. & HOLAK, T. A. (2002). Practical aspects of the 2D N-15-{H-1}-NOE experiment. *J. Biomol. NMR* **23**, 23–33.

86 LIPARI, G. & SZABO, A. (1982). Model-Free Approach to the Interpretation of Nuclear Magnetic-Resonance Relaxation in Macromolecules .1. Theory and Range of Validity. *J. Am. Chem. Soc.* **104**, 4546–4559.

87 LIPARI, G. & SZABO, A. (1982). Model-Free Approach to the Interpretation of Nuclear Magnetic-Resonance Relaxation in Macromolecules .2. Analysis of Experimental Results. *J. Am. Chem. Soc.* **104**, 4559–4570.

88 PENG, J. W. & WAGNER, G. (1992). Mapping of the spectral densities of N–H bond motions in eglin c using heteronuclear relaxation experiments. *Biochemistry* **31**, 8571–86.

89 ABRAGAM, A. (1961). *The principles of nuclear magnetism*. The International series of monographs on physics, Clarendon Press, Oxford.

90 BRUSCHWEILER, R., LIAO, X. & WRIGHT, P. E. (1995). Long-range motional restrictions in a multi-domain zinc-finger protein from anisotropic tumbling. *Science* **268**, 886–9.

91 TJANDRA, N., FELLER, S. E., PASTOR, R. W. & BAX, A. (1995). Rotational diffusion anisotropy of human ubiquitin from N-15 NMR relaxation. *J. Am. Chem. Soc.* **117**, 12562–12566.

92 LEE, L. K., RANCE, M., CHAZIN, W. J. & PALMER, A. G. (1997). Rotational diffusion anisotropy of proteins from simultaneous analysis of N-15 and C-13(alpha) nuclear spin relaxation. *J. Biomol. NMR* **9**, 287–298.

93 BUCK, M., SCHWALBE, H. & DOBSON,

C. M. (1996). Main-chain dynamics of a partially folded protein: 15N NMR relaxation measurements of hen egg white lysozyme denatured in trifluoroethanol. *J. Mol. Biol.* **257**, 669–83.

94 BUEVICH, A. V., SHINDE, U. P., INOUYE, M. & BAUM, J. (2001). Backbone dynamics of the natively unfolded pro-peptide of subtilisin by heteronuclear NMR relaxation studies. *J. Biomol. NMR* **20**, 233–249.

95 FARROW, N. A., ZHANG, O., MUHANDIRAM, R., FORMANKAY, J. D. & KAY, L. E. (1995). A comparative-study of the backbone dynamics of the folded and unfolded forms of an SH3 domain. *J. Cell. Biochem.*, 44–44.

96 YANG, D. & KAY, L. E. (1996). Contributions to conformational entropy arising from bond vector fluctuations measured from NMR-derived order parameters: application to protein folding. *J. Mol. Biol.* **263**, 369–82.

97 YANG, D., MOK, Y. K., FORMAN-KAY, J. D., FARROW, N. A. & KAY, L. E. (1997). Contributions to protein entropy and heat capacity from bond vector motions measured by NMR spin relaxation. *J. Mol. Biol.* **272**, 790–804.

98 VILES, J. H., DONNE, D., KROON, G., PRUSINER, S. B., COHEN, F. E., DYSON, H. J. & WRIGHT, P. E. (2001). Local structural plasticity of the prion protein. Analysis of NMR relaxation dynamics. *Biochemistry* **40**, 2743–53.

99 LANDRY, S. J., STEEDE, N. K. & MASKOS, K. (1997). Temperature dependence of backbone dynamics in loops of human mitochondrial heat shock protein 10. *Biochemistry* **36**, 10975–86.

100 BHATTACHARYA, S., FALZONE, C. J. & LECOMTE, J. T. (1999). Backbone dynamics of apocytochrome b5 in its native, partially folded state. *Biochemistry* **38**, 2577–89.

101 CAMPBELL, A. P., SPYRACOPOULOS, L., IRVIN, R. T. & SYKES, B. D. (2000). Backbone dynamics of a bacterially expressed peptide from the receptor binding domain of *Pseudomonas aeruginosa* pilin strain PAK from heteronuclear 1H-15N NMR spectroscopy. *J. Biomol. NMR* **17**, 239–55.

102 DEDMON, M. M., PATEL, C. N., YOUNG, G. B. & PIELAK, G. J. (2002). FlgM gains structure in living cells. *Proc. Natl. Acad. Sci. U.S.A.* **99**, 12681–12684.

103 CHOY, W. Y., MULDER, F. A., CROWHURST, K. A., MUHANDIRAM, D. R., MILLETT, I. S., DONIACH, S., FORMAN-KAY, J. D. & KAY, L. E. (2002). Distribution of molecular size within an unfolded state ensemble using small-angle X-ray scattering and pulse field gradient NMR techniques. *J. Mol. Biol.* **316**, 101–12.

104 CHOY, W. Y. & FORMAN-KAY, J. D. (2001). Calculation of ensembles of structures representing the unfolded state of an SH3 domain. *J. Mol. Biol.* **308**, 1011–1032.

105 KISSINGER, C. R., GEHLHAAR, D. K., SMITH, B. A. & BOUZIDA, D. (2001). Molecular replacement by evolutionary search. *Acta. Crystallogr. D Biol. Crystallogr.* **57**, 1474–9.

106 HUBER, R. (1987). Flexibility and rigidity, requirements for the function of proteins and protein pigment complexes. Eleventh Keilin memorial lecture. *Biochem. Soc. Trans.* **15**, 1009–20.

107 HUBER, R. & BENNETT, W. S., JR. (1983). Functional significance of flexibility in proteins. *Biopolymers* **22**, 261–279.

108 DOUZOU, P. & PETSKO, G. A. (1984). Proteins at work: "stop-action" pictures at subzero temperatures. *Adv. Protein Chem.* **36**, 245–361.

109 OBRADOVIC, Z., PENG, K., VUCETIC, S., RADIVOJAC, P., BROWN, C. J. & DUNKER, A. K. (2003). Predicting intrinsic disorder from amino acid sequence. *Proteins* **53**, 566–572.

110 HOBOHM, U. & SANDER, C. (1994). Enlarged representative set of protein structures. *Protein Sci.* **3**, 522–524.

111 BAIROCH, A. & APWEILER, R. (2000). The SWISS-PROT protein sequence database and its supplement TrEMBL in 2000. *Nucleic Acids Res.* **28**, 45–48.

112 SCHACHMAN, H. K. (1959). *Ultracentrifugation in Biochemistry*, Academic Press, New York.

113 GLATTER, O. & KRATKY, O. (1982). *Small Angle X-ray Scattering*, Academic Press, London, England.

114 DONIACH, S., BASCLE, J., GAREL, T. & ORLAND, H. (1995). Partially folded states of proteins: characterization by X-ray scattering. *J. Mol. Biol.* **254**, 960–7.

115 KATAOKA, M. & GOTO, Y. (1996). X-ray solution scattering studies of protein folding. *Fold. Des.* **1**, R107–14.

116 UVERSKY, V. N., GILLESPIE, J. R., MILLETT, I. S., KHODYAKOVA, A. V., VASILIEV, A. M., CHERNOVSKAYA, T. V., VASILENKO, R. N., KOZLOVSKAYA, G. D., DOLGIKH, D. A., FINK, A. L., DONIACH, S. & ABRAMOV, V. M. (1999). Natively unfolded human prothymosin alpha adopts partially folded collapsed conformation at acidic pH. *Biochemistry* **38**, 15009–16.

117 UVERSKY, V. N., GILLESPIE, J. R., MILLETT, I. S., KHODYAKOVA, A. V., VASILENKO, R. N., VASILIEV, A. M., RODIONOV, I. L., KOZLOVSKAYA, G. D., DOLGIKH, D. A., FINK, A. L., DONIACH, S., PERMYAKOV, E. A. & ABRAMOV, V. M. (2000). Zn(2+)-mediated structure formation and compaction of the "natively unfolded" human prothymosin alpha. *Biochem. Biophys. Res. Commun.* **267**, 663–668.

118 LI, J., UVERSKY, V. N. & FINK, A. L. (2001). Effect of familial Parkinson's disease point mutations A30P and A53T on the structural properties, aggregation, and fibrillation of human alpha-synuclein. *Biochemistry* **40**, 11604–13.

119 UVERSKY, V. N., LI, J. & FINK, A. L. (2001). Evidence for a partially folded intermediate in alpha-synuclein fibril formation. *J. Biol. Chem.* **276**, 10737–10744.

120 UVERSKY, V. N., LI, J., SOUILLAC, P., JAKES, R., GOEDERT, M. & FINK, A. L. (2002). Biophysical properties of the synucleins and their propensities to fibrillate: inhibition of alpha-synuclein assembly by beta- and gamma-synucleins. *J. Biol. Chem.* **25**, 25.

121 PERMYAKOV, S. E., MILLETT, I. S., DONIACH, S., PERMYAKOV, E. A. & UVERSKY, V. N. (2003). Natively unfolded C-terminal domain of caldesmon remains substantially unstructured after the effective binding to calmodulin. *Proteins* **53**, 855–62.

122 MUNISHKINA, L. A., FINK, A. L. & UVERSKY, V. N. (In press). Formation of amyloid fibrils from the core histones in vitro. *J. Mol. Biol.*

123 UVERSKY, V. N. (2002). What does it mean to be natively unfolded? *Eur. J. Biochem.* **269**, 2–12.

124 GUINIER, A. & FOURNET, G. (1955). *Small-angle scattering of X-rays*, John Wiley & Sons.

125 MILLETT, I. S., DONIACH, S. & PLAXCO, K. W. (2002). Toward a taxonomy of the denatured state: small angle scattering studies of unfolded proteins. *Adv. Protein Chem.* **62**, 241–62.

126 JAENICKE, R. & SECKLER, R. (1997). Protein misassembly in vitro. *Adv. Protein Chem.* **50**, 1–59.

127 ROSE, G. D. (2002). *Unfolded Proteins*. Advances in Protein Chemistry (F. M., R., S., E. D. & J., K., Eds.), 62, Academic Press, New York.

128 SEVERGUN, D. I. (1992). Determination of the regularization parameter in indirect-transform methods using perceptual criteria. *J. Appl. Cryst.* **25**, 495–503.

129 BERGMANN, A., ORTHABER, D., SCHERF, G. & GLATTER, O. (2000). Improvement of SAXS measurements on Kratky slit systems by Göbel mirrors and imaging-plate detectors. *J. Appl. Cryst.* **33**, 869–875.

130 PANICK, G., MALESSA, R., WINTER, R., RAPP, G., FRYE, K. J. & ROYER, C. A. (1998). Structural characterization of the pressure-denatured state and unfolding/refolding kinetics of staphylococcal nuclease by synchrotron small-angle X-ray scattering and Fourier-transform infrared spectroscopy. *J. Mol. Biol.* **275**, 389–402.

131 PÉREZ, J., VACHETTE, P., RUSSO, D., DESMADRIL, M. & DURAND, D. (2001). Heat-induced unfolding of neocarzinostatin, a small all-beta protein investigated by small-angle X-ray scattering. *J. Mol. Biol.* **308**, 721–43.

132 UVERSKY, V. N. (1993). Use of fast protein size-exclusion liquid chromatography to study the unfolding of proteins which denature through the molten globule. *Biochemistry* **32**, 13288–13298.

133 UVERSKY, V. N. (1994). Gel-permeation

chromatography as a unique instrument for quantitative and qualitative analysis of protein denaturation and unfolding. *Int. J. Bio-Chromatography* **1**, 103–114.

134 UVERSKY, V. N. (2003). Protein folding revisited. A polypeptide chain at the folding – misfolding – no-folding crossroads: Which way to go? *Cell. Mol. Life Sci.*

135 UVERSKY, V. N. (2002). Natively unfolded proteins: a point where biology waits for physics. *Protein Sci.* **11**, 739–756.

136 UVERSKY, V. N. & PTITSYN, O. B. (1996). Further evidence on the equilibrium "pre-molten globule state": four-state guanidinium chloride-induced unfolding of carbonic anhydrase B at low temperature. *J. Mol. Biol.* **255**, 215–28.

137 FASMAN, G. D. (1996). *Circular Dichroism and the Conformational Analysis of Biomolecules*, Plenum Press, New York.

138 ADLER, A. J., GREENFIELD, N. J. & FASMAN, G. D. (1973). Circular dichroism and optical rotatory dispersion of proteins and polypeptides. *Methods Enzymol.* **27**, 675–735.

139 UVERSKY, V. N. & FINK, A. L. (2002). The chicken-egg scenario of protein folding revisited. *FEBS Lett.* **515**, 79–83.

140 PALLEROS, D. R., REID, K. L., MCCARTY, J. S., WALKER, G. C. & FINK, A. L. (1992). DnaK, hsp73, and their molten globules. Two different ways heat shock proteins respond to heat. *J. Biol. Chem.* **267**, 5279–85.

141 FINK, A. L., OBERG, K. A. & SESHADRI, S. (1998). Discrete intermediates versus molten globule models for protein folding: characterization of partially folded intermediates of apomyoglobin. *Fold. Des.* **3**, 19–25.

142 UVERSKY, V. N., KARNOUP, A. S., SEGEL, D. J., SESHADRI, S., DONIACH, S. & FINK, A. L. (1998). Anion-induced folding of *Staphylococcal* nuclease: characterization of multiple equilibrium partially folded intermediates. *J. Mol. Biol.* **278**, 879–894.

143 KRIMM, S. & TIFFANY, M. L. (1974). The circular dichroism spectrum and structure of unordered polypeptides and proteins. *Israeli J. of Chem.* **12**, 189–200.

144 KELLY, M. A., CHELLGREN, B. W., RUCKER, A. L., TROUTMAN, J. M., FRIED, M. G., MILLER, A. F. & CREAMER, T. P. (2001). Host-guest study of left-handed polyproline II helix formation. *Biochemistry* **40**, 14376–83.

145 RUCKER, A. L. & CREAMER, T. P. (2002). Polyproline II helical structure in protein unfolded states: lysine peptides revisited. *Protein Sci.* **11**, 980–5.

146 HOUSE-POMPEO, K., XU, Y., JOH, D., SPEZIALE, P. & HOOK, M. (1996). Conformational changes in the binding MSCRAMMs are induced by ligand binding. *J. Biol. Chem.* **271**, 1379–1384.

147 GAST, K., DAMASCHUN, H., ECKERT, K., SCHULZE-FORSTER, K., MAURER, H. R., MÜLLER-FROHNE, M., ZIRWER, D., CZARNECKI, J. & DAMASCHUN, G. (1995). Prothymosin alpha: a biologically active protein with random coil conformation. *Biochemistry* **34**, 13211–13218.

148 FLAUGH, S. L. & LUMB, K. J. (2001). Effects of macromolecular crowding on the intrinscially disordered proteins c-Fos and p27(Kip 1). *Biomacromolecules* **2**, 538–540.

149 DENNING, D. P., PATEL, S. S., UVERSKY, V., FINK, A. L. & REXACH, M. (2003). Disorder in the nuclear pore complex: the FG repeat regions of nucleoporins are natively unfolded. *Proc. Natl. Acad. Sci. U.S.A.* **100**, 2450–2455.

150 BOTHNER, B., AUBIN, Y. & KRIWACKI, R. W. (2003). Peptides derived from two dynamically disordered proteins self-assemble into amyloid-like fibrils. *J. Am. Chem. Soc.* **125**, 3200–1.

151 MACKAY, J. P., MATTHEWS, J. M., WINEFIELD, R. D., MACKAY, L. G., HAVERKAMP, R. G. & TEMPLETON, M. D. (2001). The hydrophobin EAS is largely unstructured in solution and functions by forming amyloid-like structures. *Structure (Camb)* **9**, 83–91.

152 BASKAKOV, I. & BOLEN, D. W. (1998). Forcing thermodynamically unfolded proteins to fold. *J. Biol. Chem.* **273**, 4831–4834.

153 QU, Y., BOLEN, C. L. & BOLEN, D. W. (1998). Osmolyte-driven contraction of a random coil protein. *Proc. Natl. Acad. Sci. U.S.A.* **95**, 9268–9273.

154 BASKAKOV, I. V., KUMAR, R., SRINIVASAN, G., JI, Y. S., BOLEN, D. W. & THOMPSON, E. B. (1999). Trimethylamine N-oxide-induced cooperative folding of an intrinsically unfolded transcription-activating fragment of human glucocorticoid receptor. *J. Biol. Chem.* **274**, 10693–10696.

155 QU, Y. & BOLEN, D. W. (2002). Efficacy of macromolecular crowding in forcing proteins to fold. *Biophys. Chem.* **101–102**, 155–65.

156 DAVIS-SEARLES, P. R., MORAR, A. S., SAUNDERS, A. J., ERIE, D. A. & PIELAK, G. J. (1998). Sugar-induced molten-globule model. *Biochemistry* **37**, 17048–17053.

157 SAUNDERS, A. J., DAVIS-SEARLES, P. R., ALLEN, D. L., PIELAK, G. J. & ERIE, D. A. (2000). Osmolyte-induced changes in protein conformational equilibria. *Biopolymers* **53**, 293–307.

158 MORAR, A. S., OLTEANU, A., YOUNG, G. B. & PIELAK, G. J. (2001). Solvent-induced collapse of alpha-synuclein and acid-denatured cytochrome c. *Protein Sci.* **10**, 2195–2199.

159 CHIRGADZE, Y. N., SHESTOPALOV, B. V. & VENYAMINOV, S. Y. (1973). Intensities and other spectral parameters of infrared amide bands of polypeptides in the beta- and random forms. *Biopolymers* **12**, 1337–51.

160 SUSI, H., TIMASHEFF, S. N. & STEVENS, L. (1967). Infrared spectra and protein conformations in aqueous solutions. I. The amide I band in H2O and D2O solutions. *J. Biol. Chem.* **242**, 5460–6.

161 CHIRGADZE, Y. N. & BRAZHNIKOV, E. V. (1974). Intensities and other spectral parameters of infrared amide bands of polypeptides in the alpha-helical form. *Biopolymers* **13**, 1701–12.

162 MIYAZAWA, T. & BLOUT, E. R. (1961). The Infrared Spectra of Polypeptides in Various Conformations: Amide I and II Bands. *J. Am. Chem. Soc.* **83**, 712–719.

163 LACKOWICZ, J. (1999). *Principals of Fluorescence Spectroscopy*, Kluwer Academic/Plenum Publishers, New York.

164 STRYER, L. (1968). Fluorescence spectroscopy of proteins. *Science* **162**, 526–33.

165 PERMYAKOV, E. A. (1993). *Luminescence spectroscopy of proteins*, CRC Press, London.

166 EFTINK, M. R. & GHIRON, C. A. (1981). Fluorescence quenching studies with proteins. *Anal. Biochem.* **114**, 199–227.

167 LAKOWICZ, J. R. & WEBER, G. (1973). Quenching of fluorescence by oxygen. A probe for structural fluctuations in macromolecules. *Biochemistry* **12**, 4161–70.

168 LAKOWICZ, J. R. & WEBER, G. (1973). Quenching of protein fluorescence by oxygen. Detection of structural fluctuations in proteins on the nanosecond timescale. *Biochemistry* **12**, 4171–9.

169 EFTINK, M. R. & GHIRON, C. A. (1976). Exposure of tryptophanyl residues in proteins. Quantitative determination by fluorescence quenching studies. *Biochemistry* **15**, 672–80.

170 EFTINK, M. R. & GHIRON, C. A. (1975). Dynamics of a protein matrix revealed by fluorescence quenching. *Proc. Natl. Acad. Sci. U.S.A.* **72**, 3290–4.

171 EFTINK, M. R. & GHIRON, C. A. (1977). Exposure of tryptophanyl residues and protein dynamics. *Biochemistry* **16**, 5546–51.

172 CHAFFOTTE, A. F. & GOLDBERG, M. E. (1984). Fluorescence-quenching studies on a conformational transition within a domain of the beta 2 subunit of *Escherichia coli* tryptophan synthase. *Eur. J. Biochem.* **139**, 47–50.

173 VARLEY, P. G., DRYDEN, D. T. & PAIN, R. H. (1991). Resolution of the fluorescence of the buried tryptophan in yeast 3-phosphoglycerate kinase using succinimide. *Biochim. Biophys. Acta.* **1077**, 19–24.

174 WEBER, G. (1952). Polarization of the

fluorescence of macromolecules. II. Fluorescent conjugates of ovalbumin and bovine serum albumin. *Biochem. J.* **51**, 155–67.
175 WEBER, G. (1952). Polarization of the fluorescence of macromolecules. I. Theory and experimental method. *Biochem. J.* **51**, 145–55.
176 WEBER, G. (1953). Rotational Brownian motion and polarization of the fluorescence of solutions. *Adv. Protein Chem.* **8**, 415–59.
177 WEBER, G. (1960). Fluorescence-polarization spectrum and electronic-energy transfer in proteins. *Biochem. J.* **75**, 345–52.
178 SEMISOTNOV, G. V., ZIKHERMAN, K. K., KASATKIN, S. B., PTITSYN, O. B. & V., A. E. (1981). Polarized luminescence and mobility of tryptophan residues in polypeptide chains. *Biopolymers* **20**, 2287–2309.
179 DOLGIKH, D. A., ABATUROV, L. V., BOLOTINA, I. A., BRAZHNIKOV, E. V., BYCHKOVA, V. E., GILMANSHIN, R. I., LEBEDEV YU, O., SEMISOTNOV, G. V., TIKTOPULO, E. I., PTITSYN, O. B. & et al. (1985). Compact state of a protein molecule with pronounced small-scale mobility: bovine alpha-lactalbumin. *Eur. Biophys. J.* **13**, 109–21.
180 RODIONOVA, N. A., SEMISOTNOV, G. V., KUTYSHENKO, V. P., UVERSKII, V. N. & BOLOTINA, I. A. (1989). [Staged equilibrium of carbonic anhydrase unfolding in strong denaturants]. *Mol Biol (Mosk)* **23**, 683–92.
181 FÖRSTER, T. (1948). Intermolecular energy migration and fluorescence. *Ann. Physik.* **2**, 55–75.
182 LUNDBLAD, R. L. (1991). *Chemical reagents for protein modification*, CRC Press, Boca Raton.
183 RISCHEL, C. & POULSEN, F. M. (1995). Modification of a specific tyrosine enables tracing of the end-to-end distance during apomyoglobin folding. *FEBS Lett.* **374**, 105–9.
184 UVERSKY, V. N. & FINK, A. L. (1999). Do protein molecules have a native-like topology in the pre-molten globule state? *Biochemistry (Mosc)* **64**, 552–5.
185 TCHERKASSKAYA, O. & PTITSYN, O. B. (1999). Molten globule versus variety of intermediates: influence of anions on pH-denatured apomyoglobin. *FEBS Lett.* **455**, 325–31.
186 TCHERKASSKAYA, O. & PTITSYN, O. B. (1999). Direct energy transfer to study the 3D structure of non-native proteins: AGH complex in molten globule state of apomyoglobin. *Protein Eng.* **12**, 485–90.
187 TCHERKASSKAYA, O. & UVERSKY, V. N. (2001). Denatured collapsed states in protein folding: example of apomyoglobin. *Proteins* **44**, 244–254.
188 STRYER, L. (1965). The interaction of a naphthalene dye with apomyoglobin and apohemoglobin. A fluorescent probe of non-polar binding sites. *J. Mol. Biol.* **13**, 482–495.
189 TURNER, D. C. & BRAND, L. (1968). Quantitative estimation of protein binding site polarity. Fluorescence of N-arylaminonaphthalenesulfonates. *Biochemistry* **7**, 3381–90.
190 SEMISOTNOV, G. V., RODIONOVA, N. A., RAZGULYAEV, O. I., UVERSKY, V. N., GRIPAS, A. F. & GILMANSHIN, R. I. (1991). Study of the "molten globule" intermediate state in protein folding by a hydrophobic fluorescent probe. *Biopolymers* **31**, 119–128.
191 SEMISOTNOV, G. V., RODIONOVA, N. A., KUTYSHENKO, V. P., EBERT, B., BLANCK, J. & PTITSYN, O. B. (1987). Sequential mechanism of refolding of carbonic anhydrase B. *FEBS Lett.* **224**, 9–13.
192 UVERSKY, V. N., WINTER, S. & LOBER, G. (1996). Use of fluorescence decay times of 8-ANS-protein complexes to study the conformational transitions in proteins which unfold through the molten globule state. *Biophys. Chem.* **60**, 79–88.
193 UVERSKY, V. N., WINTER, S. & LOBER, G. (1998). Self-association of 8-anilino-1-naphthalene-sulfonate molecules: spectroscopic characterization and application to the investigation of protein folding. *Biochim. Biophys. Acta.* **1388**, 133–142.
194 UVERSKY, V. N. & PTITSYN, O. B. (1996). All-or-none solvent-induced transitions between native, molten

globule and unfolded states in globular proteins. *Fold. & Des.* **1**, 117–122.

195 ROBERTS, L. M. & DUNKER, A. K. (1993). Structural changes accompanying chloroform-induced contraction of the filamentous phage fd. *Biochemistry* **32**, 10479–10488.

196 PRIVALOV, P. L. (1979). Stability of proteins: small globular proteins. *Adv. Protein. Chem.* **33**, 167–241.

197 PTITSYN, O. (1995). Molten globule and protein folding. *Adv. Protein Chem.* **47**, 83–229.

198 UVERSKY, V. N. (1999). A multiparametric approach to studies of self-organization of globular proteins. *Biochemistry (Mosc)* **64**, 250–266.

199 PTITSYN, O. B. & UVERSKY, V. N. (1994). The molten globule is a third thermodynamical state of protein molecules. *FEBS Lett.* **341**, 15–18.

200 UVERSKY, V. N., PERMYAKOV, S. E., ZAGRANICHNY, V. E., RODIONOV, I. L., FINK, A. L., CHERSKAYA, A. M., WASSERMAN, L. A. & PERMYAKOV, E. A. (2002). Effect of zinc and temperature on the conformation of the gamma subunit of retinal phosphodiesterase: a natively unfolded protein. *J. Proteome Res.* **1**, 149–159.

201 TIMM, D. E., VISSAVAJJHALA, P., ROSS, A. H. & NEET, K. E. (1992). Spectroscopic and chemical studies of the interaction between nerve growth factor (NGF) and the extracellular domain of the low affinity NGF receptor. *Protein Sci.* **1**, 1023–31.

202 KIM, T. D., RYU, H. J., CHO, H. I., YANG, C. H. & KIM, J. (2000). Thermal behavior of proteins: heat-resistant proteins and their heat-induced secondary structural changes. *Biochemistry* **39**, 14839–46.

203 KONNO, T., TANAKA, N., KATAOKA, M., TAKANO, E. & MAKI, M. (1997). A circular dichroism study of preferential hydration and alcohol effects on a denatured protein, pig calpastatin domain I. *Biochim. Biophys. Acta.* **1342**, 73–82.

204 LYNN, A., CHANDRA, S., MALHOTRA, P. & CHAUHAN, V. S. (1999). Heme binding and polymerization by *Plasmodium falciparum* histidine rich protein II: influence of pH on activity and conformation. *FEBS Lett.* **459**, 267–71.

205 JOHANSSON, J., GUDMUNDSSON, G. H., ROTTENBERG, M. E., BERNDT, K. D. & AGERBERTH, B. (1998). Conformation-dependent antibacterial activity of the naturally occurring human peptide LL-37. *J. Biol. Chem.* **273**, 3718–3724.

206 ENGLANDER, S. W. & KRISHNA, M. M. (2001). Hydrogen exchange. *Nat. Struct. Biol.* **8**, 741–2.

207 LINDERSTRØM-LANG, K. (1955). Deuterium exchange between peptides and water. *Chem. Soc. (London) Spec. Publ.* **2**, 1–20.

208 HVIDT, A. & NIELSEN, S. O. (1966). Hydrogen exchange in proteins. *Adv. Protein Chem.* **21**, 287–386.

209 BAI, Y., MILNE, J. S., MAYNE, L. & ENGLANDER, S. W. (1993). Primary structure effects on peptide group hydrogen exchange. *Proteins* **17**, 75–86.

210 HAMURO, Y., COALES, S. J., SOUTHERN, M. R., NEMETH-CAWLEY, J. F., STRANZ, D. D. & GRIFFIN, P. R. (2003). Rapid analysis of protein structure and dynamics by hydrogen/deuterium exchange mass spectrometry. *J Biomol Tech* **14**, 171–82.

211 HAMURO, Y., COALES, S. J., SOUTHERN, M. R., NEMETH-CAWLEY, J. F., STRANZ, D. D. & GRIFFIN, P. R. (2003). Rapid analysis of protein structure and dynamics by hydrogen/deuterium exchange mass spectrometry. *J. Biomol. Tech.* **14**, 171–82.

212 ENGLANDER, J. J., ROGERO, J. R. & ENGLANDER, S. W. (1985). Protein hydrogen exchange studied by the fragment separation method. *Anal. Biochem.* **147**, 234–44.

213 ROSA, J. J. & RICHARDS, F. M. (1979). An experimental procedure for increasing the structural resolution of chemical hydrogen-exchange measurements on proteins: application to ribonuclease S peptide. *J. Mol. Biol.* **133**, 399–416.

214 ZHANG, Z. & SMITH, D. L. (1996). Thermal-induced unfolding domains in aldolase identified by amide

hydrogen exchange and mass spectrometry. *Protein Sci* **5**, 1282–9.

215 PANTAZATOS, D., KIM, J. S., KLOCK, H. E., STEVENS, R. C., WILSON, I. A., LESLEY, S. A. & WOODS, V. L., JR. (2004). Rapid refinement of crystallographic protein construct definition employing enhanced hydrogen/deuterium exchange MS. *Proc. Natl. Acad. Sci. U.S.A.* **101**, 751–6.

216 WU, H. (1931). Studies on denaturation of proteins XIII A theory of denaturation. *Chin. J. Physiol.* **1**, 219–234.

217 LINDERSTRØM-LANG, K. (1952). Structure and enzymatic break-down of proteins. *Lane Medical Lectures* **6**, 117–126.

218 MARKUS, G. (1965). Protein substrate conformation and proteolysis. *Proc. Natl. Acad. Sci. U.S.A.* **54**, 253–258.

219 WRIGHT, H. T. (1977). Secondary and conformational specificities of trypsin and chymotrypsin. *Eur. J. Biochem.* **73**, 567–578.

220 SWEET, R. M., WRIGHT, H. T., JANIN, J., CHOTHIA, C. H. & BLOW, D. M. (1974). Crystal structure of the complex of porcine trypsin with soybean trypsin inhibitor (Kunitz) at 2.6-A resolution. *Biochemistry* **13**, 4212–4228.

221 HUBBARD, S. J., CAMPBELL, S. F. & THORNTON, J. M. (1991). Molecular recognition. Conformational analysis of limited proteolytic sites and serine proteinase protein inhibitors. *J. Mol. Biol.* **220**, 507–530.

222 HUBBARD, S. J., BEYNON, R. J. & THORNTON, J. M. (1998). Assessment of conformational parameters as predictors of limited proteolytic sites in native protein structures. *Protein Eng.* **11**, 349–359.

223 HUBBARD, S. J., EISENMENGER, F. & THORNTON, J. M. (1994). Modeling studies of the change in conformation required for cleavage of limited proteolytic sites. *Protein Sci.* **3**, 757–768.

224 FERNANDEZ, A. & SCHERAGA, H. A. (2003). Insufficiently dehydrated hydrogen bonds as determinants of protein interactions. *Proc. Natl. Acad. Sci. U.S.A.* **100**, 113–8.

225 FONTANA, A., DE LAURETO, P. P., DE FILIPPIS, V., SCARAMELLA, E. & ZAMBONIN, M. (1997). Probing the partly folded states of proteins by limited proteolysis. *Fold. Des.* **2**, R17–R26.

226 FONTANA, A., ZAMBONIN, M., POLVERINO DE LAURETO, P., DE FILIPPIS, V., CLEMENTI, A. & SCARAMELLA, E. (1997). Probing the conformational state of apomyoglobin by limited proteolysis. *J. Mol. Biol.* **266**, 223–230.

227 UVERSKY, V. N., GILLESPIE, J. R. & FINK, A. L. (2000). Why are "natively unfolded" proteins unstructured under physiologic conditions? *Proteins* **41**, 415–27.

228 WILLIAMS, R. J. (1978). Energy states of proteins, enzymes and membranes. *Proc. R. Soc. Lond. B Biol. Sci.* **200**, 353–389.

229 LINDING, R., RUSSELL, R. B., NEDUVA, V. & GIBSON, T. J. (2003). GlobPlot: Exploring protein sequences for globularity and disorder. *Nucleic Acids Res.* **31**, 3701–8.

230 LINDING, R., JENSEN, L. J., DIELLA, F., BORK, P., GIBSON, T. J. & RUSSELL, R. B. (2003). Protein disorder prediction: implications for structural proteomics. *Structure (Camb)* **11**, 1453–1459.

231 ROMERO, P., OBRADOVIC, Z. & DUNKER, A. K. (1997). Sequence data analysis for long disordered regions prediction in the calcineurin family. *Genome Inform. Ser. Workshop Genome Inform.* **8**, 110–124.

232 ROMERO, P., OBRADOVIC, Z., LI, X., GARNER, E. C., BROWN, C. J. & DUNKER, A. K. (2001). Sequence complexity of disordered protein. *Proteins* **42**, 38–48.

233 JONES, D. T. & WARD, J. (2003). Prediction of disordered regions in proteins from position specific score matrices. *Proteins* **53**, 573–578.

234 MELAMUD, E. & MOULT, J. (2003). Evaluation of disorder predictions in CASP5. *Proteins* **53 Suppl 6**, 561–5.

235 YANG, S. A. & KLEE, C. B. (2000). Low affinity Ca2+-binding sites of calcineurin B mediate conformational changes in calcineurin A. *Biochemistry* **39**, 16147–16154.

236 BRANDSTETTER, H., BAUER, M., HUBER, R., LOLLAR, P. & BODE, W. (1995). X-ray structure of clotting factor IXa: active site and module structure related to Xase activity and hemophilia B. *Proc. Natl. Acad. Sci. U.S.A.* **92**, 9796–9800.

237 BRANDSTETTER, H., KUHNE, A., BODE, W., HUBER, R., VON DER SAAL, W., WIRTHENSOHN, K. & ENGH, R. A. (1996). X-ray structure of active site-inhibited clotting factor Xa. Implications for drug design and substrate recognition. *J. Biol. Chem.* **271**, 29988–29992.

238 PADMANABHAN, K., PADMANABHAN, K. P., TULINSKY, A., PARK, C. H., BODE, W., HUBER, R., BLANKENSHIP, D. T., CARDIN, A. D. & KISIEL, W. (1993). Structure of human des(1–45) factor Xa at 2.2 A resolution. *J. Mol. Biol.* **232**, 947–966.

239 AVILES, F. J., CHAPMAN, G. E., KNEALE, G. G., CRANE-ROBINSON, C. & BRADBURY, E. M. (1978). The conformation of histone H5. Isolation and characterisation of the globular segment. *Eur. J. Biochem.* **88**, 363–371.

240 RAMAKRISHNAN, V., FINCH, J. T., GRAZIANO, V., LEE, P. L. & SWEET, R. M. (1993). Crystal structure of globular domain of histone H5 and its implications for nucleosome binding. *Nature* **362**, 219–223.

241 GRAZIANO, V., GERCHMAN, S. E., WONACOTT, A. J., SWEET, R. M., WELLS, J. R., WHITE, S. W. & RAMAKRISHNAN, V. (1990). Crystallization of the globular domain of histone H5. *J. Mol. Biol.* **212**, 253–257.

242 SHAIU, W. L., HU, T. & HSIEH, T. S. (1999). The hydrophilic, protease-sensitive terminal domains of eucaryotic DNA topoisomerases have essential intracellular functions. *Pac. Symp. Biocomput.* **4**, 578–589.

243 BERGER, J. M., GAMBLIN, S. J., HARRISON, S. C. & WANG, J. C. (1996). Structure and mechanism of DNA topoisomerase II. *Nature* **379**, 225–232.

244 CARON, P. R., WATT, P. & WANG, J. C. (1994). The C-terminal domain of Saccharomyces cerevisiae DNA topoisomerase II. *Mol. Cell. Biol.* **14**, 3197–3207.

245 MUCHMORE, S. W., SATTLER, M., LIANG, H., MEADOWS, R. P., HARLAN, J. E., YOON, H. S., NETTESHEIM, D., CHANG, B. S., THOMPSON, C. B., WONG, S. L., NG, S. L. & FESIK, S. W. (1996). X-ray and NMR structure of human Bcl-x_L, an inhibitor of programmed cell death. *Nature* **381**, 335–341.

246 YAMAMOTO, K., ICHIJO, H. & KORSMEYER, S. J. (1999). BCL-2 is phosphorylated and inactivated by an ASK1/Jun N-terminal protein kinase pathway normally activated at G(2)/M. *Mol. Cell. Biol.* **19**, 8469–8478.

247 CHENG, E. H., KIRSCH, D. G., CLEM, R. J., RAVI, R., KASTAN, M. B., BEDI, A., UENO, K. & HARDWICK, J. M. (1997). Conversion of Bcl-2 to a Bax-like death effector by caspases. *Science* **278**, 1966–1968.

248 CHANG, B. S., MINN, A. J., MUCHMORE, S. W., FESIK, S. W. & THOMPSON, C. B. (1997). Identification of a novel regulatory domain in Bcl-X(L) and Bcl-2. *EMBO J.* **16**, 968–977.

249 HOLLIGER, P., RIECHMANN, L. & WILLIAMS, R. L. (1999). Crystal structure of the two N-terminal domains of g3p from filamentous phage fd at 1.9 A: evidence for conformational lability. *J. Mol. Biol.* **288**, 649–657.

250 HOLLIGER, P. & RIECHMANN, L. (1997). A conserved infection pathway for filamentous bacteriophages is suggested by the structure of the membrane penetration domain of the minor coat protein g3p from phage fd. *Structure* **5**, 265–275.

251 NILSSON, N., MALMBORG, A. C. & BORREBAECK, C. A. (2000). The phage infection process: a functional role for the distal linker region of bacteriophage protein 3. *J. Virol.* **74**, 4229–4235.

252 LEE, C. H., SAKSELA, K., MIRZA, U. A., CHAIT, B. T. & KURIYAN, J. (1996). Crystal structure of the conserved core of HIV-1 Nef complexed with a Src family SH3 domain. *Cell* **85**, 931–942.

253 Arold, S., Franken, P., Strub, M. P., Hoh, F., Benichou, S., Benarous, R. & Dumas, C. (1997). The crystal structure of HIV-1 Nef protein bound to the Fyn kinase SH3 domain suggests a role for this complex in altered T cell receptor signaling. *Structure* **5**, 1361–1372.

254 Geyer, M., Munte, C. E., Schorr, J., Kellner, R. & Kalbitzer, H. R. (1999). Structure of the anchor-domain of myristoylated and non-myristoylated HIV-1 Nef protein. *J. Mol. Biol.* **289**, 123–138.

255 Arold, S. T. & Baur, A. S. (2001). Dynamic Nef and Nef dynamics: how structure could explain the complex activities of this small HIV protein. *Trends Biochem. Sci.* **26**, 356–363.

256 Karlin, D., Longhi, S., Receveur, V. & Canard, B. (2002). The N-terminal domain of the phosphoprotein of *morbilliviruses* belongs to the natively unfolded class of proteins. *Virology* **296**, 251–262.

257 Karlin, D., Ferron, F., Canard, B. & Longhi, S. (2003). Structural disorder and modular organization in Paramyxovirinae N and P. *J. Gen. Virol.* **84**, 3239–52.

258 Longhi, S., Receveur-Brechot, V., Karlin, D., Johansson, K., Darbon, H., Bhella, D., Yeo, R., Finet, S. & Canard, B. (2003). The C-terminal domain of the measles virus nucleoprotein is intrinsically disordered and folds upon binding the C-terminal moiety of the phosphoprotein. *J. Biol. Chem.* **278**, 18638.

259 Lesk, A. M., Levitt, M. & Chothia, C. (1986). Alignment of the amino acid sequences of distantly related proteins using variable gap penalties. *Protein Eng.* **1**, 77–8.

260 Lesk, A. M. & Chothia, C. (1980). How different amino acid sequences determine similar protein structures: the structure and evolutionary dynamics of the globins. *J. Mol. Biol.* **136**, 225–70.

261 Chothia, C. & Lesk, A. M. (1987). The evolution of protein structures. *Cold Spring Harb. Symp. Quant. Biol.* **52**, 399–405.

262 Harrison, R. W., Chatterjee, D. & Weber, I. T. (1995). Analysis of six protein structures predicted by comparative modeling techniques. *Proteins* **23**, 463–71.

263 Skolnick, J. & Fetrow, J. S. (2000). From genes to protein structure and function: novel applications of computational approaches in the genomic era. *Trends Biotechnol.* **18**, 34–9.

264 Skolnick, J., Fetrow, J. S. & Kolinski, A. (2000). Structural genomics and its importance for gene function analysis. *Nat. Biotechnol.* **18**, 283–287.

265 Thornton, J. M., Orengo, C. A., Todd, A. E. & Pearl, F. M. (1999). Protein folds, functions and evolution. *J. Mol. Biol.* **293**, 333–42.

266 Dunker, A. K., Obradovic, Z., Romero, P., Garner, E. C. & Brown, C. J. (2000). Intrinsic protein disorder in complete genomes. *Genome Inform. Ser. Workshop Genome Inform.* **11**, 161–171.

267 Daughdrill, G. W., Chadsey, M. S., Karlinsey, J. E., Hughes, K. T. & Dahlquist, F. W. (1997). The C-terminal half of the anti-sigma factor, FlgM, becomes structured when bound to its target, sigma 28. *Nat. Struct. Biol.* **4**, 285–291.

268 Donne, D. G., Viles, J. H., Groth, D., Mehlhorn, I., James, T. L., Cohen, F. E., Prusiner, S. B., Wright, P. E. & Dyson, H. J. (1997). Structure of the recombinant full-length hamster prion protein PrP(29–231): the N terminus is highly flexible. *Proc. Natl. Acad. Sci. U.S.A.* **94**, 13452–13457.

269 Jacobs, D. M., Lipton, A. S., Isern, N. G., Daughdrill, G. W., Lowry, D. F., Gomes, X. & Wold, M. S. (1999). Human replication protein A: global fold of the N-terminal RPA-70 domain reveals a basic cleft and flexible C-terminal linker. *J. Biomol. NMR* **14**, 321–331.

270 Lee, H., Mok, K. H., Muhandiram, R., Park, K. H., Suk, J. E., Kim, D. H., Chang, J., Sung, Y. C., Choi, K. Y. & Han, K. H. (2000). Local structural elements in the mostly unstructured transcriptional activation domain of human p53. *J. Biol. Chem.* **275**, 29426–29432.

271 TOMPA, P. (2002). Intrinsically unstructured proteins. *Trends Biochem. Sci.* **27**, 527–33.

272 DAUGHDRILL, G. W., ACKERMAN, J., ISERN, N. G., BOTUYAN, M. V., ARROWSMITH, C., WOLD, M. S. & LOWRY, D. F. (2001). The weak interdomain coupling observed in the 70 kDa subunit of human replication protein A is unaffected by ssDNA binding. *Nucleic Acids Res.* **29**, 3270–3276.

273 CHOTHIA, C. & LESK, A. M. (1986). The relation between the divergence of sequence and structure in proteins. *EMBO J.* **5**, 823–826.

274 PHILLIPS, A., JANIES, D. & WHEELER, W. (2000). Multiple sequence alignment in phylogenetic analysis. *Mol. Phylogenet. Evol.* **16**, 317–30.

275 BROWN, C. J., TAKAYAMA, S., CAMPEN, A. M., VISE, P., MARSHALL, T., OLDFIELD, C. J., WILLIAMS, C. J. & DUNKER, A. K. (2002). Evolutionary rate heterogeneity in proteins with long disordered regions. *J. Mol. Evol.* **55**, 104–110.

276 SHEN, J. C., LAO, Y., KAMATH-LOEB, A., WOLD, M. S. & LOEB, L. A. (2003). The N-terminal domain of the large subunit of human replication protein A binds to Werner syndrome protein and stimulates helicase activity. *Mech. Ageing. Dev.* **124**, 921–30.

277 LONGHESE, M. P., PLEVANI, P. & LUCCHINI, G. (1994). Replication factor A is required in vivo for DNA replication, repair, and recombination. *Mol. Cell. Biol.* **14**, 7884–90.

278 UMEZU, K., SUGAWARA, N., CHEN, C., HABER, J. E. & KOLODNER, R. D. (1998). Genetic analysis of yeast RPA1 reveals its multiple functions in DNA metabolism. *Genetics* **148**, 989–1005.

279 WOLD, M. S. (1997). Replication protein A: a heterotrimeric, single-stranded DNA-binding protein required for eukaryotic DNA metabolism. *Annu. Rev. Biochem.* **66**, 61–92.

280 GOLDMAN, N., THORNE, J. L. & JONES, D. T. (1998). Assessing the impact of secondary structure and solvent accessibility on protein evolution. *Genetics* **149**, 445–58.

281 LIO, P., GOLDMAN, N., THORNE, J. L. & JONES, D. T. (1998). PASSML: combining evolutionary inference and protein secondary structure prediction. *Bioinformatics* **14**, 726–33.

282 THORNE, J. L., GOLDMAN, N. & JONES, D. T. (1996). Combining protein evolution and secondary structure. *Mol. Biol. Evol.* **13**, 666–73.

283 DAYHOFF, M. O., SCHWARTZ, R. M. & ORCUTT, B. C. (1978). A model of evolutionary change in proteins. *Atlas of Protein Sequence and Structure* **5**, 345–352.

284 HENIKOFF, S. & HENIKOFF, J. G. (1992). Amino acid substitution matrices from protein blocks. *Proc. Natl. Acad. Sci. U.S.A.* **89**, 10915–9.

285 RADIVOJAC, P., OBRADOVIC, Z., BROWN, C. J. & DUNKER, A. K. (2002). Improving sequence alignments for intrinsically disordered proteins. *Pac. Symp. Biocomput.* **7**, 589–600.

286 RADIVOJAC, P., OBRADOVIC, Z., BROWN, C. J. & DUNKER, A. K. (2002). *Pac. Symp. Biocomput.*

287 BROWN, C. J., TAKAYAMA, S., CAMPEN, A. M., VISE, P., MARSHALL, T. W., OLDFIELD, C. J., WILLIAMS, C. J. & DUNKER, A. K. (2002). Evolutionary rate heterogeneity in proteins with long disordered regions. *J. Mol. Evol.* **55**, 104–10.

288 LUBY-PHELPS, K. (2000). Cytoarchitecture and physical properties of cytoplasm: volume, viscosity, diffusion, intracellular surface area. *Int. Rev. Cytol.* **192**, 189–221.

289 ZIMMERMAN, S. B. & MINTON, A. P. (1993). Macromolecular crowding: biochemical, biophysical and physiological consequences. *Annu. Rev. Biophys. Biomol. Struct.* **22**, 27–65.

290 FULTON, A. B. (1982). How crowded is the cytoplasm? *Cell* **30**, 345–347.

291 DAVIS-SEARLES, P. R., SAUNDERS, A. J., ERIE, D. A., WINZOR, D. J. & PIELAK, G. J. (2001). Interpreting the effects of small uncharged solutes on protein-folding equilibria. *Annu. Rev. Biophys. Biomol. Struct.* **30**, 271–306.

292 Minton, A. P. (2001). The influence of macromolecular crowding and macromolecular confinement on biochemical reactions in physiological media. *J. Biol. Chem.* **276**, 10577–80.

293 Sasahara, K., McPhie, P. & Minton, A. P. (2003). Effect of dextran on protein stability and conformation attributed to macromolecular crowding. *J. Mol. Biol.* **326**, 1227–37.

294 Elston, T. C. & Oster, G. (1997). Protein turbines. I: The bacterial flagellar motor. *Biophys. J.* **73**, 703–21.

295 Hughes, K. T., Gillen, K. L., Semon, M. J. & Karlinsey, J. E. (1993). Sensing structural intermediates in bacterial flagellar assembly by export of a negative regulator. *Science* **262**, 1277–80.

296 Schulman, B. A., Kim, P. S., Dobson, C. M. & Redfield, C. (1997). A residue-specific NMR view of the non-cooperative unfolding of a molten globule. *Nat. Struct. Biol.* **4**, 630–634.

297 Redfield, C., Schulman, B. A., Milhollen, M. A., Kim, P. S. & Dobson, C. M. (1999). Alpha-lactalbumin forms a compact molten globule in the absence of disulfide bonds. *Nat. Struct. Biol.* **6**, 948–52.

298 McParland, V. J., Kalverda, A. P., Homans, S. W. & Radford, S. E. (2002). Structural properties of an amyloid precursor of beta(2)-microglobulin. *Nat. Struct. Biol.* **9**, 326–331.

299 Shimba, N., Serber, Z., Ledwidge, R., Miller, S. M., Craik, C. S. & Dötsch, V. (2003). Quantitative identification of the protonation state of histidines in vitro and in vivo. *Biochemistry* **42**, 9227–34.

300 Serber, Z. & Dötsch, V. (2001). In-cell NMR spectroscopy. *Biochemistry* **40**, 14317–23.

301 Serber, Z., Ledwidge, R., Miller, S. M. & Dötsch, V. (2001). Evaluation of parameters critical to observing proteins inside living *Escherichia coli* by in-cell NMR spectroscopy. *J. Am. Chem. Soc.* **123**, 8895–901.

302 Honnappa, S., Cutting, B., Jahnke, W., Seelig, J. & Steinmetz, M. O. (2003). Thermodynamics of the Op18/stathmin-tubulin interaction. *J. Biol. Chem.* **278**, 38926–34.

303 Demchenko, A. P. (2001). Recognition between flexible protein molecules: induced and assisted folding. *J. Mol. Recognit.* **14**, 42–61.

304 Dyson, H. J. & Wright, P. E. (2002). Coupling of folding and binding for unstructured proteins. *Curr. Opin. Struct. Biol.* **12**, 54–60.

305 Spolar, R. S. & Record, M. T., Jr. (1994). Coupling of local folding to site-specific binding of proteins to DNA. *Science* **263**, 777–84.

306 Uversky, V. N. & Narizhneva, N. V. (1998). Effect of natural ligands on the structural properties and conformational stability of proteins. *Biochemistry (Mosc)* **63**, 420–33.

307 Uversky, V. N. (2003). A rigidifying union: The role of ligands in protein structure and stability. In *Recent Research Developments in Biophysics & Biochemistry* (Pandalai, S. G., ed.), Vol. 3, pp. 711–745. Transworld Research Network, Kerala, India.

308 Zwahlen, C., Li, S. C., Kay, L. E., Pawson, T. & Forman-Kay, J. D. (2000). Multiple modes of peptide recognition by the PTB domain of the cell fate determinant Numb. *EMBO J.* **19**, 1505–15.

309 Li, S. C., Zwahlen, C., Vincent, S. J., McGlade, C. J., Kay, L. E., Pawson, T. & Forman-Kay, J. D. (1998). Structure of a Numb PTB domain-peptide complex suggests a basis for diverse binding specificity. *Nat. Struct. Biol.* **5**, 1075–83.

310 Jen-Jacobson, L., Engler, L. E. & Jacobson, L. A. (2000). Structural and thermodynamic strategies for site-specific DNA binding proteins. *Structure Fold. Des.* **8**, 1015–23.

311 Wester, M. R., Johnson, E. F., Marques-Soares, C., Dansette, P. M., Mansuy, D. & Stout, C. D. (2003). Structure of a substrate complex of mammalian cytochrome P450 2C5 at 2.3 A resolution: evidence for multiple substrate binding modes. *Biochemistry* **42**, 6370–9.

312 Furuke, K., Shiraishi, M.,

Mostowski, H. S. & Bloom, E. T. (1999). Fas ligand induction in human NK cells is regulated by redox through a calcineurin-nuclear factors of activated T cell-dependent pathway. *J. Immunol.* **162**, 1988–93.

313 Liu, J., Farmer, J. D., Jr., Lane, W. S., Friedman, J., Weissman, I. & Schreiber, S. L. (1991). Calcineurin is a common target of cyclophilin-cyclosporin A and FKBP-FK506 complexes. *Cell* **66**, 807–815.

314 O'Day, D. H. (2003). CaMBOT: profiling and characterizing calmodulin-binding proteins. *Cell Signal.* **15**, 347–54.

315 Zhang, L. & Lu, Y. T. (2003). Calmodulin-binding protein kinases in plants. *Trends Plant Sci.* **8**, 123–7.

316 Urbauer, J. L., Short, J. H., Dow, L. K. & Wand, A. J. (1995). Structural analysis of a novel interaction by calmodulin: high-affinity binding of a peptide in the absence of calcium. *Biochemistry* **34**, 8099–8109.

317 Sandak, B., Wolfson, H. J. & Nussinov, R. (1998). Flexible docking allowing induced fit in proteins: insights from an open to closed conformational isomers. *Proteins* **32**, 159–74.

318 Yang, S. A. & Klee, C. (2002). Study of calcineurin structure by limited proteolysis. *Methods Mol. Biol.* **172**, 317–34.

319 Manalan, A. S. & Klee, C. B. (1983). Activation of calcineurin by limited proteolysis. *Proc. Natl. Acad. Sci. U.S.A.* **80**, 4291–4295.

320 Zidek, L., Novotny, M. V. & Stone, M. J. (1999). Increased protein backbone conformational entropy upon hydrophobic ligand binding. *Nat. Struct. Biol.* **6**, 1118–21.

321 Loh, A. P., Pawley, N., Nicholson, L. K. & Oswald, R. E. (2001). An increase in side chain entropy facilitates effector binding: NMR characterization of the side chain methyl group dynamics in Cdc42Hs. *Biochemistry* **40**, 4590–600.

322 Leontiev, V. V., Uversky, V. N., Permyakov, E. A. & Murzin, A. G. (1993). Introduction of Ca(2+)-binding amino-acid sequence into the T4 lysozyme. *Biochim. Biophys. Acta.* **1162**, 84–8.

323 VanScyoc, W. S. & Shea, M. A. (2001). Phenylalanine fluorescence studies of calcium binding to N-domain fragments of Paramecium calmodulin mutants show increased calcium affinity correlates with increased disorder. *Protein Sci.* **10**, 1758–68.

324 Sorensen, B. R. & Shea, M. A. (1998). Interactions between domains of apo calmodulin alter calcium binding and stability. *Biochemistry* **37**, 4244–53.

325 Fredricksen, R. S. & Swenson, C. A. (1996). Relationship between stability and function for isolated domains of troponin C. *Biochemistry* **35**, 14012–26.

326 Li, Z., Stafford, W. F. & Bouvier, M. (2001). The metal ion binding properties of calreticulin modulate its conformational flexibility and thermal stability. *Biochemistry* **40**, 11193–201.

327 Eakin, C. M., Knight, J. D., Morgan, C. J., Gelfand, M. A. & Miranker, A. D. (2002). Formation of a copper specific binding site in non-native states of beta-2-microglobulin. *Biochemistry* **41**, 10646–10656.

328 Opitz, U., Rudolph, R., Jaenicke, R., Ericsson, L. & Neurath, H. (1987). Proteolytic dimers of porcine muscle lactate dehydrogenase: characterization, folding, and reconstitution of the truncated and nicked polypeptide chain. *Biochemistry* **26**, 1399–406.

329 Namba, K. (2001). Roles of partly unfolded conformations in macromolecular self-assembly. *Genes Cells* **6**, 1–12.

330 Hoh, J. H. (1998). Functional protein domains from the thermally driven motion of polypeptide chains: a proposal. *Proteins* **32**, 223–228.

331 Hupp, T. R., Meek, D. W., Midgley, C. A. & Lane, D. P. (1992). Regulation of the specific DNA binding function of p53. *Cell* **71**, 875–86.

332 Brown, H. G. & Hoh, J. H. (1997). Entropic exclusion by neurofilament sidearms: a mechanism for maintaining interfilament spacing. *Biochemistry* **36**, 15035–15040.

333 KUMAR, S., YIN, X., TRAPP, B. D., HOH, J. H. & PAULAITIS, M. E. (2002). Relating interactions between neurofilaments to the structure of axonal neurofilament distributions through polymer brush models. *Biophys. J.* **82**, 2360–2372.

334 KUMAR, S., YIN, X., TRAPP, B. D., PAULAITIS, M. E. & HOH, J. H. (2002). Role of long-range repulsive forces in organizing axonal neurofilament distributions: evidence from mice deficient in myelin-associated glycoprotein. *J. Neurosci. Res.* **68**, 681–690.

335 WISSMANN, R., BAUKROWITZ, T., KALBACHER, H., KALBITZER, H. R., RUPPERSBERG, J. P., PONGS, O., ANTZ, C. & FAKLER, B. (1999). NMR structure and functional characteristics of the hydrophilic N terminus of the potassium channel beta-subunit Kvbeta1.1. *J. Biol. Chem.* **274**, 35521–35525.

336 ARMSTRONG, C. M. & BEZANILLA, F. (1977). Inactivation of the sodium channel. II. Gating current experiments. *J. Gen. Physiol.* **70**, 567–590.

337 HOSHI, T., ZAGOTTA, W. N. & ALDRICH, R. W. (1990). Biophysical and molecular mechanisms of *Shaker* potassium channel inactivation. *Science* **250**, 533–538.

338 HELMES, M., TROMBITAS, K., CENTNER, T., KELLERMAYER, M., LABEIT, S., LINKE, W. A. & GRANZIER, H. (1999). Mechanically driven contour-length adjustment in rat cardiac titin's unique N2B sequence. *Circ. Res.* **84**, 1339–1352.

339 FONTANA, A., FASSINA, G., VITA, C., DALZOPPO, D., ZAMAI, M. & ZAMBONIN, M. (1986). Correlation between sites of limited proteolysis and segmental mobility in thermolysin. *Biochemistry* **25**, 1847–1851.

340 FONTANA, A., DE LAURETO, P. P., DE FILIPPIS, V., SCARAMELLA, L. & ZAMBONIN, M. (1999). Limited proteolysis in the study of protein conformation. In *Proteolytic Enzymes: Tool and Targets*, pp. 257–284, Springer Verlag.

341 MARKS, F. (1996). *Protein Phosphorylation*, VCH Weinheim, New York, Basel, Cambridge, Tokyo.

342 JOHNSON, L. N. & LEWIS, R. J. (2001). Structural basis for control by phosphorylation. *Chem. Rev.* **101**, 2209–2242.

343 IAKOUCHEVA, L. M., RADIVOJAC, P., BROWN, C. J., O'CONNOR, T. R., SIKES, J. G., OBRADOVIC, Z. & DUNKER, A. K. (2004). The importance of intrinsic disorder for protein phosphorylation. *Nucleic Acids Res.* **32**, 1037–49.

344 KWONG, P. D., WYATT, R., DESJARDINS, E., ROBINSON, J., CULP, J. S., HELLMIG, B. D., SWEET, R. W., SODROSKI, J. & HENDRICKSON, W. A. (1999). Probability analysis of variational crystallization and its application to gp120, the exterior envelope glycoprotein of type 1 human immunodeficiency virus (HIV-1). *J. Biol. Chem.* **274**, 4115–23.

345 BOSSEMEYER, D., ENGH, R. A., KINZEL, V., PONSTINGL, H. & HUBER, R. (1993). Phosphotransferase and substrate binding mechanism of the cAMP-dependent protein kinase catalytic subunit from porcine heart as deduced from the 2.0 A structure of the complex with Mn^{2+} adenylyl imidodiphosphate and inhibitor peptide PKI(5–24). *EMBO J.* **12**, 849–59.

346 NARAYANA, N., COX, S., SHALTIEL, S., TAYLOR, S. S. & XUONG, N. (1997). Crystal structure of a polyhistidine-tagged recombinant catalytic subunit of cAMP-dependent protein kinase complexed with the peptide inhibitor PKI(5–24) and adenosine. *Biochemistry* **36**, 4438–4448.

347 LOWE, E. D., NOBLE, M. E., SKAMNAKI, V. T., OIKONOMAKOS, N. G., OWEN, D. J. & JOHNSON, L. N. (1997). The crystal structure of a phosphorylase kinase peptide substrate complex: kinase substrate recognition. *Embo. J.* **16**, 6646–6658.

348 TER HAAR, E., COLL, J. T., AUSTEN, D. A., HSIAO, H. M., SWENSON, L. & JAIN, J. (2001). Structure of GSK3beta reveals a primed phosphorylation mechanism. *Nat. Struct. Biol.* **8**, 593–596.

349 HUBBARD, S. R. (1997). Crystal structure of the activated insulin receptor tyrosine kinase in complex with peptide substrate and ATP analog. *Embo. J.* **16**, 5572–5581.

350 MCDONALD, I. K. & THORNTON, J. M. (1994). Satisfying hydrogen bonding potential in proteins. *J. Mol. Biol.* **238**, 777–793.

351 TANFORD, C. (1968). Protein denaturation. *Adv. Protein Chem.* **23**, 121–282.

352 FLORY, P. J. (1969). *Statistical Mechanics of Chain Molecules*, John wiley.

353 BALDWIN, R. L. (2002). A new perspective on unfolded proteins. *Adv Protein Chem* **62**, 361–7.

354 PAPPU, R. V., SRINIVASAN, R. & ROSE, G. D. (2000). The Flory isolated-pair hypothesis is not valid for polypeptide chains: implications for protein folding. *Proc. Natl. Acad. Sci. U.S.A.* **97**, 12565–12570.

355 ZHOU, H. X. (2004). Polymer models of protein stability, folding, and interactions. *Biochemistry* **43**, 2141–54.

356 HUBER, R. (1979). Conformational flexibility in protein molecules. *Nature* **280**, 538–539.

357 KOSSIAKOFF, A. A., CHAMBERS, J. L., KAY, L. M. & STROUD, R. M. (1977). Structure of bovine trypsinogen at 1.9 A resolution. *Biochemistry* **16**, 654–664.

358 BENNETT, W. S. & HUBER, R. (1984). Structural and functional aspects of domain motions in proteins. *Crit. Rev. Biochem.* **15**, 291–384.

359 ROMERO, P., OBRADOVIC, Z., LI, X., GARNER, E. C., BROWN, C. J. & DUNKER, A. K. (2001). Sequence complexity and disordered protein. *Proteins: Structure, Function, Genetics* **42**, 38–49.

360 PLAXCO, K. W., SIMONS, K. T. & BAKER, D. (1998). Contact order, transition state placement and the refolding rates of single domain proteins. *J Mol Biol* **277**, 985–94.

361 GROMIHA, M. M. & SELVARAJ, S. (2001). Comparison between long-range interactions and contact order in determining the folding rate of two-state proteins: application of long-range order to folding rate prediction. *J Mol Biol* **310**, 27–32.

362 MIRNY, L. & SHAKHNOVICH, E. (2001). Protein folding theory: from lattice to all-atom models. *Annu Rev Biophys Biomol Struct* **30**, 361–96.

363 ZHOU, H. & ZHOU, Y. (2002). Folding rate prediction using total contact distance. *Biophys J* **82**, 458–63.

364 MAKAROV, D. E. & PLAXCO, K. W. (2003). The topomer search model: A simple, quantitative theory of two-state protein folding kinetics. *Protein Sci* **12**, 17–26.

365 MAKAROV, D. E., KELLER, C. A., PLAXCO, K. W. & METIU, H. (2002). How the folding rate constant of simple, single-domain proteins depends on the number of native contacts. *Proc Natl Acad Sci USA* **99**, 3535–9.

366 DEBE, D. A. & GODDARD, W. A., 3rd. (1999). First principles prediction of protein folding rates. *J Mol Biol* **294**, 619–25.

367 DEBE, D. A., CARLSON, M. J. & GODDARD, W. A., 3rd. (1999). The topomer-sampling model of protein folding. *Proc Natl Acad Sci USA* **96**, 2596–601.

368 KAY, L. E., KEIFER, P. & SAARINEN, T. (1992). Pure absorption gradient enhanced heteronuclear single quantum correlation spectroscopy with improved sensitivity. *J. Am. Chem. Soc.* **114**, 10663.

369 MUHANDIRAM, D. R. & KAY, L. E. (1994). Gradient-enhanced triple-resonance three-dimensional NMR experiments with improved sensitivity. *J. of Magn. Reson. B* **3**, 203–216.

370 WITTEKIND, M. & MUELLER, L. (1993). HNCACB, a high-sensitivity 3D NMR experiment to correlate amide-proton and nitrogen resonances with the alpha- and beta-carbon resonances in proteins. *J. of Magn. Reson. B* **2**, 201–205.

371 DINGLEY, A. J., MACKAY, J. P., CHAPMAN, B. E., MORRIS, M. B., KUCHEL, P. W., HAMBLY, B. D. & KING, G. F. (1995). Measuring protein self-association using pulsed-field-

gradient NMR spectroscopy: application to myosin light chain 2. *J. Biomol. NMR* **6**, 321–8.

372 STEJSKAL, E. O. & TANNER, J. E. (1965). Spin diffusion measurements spin echoes in the presence of a time dependent field gradient. *J. Chem. Phys.* **42**, 288–292.

373 PENG, J. W. & WAGNER, G. (1995). Frequency spectrum of NH bonds in eglin c from spectral density mapping at multiple fields. *Biochemistry* **34**, 16733–52.

374 FARROW, N. A., ZHANG, O. W., SZABO, A., TORCHIA, D. A. & KAY, L. E. (1995). Spectral density-function mapping using N-15 relaxation data exclusively. *J. Biomol. NMR* **6**, 153–162.

375 PLANSON, A. G., GUIJARRO, J. I., GOLDBERG, M. E. & CHAFFOTTE, A. F. (2003). Assistance of maltose binding protein to the in vivo folding of the disulfide-rich C-terminal fragment from Plasmodium falciparum merozoite surface protein 1 expressed in *Escherichia coli*. *Biochemistry* **42**, 13202–11.

376 STOUT, G. H. & JENSEN, L. H. (1989). *X-Ray Structure Determination: A Practical Guide*, Wiley-Interscience, New York.

377 MCREE, D. E. (1997). *Practical Protein Crystallography*. 2nd edit, Academic Press, New York.

378 IKEMOTO, N., NAGY, B., BHATNAGAR, G. M. & GERGELY, J. (1974). Studies on a metal-binding protein of the sarcoplasmic reticulum. *J. Biol. Chem.* **249**, 2357–65.

379 OSTVALD, T. V., MACLENNON, D. H. & DORRINGTON, K. J. (1974). Effects of Cation Binding on the Conformation of Calsequestrin and the High Affinity Calcium-binding Protein of Sarcoplasmic Reticulum. *J. Biol. Chem.* **249**, 567–5871.

380 AARON, B. M., OIKAWA, K., REITHMEIER, R. A. & SYKES, B. D. (1984). Characterization of skeletal muscle calsequestrin by 1H NMR spectroscopy. *J. Biol. Chem.* **259**, 11876–11881.

381 HE, Z., DUNKER, A. K., WESSON, C. R. & TRUMBLE, W. R. (1993). Ca(2+)-induced folding and aggregation of skeletal muscle sarcoplasmic reticulum calsequestrin. The involvement of the trifluoperazine-binding site. *J. Biol. Chem.* **268**, 24635–24641.

382 IKEMOTO, N., BHATNAGAR, G. M., NAGY, B. & GERGELY, J. (1972). Interaction of divalent cations with the 55,000-dalton protein component of the sarcoplasmic reticulum. Studies of fluorescence and circular dichroism. *J. Biol. Chem.* **247**, 7835–7.

383 COZENS, B. & REITHMEIER, R. A. (1984). Size and shape of rabbit skeletal muscle calsequestrin. *J. Biol. Chem.* **259**, 6248–6252.

384 OHNISHI, M. & REITHMEIER, R. A. (1987). Fragmentation of rabbit skeletal muscle calsequestrin: spectral and ion binding properties of the carboxyl-terminal region. *Biochemistry* **26**, 7458–65.

385 MITCHELL, R. D., SIMMERMAN, H. K. & JONES, L. R. (1988). Ca2+ binding effects on protein conformation and protein interactions of canine cardiac calsequestrin. *J Biol. Chem.* **263**, 1376–81.

386 WILLIAMS, R. W. & BEELER, T. J. (1986). Secondary structure of calsequestrin in solutions and in crystals as determined by Raman spectroscopy. *J. Biol. Chem.* **261**, 12408–12413.

387 MAURER, A., TANAKA, M., OZAWA, T. & FLEISCHER, S. (1985). Purification and crystallization of the calcium binding protein of sarcoplasmic reticulum from skeletal muscle. *Proc. Natl. Acad. Sci. U.S.A.* **82**, 4036–40.

388 FRANZINI-ARMSTRONG, C., KENNEY, L. J. & VARRIANO-MARSTON, E. (1987). The structure of calsequestrin in triads of vertebrate skeletal muscle: a deep-etch study. *J. Cell Biol.* **105**, 49–56.

389 KANG, C. H., TRUMBLE, W. R. & DUNKER, A. K. (2001). Crystallization and structure-function of calsequestrin. In *Calcium Binding Protein Protocols: volume 1 reviews and case studies* (VOGEL, H. J., ed.), Vol. 172, pp. 281–294. Humana Press, Totowa, New Jersey.

390 WANG, S., TRUMBLE, W. R., LIAO, H., WESSON, C. R., DUNKER, A. K. & KANG, C. H. (1998). Crystal structure of calsequestrin from rabbit skeletal muscle sarcoplasmic reticulum. *Nat. Struct. Biol.* **5**, 476–483.

391 MCPHERSON, A. (1990). Current approaches to macromolecular crystallization. *Eur. J. Biochem.* **189**, 1–23.

392 JANCARIK, J. & KIM, S.-H. (1991). Sparse matrix sampling: a screening method for crystallization of proteins. *J. Appl. Cryst.* **24**, 409–411.

393 FLIEGEL, L., OHNISHI, M., CARPENTER, M. R., KHANNA, V. K., REITHMEIER, R. A. F. & MACLENNAN, D. H. (1987). Amino acid sequence of rabbit fast-twitch skeletal muscle calsequestrin deduced from cDNA and peptide sequencing. *Proc. Natl. Acad. Sci. U.S.A.* **84**, 1167–1171.

394 SCOTT, B. T., SIMMERMAN, H. K., COLLINS, J. H., NADAL-GINARD, B. & JONES, L. R. (1988). Complete amino acid sequence of canine cardiac calsequestrin deduced by cDNA cloning. *J Biol. Chem.* **263**, 8958–64.

395 GILL, S. C. & VON HIPPEL, P. H. (1989). Calculation of protein extinction coefficients from amino acid sequence data. *Anal Biochem* **182**, 319–26.

396 ANTZ, C., GEYER, M., FAKLER, B., SCHOTT, M. K., GUY, H. R., FRANK, R., RUPPERSBERG, J. P. & KALBITZER, H. R. (1997). NMR structure of inactivation gates from mammalian voltage-dependent potassium channels. *Nature* **385**, 272–275.

397 KORZHNEV, D. M., SALVATELLA, X., VENDRUSCOLO, M., DI NARDO, A. A., DAVIDSON, A. R., DOBSON, C. M. & KAY, L. E. (2004). Low-populated folding intermediates of Fyn SH3 characterized by relaxation dispersion NMR. *Nature* **430**, 586–590.

398 SPIRO, T. G., GABER, B. P. (1977). Laser Raman scattering as a probe of protein structure. *Annu. Rev. Biochem.* **46**, 553–572.

399 MAITI, N. C., APETRI, M. M., ZAGORSKI, M. G., CAREY, P. R. & ANDERSON, V. E. (2004). Raman spectroscopic characterization of secondary structure in natively unfolded proteins: alpha-synuclein. *J. Am. Chem. Soc.* **126**, 2399–2408.

400 SYME, C. D., BLANCH, E. W., HOLT, C., JAKES, R., GOEDERT, M., HECHT, L. & BARRON, L. D. (2002). A Raman optical activity study of rheomorphism in caseins, synucleins and tau. New insight into the structure and behaviour of natively unfolded proteins. *Eur. J. Biochem.* **269**, 148–156.

401 DAUGHDRILL, G. W., VISE, P. D., ZHOU, H., YANG, X., YU, W. F., TASAYCO, M. L. & LOWRY, D. F. (2004). Reduced spectral density mapping of a partially folded fragment of E. coli thioredoxin. *J. Biomol. Struct. Dyn.* **21**, 663–670.

402 KRIWACKI, R. W., HENGST, L., TENNANT, L., REED, S. I. & WRIGHT, P. E. (1996). Structural studies of p21Waf1/Cip1/Sdi1 in the free and Cdk2-bound state: conformational disorder mediates binding diversity. *Proc. Natl. Acad. Sci. U.S.A.* **93**, 11504–11509.

403 KRIWACKI, R. W., WU, J., TENNANT, J., WRIGHT, P. E. & SIUZDAK, G. (1997). Probing protein structure using biochemical and biophysical methods. Proteolysis, matrix-assisted laser desorption/ionization mass spectrometry, high-performance liquid chromatography and size-exclusion chromatography of p21Waf1/Cip1/Sdi1. *J. Chromatogr. A* **777**, 23–30.

404 SADQI, M., HERNANDEZ, F., PAN, U., PEREZ, M., SCHAEBERLE, M. D., AVILA, J. & MUNOZ, V. (2002). Alpha-helix structure in Alzheimer's disease aggregates of tau-protein. *Biochemistry* **41**, 7150–7155.

405 GHAEMMAGHAMI, S., FITZGERALD, M. C. & OAS, T. G. (2000). A quantitative, high-throughput screen for protein stability. *Proc. Natl. Acad. Sci. USA* **97**, 8296–8301.

406 GONG, H., ISOM, D. G., SRINIVASAN, R. & ROSE, G. D. (2003). Local secondary structure content predicts folding rates for simple, two-state proteins. *J. Mol. Biol.* **327**, 1149–1154.

9
The Catalysis of Disulfide Bond Formation in Prokaryotes

Jean-Francois Collet and James C. Bardwell

9.1
Introduction

Oxidation of two cysteine residues leads to the formation of a disulfide bond and the concomitant release of two electrons. The formation of disulfide bonds is a required step in the folding pathway of many secreted proteins. It takes place in the eukaryotic endoplasmic reticulum or in the bacterial periplasm. In contrast, this oxidation reaction is harmful to most cytoplasmic proteins and may lead to protein misfolding and aggregation. Both eukaryotic and prokaryotic cells possess mechanisms to ensure that cytoplasmic cysteines are kept reduced. These mechanisms involve enzymes of the thioredoxin and glutaredoxin systems.

This review will focus on disulfide bond formation in bacteria, taking *Escherichia coli* as a model. In the first two sections we will discuss disulfide bond formation and isomerization in the periplasm. In the third section, we will present some of the techniques that have been used to study thiol-disulfide chemistry.

9.2
Disulfide Bond Formation in the *E. coli* Periplasm

9.2.1
A Small Bond, a Big Effect

In both eukaryotes and prokaryotes, structural disulfide bonds are found in extra-cytoplasmic proteins. The presence of a covalent bond between two cysteine residues confers a higher stability to these proteins and is often required for their correct folding. One striking example of the importance of disulfide bonds for the correct folding of secreted proteins is the *E. coli* flagellar protein FlgI, a component of the bacterial flagellum. The folding of FlgI requires a disulfide bond. When this disulfide is not introduced in FlgI, FlgI cannot be properly folded and a functional flagellum cannot be assembled [1]. This results in the complete loss of motility as shown in Figure 9.1.

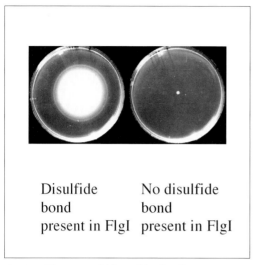

Fig. 9.1. The correct folding of the flagellum component FlgI requires a disulfide bond.

9.2.2
Disulfide Bond Formation Is a Catalyzed Process

Disulfide bonds can be formed spontaneously by molecular oxygen according to the following equation:

$$2\ \text{R-SH} + 1/2\ \text{O}_2 \rightarrow \text{RS-SR} + \text{H}_2\text{O}$$

However, this reaction is very slow, occurring on the hour timescale in vitro. Such reaction times are unrealistic in vivo, where proteins are synthesized and folded within minutes. This discrepancy led Anfinsen in the early 1960s to postulate the existence of catalysts of disulfide bond formation. In 1964, he discovered the first catalyst for disulfide bond formation, the eukaryotic protein disulfide isomerase (PDI). PDI is a multi-domain enzyme that is active in the eukaryotic endoplasmic reticulum. The eukaryotic disulfide formation pathway has been recently reviewed by Sevier and Kaiser and will not be discussed here [2].

In *E. coli*, disulfide bonds are introduced exclusively in the periplasm. The enzymes responsible are members of the Dsb (disulfide bond) protein family: DsbA and DsbB, which are involved in disulfide bond formation, and DsbC, DsbG, and DsbD, which are involved in disulfide bond isomerization.

9.2.3
DsbA, a Protein-folding Catalyst

DsbA is a 21-kDa soluble protein that is required for disulfide bond formation in the *E. coli* periplasm. In agreement with the central role of this protein in the oxi-

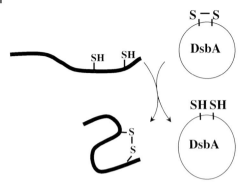

Fig. 9.2. Disulfide bond transfer from DsbA to the target protein. Oxidized DsbA is very unstable and reacts rapidly with unfolded proteins entering the periplasm. The disulfide bond is transferred onto the target protein and DsbA is reduced.

dative protein-folding pathways, $dsbA^-$ strains exhibit pleiotropic phenotypes [3]. For instance, $dsbA^-$ mutants are hypersensitive to benzylpenicillin, dithiothreitol, and metals [4, 5]. They show reduced levels of several secreted proteins such as the outer-membrane OmpA and alkaline phosphatase [3].

Like many other oxidoreductases, DsbA has a thioredoxin-like fold and a catalytic CXXC motif. In agreement with DsbA acting as a thiol-oxidase, the CXXC motif is found to be predominantly oxidized in vivo [6]. Biochemical and biophysical studies have shown that the disulfide bond present in DsbA is very unstable [7] and can be rapidly transferred to newly translocated proteins (Figure 9.2).

With a redox potential around -120 mV [7], DsbA is the second most oxidizing protein after DsbB (see below). The oxidizing power of DsbA originates from the low pK_a of Cys30, the first cysteine of the CXXC motif [8]. This residue has a pK_a of ≈ 3, which is much more acidic than the usual pK_a of cysteine residues (≈ 9). Due to its low pK_a value, Cys30 is in the thiolate-anion state at physiological pH. Structural studies have shown that this thiolate anion, which is highly accessible in the folded protein, is stabilized by electrostatic interactions [9]. The most important electrostatic interactions take place between Cys30 and the third residue of the CXXC motif, His32. Mutations of this single His32 residue increase the stability of the oxidized protein, thereby decreasing the redox potential of DsbA and the reactivity of the enzyme [10]. The stabilization of the thiolate anion is a key characteristic of DsbA, as it drives the reaction towards the reduction of DsbA and the transfer of the disulfide bond to the folding protein.

In addition to the thioredoxin-like domain, DsbA contains a second compactly folded helical domain [9]. This domain covers the CXXC active site and contains several residues that form a hydrophobic patch. This patch, together with a long and relatively deep groove running in the thioredoxin domain below the active site, is likely to interact in a hydrophobic manner with unfolded protein substrates

[11]. Such hydrophobic interactions have indeed been reported to take place. Using NMR, Couprie et al. showed that DsbA binds a model peptide in a hydrophobic way [12]. In a mixed disulfide between DsbA and the substrate protein Rnase T1, the conformational stability of Rnase T1 is decreased by 5 kJ mol^{-1} and the conformational stability of DsbA is increased by 5 kJ mol^{-1}, suggesting that DsbA interacts with the unfolded protein by preferential noncovalent interactions [13]. Proteolysis experiments also showed that oxidized DsbA is more flexible than reduced DsbA. This flexibility might help the oxidized protein to establish interactions with different partner proteins by accommodating the peptide-binding groove to various substrates. On the other hand, the increased rigidity of the reduced form could facilitate the release of oxidized products [14].

9.2.4
How is DsbA Re-oxidized?

After the transfer of the disulfide bond to target proteins, DsbA is reduced and therefore incapable of donating a disulfide bond. The protein responsible for the re-oxidation of DsbA is an inner-membrane protein called DsbB [4, 15]. Mutants lacking DsbB accumulate DsbA in the reduced state and exhibit the same pleiotropic phenotypes as *dsbA*$^-$ mutants, indicating that DsbA and DsbB are on the same pathway [15]. This was confirmed in vitro when we purified DsbB to homogeneity and showed that it is able to catalytically oxidize DsbA in the presence of molecular oxygen [16].

DsbB is a 21-kDa protein that is predicted to have four transmembrane segments and two loops protruding into the periplasm. Each of these periplasmic domains contains a conserved pair of cysteine residues: Cys41 and Cys44 in domain 1 and Cys104 and Cys130 in domain 2. These cysteines can undergo oxidation-reduction cycles. They are essential for activity, as removal of any one of these cysteines causes the accumulation of reduced DsbA [17–19]. The first pair of cysteine residues (Cys41–Cys44) is present in a CXXC motif, a motif that is often found in thioredoxin-like folds. However, the first periplasmic loop of DsbB contains no other recognizable similarity to thioredoxin, and because it is predicted to be only $\approx$17 residues long, it is far too short to contain a thioredoxin fold.

9.2.5
From Where Does the Oxidative Power of DsbB Originate?

In vivo data obtained by Kobayashi et al. suggested that DsbB is recycled by the electron transport chain [20]. We confirmed this hypothesis by successfully reconstituting the complete disulfide bond formation system in vitro [16, 21]. Using purified components, we showed that electrons flow from DsbB to ubiquinone, then onto cytochrome oxidases *bd* and *bo*, and finally to molecular oxygen. Under anaerobic conditions, electrons are transferred from DsbB to menaquinone then to fumarate or nitrate reductase (see Figure 9.3).

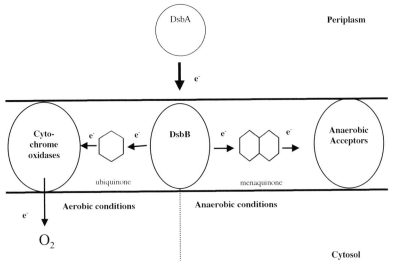

Fig. 9.3. Oxidation pathway in the disulfide periplasm. Reduced proteins are oxidized by DsbA. DsbA is re-oxidized by the inner-membrane protein DsbB. Under aerobic conditions, DsbB is re-oxidized by molecular oxygen in a ubiquinone- and cytochrome oxidase-dependent reaction. Under anaerobic conditions, electrons flow to menaquinone and then to alternative terminal acceptors.

DsbB is a key element of the disulfide bond formation pathway, as it uses the oxidizing power of quinones to generate disulfides de novo [22]. This is a novel catalytic activity that seems to be the major source of disulfide bonds in prokaryotes. We showed that purified DsbB contains a single, highly specific quinone-binding site and has a K_M of 2 µM for quinones [22]. The quinone-binding domain of DsbB seems to be in close proximity to residues 91–97 of the protein [23]. This segment is in the second periplasmic domain of DsbB, in a loop connecting transmembrane helices 3 and 4. Another residue, R48, is also likely to be involved in quinone binding. R48 mutants exhibit a major defect in catalyzing disulfide bond formation in vivo [24]. We purified some of these DsbB mutant proteins and showed that the R48H mutant has a K_M value for quinones that is seven times greater than that of wild-type DsbB, indicating that R48 is involved in quinone binding [24].

9.2.6
How Are Disulfide Bonds Transferred From DsbB to DsbA?

An interesting question is how disulfide bonds are transferred from DsbB to DsbA. Under aerobic conditions, the CXXC motif of DsbB is very difficult to reduce with dithiothreitol when quinones are present in the preparation [25]. This is one of the major arguments in favor of a direct oxidation of Cys41 and Cys44 by quinones.

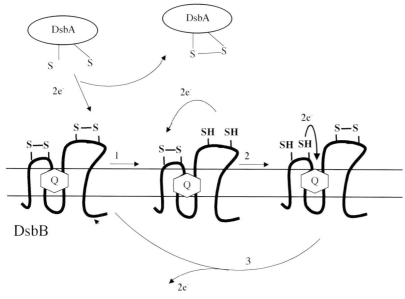

Fig. 9.4. DsbB transfers electrons by a succession of thiol-disulfide exchange reactions. Step 1: a disulfide bond is transferred from the C-terminal cysteine residues of DsbB (Cys104 and Cys130) to DsbA. Cys104 and Cys130 are reduced and DsbA is oxidized. Step 2: a disulfide bond is transferred from the N-terminal cysteine residues of DsbB (Cys41 and Cys44) to the C-terminal cysteine residues. Cys104–Cys130 are oxidized and Cys41–Cys44 are reduced. Step 3: Cys41 and Cys44 are re-oxidized by quinone reduction. Electrons are transferred to the electron transport system.

Then, according to a model first proposed by Ito and coworkers [18, 19, 26], the generated disulfide bond is transferred to the cysteine pair Cys104–Cys130 and then onto DsbA (Figure 9.4). Indeed, a mixed disulfide between Cys30 of DsbA and Cys104 of DsbB has been isolated [18]. This model has been a matter of controversy, and results obtained recently by us and two other labs do not entirely agree with it [26–28]. However, in a paper published recently, Glockshuber and coworkers reported strong arguments in favor of the proposed original mechanism [29]. They showed that Cys104–Cys130 of a quinone-depleted DsbB can indeed directly transfer a disulfide to DsbA and is then re-oxidized by the Cys41–Cys44 disulfide bond. They determined the redox potential of Cys41–Cys44 and Cys104–Cys130 disulfide bonds of DsbB to be −69 mV and −186 mV, respectively. This high redox potential of the Cys41–Cys44 disulfide bond provides strong support for the above-mentioned model, because this redox potential lies between that of ubiquinone (+100 mV) and that of DsbA. This redox potential makes the Cys41–Cys44 disulfide bond the most oxidizing disulfide ever identified in proteins.

The redox potential of Cys104–Cys130 is about 40 mV more negative than the redox potential of DsbA, but the very oxidizing Cys41–Cys44 disulfide bond could provide the thermodynamic driving force that allows the energetically uphill oxidation of DsbA by the Cys104–Cys130 disulfide in DsbB to occur.

9.2.7
How Can DsbB Generate Disulfide by Quinone Reduction?

Another very intriguing and still unanswered question is, how can DsbB form disulfide bonds by reducing quinone molecules? We recently showed that purified DsbB has a purple color, presumably due to the presence of a quinhydrone, a purple charge-transfer complex consisting of a hydroquinone and a quinone in a stacked conformation [30]. This finding suggested that the DsbB mechanism involves two quinone molecules. According to our model, one of these quinones is very tightly bound to DsbB, whereas the other one is exchangeable. The resident quinone would be directly involved in disulfide bond formation and would undergo oxidation-reduction cycles: it would be reduced during the generation of a disulfide bond and then re-oxidized by an exchangeable quinone derived from the oxidized quinones pool (see Figure 9.5).

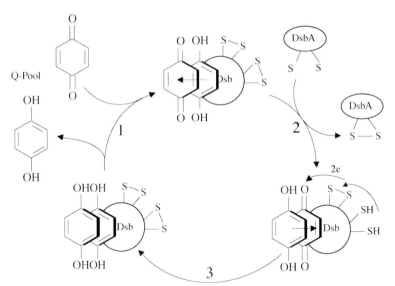

Fig. 9.5. Model for the quinone reductase activity of DsbB. The quinone reductase activity of DsbB involves two quinone molecules: a tightly bound quinone and an exchangeable quinone. The two quinones are initially reduced. Step 1: the reduced exchangeable quinone is replaced by an oxidized quinone derived from the quinone pool. A purple quinhydrone-like charge-transfer complex is formed between the reduced and the oxidized quinone molecules. Step 2: Electrons are transferred from DsbA to the C-terminal disulfide in DsbB. Then, electrons are transferred to the N-terminal disulfide of DsbB. Step 3: the N-terminal cysteine residues reduce the resident quinone. The two reduced quinone molecules do not form a stable complex, and the transient quinone is exchanged with an oxidized one.

9.3 Disulfide Bond Isomerization

9.3.1 The Protein Disulfide Isomerases DsbC and DsbG

The powerful oxidant DsbA can introduce nonnative disulfide bonds into proteins with more than two cysteines. It is important to quickly resolve these incorrect disulfides and thus prevent the accumulation of misfolded proteins. Cells therefore contain a "disulfide correction system," which, in E. coli, involves the two protein disulfide isomerases DsbC and DsbG. The proteins are homologous ($\approx$ 30% amino acid identity) and share many common properties. They seem to differ, however, in their substrate specificity. We showed, for instance, that RNAse I and MepA, which contain eight and six conserved cysteines, respectively, are substrates for DsbC but not for DsbG [31].

DsbC and DsbG are dimeric proteins with a thioredoxin-like domain and a CXXC active site motif. In contrast to DsbA, the two cysteine residues of these motifs are kept reduced in the periplasm [32]. This allows the proteins to reduce nonnative disulfides and to isomerize them to the correct form. According to the proposed mechanism for DsbC and DsbG action (Figure 9.6), the first cysteine res-

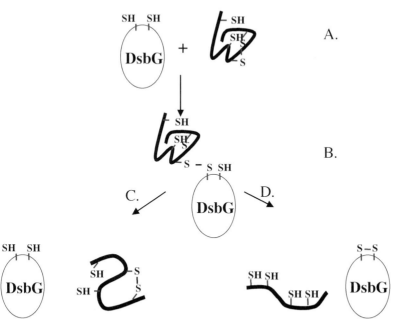

Fig. 9.6. Disulfide bond isomerization by DsbG. (A) Reduced DsbG reacts with a nonnative disulfide. (B) A mixed disulfide is formed between DsbG and the target protein. The mixed disulfide is resolved either by attack of another cysteine of the substrate (C) or by attack of the second cysteine of DsbG (D).

idue attacks a nonnative disulfide bond in the substrate protein, leading to the formation of a mixed disulfide. This mixed disulfide can then be resolved either by the attack of the second active site cysteine of the isomerase or by attack of a third cysteine of the target protein. In the first case, the nonnative disulfide in the substrate protein is reduced and the isomerase is released in the oxidized state. In the second case, alternative disulfide bonds form in the substrate protein and the isomerase is released in the reduced state.

DsbC and DsbG have been shown to be able to correct nonnative disulfide bonds both in vitro and in vivo. They have been found to assist in the refolding of eukaryotic recombinant proteins expressed in *E. coli*, such as bovine pancreatic trypsin inhibitor (three disulfide bonds), RNAse A (four disulfide bonds), and mouse urokinase (12 disulfide bonds) [22, 33–35]. We have recently shown that DsbC is required to fold at least two native *E. coli* proteins, RNAse I (four disulfides) and MepA (six conserved cysteines) [31]. Curiously, so far in vivo substrates of DsbG are restricted to eukaryotic proteins expressed in *E. coli*, and its physiological substrates are still unknown. It seems likely that DsbG has a very limited substrate specificity, which makes the identification rather difficult. Alternatively, it is conceivable that DsbG has an in vivo function that is not related to its ability to act on disulfide bonds. Further work is needed to further characterize the function and substrate specificity of DsbC and DsbG. This should help us to understand why *E. coli* has evolved two homologous protein disulfide isomerases.

9.3.2
Dimerization of DsbC and DsbG Is Important for Isomerase and Chaperone Activity

An important feature of DsbC and DsbG is the presence of a dimerization domain at the amino-terminus of the protein [36]. Although no catalytically active cysteine residues are present in this domain, this region is important for the isomerase activity of DsbC. This was suggested by limited proteolysis experiments performed on DsbC, which showed that monomeric DsbC lacking the amino-terminal domain is inactive as an isomerase and cannot catalyze the formation of correct disulfide bonds in proteins such as RNase A and bovine pancreatic trypsin inhibitor [36]. In particular, this amino-terminal domain is thought to be important for peptide binding. This was suggested by the crystal structure of DsbC, which showed that DsbC is a V-shaped dimer, with each monomer forming one arm of the V [37]. The dimerization domains form an extended cleft in the DsbC structure. The surface of this cleft is predominantly composed of hydrophobic residues, which are predicted to bind misfolded substrates. These hydrophobic interactions are proposed to stabilize the complex formed between DsbC and the target protein until the target protein reaches a more stable conformation. Depending on the redox state of the CXXC motif, the hydrophobic cleft switches between an open and a closed conformation [38]. This allows DsbC to adapt to the size and shape of the substrate proteins. In addition to the isomerase activity, both DsbC and DsbG have been reported to have a chaperone activity. DsbC can assist in the refolding of lysozyme and promotes the in vitro reactivation of denatured D-glyceraldehyde-3-phosphate dehydrogenase [39]. DsbG is able to prevent the aggregation of citrate synthase and luciferase in vitro [40]. Dimerization seems to be required for this activity, as truncated DsbC variants that lack the dimerization domain have no chaperone activity [36]. Furthermore, addition of the DsbC amino-terminal domain to DsbA or thioredoxin, two oxidoreductases that do not exhibit chaperone activity, causes their dimerization and confers some chaperone activity [41]. The formation of similar V-shaped dimers with a hydrophobic cleft by these hybrid proteins is likely to be the cause of this new activity.

9.3.3
Dimerization Protects from DsbB Oxidation

If DsbB were able to oxidize the active-site cysteines of DsbC and DsbG, the proteins would no longer be able to act as disulfide isomerases. However, we and others have found that DsbB is unable to oxidize DsbC and DsbG either in vivo or in vitro [22, 32, 35, 42]. The observed dimerization of DsbC and DsbG appears to be the molecular basis for this discrimination, as disruption of the dimer interface produces a monomeric DsbC protein that is now oxidized by DsbB both in vivo and in vitro and that can substitute for DsbA in the cell [43].

9.3.4
Import of Electrons from the Cytoplasm: DsbD

In order to stay reduced and active in the oxidizing environment of the periplasm, DsbC and DsbG have to be kept reduced. This reduction of DsbC and DsbG depends on the protein DsbD (Figure 9.7). DsbD is an inner-membrane protein of 59 kDa. It is composed of three domains (Figure 9.7): a periplasmic amino-terminal domain (α), a central membranous domain with eight transmembrane segments (β), and a periplasmic carboxy-terminal domain with a thioredoxin-like fold (γ). Each of these domains harbors a conserved pair of cysteine residues. These six conserved cysteine residues are important for DsbD activity, as removal of any one of them leads to the accumulation of oxidized DsbC and DsbG in the E. coli periplasm [44, 45].

Rietsche et al. showed that the cytoplasmic thioredoxin system is the source of reducing equivalents that keep DsbD reduced [32, 34]. The thioredoxin system, which consists of thioredoxin (TrxA), thioredoxin reductase (TrxB), and NADPH, is responsible for maintaining cytoplasmic proteins in a reduced state. Thus, the function of DsbD is to transfer electrons across the inner membrane, using the cytoplasmic pool of NADPH as a source of electrons.

The mechanism by which DsbD transfers electrons across the membrane is still unclear. In vivo and in vitro experiments suggest that electrons are transferred via a cascade of disulfide bond exchange reactions. According to the current model [46], the first step is the reduction of the two conserved cysteine residues within the transmembrane β domain by cytoplasmic TrxA. Then, electrons are successively transferred to the γ and α domains and finally onto DsbC or DsbG. Most of the steps that have been proposed by this model are well characterized, especially the reactions taking place in the cytoplasm and in the periplasm. The isolation of a

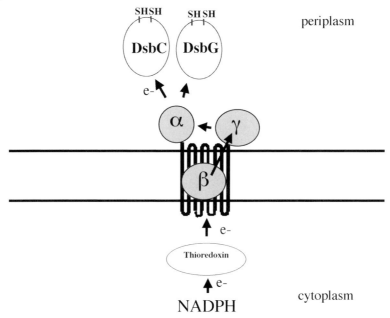

Fig. 9.7. Disulfide bond isomerization pathway in the periplasm. DsbC and DsbG are kept reduced by DsbD. DsbD is kept reduced by the thioredoxin system in the cytoplasm. The proposed direction for the electron flow is shown by the arrows.

mixed disulfide between TrxA and one of the conserved cysteines of the β domain is an argument in favor of a direct interaction between TrxA and the β domain [46]. Using purified components, we showed that electrons flow from the cysteines of the γ domain to the α domain and then to DsbC or DsbG [47]. This was further confirmed by the determination of the redox potential of the cysteine residues in the γ domain (-241 mV) and in the α domain (-229 mV). These lie midway between the redox potential of thioredoxin (-270 mV) and that of DsbC (-134 mV) [47]. The fact that the cysteine residues of the γ domain are more reducing than the cysteine residues of the α domain is consistent with the reaction being thermodynamically driven.

The structure of the γ domain has been determined, and residues that might be important for the interaction with the α domain have been identified [48]. The interaction between the α domain and DsbC or DsbG is also well characterized. In vitro, cysteines of the α domain can efficiently reduce DsbC and DsbG with a K_M of 5 μM, and the structure of a complex between the α domain and DsbC has been solved [38].

The membranous part of the DsbD reaction is still obscure. Removal of the β domain leads to the accumulation of oxidized γ and α domains, which is a strong argument that the β domain is required to transfer electrons to the periplasmic domains [46]. However, *how* electrons are transferred from the β domain to the γ domain is still unclear. Is it by pure disulfide exchange? Is a cofactor, such as a quinone, required as with DsbB? So far, there is no evidence that the cysteine residues of the β domain interact directly with the cysteine residues of the γ domain, and attempts to isolate mixed disulfides between the two domains have been unsuccessful. Moreover, according to a recent report, both cysteine residues of the β domain are on the cytoplasmic side of the membrane [49]. If this is the case, how can they reduce the cysteines of the γ domain, which are in the periplasm? We purified the β domain and showed that it can transfer electrons from TrxA to the cysteines of the γ domain in vitro [47]. However, the small amount of pure β domain that we were able to produce and the relatively low activity of the preparation prevented us from either proving or ruling out the presence of a cofactor. The important question of how electrons are transferred across the membrane is still unanswered and further work, such as crystallization studies and biochemical characterization of the β domain, is required.

9.3.5
Conclusions

Before the discovery of DsbA in 1991, it was thought that disulfide bond formation in the periplasm was an uncatalyzed process. Since the discovery of this enzyme, much work has been done and most of the important steps in the pathways of disulfide bond formation and isomerization in the *E. coli* periplasm have been unraveled. However, some very interesting questions remain unanswered. These mainly concern the mechanism of action of the two membrane proteins of the Dsb protein family, DsbB and DsbD. DsbB generates disulfides de novo via quinone reduction. More work is required to establish exactly how it performs this key reaction, which appears to be the source of the vast majority of disulfides in *E. coli*. DsbD acts to transport disulfides across the membrane, but how it does so is almost entirely unclear. Characterization of the membranous domain of DsbD will help us to understand how DsbD solves this unique problem. There are many examples of proteins involved in the transport of metals, sugars, or even complex polypeptides across the membranes, and some of these proteins are very well characterized. In contrast, DsbD and its homologues are the only proteins known to transport disulfides, and their mechanism is still obscure. It is not surprising that these two membrane proteins are the least understood of the family of Dsb proteins, as membrane proteins are traditionally difficult to work with. However, both DsbB and DsbD can be purified in large amounts, so further biochemical progress is likely.

9.4
Experimental Protocols

9.4.1
Oxidation-reduction of a Protein Sample

The common way to reduce or oxidize a protein is to incubate the protein with reduced dithiothreitol (DTT) or oxidized glutathione (GSSG), respectively. Then, the excess redox agents are removed by gel filtration or dialysis.

Procedure

1. Incubate the protein with DTT or GSSG for 30 min at 30 °C. DTT and GSSG concentrations are chosen so that they are in large excess compared to the protein. Typical DTT and GSSG concentrations vary between 10 and 20 mM, whereas protein concentrations are usually in the micromolar range. It is often overlooked that the pH for the incubation should be alkaline (8.0–9.0), as most thiols have pK_a values in this range. Thiol disulfide exchange reactions are dependent on the presence of a thiolate ion attacking a disulfide. Thus, reactions are decreased about 10-fold with each pH unit below the pK_a of the relevant thiol.
2. After the oxidation/reduction reaction, the protein is separated from excess DTT and GSSG by dialysis or gel filtration on Sephadex G25 columns. If using prepacked NAP5 or PD10 columns (Amersham), it is important to choose an elution volume allowing a complete separation of the protein from the small redox molecules.

9.4.2
Determination of the Free Thiol Content of a Protein

A simple method to determine whether the cysteine residues present in a protein are reduced or oxidized is to perform an Ellman assay. This assay relies on the reaction between dithionitrobenzoate (DTNB) and reduced thiols (Figure 9.8).

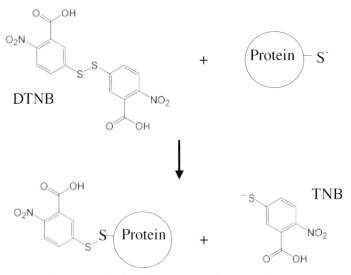

Fig. 9.8. Ellman assay for free thiols. Reaction of DTNB with thiols releases thionitrobenzoate (TNB), which absorbs at 412 nm.

The released TNB anion absorbs at 412 nm ($\varepsilon = 13\,700$ M^{-1} or $14\,150$ M^{-1} with or without 6 M guanidinium chloride, respectively), which allows the determination of the concentration of reduced cysteine residues in the protein sample.

Procedure

1. Prepare a 3-mM DNTB solution in 100 mM phosphate buffer, pH 7.4. Keep in the dark at 4 °C.
2. Prepare the protein in 100 mM phosphate buffer, pH 7.4, containing 6 M guanidinium chloride and 100 mM EDTA. Guanidinium is used to denature the protein, which increases the accessibility of all thiol groups to DTNB. Choose a protein concentration high enough to allow detection of all free thiols (absorbance of 1 µM TNB anion = 0.0137 unit). Prepare a reference sample without the protein.
3. Measure the absorbance at 412 nm of both the reference and the protein samples.
4. Add 50 µL of the DTNB solution to 1 mL of each sample.
5. Follow the increase in absorbance at 412 nm until it stops increasing.
6. Subtract the absorbance of the reference sample from the absorbance of the protein sample and calculate the molar concentration of free thiols present in the protein using the absorption coefficient (ε). This methodology performed in the absence of denaturants can also allow one to measure the accessibility and reactivity of thiol groups [50].

9.4.3
Separation by HPLC

Reverse-phase HPLC has been used to separate the reduced and oxidized forms of several redox proteins such as TrxA, DsbA, or the periplasmic domains of DsbD [47, 51]. First, the thiol-disulfide exchange reactions are stopped by acidification and the samples are loaded onto a hydrophobic HPLC column. Then, elution conditions are optimized to allow the separation of the oxidized form from the reduced form. The absorbance at 280 nm (aromatic residues) or 214 nm (peptide bond) is recorded and the respective amounts of both forms are determined by integration of the chromatogram.

The time-consuming part of this technique is the optimization of the elution conditions, as it often varies from protein to protein. It is often necessary to experiment with different column types (C4, C8, or C18 columns), to vary the elution buffer's composition, or to vary the type of elution (gradient or isocratic). Once appropriate elution conditions are found, separation of the two redox forms of a protein by HPLC is an accurate, sensitive, fast, and reliable technique. We describe below the conditions that we used to separate oxidized and reduced forms of the γ and α domains of DsbD.

Procedure

1. Add 25 µL of 1% H_3PO_4 to 25 µL of a 50 µM protein sample.
2. Add 50 µL of 50% acetonitrile in 0.1% trifluoroacetic acid (TFA).
3. Load 80 µL on a C18 column (for instance, a Phenomenex Primesphere, 250 × 4.60 mm).

A. Separation of the redox forms of the γ domain was performed using a linear aqueous 39–51% acetonitrile gradient in 0.1% trifluoroacetic acid over 25 min at a flow rate of 1 mL min^{-1}.

B. The oxidized and reduced forms of the α domain were separated using an isocratic elution with 15% methanol and 28% acetonitrile in 0.1% trifluoroacetic acid over 20 min at a flow rate of 0.5 mL min^{-1} [47].

9.4.4
Tryptophan Fluorescence

Protein tryptophan fluorescence is sensitive to changes in the environment of the chromophore residue. Formation of a disulfide bond between two cysteine residues that are in proximity to a tryptophan usually quenches its fluorescence signal. For instance, the fluorescence signal of DsbA is reduced by more than threefold upon disulfide bond formation. This can be used to determine the redox state of the protein of interest. For example, the catalytic oxidation reaction of DsbA by DsbB was characterized by following the fluorescence signal of DsbA [16].

Procedure

1. Prepare a protein solution ($\approx$1–3 µM) in a quartz cuvette. Use sodium phosphate as a buffer, pH 7.0. The fluorescence signal decreases with increasing temperature, so the buffer solution should be equilibrated at the desired temperature.
2. Use 278 nm as the excitation wavelength and record the emission signal between 300 and 400 nm. The peak of the signal should be around 350 nm.

9.4.5
Assay of Disulfide Oxidase Activity

The ability of a protein to form disulfide bonds can be assayed by hirudin folding experiments [52]. Hirudin is a 65-residue protein, purified from the medicinal leech, with three disulfide bonds in the native state. It is a well-characterized substrate for DsbA. The stoichiometric reaction between oxidized DsbA and reduced hirudin is characterized by a rapid random oxidation of hirudin by DsbA. Then, reduced DsbA, formed in the course of hirudin oxidation, catalyzes the slow isomerization of the nonnative disulfide bonds to form native hirudin. The formation of native hirudin can be analyzed by reverse-phase HPLC.

Procedure

1. Mix completely reduced and unfolded hirudin (≈ 25 µM) with three molar equivalents of oxidized DsbA in 100 mM sodium phosphate pH 7.0, 0.2 M KCl, and 1 mM EDTA.
2. Incubate at 25 °C. At different times, take 50-µL aliquots and quench disulfide exchange reactions by addition of 50 µL formic acid (final pH < 2).
3. Separate the hirudin folding intermediates at 55 °C on a C18 reversed-phase HPLC column using a gradient from 20% to 24% (v/v) acetonitrile in 0.1% (v/v) trifluoroacetic acid at a flow rate of 0.5 mL min^{-1}.

References

1 DAILEY, F. E. & BERG, H. C. (1993). Mutants in disulfide bond formation that disrupt flagellar assembly in Escherichia coli. *Proc Natl Acad Sci USA* **90**, 1043–7.
2 SEVIER, C. S. & KAISER, C. A. (2002). Formation and transfer of disulphide bonds in living cells. *Nat Rev Mol Cell Biol* **3**, 836–47.
3 BARDWELL, J. C., McGOVERN, K. & BECKWITH, J. (1991). Identification of a protein required for disulfide bond formation in vivo. *Cell* **67**, 581–9.
4 MISSIAKAS, D., GEORGOPOULOS, C. & RAINA, S. (1993). Identification and characterization of the Escherichia coli gene dsbB, whose product is involved in the formation of disulfide bonds in vivo. *Proc Natl Acad Sci USA* **90**, 7084–8.
5 STAFFORD, S. J., HUMPHREYS, D. P. & LUND, P. A. (1999). Mutations in dsbA and dsbB, but not dsbC, lead to an enhanced sensitivity of Escherichia coli to Hg2+ and Cd2+. *FEMS Microbiol Lett* **174**, 179–84.
6 KISHIGAMI, S., AKIYAMA, Y. & ITO, K. (1995). Redox states of DsbA in the periplasm of Escherichia coli. *FEBS Lett* **364**, 55–8.
7 ZAPUN, A., BARDWELL, J. C. & CREIGHTON, T. E. (1993). The reactive and destabilizing disulfide bond of DsbA, a protein required for protein disulfide bond formation in vivo. *Biochemistry* **32**, 5083–92.
8 GRAUSCHOPF, U., WINTHER, J. R., KORBER, P., ZANDER, T., DALLINGER, P. & BARDWELL, J. C. (1995). Why is DsbA such an oxidizing disulfide catalyst? *Cell* **83**, 947–55.
9 MARTIN, J. L., BARDWELL, J. C. & KURIYAN, J. (1993). Crystal structure of the DsbA protein required for disulfide bond formation in vivo. *Nature* **365**, 464–8.
10 GUDDAT, L. W., BARDWELL, J. C., GLOCKSHUBER, R., HUBER-WUNDERLICH, M., ZANDER, T. & MARTIN, J. L. (1997). Structural analysis of three His32 mutants of DsbA: support for an electrostatic role of His32 in DsbA stability. *Protein Sci* **6**, 1893–900.
11 GUDDAT, L. W., BARDWELL, J. C., ZANDER, T. & MARTIN, J. L. (1997). The uncharged surface features surrounding the active site of Escherichia coli DsbA are conserved and are implicated in peptide binding. *Protein Sci* **6**, 1148–56.
12 COUPRIE, J., VINCI, F., DUGAVE, C., Qu inverted question markem inverted question markeneur, E. & MOUTIEZ, M. (2000). Investigation of the DsbA mechanism through the

synthesis and analysis of an irreversible enzyme-ligand complex. *Biochemistry* **39**, 6732–42.

13 FRECH, C., WUNDERLICH, M., GLOCKSHUBER, R. & SCHMID, F. X. (1996). Preferential binding of an unfolded protein to DsbA. *EMBO J* **15**, 392–98.

14 VINCI, F., COUPRIE, J., PUCCI, P., QUEMENEUR, E. & MOUTIEZ, M. (2002). Description of the topographical changes associated to the different stages of the DsbA catalytic cycle. *Protein Sci* **11**, 1600–12.

15 BARDWELL, J. C., LEE, J. O., JANDER, G., MARTIN, N., BELIN, D. & BECKWITH, J. (1993). A pathway for disulfide bond formation in vivo. *Proc Natl Acad Sci USA* **90**, 1038–42.

16 BADER, M., MUSE, W., ZANDER, T. & BARDWELL, J. (1998). Reconstitution of a protein disulfide catalytic system. *J Biol Chem* **273**, 10302–7.

17 GUILHOT, C., JANDER, G., MARTIN, N. L. & BECKWITH, J. (1995). Evidence that the pathway of disulfide bond formation in Escherichia coli involves interactions between the cysteines of DsbB and DsbA. *Proc Natl Acad Sci USA* **92**, 9895–9.

18 KISHIGAMI, S., KANAYA, E., KIKUCHI, M. & ITO, K. (1995). DsbA-DsbB interaction through their active site cysteines. Evidence from an odd cysteine mutant of DsbA. *J Biol Chem* **270**, 17072–4.

19 KISHIGAMI, S. & ITO, K. (1996). Roles of cysteine residues of DsbB in its activity to reoxidize DsbA, the protein disulphide bond catalyst of Escherichia coli. *Genes Cells* **1**, 201–8.

20 KOBAYASHI, T., KISHIGAMI, S., SONE, M., INOKUCHI, H., MOGI, T. & ITO, K. (1997). Respiratory chain is required to maintain oxidized states of the DsbA-DsbB disulfide bond formation system in aerobically growing Escherichia coli cells. *Proc Natl Acad Sci USA* **94**, 11857–62.

21 BADER, M., MUSE, W., BALLOU, D. P., GASSNER, C. & BARDWELL, J. C. (1999). Oxidative protein folding is driven by the electron transport system. *Cell* **98**, 217–27.

22 BADER, M. W., XIE, T., YU, C. A. & BARDWELL, J. C. (2000). Disulfide bonds are generated by quinone reduction. *J Biol Chem* **275**, 26082–8.

23 XIE, T., YU, L., BADER, M. W., BARDWELL, J. C. & YU, C. A. (2002). Identification of the ubiquinone-binding domain in the disulfide catalyst disulfide bond protein B. *J Biol Chem* **277**, 1649–52.

24 KADOKURA, H., BADER, M., TIAN, H., BARDWELL, J. C. & BECKWITH, J. (2000). Roles of a conserved arginine residue of DsbB in linking protein disulfide-bond-formation pathway to the respiratory chain of Escherichia coli. *Proc Natl Acad Sci USA* **97**, 10884–9.

25 KOBAYASHI, T. & ITO, K. (1999). Respiratory chain strongly oxidizes the CXXC motif of DsbB in the Escherichia coli disulfide bond formation pathway. *EMBO J* **18**, 1192–8.

26 INABA, K. & ITO, K. (2002). Paradoxical redox properties of DsbB and DsbA in the protein disulfide-introducing reaction cascade. *EMBO J* **21**, 2646–54.

27 REGEIMBAL, J. & BARDWELL, J. C. (2002). DsbB catalyzes disulfide bond formation de novo. *J Biol Chem* **277**, 32706–13.

28 KADOKURA, H. & BECKWITH, J. (2002). Four cysteines of the membrane protein DsbB act in concert to oxidize its substrate DsbA. *EMBO J* **21**, 2354–63.

29 GRAUSCHOPF, U., FRITZ, A. & GLOCKSHUBER, R. (2003). Mechanism of the electron transfer catalyst DsbB from Escherichia coli. *EMBO J* **22**, 3503–13.

30 REGEIMBAL, J., GLEITER, S., TRUMPOWER, B. L., YU, C. A., DIWAKAR, M., BALLOU, D. P. & BARDWELL, J. C. (2003). Disulfide bond formation involves a quinhydrone-type charge-transfer complex. *Proc Natl Acad Sci USA* **100**, 13779–84.

31 HINIKER, A. & BARDWELL, J. C. (2004). In vivo substrate specificity of periplasmic disulfide oxidoreductases. *J Biol Chem* **279**, 12967–73.

32 Rietsch, A., Bessette, P., Georgiou, G. & Beckwith, J. (1997). Reduction of the periplasmic disulfide bond isomerase, DsbC, occurs by passage of electrons from cytoplasmic thioredoxin. *J Bacteriol* **179**, 6602–8.

33 Zapun, A., Missiakas, D., Raina, S. & Creighton, T. E. (1995). Structural and functional characterization of DsbC, a protein involved in disulfide bond formation in Escherichia coli. *Biochemistry* **34**, 5075–89.

34 Rietsch, A., Belin, D., Martin, N. & Beckwith, J. (1996). An in vivo pathway for disulfide bond isomerization in Escherichia coli. *Proc Natl Acad Sci USA* **93**, 13048–53.

35 Bessette, P. H., Cotto, J. J., Gilbert, H. F. & Georgiou, G. (1999). In vivo and in vitro function of the Escherichia coli periplasmic cysteine oxidoreductase DsbG. *J Biol Chem* **274**, 7784–92.

36 Sun, X. X. & Wang, C. C. (2000). The N-terminal sequence (residues 1–65) is essential for dimerization, activities, and peptide binding of Escherichia coli DsbC. *J Biol Chem* **275**, 22743–9.

37 McCarthy, A. A., Haebel, P. W., Torronen, A., Rybin, V., Baker, E. N. & Metcalf, P. (2000). Crystal structure of the protein disulfide bond isomerase, DsbC, from Escherichia coli. *Nat Struct Biol* **7**, 196–9.

38 Haebel, P. W., Goldstone, D., Katzen, F., Beckwith, J. & Metcalf, P. (2002). The disulfide bond isomerase DsbC is activated by an immunoglobulin-fold thiol oxidoreductase: crystal structure of the DsbC-DsbDalpha complex. *EMBO J* **21**, 4774–84.

39 Chen, J., Song, J. L., Zhang, S., Wang, Y., Cui, D. F. & Wang, C. C. (1999). Chaperone activity of DsbC. *J Biol Chem* **274**, 19601–5.

40 Shao, F., Bader, M. W., Jakob, U. & Bardwell, J. C. (2000). DsbG, a protein disulfide isomerase with chaperone activity. *J Biol Chem* **275**, 13349–52.

41 Zhao, Z., Peng, Y., Hao, S. F., Zeng, Z. H. & Wang, C. C. (2003). Dimerization by domain hybridization bestows chaperone and isomerase activities. *J Biol Chem* **278**, 43292–8.

42 Joly, J. C. & Swartz, J. R. (1997). In vitro and in vivo redox states of the Escherichia coli periplasmic oxidoreductases DsbA and DsbC. *Biochemistry* **36**, 10067–72.

43 Bader, M. W., Hiniker, A., Regeimbal, J., Goldstone, D., Haebel, P. W., Riemer, J., Metcalf, P. & Bardwell, J. C. (2001). Turning a disulfide isomerase into an oxidase: DsbC mutants that imitate DsbA. *EMBO J* **20**, 1555–62.

44 Stewart, E. J., Katzen, F. & Beckwith, J. (1999). Six conserved cysteines of the membrane protein DsbD are required for the transfer of electrons from the cytoplasm to the periplasm of Escherichia coli. *EMBO J* **18**, 5963–71.

45 Chung, J., Chen, T. & Missiakas, D. (2000). Transfer of electrons across the cytoplasmic membrane by DsbD, a membrane protein involved in thiol-disulphide exchange and protein folding in the bacterial periplasm. *Mol Microbiol* **35**, 1099–109.

46 Katzen, F. & Beckwith, J. (2000). Transmembrane electron transfer by the membrane protein DsbD occurs via a disulfide bond cascade. *Cell* **103**, 769–79.

47 Collet, J. F., Riemer, J., Bader, M. W. & Bardwell, J. C. (2002). Reconstitution of a disulfide isomerization system. *J Biol Chem* **277**, 26886–92.

48 Kim, J. H., Kim, S. J., Jeong, D. G., Son, J. H. & Ryu, S. E. (2003). Crystal structure of DsbDgamma reveals the mechanism of redox potential shift and substrate specificity(1). *FEBS Lett* **543**, 164–9.

49 Katzen, F. & Beckwith, J. (2003). Role and location of the unusual redox-active cysteines in the hydrophobic domain of the transmembrane electron transporter DsbD. *Proc Natl Acad Sci USA* **100**, 10471–6.

50 Wunderlich, M., Otto, A., Maskos, K., Mucke, M., Seckler, R. & Glockshuber, R. (1995). Efficient

catalysis of disulfide formation during protein folding with a single active-site cysteine. *J Mol Biol* **247**, 28–33.

51 ASLUND, F., BERNDT, K. D. & HOLMGREN, A. (1997). Redox potentials of glutaredoxins and other thiol-disulfide oxidoreductases of the thioredoxin superfamily determined by direct protein-protein redox equilibria. *J Biol Chem* **272**, 30780–6.

52 WUNDERLICH, M., OTTO, A., SECKLER, R. & GLOCKSHUBER, R. (1993). Bacterial protein disulfide isomerase: efficient catalysis of oxidative protein folding at acidic pH. *Biochemistry* **32**, 12251–6.

10
Catalysis of Peptidyl-prolyl *cis/trans* Isomerization by Enzymes

Gunter Fischer

10.1
Introduction

The proper function of proteins within living cells depends on how those macromolecules adopt a unique three-dimensional structure and on how spatial-temporal control underlies the mechanism of protein folding in vivo. The numerous contact points formed and released in the course of protein folding are based on countless rotational movements. Most rotations of covalent bonds have proven to be intrinsically very fast and do not need further rate acceleration by either external factors or intramolecular assistance. Potentially, external influences on rotational rates could be mediated by a wide variety of physical and chemical means such as enzyme catalysis, catalysis by low-molecular-mass compounds, heat, mechanical forces, and supportive microenvironments. Acid-base catalysis, torsional strain, and proximity and field effects are major forces with the potential to act in intramolecular assistance. Submillisecond burst phase signals measured in the kinetics of protein-folding experiments and triplet-triplet energy transfer rates measured for side chain contacts of oligopeptides have been widely interpreted in terms of the fast formation of collapsed folding intermediates and secondary structures formation, both involving countless conformational interconversions [1–3]. Thus, in the evolutionary history, attempts to invent mechanisms of rate acceleration for the formation of collapsed polypeptides by accessory molecules appeared to be futile. However, in contrast to the information provided by its graphical representation, a protein backbone possesses a considerable degree of rigidity that exists independently of the three-dimensional fold. Despite the formal single-bond, fast-rotation nature of its covalent linkages, the repeating unit of the polypeptide backbone suffers from at least one slow bond rotation. It is the high rotational barrier of the carbon-nitrogen linkage of the peptide bond (angle ω) that casts serious doubt on the validity of the structure-based treatment of a polypeptide chain as an array of single-bond units. Considering the influences on the dynamics of biochemical reactions such as protein folding/restructuring and protein-ligand/protein-protein interactions, the electronic structure of the C–N bond is best described as a partial double bond. Under this classical view, the peptide bond exhibits characteristic features of E-Z geometrical isomerism: the existence of chemically and physically distinct *cis/trans* isomers, a high barrier to rotation, and a distinct chemistry of the *cis/trans* interconversion [4, 5]. By virtue of having two thermodynamically stable chain arrangements of different chemical reactivity separated by an energetic barrier, peptide bond *cis/trans* isomerization resembles a molecular switch. Switching is manifested in the form of directed mechanical movement of the polypeptide chain. This fact is best illustrated by analysis of the protein crystal structure database for homologous proteins with a different isomeric state of a certain prolyl residue. A subset of structures has been created that has a particular proline residue either in the *cis* or the *trans* isomeric state for similar proteins [6]. Structural alignment of matched pairs of isomeric proteins generates three classes with respect to position-specific distribution of Cα-atom displacements around the isomeric proline imide bond. Besides small bidirectional chain displacements, there is a class of 12 protein pairs in which the structural changes are unidirectional relative to the isomerizing bond. The magnitude of the isomer-specific distance effect exceeds 3.0 ± 2.0 Å at sequence positions remote from the prolyl bond. Interestingly, the magnitude of the intramolecular isomer-specific Cα atom displacements reveals a lever-arm amplification of distance effects because there is only a *cis*-specific chain contraction of 0.8 Å for the Cα atoms directly involved in the prolyl bond. A type of directed oscillating movement within the backbone in its native conformation, if a fast kinetic step precedes it, offers a structural basis for the explanation of the isomer-specific control of the bioactivity. The findings indicate that cellular folding catalysts may provide a previously unrecognized level of control for the biological activity of the equilibrium-folding pathway of native proteins, which exists independently of the de novo folding pathway.

Chemically, peptide bond *cis/trans* isomerization is reminiscent of other important biological events, such as the retinal-mediated photocycle, carotenoids in antioxidant processes, and phytochrome-based reactions, which, like retinal-based, display isomerization-linked geometrical distortions that are transmitted to amino acid residues of their respective binding protein [7]. Whereas visible light absorption gives rise to the very fast *cis/trans* isomerization-mediated photocycle of the protein-bound 11-*cis*-retinal chromophore, which finally results in the formation of a G-protein-interacting state of rhodopsin, the rate of peptide bond *cis/trans* isomerization is naturally insensitive to irradiation at wavelengths > 220 nm [8]. Thus, light irradiation is also unable to decrease the double-bond character of the bonds determining protein backbone rigidity, with the consequent lack of directed atomic displacements as well as structural "backbone liquification" or facilitating torsional movements by "lubrication." However, specific peptide bond *cis/trans* isomerases have evolved to trigger a similar type of structural response and, thereby, to have a specific contribution to lowering of the high rotational barrier of peptide bonds [9, 10]. Unlike photochemical triggering, where a very slow dark reaction at the carbon-carbon double bond after switching off the light is to occur, enzyme-catalyzed peptide *cis/trans* isomerizations tend to be both fast and reversible under most conditions.

It should be noted that the peptide bond *cis/trans* isomerases identified up to now could not shift a *cis/trans* equilibrium but could accelerate re-equilibration after transient chemical or physical perturbbations. Except for enzyme concentrations much larger than substrate concentrations, where the Michaelis complexes of the substrate and the product will confer an individual isomer ratio, peptide bond *cis/trans* isomerases must be considered simple catalysts in bioreactions. Obviously, enzymatic coupling to an energy-consuming process would allow generating a permanent non-equilibrium composition of *cis/trans* isomers free in solution. Such multifunctional peptide bond *cis/trans* isomerases have not yet been found but are likely to exist as isolated enzymes or hetero-oligomeric complexes.

From the point of view of biological importance, when a polypeptide binds to a native protein, the resulting complex forms a global ensemble that must seek its own minimum free-energy fold. Consequently, the rules underlying the process of de novo protein folding do not change, in principle, for chain rearrangements induced in native proteins during ligand association, enzyme catalysis, and other reactions. Since peptide bond *cis/trans* isomers are frequently found to exist in a dynamic equilibrium in the context of a native fold, the data collected for the isomerization process in the refolding of denatured proteins are principally applicable for the analysis of native-state *cis/trans* isomerizations. Most importantly, native-state isomers might exhibit significant differences in their biological activity [6, 11–13]. In an approach to understanding the nature of catalyzed *cis/trans* isomerizations in cells, two major effects have to be assumed: (1) accelerated re-equilibration of a transiently perturbed *cis/trans* isomer ratio and (2) enhanced conformational dynamics in the presence of a catalyst in an already equilibrated mixture of *cis/trans* isomers. The increase in the rate of re-equilibration brought about by the catalyst according to effect (1) requires a fast-equilibrium, perturbing-type reaction prior to the *cis/trans* isomerization in order for isomer-specific reaction control to occur. In contrast, effect (2) permanently enables a small segment of the polypeptide backbone to escape the steric restrictions imposed by the partial double-bond character of the C–N linkage. Concomitantly, the position of the substrate chain oscillates in a directed manner with a frequency determined by k_{cat} values of the reversible enzyme reaction. Because of dynamic coupling within the enzyme-substrate or enzyme-product complexes, specific amino acid residues of the enzyme might oscillate with the same frequency [14].

10.2
Peptidyl-prolyl *cis/trans* Isomerization

Proline is the only one of the 20 canonical amino acids that is involved in the formation of an imidic peptide bond at its N-terminal tail because the peptide bond nitrogen is in an *N*-alkylated state (Figure 10.1). Essential features of the peptidyl-prolyl peptide bond and the peptidyl-prolyl bond's *cis/trans* isomerization, which we subsequently refer to as prolyl bond and prolyl isomerization, respectively, are dynamics and unique conformational constraints. Among all peptide bonds, prolyl bonds should be afforded special attention regarding conformational polymor-

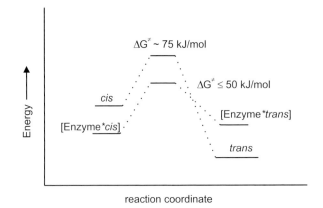

Fig. 10.1. Cis/trans isomerization of a prolyl bond and free-energy profiles of the spontaneous and enzyme-catalyzed reactions.

phism of proline-containing polypeptides. Regardless of their sequential context, the *cis* and *trans* prolyl conformers in unstructured or partially structured polypeptide chains exhibit energetic differences that tend to be small, thus leading to a comparable level of isomers in solution. The slow rate of *cis/trans* interconversion obtained in refolding studies plays a significant role in determining the uniqueness of prolyl isomerization because just one covalent bond serves to control an enormous body of structural changes from the seconds time range of slow folding steps in small globular proteins [15] to the hours time range for the zipper-like mechanism of the triple-helix formation of collagen [16, 17] or protein-protein associations [18]. The classical experiment conducted by Brandts et al. led us to realize that a prolyl isomerization is also involved in protein denaturation [19]. The experimental data reveal that the relatively few examples of functional coupling between protein folding and prolyl isomerization are seemingly at odds with the large number of proline residues in proteins. According to the marked differences of the cis/trans isomer ratios between the unfolded and the folded states, every proline residue might cause slow folding events unless intramolecular assistance greatly enhances the isomerization rate. This ambiguity is of more than theoretical interest since the whole body of slowly isomerizing prolyl bonds can be seen as a potential molecular basis of metastable states of proteins and classed as potential targets for enzyme catalysis. Obviously, the existence of non-rate-limiting prolyl isomerizations must take into account the potential impact of intramolecular catalysis by amino acid side chains and steric strain in polypeptides, but only a few examples have been identified [20, 21]. An alternative explanation has been to suggest that the prolyl isomerization is silent in common folding probes. This case is easily conceivable, but nevertheless, it is likely that formation of bioactivity might be affected in many cases [22–24].

In principle, peptides or proteins with n proline residues can form 2^n isomers unless structural constraints (such as the three-dimensional fold of proteins) stabilize one isomer strongly relative to the others. While being a highly abundant motif in proteins, a polyproline stretch tends to be a conformationally uniform segment but is largely undefined in its isomerization dynamics [25, 26]. For example, all theoretically predicted cis/trans isomers could be identified by ^{1}H-NMR spectroscopy for the octapeptide Ile-Phe-Pro-Pro-Val-Pro-Gly-Pro-Gly, which is derived from the prolactin receptor [27]. Counter to the expectations drawn from statistics, a decreasing fractional isomer probability among the 16 isomeric forms does not correlate with a decreasing number of trans peptide bonds in the respective isomer. In many cases among prolyl isomers presented to other proteins for interactions to occur, some isomers might lack productive binding. In fact, proteins have evolved highly efficient means of selecting a particular structure among an ensemble of conformers of ligands.

Consequently, isomer-specific enzymes and transporters, such as proteases [28], protein phosphatases [29], proline-directed protein kinases [30], and the peptide transporter Pept1 [31], that favor molecules with trans prolyl bonds at critical positions discriminating the coexisting cis isomer have been found. However, bioactivity is not stringently linked with chains with all-trans peptide bond conformations. In several cases the optimal receptor-ligand interaction requires appropriately positioned cis prolyl bonds [32–35].

Since bioactivity of polypeptides exhibits specificity for the slowly interconverting prolyl isomers, enzymes have evolved (1) to accelerate the interconversion of the isomers and (2) to establish a new level of conformational control.

In the years following the discovery of the first peptidyl-prolyl cis/trans isomerase (PPIase) in pig kidney in 1984, three different families of PPIases have been characterized, which now collectively form the enzyme class EC 5.2.1.8 [9, 28, 36]. In humans, the families of FK506-binding proteins (FKBPs) and cyclophilins have many members, whereas the human genome probably restricts the total number of the parvulins to not more than two members (Figure 10.2). Sequence similarity of the catalytic domains forms the signature of family membership. Across families, the catalytic domains are unrelated to each other in their amino acid se-

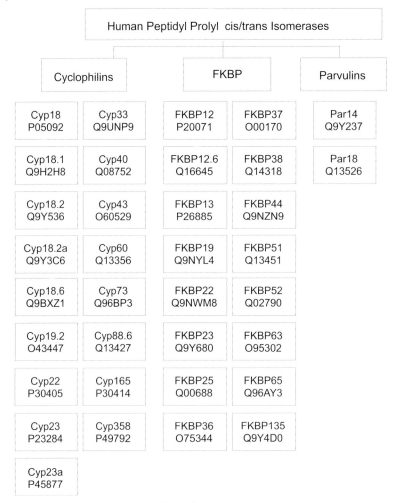

Fig. 10.2. Presently known peptidyl-prolyl *cis/trans* isomerase in humans classified according to amino acid sequence-based similarities. The individual enzymes have been identified by the SWISS-Prot/TREMBL accession numbers.

quences, confer distinct substrate specificities, and prove to be binders for different types of inhibitors. In their three-dimensional structures, parvulins appeared to be closely related to FKBPs in that, by comparison, a FKBP superfold could be extracted [37]. Family members of the ubiquitous enzyme class of PPIases are found in abundance in virtually all organisms and subcellular compartments. Within families, the amino acid sequences are highly conserved phylogenetically. Organisms simultaneously express paralogs within the families that encompass one or more PPIase domains complemented with other functional polypeptide segments.

Except for targeting motifs and metal-ion-binding polypeptide stretches, the functional interplay between the supplementing domains of the larger PPIases and the catalytic domain(s) is still unclear.

Since some prototypic PPIases are proven drug targets either by blockage of the active site or by utilizing the PPIase as a presenter protein of the drug [38], knowledge about functional consequences of the supplementary domains in the multi-domain-like PPIases is urgently needed.

Recently, the secondary amide peptide bond *cis/trans* isomerases (abbreviated APIases) were added to the class of peptide bond *cis/trans* isomerases, indicating that an expanded set of polypeptide substrates is potentially sensitive to biocatalysis of a conformational interconversion in the cell. This enzyme activity is discussed in chapter 11.

10.3 Monitoring Peptidyl-prolyl *cis/trans* Isomerase Activity

The chemical and physical principles described for the detection of isomer-specific reactions and native-state isomerizations of prolyl bond-containing oligopeptide chains apply, in general, to all PPIase assays. In all cases, the rate enhancement in the presence of the PPIase quantitatively reflects the enzyme activity. As a monomolecular reaction, the spontaneous *cis/trans* isomerization conforms to first-order kinetics with a specific rate constant k_o (s^{-1}) that is independent of the substrate concentration. With some exceptions, the isomer ratio cannot be extracted from the data of the spectrometric PPIase assays, but NMR-based methods can be utilized.

Under favorable conditions, where the concentration of the active PPIase is much smaller than the total substrate concentration and the total substrate concentration is much smaller than the K_m value, Eq. (1) provides a quantitative estimate of enzyme catalysis. Such conditions also include first-order kinetics of the PPIase-catalyzed progress curves.

$$k_{cat}/K_m = (k_{ob} - k_o)/E_o \tag{1}$$

The k_{cat}/K_m (L mol^{-1}s^{-1}) value is the specificity constant for the enzymatic catalysis of the interconversion of the *cis/trans* isomers of the substrate, k_{ob} (s^{-1}) is the first-order rate constant of the catalyzed interconversion, k_o (s^{-1}) is the first-order rate constant of the spontaneous *cis/trans* interconversion, and E_o (mol L^{-1}) is the concentration of active enzyme molecules. However, active-site titration methods cannot be employed for PPIases except for cyclophilins. Cyclosporin A proved to be useful for fluorescence-based active-site titration of those cyclophilins with a tryptophan residue corresponding to position 121 of Cyp18 [39]. Detailed inspection of the catalyzed reaction will reveal the specific meaning of the k_{cat}/K_m value: either that the reversible *cis/trans* interconversion determines this constant or that there is a direct coupling to an irreversible process, with preceding *trans* to *cis* or *cis* to *trans* interconversions, that plays a role in determining the available enzymatic parameters. The presence of PPIases with high affinity for a substrate ($K_M \ll [S]$) is indicated by deviations of the experimental data from strict first-order kinetics in the initial phase of the progress curve for the catalyzed interconversion.

Several assays make use of the transient production of pure *cis* or *trans* isomers or at least of an isomer ratio different from that expected based on the steady-state reaction conditions. A comparison between the PPIase assays revealed that the higher the deviation from the equilibrium ratio of isomers, the higher the signal-to-noise ratio of the assay.

Rate enhancement measurements usually require monitoring of reaction kinetics over an extended period of time. It is vital to reliable activity measurements that kinetic traces be followed over at least three half-times of the reaction. Use of the assays as endpoint-determination methods has not yet been established.

Assaying PPIases with protein substrates has proven to be more difficult. Three-dimensional protein structures derived from crystals reveal a defined isomeric state for every prolyl peptide bond being in either *cis* or *trans* conformation, with the majority of prolyl bonds in *trans* form [6, 40]. In some cases the crystallization process was found to be a crucial determinant of the conformational state of prolyl bonds in the sample [6]. The seeming homogeneity results only partially from isomer-specific stabilization effects by intramolecular long-range interactions or isomer-specific crystal lattice sampling. When probed by NMR spectroscopy in solution, many proteinaceous prolyl bonds show conformational polymorphism, where the isomeric proteins interconvert in a reversible reaction. Typically, these native-state isomerizations exhibit a wide range of isomer ratio values. The rate of the spontaneous interconversion of isomers is usually found to be slow on the NMR timescale. Due to a considerable isomer-specific dispersion of proton signals adjacent to the isomeric bond, dynamic ^{1}H-NMR methods provide powerful tools for locating isomerizing peptide bonds in polypeptides. Rate constants of interconversion are also available from this probe, but NMR methods are limited to rather high substrate concentrations. Dynamic NMR methods generally do not require prior chemical perturbation of isomer composition by chemical means. Instead, transferred magnetization to isomers is detected. However, one-dimensional ^{1}H-NMR spectroscopy also has been used in combination with classical solvent-jump techniques when subsequent monitoring of the time course of an isomer-specific signal intensity is to occur [41].

^{19}F nuclei in labeled amino acids can also serve as a probe, as was shown for a PPIase-catalyzed isomerization reaction in a derivative of the tissue hormone bradykinin [42].

Based on the combination of dynamic ^{1}H-NMR spectroscopy and conformational polymorphism of proteins, PPIase activity can be easily determined if the number of isomerizing prolyl bonds is limited (preferably to one or two prolyl bonds) and the isomer ratio does not deviate much from unity. Because one-dimensional ^{1}H-NMR experiments reveal slow dynamics in spontaneous prolyl isomerization, line broadening is not significant for the *cis* and *trans* signals. However, in the presence of PPIases, signals experience line broadening and coalescence [43] that can be quantified by comparison with simulated line shapes based on quantum-mechanical density matrix formalism [44, 45]. The two-dimensional NOESY experiments are more sensitive to catalytically enhanced interconversion rates than are one-dimensional NMR techniques. Exchange cross-peak intensity at different mixing times gives access to the rate constants of interconversion [46, 47].

Isomer-specific proteolysis has been widely employed to assay PPIase activity with oligopeptide substrates. This robust method can be considered as the standard test for routine determination of enzyme activity for all families of PPIases in their pure state as well as in crude cell extracts and biological fluids [9, 48]. The assay is based on the fact that a linear oligopeptide containing a single prolyl bond usually consists of a mixture of two slowly interconverting conformers, the *cis* and the *trans* prolyl isomers, in solution. Depending on the nature of the amino acid flanking the proline residue of substrates in the -Xaa-Pro-Ala- moiety, the percentage of *cis* conformer is in the range of 6% to 38% in aqueous solutions [41].

Slow re-equilibration after rapid disturbance of the conformational equilibrium is a common feature of most prolyl isomerizations. Proteases such as chymotrypsin, various elastases, trypsin, subtilisin, clostripain, and others liberate 4-nitroaniline from -Pro-Yaa-4-nitroanilides (typically, Yaa = Phe, Tyr, Leu, Arg, Lys, Val, Ala) only if a *trans* prolyl bond exists in the penultimate amino acid position from the scissile linkage. Isomer-specific hydrolysis by a protease can serve as a chemical means of displaycing the isomer distribution in a very efficient manner, resulting in a transient population of about 100% *cis* isomer. A protease has not yet been detected that discriminates to the isomer level, hydrolyzing substrates with a *cis* prolyl bond exclusively. To ensure rapid cleavage of the entire population of the *trans* isomeric substrate, the protease concentration is set at a high value. The k_{cat} value of the proteolytic reaction determines the limiting cleavage rate at very high protease concentrations and defines an upper limit estimate of isomerization rates accessible to this assay. Thus, after consumption of the entire population of the *trans* isomer using this strong proteolytic pulse, the *cis* isomer of the peptide is left over. Such a reaction cannot continue to completeness, which is defined by consumption of the total amount of substrate, unless a parallel reaction couples. The coupled reaction is the first-order *cis* to *trans* interconversion, which precedes the rapid proteolytic consumption of the *trans* isomer formed during the conformational interconversion as a rate-limiting step after the initial phase of the reaction. Biphasic kinetics can be monitored under these conditions, where the PPIase-mediated rate acceleration will reside in the slow phase of the reaction (Figure 10.3). It is clear that only a small fraction of the total absorbance of a protease-coupled reaction can be utilized for assaying PPIases. Due to the distinctive characteristics of the protease-coupled assay, the part of absorbance released in the dead time of mixing at 390 nm contributes to the unwanted background signal. Substitution of 4-nitroaniline by fluorogenic leaving groups or the use of intramolecularly quenched fluorescent protease substrates allows for fluorescence-based protease-coupled PPIase assays [49, 50]. The signal-to-noise ratio of this modified assay also suffers the limitations of UV/Vis detection.

Deficits of the protease-coupled assay such as proteolytic degradation of PPIases

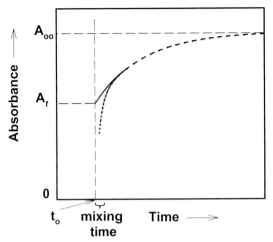

Fig. 10.3. First-order progress curve of a typical proteolytic PPIase assay. Zero-time t_o is taken at the time point of injection of the assay peptide. The time period required for sample mixing is depicted on the time axis. By extrapolation, the time zero intercept value A_t and the absorbance at infinite reaction time A_{oo} become available, the difference of which gives an estimate of the cis/trans ratio if it is set relative to the amplitude of absorbance of the very rapid kinetic phase visible at zero time.

and substrates are partially avoided by carefully selecting the helper protease. However, most PPIases may have evolved to accelerate cis/trans interconversions at or near the cis/trans equilibrium of a cellular substrate [44]. The protease-coupled assay utilizes reaction conditions that necessarily lead to a quasi-irreversible isomerization under non-equilibrium conditions. In contrast, small physical perturbations of a cis/trans equilibrium can serve to realize PPIase assays under more natural conditions

Provided that isomer-specific differences of UV/Vis spectra, NMR signals, fluorescence quenching, or chemical properties (pK_a values) exist, the time courses of re-equilibration of prolyl isomers may provide a signal that is determined in rate by the sum of the first-order rate constants for the cis to trans and trans to cis interconversion. Transient deviations from the actual equilibrium conditions have been achieved mainly by pH and solvent jumps [51–53].

Given the experimental fact that signal amplitudes arising from solvent perturbations are usually very small, improvement in signal quality of the assays is expected to be associated with covalent modifications that serve as a means for potentially increasing the cis/trans isomer ratio. By intramolecular disulfide bridging, 2-aminobenzoyl-Cys-Phe-Pro-Ala-Cys-Ala-4-nitroanilide or 2-aminobenzoyl-Cys-Phe-Pro-Ala-Cys-3-nitro-tyrosineamide experienced about 100% cis isomer for the Phe-Pro moiety, which declined after disulfide ring reduction to an open-chain cis isomer level of about 10% [54]. In the presence of high concentrations of reductive agent, such as dithiothreitol or tris-(2-carboxyethyl)-phosphine, ring opening is a rapid reaction followed by the slow equilibration of the Phe-Pro moiety. Efficient fluorescence quenching of the 2-aminobenzoyl residue by the 4-nitroanilide moiety requires the *cis* conformation. Thus, subsequent to the reductive pulse, the time course of the fluorescence increase corresponds to the *cis/trans* interconversion in the open-chain peptide that is able to reveal the presence of a PPIase [55]. Fortunately, most substrates and enzymes are not susceptible to interference from the reductive agents.

Quite early in the development of PPIase assays, it was noted that slow kinetic phases of refolding of chemically denatured RNases are sensitive to PPIase catalysis [56–58]. The reaction progress curves may in turn serve as a tool for assaying PPIase activity. For this to occur, the PPIase must recognize partially folded intermediates that result in limited access to the catalyzable prolyl bond. Both wild-type ribonuclease T1 (abbreviated wtRNaseT1) and its site-directed mutagenized protein variants turned out to be especially suitable substrates for peptide bond *cis/trans* isomerases.

Two prolyl bonds of RNase T1 (Tyr38-Pro39 and Ser54-Pro55) adopt the *cis* conformation in the native state but display a diminished *cis* population in the unfolded state. Thus, the *trans* to *cis* isomerization is a quasi-irreversible process because the reverse isomerization is disfavored under strongly native refolding conditions. The refolding kinetics can be probed by fluorescence, CD, UV/Vis, and real-time NMR spectroscopy and the regain of enzymatic activity [58–60]. Recently, time-resolved FTIR-difference spectroscopy has also been applied successfully for the spontaneous reaction [61]. A complete kinetic folding model has been established by unfolding-refolding double mixing experiments [62, 63]. If RNase T1 is denatured and reduced and the cysteine side chains are carboxymethylated, it is still able to adopt an active, native-like conformation, but only if high ionic strength applies. Thus, instead of utilizing denaturing agents or extreme thermal conditions, variable concentrations of NaCl allow for shuttling between native and unfolding conditions. In general, the PPIase activity does not reflect significant interference with the shuttling conditions. Using the remarkable features of RNaseT1, this refolding reaction is the established standard assay for PPIase catalysis on proteinaceous substrates in the refolding and unfolding directions [64].

Tendamistat provides an example where PPIase catalysis of native-state prolyl isomerization can be probed by fluorescence spectroscopy using double-jump techniques. In interrupted refolding experiments, a very slow folding reaction, with an amplitude of about 12%, was detected that is still present, although at a reduced level of amplitude, in the native state. It indicates conformational heterogeneity at the proline positions 7 and 9. Catalysis by cyclophilin identifies this very slow reaction as a limiting prolyl isomerization. The very low *cis/trans* ratio of about 0.02 for these prolyl bonds suggests that the native-state isomerization would probably escape detection by NMR spectroscopy [65].

Accelerated recovery of enzyme activity can be used to identify and quantitate PPIase catalysis on prolyl bond-limited refolding phases that are spectroscopically silent. Monitoring the time-resolved appearance of native muscle acyl phosphatase exemplifies this approach [22].

10.4
Prototypical Peptidyl-prolyl cis/trans Isomerases

10.4.1
General Considerations

This chapter addresses some fundamental principles underlying the biochemical functions of PPIases based on the discussion of prototypical enzymes. Prototypical PPIases are monomerically active enzymes ranging in size from 92 to 177 amino acids, which, per definition, do not contain any targeting signal or supplementary domain. Functional and structural studies have been performed mostly for the prototypical enzymes, whereas present knowledge of the enzymatic properties of multidomain PPIases is rather limited.

Four different molecular mechanisms can be hypothesized to define the biochemical basis of the physiological effects known to be exerted by the members of the three families of PPIases: (1) the catalysis of prolyl isomerization, (2) a holding function for unfolded polypeptides, (3) a presenter-protein function for physiological ligands, and (4) a proline-directed binding function for other proteins. The underlying reactions might control, either in isolation or collectively, the physiological role of prototypical and domain-complemented enzymes. A general consensus has not yet been reached to which extent the biochemical principles above contribute to the generation of physiological signals by PPIases [36].

In vitro experiments frequently reveal that peptide bond cis/trans isomerases encompassing more domains than just the catalytic module are able to prevent chain aggregation in typical chaperone assays [66–69]. Proline residues are not essential for sequestration of the polypeptide chain. However, polypeptide sequestration by PPIases is due to the presence of an extended stretch of catalytic subsites, also termed secondary binding sites. Classically, many polypeptide-utilizing enzymes form extensive contacts both in the active site and through secondary binding sites. All these interacting sites are closely linked to the activation of the catalytic machinery, but they will not give rise to physiological functions independent of the catalyzed reaction. More specifically, the properties of maintaining and catalyzing unfolded chains are not mutually exclusive. Therefore, although an important determinant of catalysis, the holding function plays an auxiliary role in contributing to the physiological effects of PPIases [36].

Immunosuppressive PPIase inhibitors produce a multitude of effects in transplantation medicine, reflecting the causal relationship between physiological effects and the multifunctional nature of the biochemical platform of PPIases [70–72]. The molecular mechanism underlying immunosuppression mediated by PPIase inhibitors was found to be associated with a gain of function in the form of drug activation by PPIase/inhibitor complex formation. This example unequivocally demonstrates the importance of drug target control by presenter molecules (according to the biochemical principle 3). Consequently, the most prominent effect of the Cyp18 and FKBP inhibitors CsA and FK506, respectively, i.e., the prevention of clonal expansion after antigenic stimulation of T cells, has been discussed exclusively in terms of this presenter-protein strategy. The active complex is targeted to the inhibition of the protein phosphatase calcineurin. Structurally the PPIase/drug complex was thought to present a composite surface, which is next to active-site residues of the PPIase, to other cytosolic binding partners, whereas no independent role could be elucidated for the simultaneously occurring blockage of the catalytic machinery of the PPIase [73]. Care must be taken, however, when interpreting the drug-mediated immunosuppression as a model for physiological PPIase function in humans, as the microbial drugs do not belong to the normal constituents of mammalian cells. The fact of the matter is that an authentic polypeptide ligand utilizing the presenter-protein function of a PPIase has not yet been identified, thus challenging the view of a presenter-protein function of PPIases in normal cell life. Moreover, gain of function appears to be released in a few special cases, such as the Cyp18/cyclosporin A complex. When substituted in the 3 position, cyclosporin derivatives inhibit calcineurin on their own but are inactive in the Cyp18/drug complexes. Of the most important structural parameters, only the ring conformation of the cyclosporins proved to be a decisive feature [74].

Only recently has attention turned to the role of inhibiting the catalytic function of PPIases for the interpretation of drug effects under physiological and pathophysiological conditions [75–78]. Small cell-permeable effectors represent in principle powerful tools for identification of functional links between catalyzed prolyl isomerization and physiological processes in living cells. Most attempts to unravel physiological functions of PPIases in vivo have been done by this pharmacological approach. In particular, the immunosuppressants cyclosporin A (CsA), FK506, and rapamycin, as well as their non-immunosuppressive derivatives, have been proposed as biochemical markers, which may provide results useful for dissecting functions in vivo. These compounds are competitive inhibitors of many cyclophilins or FKBPs present in human tissues, but to a different extent. There are only a few inhibitors available for the enzymes of the parvulin family [53, 79, 80]. A long-standing aspiration of researchers in the field has been the design and synthesis of inhibitors capable of selectively inhibiting closely related PPIase paralogs, but success is still limited at this stage.

Among all known enzymes, PPIases exhibit a unique reaction profile with a large number of potentially reactive sites in a polypeptide substrate and a low degree of electronic differences between substrate and product. Regarding a microscopic description, the minimal kinetic mechanism of the enzyme-catalyzed prolyl isomerization of protein and peptide substrates operates on the basis of three chemical transitions. Accordingly, the two Michaelis complexes, the *trans* and *cis* isomers in their enzyme-bound state, are able to interconvert to each other with or without forming a non-Michaelis complex-like intermediate. In the simplest case, the catalytic rate enhancement is due to enzyme-mediated stabilization of the partially rotated state of the prolyl bond. However, the limited access to transition-state analogue inhibitors and the minimal activities of catalytic antibodies generated against surrogates of twisted peptide bonds [81, 82] do not support an exclusive role of noncovalent transition-state stabilization for catalysis. Large inverse solvent deuterium isotope effects (J. Fanghänel and G. Fischer, in preparation), the pH dependence of k_{cat}/K_m values [83, 84], and the results of site-directed mutagenesis [85–87] point to the involvement of distinct catalytic groups in the mechanism of rate enhancement. Therefore, it is likely that the desolvation, the twisted transition-state stabilization, and the acid-base catalysis are acting in combination rather than as isolated routes to enzymatic catalysis [88]. It appears likely that only physiological substrates correctly recruiting the complete ensemble of secondary binding sites provided by the PPIase are able to make the catalytic machinery fully active [84].

PPIases have been shown to exhibit pronounced subsite recruitment, supporting the presence of many enzyme–substrate-contacting sites in the transition state of catalysis, but for most PPIases knowledge remains limited and fragmentary. In the reported cases, this secondary specificity contributes much to efficient catalysis [9, 53]. With a few exceptions, the parent five-membered proline ring represents a major determinant of catalytic specificity [89]. The active sites of some eukaryotic members of the parvulin family impose considerable constraints on substrate selection. The marked preference for substrates containing the (PO_3H_2)Thr-Pro-/(PO_3H_2)Ser-Pro-moiety illustrate a unique ability of several parvulins to isomerize polypeptides targeted by the large superfamily of proline-directed protein kinases, including cyclin-dependent and mitogen-activated protein kinases [90–92]. As a common property of PPIases, stereospecificity causes detrimental effects on catalysis when D-amino acid substitution in the P1 and P1' positions is to occur [93].

10.4.2
Prototypic Cyclophilins

Prototypic cyclophilins consist of a single catalytic domain with no N-terminal and C-terminal extensions bearing a discrete biochemical function. Their amino acid sequences are highly conserved from yeast to humans, with a smaller degree of conservation across prokaryotic taxa [94]. Currently, this group of cyclophilins comprises six members in human cells, with polypeptide chains of 160–177 amino acids long whose biochemical functions have been at least partially characterized. Many additional prototypic cyclophilins can be derived from the complete human genome, but their protein expression must await characterization. Cytosolic Cyp18 (CypA) is the most abundant constitutively expressed cyclophilin in the cytosol of mammal cells. Unexpectedly, Cyp18 can be extracellularly secreted in response to inflammatory stimuli [95] and is recognized by the chemotaxis cell-surface receptor CD147 only if the Cyp18 active site is functional [96]. Other members are the nuclearly localized Cyp19.2 (CypH, USA-Cyp, Snu-Cyp, Cyp20) [97]; the cancer-related Cyp18.1a (COAS2-Cyp) [98]; the two splicing variants of the gene *PPIL3*, Cyp18.6 and Cyp18.1 (CypJ) [99]; and Cyp18.2a (CypZ) [100]. Among the best-characterized prototypic cyclophilins, Cyp19.2 is biochemically dissected along with resolved crystal structures [101–103].

Cyp18 represents a thermally and chemically rather stable protein that contains a well-defined PPIase active site, a hidden catalytic triad typical of serine proteases [104], and a poorly defined second peptide ligand-binding site [105]. The calcium/

magnesium-dependent nuclease activity reported for cyclophilins [106, 107] is not a property of the native Cyp18 structure but might result from a nuclease contamination of Cyp18 preparations [108].

The binding partner of U4/U6-60K and hPrp18, the spliceosomal Cyp19.2, contains two well-defined protein-protein interaction sites: the PPIase site and an oppositely located peptide-binding site. Both sites are believed to act in conjunction [102].

The Cyp18 concentration approaches high levels in the brain but is also remarkable in other tissues and cells, as could be exemplified by the measurement of 5–10 µg/mg^{-1} total protein in kidney tubules and endothelial cells [109–111].

In contrast to FKBP12, Cyp18 is rather promiscuous toward substrates of a series of -Xaa-Pro- with comparable k_{cat}/K_m values for tetrapeptides, while the Xaa position varied within the natural amino acids [112, 113]. Specificity constants approach high values of about $>10^7$ M^{-1} s^{-1}, implying a nearly diffusion-controlled bimolecular reaction, and the energy barrier of Cyp18-catalyzed isomerization is almost entirely determined by the activation entropy.

General acid catalysis by Arg55 has been argued to be the major driving force of catalyzed isomerization for Cyp18 [85, 114]. Another critical residue in hydrogen bonding distance to the rotating prolyl bond is His126, but its specific function has not yet been explored. Kinetic secondary deuterium isotope effects for substrates with an H/D exchange at the Cα position of the amino acid preceding proline are found to be similar between the spontaneous and the Cyp18-catalyzed reaction. The occurrence of a tetrahedral intermediate on the catalytic pathway is ruled out by the isotope effect $k_H/k_D > 1$ for these substrates [83]. However, a detailed picture of Cyp18 catalysis has not yet begun to emerge.

Most strikingly, all genes encoding cyclophilins could be deleted without seriously affecting viability of *Saccharomyces cerevisiae* [115]. Similarly, disruption of both copies of the mouse gene encoding for Cyp18 indicates the nonessential character of this protein [116]. On the other hand, mutations of the Cpa1 and Cpa2 proteins alone, the Cyp18 homologues in *Cryptococcus neoformans*, conferred dramatic phenotypes in that cell growth, mating, virulence, and cyclosporin A toxicity were affected. Cpa1/Cpa2 double mutants exhibited severe synthetic defects in growth and virulence [117].

Cyp18 itself has pronounced biochemical and physiological effects on living cells. When Cyp18 is secreted by vascular smooth muscle cell in response to oxidative stress, it mediates typical effects of reactive oxygen species: the activation of extracellular signal-regulated kinase (ERK1/2) and stimulation of cell growth [118]. Whole-cell experiments assaying the phosphorylation status of extracellular signal-regulated protein kinases in the presence of a panel of wild-type Cyp18 and Cyp18 variants served to display possible enzymatically controlled initiation of downstream signaling events. The extracellular part of the multifunctional transmembrane glycoprotein CD147 (basigin) is considered to be the docking site of Cyp18 [95]. Whereas externally added wild-type Cyp18 fully enhances the level of ERK phosphorylation in CHO.CD147 cells, but not in CHO cells lacking induced CD147 expression, Cyp18 variants are effective only according to their level of residual PPIase activity. Site-directed mutagenesis of the CD147 extracellular domain revealed that the Asp179-Pro180 segment was the Cyp18-directed substrate region. Neutrophil chemotaxis, a downstream signaling effect of CD147 stimulation by Cyp18, is also dependent on the intact Asp179-Pro180 recognition segment [96]. Another topic concerns the intracellularly localized fraction of Cyp18 that controls the functional expression of homo-oligomeric $\alpha 7$ neuronal nicotinic and type 3 serotonin receptors in a manner dependent on its PPIase activity in *Xenopus laevis* oocytes [119].

An antisense deoxynucleotide effectively reduced Cyp18 expression in a rat neuroblastoma cell line and provided the first indication that this enzyme participates in the activation of the caspase cascade in neuronal cells [120].

Functional analysis revealed that overexpression of Cyp18 inhibited prolactin-induced Rac activation, while simultaneously prolonging Jak2 phosphorylation. This effect indicates that Cyp18 might be involved in the various prolactin receptor-mediated signaling pathways [121].

Cyp18 physically interacts with many protein ligands in cells, some of which may involve substantial chain reorganization during functional interaction affecting their bioactivity. If ligand binding is strong enough, as realized by the retinoblastoma gene product RB, it can interfere with the blockage of CsA-mediated NF-AT dephosphorylation if the Cyp18 concentration is limiting [122]. Direct evidence for the functional role of Cyp18 has come from signals associated with oxidative stress. In particular, stimulation of the ligand activity was found to occur when Cyp18 binds to the thiol-specific antioxidant protein Aop1 [123]. The familial amyotrophic lateral sclerosis (FALS)-associated mutant Cu/Zn superoxide dismutase-1 (SOD) induces apoptosis of neuronal cells in culture associated with high levels of reactive oxygen species. An increased protein turnover might result from altered cellular free radical balance that requires an increased level of folding-supportive enzymes. Thus, the apoptosis-enhancing effect of cyclosporin A and FK506, as well as non-immunosuppressive analogues of cyclosporin A, in PC12 cells infected with the AdV-SOD Val148Gly- and SOD Ala4Val variants might result from the increased need for PPIase activity in this oxidative stress situation. The importance of enzymatic activity in mutant SOD-mediated apoptosis was supported by experiments showing that overexpressed wild-type Cyp18, but not Cyp18 Arg55Ala variant with a reduced catalytic activity of about 0.1% of the wild-type enzyme, protected cells from death after SOD Val148Gly expression [124].

Interleukin-2 tyrosine kinase catalytic activity is inhibited according to Cyp18 PPIase activity level. NMR structural studies combined with mutational analysis show that a proline-dependent conformational switch within the Itk SH2 domain regulates substrate recognition and mediates regulatory interactions with the active site of the PPIase. Cyp18 and Itk form a stable complex in Jurkat T cells that is disrupted by treatment with cyclosporin A [125].

Despite the fact that the Cyp18 gene is regarded as a housekeeping gene, the regulated expression of Cyp18 mRNA in the brain points to stress control of expression [126]. Many Cyp18 homologues are subject to stress regulation [127–129]. Similarly, proteome analyses detected upregulation of Cyp18 in the higher passages of fetal skin cells and downregulation in the fibroblasts of older adults [130, 131]. A proteome-based analysis of potential mitogens to human bone marrow stromal cells secreted by PC3 cancer cells identified Cyp18 [132]. Both the prototypic COAS-Cyp and Cyp18 belong to the highly overexpressed proteins in human tumors [98, 133, 134]. Secretion of prototypic cyclophilins appears to be involved in host cell defense against parasitic invasions, because a *Toxoplasma gondii* Cyp18 homologue induces IL-12 production in dendritic cells via chemokine receptor signaling [135]. In other parasites, Cyp18 inactivation causes anti-parasite effects [136].

Cyclophilins and FKBPs have frequently been found to be involved in protein trafficking. Yeast Cyp17 (Cpr1p) was identified as the protein that is required for fructose-1,6-bisphosphatase import into intermediate transport vesicles. Mutants lacking Cpr1p were defective in this import function, but addition of the purified Cyp17 restored import function in the ΔCpr1p mutants [137].

The three-dimensional structure of human Cyp18 revealed an eight-stranded, antiparallel β-barrel capped by two short helices, with the active-site groove located on one side of the barrel. Many three-dimensional structures of Cyp18 exist either in the free state or in complex with a variety of inhibitors and oligopeptide substrates, and the structure of a ternary complex of Cyp18/CsA/calcineurin has also been resolved [138–140]. Generally, the Cyp18 fold does not change much in response to ligand binding.

Complexes of Cyp18 with physiological ligands have been realized for human Cyp18 bound to HIV-1 capsid protein fragments [141, 142]. HIV-1 replication requires capsid protein-mediated host cell Cyp18 packaging [143]. The binding portion of the capsid protein fragments extend over nine residues, at least, with a Gly89-Pro90-Ile91 moiety bound to the primary catalytic site of Cyp18. As already discussed above, such extended ground-state binding sites may have the potential to sequester unfolded polypeptide chains, thus realizing the properties of a holding chaperone. The reaction conditions, however, determine whether Cyp18 is able to prevent aggregation [144] or remains refractory to unfolded chains [145]. In response to their additional domains, larger cyclophilins frequently exhibit chaperone-like function. The mobility of the binding loop of the capsid protein fragments causes native-state isomers to exist in the unbound state. Cyp18 effectively exerts control on the dynamics of native-state prolyl isomerization by increasing the isomerization rate [146, 147]. It can be hypothesized that the Cyp18/capsid protein interaction protects HIV-1 from the host cell restriction factor Ref-1 by a catalyzed distortion of a capsid protein segment representing the Ref-1 docking site [148].

A Cyp18-mediated rate enhancement has been established as a standard probe for the identification of prolyl-bond-limited steps in refolding [149–154]. This enzyme is also a major component of the mixture of folding helper proteins of the oxidative refolding chromatography used, for example, for "empty" MHC class I-like molecules to assemble at physiological temperatures in the absence of ligand [155–157].

Cyp18 constitutes a common target protein for the immunosuppressants cyclosporin A and sanglifehrin, which both inhibit the enzyme by binding to the deep hydrophobic pocket of the PPIase active site [158–161].

Besides cyclosporin A and its derivatives, only a few other Cyp18 inhibitors have been reported, but at a reduced level of inhibitory potency [162–164]. To date the Cyp18/cyclosporin A complex forms a sole example among the Cyp18/inhibitor complexes in that complex formation results in a gain of function that is exclusively targeted to the Ser/Thr protein phosphatase calcineurin [71].

10.4.3
Prototypic FK506-binding Proteins

The FK506-binding proteins (FKBPs) are a family of PPIases in which the prototypic enzymes have been found to bind to and become inhibited by the peptidomacrolides FK506 and rapamycin. The group of prototypic FKBPs is much smaller than those of Cyp18 because FKBP12, FKBP12.6, and a third, uncharacterized FKBP12 encoded on human chromosome 6 are the only members. FKBP12 differs from FKBP12 in 16 out of 107 amino acids. The genomes of lower eukaryotes, eubacteria, and archaea encode homologues that have, in some cases, a bulge insertion in the flap region of FKBP12, conferring holding chaperone properties to the prototypic enzymes [165]. The property is most easily explained with largely extended catalytic subsites required for optimal recognition of the cellular substrates in these organisms.

In the catalytic modules of FKBPs, in view of the sequence similarity among members of the FKBP family, affinity to FK506 is less well conserved [87]. As was already found for cyclophilins, sequence variations at the positions that are crucial for binding the inhibitors and enzyme activity are relatively small. The human members of the FKBP family are depicted in Figure 10.2.

The concentration of FKBP12 in T cells is extraordinarily high for an enzyme (about 20 µM) [28]. Furthermore, there is an increased level of FKBP12 in neurons situated in areas of degeneration [166].

Mimicking the intrinsically high concentrations of authentic FKBP12 in the male reproductive tract, recombinant FKBP12 enhances the curvilinear velocity of immature sperm, suggesting a role for FKBP12 in motility initiation [167]. FKBP12 also possesses chemotactic effects on neutrophils unless the active site is blocked by FK506 [168].

Native FKBP12 contains seven *trans*-prolyl peptide bonds, and the *cis* to *trans* isomerizations of some or all of them constitute the slow, rate-limiting event in folding. Its refolding process from a chemically denatured state constituted the first example of an autocatalytic formation of a native protein from kinetically trapped species with nonnative proline isomers [169, 170].

Knowledge of FKBP12 and FKBP12.6 structures either alone or complexed with ligands exists, but the structure of an FKBP12/oligopeptide substrate complex is still missing [171]. Thus, the FK506-binding site was taken to represent the catalytic site. A docking procedure to FKBP12 was used for a potential substrate: the proline-rich prolactin receptor-derived octapeptide. When bound to FKBP12, the docked peptide reveals a type I β-turn conformation and many hydrophobic contacting sites [172]. The major structural elements surrounding the hydrophobic catalytic cleft are a concave five-stranded antiparallel β-sheet, wrapping around a short alpha-helix, and a large flap region. In most preferred substrates, hydrophobic side chains are positioned toward the N-terminus of the reactive prolyl bond [83]. Unlike Cyp18 and Par10, however, k_{cat}/K_m values of FKBP12 do not achieve the diffusion-controlled limit. Both PPIase activity and FK506 inhibition proved to be similar for FKBP12.6 and FKBP12 [173]. Consequently, steady-state catalytic data of prototypic FKBPs suggest an unassisted conformational twist mechanism with rate enhancement due in part to desolvation of the peptide bond at the active site [88, 174].

Comparison of the effects of point mutations on k_{cat}/K_m values for tetrapeptide substrates revealed catalytically important side chains of FKBP12 [86]. Several FKBP12 variants (Tyr82Phe, Trp59Ala, and Asp37Val) experienced marked reduction (about 10-fold) in the specificity constant k_{cat}/K_m, with the k_{cat} value being the most severely affected parameter. None of the point mutations confers complete catalytic inactivity to the enzyme variants.

In comparison to Cyp18, FKBP12 confers a more diverse gain of function because the FKBP12/FK506 complex inhibits the protein phosphatase calcineurin, but the structurally related FKBP12/rapamycin complex inactivates the human target of rapamycin (mTOR) protein, a cell-cycle-specific protein kinase known to be involved in signaling pathways that promote protein synthesis and cell growth [175]. The downstream events that follow the inactivation of TOR result in the blockage of cell-cycle progression at the juncture of the G1 and S phases.

Besides the natural products FK506 and rapamycin and their derivatives, many other inhibitors of several totally unrelated substance classes have been developed, indicating a highly promiscuous binding affinity of the active site of FKBP12 [176–178]. That is, a rather limited number of contacting sites suffice to account for specificity and affinity. The dissociation constants for the FKBP12 complexes of dimethyl sulfoxide and 5-diethylamino-2-pentanone of 20 mM and 500 µM, respectively, highlight the unique ligand-affinity pattern of the FKBP12 catalytic cleft [179].

Most potent compounds are derived from α-ketoamides, from which the potent neuroactive inhibitor GPI 1046 (3-(3-pyridyl)-1-propyl (2S)-1-(3,3-dimethyl-1,2-dioxopentyl)-2-pyrrolidinecarboxylate) was derived [180–183]. Displacement of enzyme-bound water molecules upon inhibitor binding is a major driving force for the tight binding of α-ketoamides to the active site [184].

Some FKBPs, notably FKBP12, have defined roles in regulating ion channels, protein transport, or cell signaling. Using cells from FKBP12-deficient (FKBP12$^{(-/-)}$) mice, cell-cycle arrest in the G(1) phase could be detected. These cells can be rescued by FKBP12 transfection [185]. The transgenic mice do not show abnormal development of skeletal muscle but have severe dilated cardiomyopathy and ventricular septal defects that mimic a human congenital heart disorder [186]. In contrast, disruption of the FKBP12.6 gene in mice results in cardiac hypertrophy in male mice, but not in females. However, female FKBP12.6$^{(-/-)}$ mice treated with an estrogen receptor antagonist develop cardiac hypertrophy similar to that of male mice [187].

Physically interacting proteins of FKBP12 and FKBP12.6 include the intracellular calcium release channels [188, 189], the type II TGF-β receptor [190], the EGF receptor [191], the HMG-box proteins 1 and 2 [192], aspartokinase [193], FAP48, a fragment of the glomuvenous malformation-related glomulin [194, 195], and the transcription factor YY1 [196]. An interacting protein similar to FAP48, the FAP48 homologue FIP37, has been found for the prototypic plant FKBP12 in *Arabidopsis thaliana* [197].

Two transmembrane receptor protein kinases have been shown to be downregulated in their activating phosphorylation in the presence of FKBP12. FKBP12 can affect TGF-β signaling by preventing Ser/Thr phosphorylation of the type I TGF-β receptor by the type II TGF-β receptor via binding to a Leu-Pro motif of the type I TGF-β receptor [190]. Type I TGF-β receptor affinity for FKBP12 is abolished for a receptor molecule phosphorylated on the GS box. The modulation of the EGF receptor signaling by FKBP12 is directed against the tyrosine autophosphorylation of the cytoplasmic domain of the receptor [191]. FK506 and rapamycin abolish the FKBP12 effect to the same extent in EGF receptor-rich A431 fibroblasts, leading to an increase in cell growth. In both cases, by impeding the activation of receptors in the absence of native ligands, FKBP12 may provide a safeguard against leaky signaling, resulting in a basal level of receptor phosphorylation. In cell signaling, the function of prototypic FKBPs emerges as catalysts affecting the rate of decay of metastable states of receptor proteins (C. Schiene-Fischer, in preparation).

The sarco(endo)plasmic reticulum, the main Ca^{2+} storage compartment, contains two different families of Ca^{2+} release channels: the ryanodine receptors (RyR) and the inositol 1,4,5-trisphosphate receptors (IP$_3$R), both of which share affinity binding to prototypic FKBPs.

Mammalian tissues express three large, high-conductance ryanodine receptors known as skeletal muscle (RyR1), cardiac muscle (RyR2), and brain (RyR3). They are able to control dynamically the level of intracellular Ca^{2+} by releasing Ca^{2+} from the intracellular sarco(endo)plasmic reticulum.

The skeletal muscle ryanodine receptor (RyR1) can be isolated as a hetero-oligomer containing FKBP12 whereas the cardiac ryanodine receptor (RyR2) selectively associates with FKBP12.6. However, RyR2 isolated from many vertebrates contains both FKBP12 and FKBP12.6 [198]. The structural response to mutation suggests that considerable hydrophobicity of tryptophan 59 (which is a phenylalanine in FKBP12.6) contributes to the specificity of binding between FKBP12 isoforms and ryanodine receptors. More specifically, the triple mutant (Gln31Glu/Asn32Asp/Phe59Trp) of FKBP12.6 was found to lack selective binding to RyR2, comparable to that of FKBP12. In complementary studies, mutations of FKBP12 to the three critical amino acids of FKBP12.6 conferred selective binding to RyR2 [199].

The 4:1 molar ratio of FKBP12/tetrameric RyR indicates the stoichiometric binding of one FKBP12 molecule to one RyR subunit. The heterocomplex dissociation by titration with FK506 reveals that the interaction involves the active site of FKBP12.

The prototypic FKBPs are regulatory constituents of the RyR that can dramatically affect RyR function, e.g., by overexpression [200] or deletion [186, 200]. For example, when FKB12 is bound to RyR1, it inhibits the channel by stabilizing its closed state. Taking a more detailed view, FKBP12 deficiency causes longer mean open times and greater open probability in cardiac myocytes of the ryanodine receptor when compared with reference cells. Generally, the Ca^{2+} leak rate increases considerably [187]. Marks et al. reported that the FKBP12 Phe36Tyr variant is still able to bind but not to modulate RyR channeling, pointing to a crucial role of the unperturbed enzyme activity of FKBP12 [201].

In a canine model of pacing-induced heart failure, the ratio of FKBP12.6 per failing RyR2 is significantly decreased. A conformational change was suggested for RyR2 chains concomitant with abnormal Ca^{2+} leak of the channel [202].

Ca^{2+} stores of the endoplasmic reticulum become superfilled after co-expression of FKBP12.6, but not of FKBP12, in CHOhRyR2 cells. The effects of FKBP12.6 on hRyR2-mediated intracellular Ca^{2+} handling could be antagonized using the FKBP inhibitor rapamycin [203].

10.4.4
Prototypic Parvulins

The parvulins represent the most recently discovered PPIase family [204, 205]. The most typical parvulin known to date is the *E. coli* Par10, a monomerically active enzyme comprising 92 amino acids in its mature form. Parvulin sequences are characterized by two conserved histidine motifs (three to five amino acids in each segment), which are separated by a stretch of about 70–85 amino acids containing a few other conserved residues. There is no sequence similarity to either cyclophilins or FKBPs. In various plants, a parvulin member corresponds to the prototypical form (it is 117–119 amino acids in length) because it does not show an N-terminal or C-terminal extension when compared to Par10 [47, 91, 206]. The difference of Par10 in chain length is due to a single polypeptide segment of 23 amino acids inserted in the N-terminal half of the protein that adopt a loop conformation in the $\alpha 1/\beta 1$ region [207]. In two other eukaryotic parvulins, functional domains complement N-terminally the catalytic domain, a DNA-binding domain, and a (PO_3H_2)Ser/Thr-directed type IV WW domain in Par14 and Pin1, respectively. So far, the existence of only these two genes encoding parvulin proteins has been described in the human genome, i.e., they have a low frequency compared with both Cyp and FKBP sequences (Figure 10.2). Originally, an activity-based screening approach for cytosolic extracts identified Par10, and it was not shown to be a predicted ORF of the *E. coli* genome. Besides plants, no other organism was reported to possess a prototypic parvulin. However, the concept of ORF identification is confounded by several factors. The procedure may permit small proteins to escape detection due to their size falling below some arbitrary researcher-defined minimum cutoff, or there may be an inability to precisely define a promoter or translational start [208]. Thus, future identification of new prototypic enzymes is predicted. Many larger parvulins have been characterized in prokaryotes.

It is of interest that Par10 is as active as Cyp18 toward oligopeptide substrates but that proteins are catalyzed less efficiently. When compared to Cyp18, Par10 was found to have a much narrower substrate specificity at the position preceding proline (measured by k_{cat}/K_m), a finding reminiscent of FKBP12 [204]. However, unlike other PPIase families, the parvulins exhibit an extraordinary pattern of substrate specificity across the family members, ranging from the positively charged arginine side chain to the double-negative phosphoserine (phosphothreonine) side chain for Par14 and plant Par13 or Pin1, respectively.

Like FKBP12, Par10 accelerates its own refolding in an autocatalytic fashion, and the rate of refolding increases 10-fold when the concentration of parvulin is increased from 0.5 to 3.0 μM [209]. In larger parvulins, such as *E. coli* SurA, two parvulin-like domains exist, one of which, by itself, is devoid of PPIase activity [210]. The enzymatically active parvulin-like domain protrudes out from the core domain about 25–30 Å in the three-dimensional structure, implicating an isolated action onto substrates [211]. The three-dimensional structure of Par10 is not yet published, but sequence similarity suggests a common structural pattern shared by human Par14, Pin1, the active domain of SurA, and *Arabidopsis thaliana* Par13. Analyses of the tertiary structures of the parvulins revealed the existence of an FKBP-like superfold that is also found for the *E. coli* GreA transcript cleavage factor of 17.6 kDa [37]. When measuring 2D-^{1}H-^{15}N HSQC NMR spectra of human Par14 in the presence of a substrate, site-directed chemical shift changes in the protein were identified [212]. Hydrophobic amino acids in positions Met90, Val91, and Phe94 have been identified to be substrate-sensitive residues that match those of the active site of Pin1 [213]. However, the basic cluster of two neighboring arginines and a lysine, which collectively determine the -(PO$_3$H$_2$)Ser/Thr-Pro- directed substrate specificity of Pin1 and plant Par13, is absent in Par10 and Par14.

The putative biological functions of human Pin1, which has highly conserved homologues in eukaryotes, are best considered in the context of the manifold of known phosphoprotein targets [214]. Cellular processes include cell cycle progression, transcriptional regulation, cell proliferation and differentiation, DNA damage response, p53 functions, Alzheimer's disease, and other tauopathies. On these Pin1 topics, excellent comprehensive reviews have been compiled recently [24, 214, 215].

Although plant prototypic parvulins lack the type IV WW domain, they are able to rescue the lethal mitotic phenotype of a temperature-sensitive mutation in the Pin1 homologue ESS1/PTF1 gene in *Saccharomyces* [91, 206]. Plant Par13 proteins are expressed in high abundance, predominantly in rapidly dividing cells. Expression of apple Par13 is tightly associated with cell division during both apple fruit development in vivo and cell cultures in vitro.

An oxidative stress phenotype could be observed for *E. coli* Par10 deletion strains. The results of complementation experiments with several Par10 variants suggest a correlation between residual enzyme activity and the capability to rescue protein most strongly associated with the putative Par10 function (B. Hernandez-Alavarez, unpublished results).

10.5
Concluding Remarks

Elucidation of the molecular mechanisms active in conformational regulation of bioactivity as applied in catalytic fashion to polypeptide chains of various folding states by peptide bond *cis/trans* isomerases is an emerging research field. Regarding safety of conformational control, the high-resistance on/off switch of a prolyl bond offers unique advantages over other conformational rearrangements within the polypeptide backbone. Thus, the topic extends from de novo protein folding to allosteric control and protein metastability and includes processes that govern the spatial-temporal control of bioactivity in cell life. Research should make use of highly multidisciplinary approaches involving physicochemical techniques and bioorganic chemistry as well as molecular and cell biology. Fundamental probes characterizing prolyl isomerizations in larger polypeptide chains are complicated, are in part insufficiently developed, and suffer from severe limitations regarding sensitivity and selectivity. Advances in current detection methods on the one-bond level of conformational couplings for proteins in vivo will be of essence if the molecular aspect of regulation is to be explored. Experiments combining gene deletion with variant enzyme gene transfection and the small-molecule approach with monofunctional PPIase inhibitors must be supplemented with new techniques allowing the externally controlled switching of a peptide bond conformation.

10.6
Experimental Protocols

10.6.1
PPIase Assays: Materials

Pancreatic α-chymotrypsin, subtilisin Carlsberg type VIII, pancreatic elastase, and trypsin are available from a number of vendors. In the substrate, the position succeeding proline determines the type of protease required: α-chymotrypsin for -Pro-Phe(Tyr)-, trypsin and clostripain for Pro-Lys(Arg)-, and pancreatic elastase for Pro-Val-. Because defined reaction conditions are essential, determination of the active protease concentration is required using an appropriate method for the respective enzyme [216]. Stock solutions of the protease (typically 40 mg mL^{-1} active enzyme solutions) are prepared in assay buffer subsequent to the addition of a small amount of 1-mM HCl to improve dissolution and can be stored at 0 °C for several days.

Typical PPIase substrates such as succinyl-Ala-Ala-Pro-Phe-4-nitroanilide and succinyl-Ala-Leu-Pro-Phe-4-nitroanilide can be purchased from Bachem (Heidelberg, Germany). The purity of the substrate peptides should be checked by HPLC at 330 nm because the clear separation of the kinetic phases of the assay stringently relates to the absence of chromogenic impurities in the substrates. Stock solutions are prepared in analytical grade DMSO mixtures or 0.2–0.5 M LiCl dissolved in anhydrous trifluoroethanol. Typically, a sample of 25 mg peptide is dissolved in 1 mL DMSO for the stock solution. Human PPIases, such as Cyp18 and FKBP12, and *E. coli* Par10 can be easily prepared recombinantly in *E. coli* by standard procedures or are present as a mixture of many PPIases in most cytosolic extracts of tissues and cells. Except for human Pin1 [90], PPIase mixtures may be difficult to differentiate based solely on a panel of different substrates. HEPES buffer is degassed prior to use.

10.6.2
PPIase Assays: Equipment

Measurements can be performed by monitoring the UV/Vis absorbance of released 4-nitroaniline at 390 nm ($\varepsilon = 11\,800$ mM cm^{-2} on a Hewlett-Packard 8452 diode array spectrophotometer (Hewlett Packard, USA) equipped with a thermostated cell holder and a stirring device within the sample cell ($d = 1$ cm, silica cell). The reference wavelength is set to 510 nm. Constant temperature of 10 °C is maintained within the cell by water circulated from a Cryostat Haake D8 (Haake-Fisons, Germany). The Sigma-Plot Scientific Graphing System Vers. 8.0 (SPSS Science, USA) is used for data analysis. The possibility of monitoring of the reaction kinetics over three or more half-times of the reaction is the necessary precondition to reliable PPIase assays.

10.6.3
Assaying Procedure: Protease-coupled Spectrophotometric Assay

The example given here is for the chymotrypsin-coupled Cyp18 assay [217]. HEPES buffer 0.035 M, pH 7.8 (1.19 mL) and 10 µL of the stock solution of pancreatic α-chymotrypsin are preincubated at 10 °C in the cell holder for 5 min. In a typical PPIase assay, recombinant human Cyp18 is added to a final concentration of 2–5 nM. In our hand-mixing experiment, too-high isomerization rates at high PPIase concentrations result in datasets that are incompatible with the fitting procedure of the first-order calculations. Assays in the absence of a PPIase yield rate constants that characterize the spontaneous rates of prolyl isomerization of substrates. Continuous stirring is necessary during the assay. After the solution has reached thermal equilibrium, the reaction is initiated by the addition of 2 µL of the stock solution of the succinyl-Ala-Ala-Pro-Phe-4-nitroanilide, and time-absorbance readings are stored. After the reaction is finished, the data list is transferred to the first-order routine of the Sigma-Plot program for calculating the first-order rate constant and the amplitude of the slow kinetic phase. Prior to calculation, sample-mixing-corrupted data points (the first 5–15 s after reaction start) are removed from the data table by visual inspection (Figure 10.2).

A considerable improvement of the signal-to-noise ratio of the assay is achieved with the peptide dissolved in 0.2 M LiCl/trifluoroethanol [218]. Due to an increase in the equilibrium population of the *cis* isomer from about 5% to about 50%, the amplitude of the kinetic phase of the *cis* to *trans* isomerization increases in a similar manner. The residual water content in the TFE/LiCl/peptide mixture determines the percentage of *cis* isomer.

10.6.4
Assaying Procedure: Protease-free Spectrophotometric Assay

For assaying prolyl isomerization with the protease-free method, progress curves are recorded, applying the experimental conditions already described above, with the exception that a protease is omitted from the assay mixture, and a different detection wavelength [51]. The slight difference of 600 mM^{-1} cm^{-2} of the UV/Vis absorption coefficients at 330 nm between the *cis* and the *trans* isomer of many standard peptide 4-nitroanilide substrates (or other ring-substituted peptide anilides) is utilized. At 330 nm a time-dependent decrease of absorbance is recorded subsequent to a rapid 88-fold dilution from a dry stock solution of succinyl-Ala-Ala-Pro-Phe-4-nitroanilide in 0.47 M LiCl/trifluoroethanol into 1.2 mL of 0.0035-M HEPES buffer, pH 7.8. The readout of time-absorbance data pairs provides the basis for the calculation of first-order rate constants according to the procedure described above.

Besides a nanomolar concentration of Cyp18, the final reaction mixture should contain bovine serum albumin (1 µM) to avoid absorption of the PPIase to the surface of the silica cell. The first-order rate constant for approach to *cis/trans* equilibrium equals $k_{obs} = k_{cis\ to\ trans} + k_{trans\ to\ cis}$.

References

1 EATON, W. A., MUNOZ, V., HAGEN, S. J., JAS, G. S., LAPIDUS, L. J., HENRY, E. R. & HOFRICHTER, J. (2000). Fast kinetics and mechanisms in protein folding. *Annu. Rev. Biophys. Biomol. Struct* **29**, 327–359.

2 BIERI, O., WIRZ, J., HELLRUNG, B., SCHUTKOWSKI, M., DREWELLO, M. & KIEFHABER, T. (1999). The speed limit for protein folding measured by triplet-triplet energy transfer. *Proc. Natl. Acad. Sci. USA* **96**, 9597–9601.

3 KRIEGER, F., FIERZ, B., BIERI, O., DREWELLO, M. & KIEFHABER, T. (2003). Dynamics of unfolded polypeptide chains as model for the earliest steps in protein folding. *J. Mol. Biol.* **332**, 265–274.

4 DUGAVE, C. & DEMANGE, L. (2003). Cis/trans isomerization of organic molecules and biomolecules: implications and applications. *Chem. Rev.* **103**, 2475–2532.

5 FISCHER, G. (2000). Chemical aspects of peptide bond isomerization. *Chem. Soc. Rev.* **29**, 119–127.

6 REIMER, U. & FISCHER, G. (2002). Local structural changes caused by peptidyl-prolyl *cis/trans* isomerization

in the native state of proteins. *Biophys. Chem.* **96**, 203–212.

7 NEUTZE, R., PEBAY PEYROULA, E., EDMAN, K., ROYANT, A., NAVARRO, J. & LANDAU, E, M. (2002). Bacteriorhodopsin: a high-resolution structural view of vectorial proton transport. *Biochim. Biophys. Acta* **1565**, 144–167.

8 SONG, S. H., ASHER, S. A., KRIMM, S. & SHAW, K. D. (1991). Ultraviolet resonance raman studies of *trans*-peptides and *cis*-peptides – photochemical consequences of the twisted pi-star excited state. *J. Am. Chem. Soc.* **113**, 1155–1163.

9 FISCHER, G., BANG, H. & MECH, C. (1984). Determination of enzymatic catalysis for the *cis/trans* isomerization of peptide bonds in proline-containing peptides. *Biomed. Biochim. Acta* **43**, 1101–1111.

10 SCHIENE-FISCHER, C., HABAZETTL, J., SCHMID, F. X. & FISCHER, G. (2002). The hsp70 chaperone DnaK is a secondary amide peptide bond *cis/trans* isomerase. *Nat. Struct Bio.l* **9**, 419–424.

11 ANDREOTTI, A. H. (2003). Native state proline isomerization. *Biochemistry* **42**, 9515–9524.

12 STUKENBERG, P. T., KIRSCHNER, M. W. (2001). Pin1 acts catalytically to promote a conformational change in Cdc25. *Mo.l Cell* **7**, 1071–1083.

13 NG, K. K. S. & WEIS, W. I. (1998). Coupling of prolyl peptide bond isomerization and Ca^{2+} binding in a C-type mannose-binding protein. *Biochemistry* **37**, 17977–17989.

14 EISENMESSER, E. Z., BOSCO, D. A., AKKE, M. & KERN, D. (2002). Enzyme dynamics during catalysis. *Science* **295**, 1520–1523.

15 SCHMID, F. X. (1992). The mechanism of protein folding. *Curr. Opin. Struct. Biol.* **2**, 21–25.

16 BÄCHINGER, H. P. (1987). The influence of peptidyl-prolyl *cis/trans* isomerase on the in vitro folding of type III collagen. *J. Biol. Chem.* **262**, 17144–17148.

17 STEINMANN, B., BRUCKNER, P. & SUPERTI FURGA, A. (1991). Cyclosporin A slows collagen triple-helix formation in vivo: indirect evidence for a physiologic role of peptidyl-prolyl *cis/trans*-isomerase. *J. Biol. Chem.* **266**, 1299–1303.

18 THIES, M. J. W., MAYER, J., AUGUSTINE, J. G., FREDERICK, C. A., LILIE, H. & BUCHNER, J. (1999). Folding and association of the antibody domain C(H)3: Prolyl isomerization preceeds dimerization. *J. Mol. Biol.* **293**, 67–79.

19 BRANDTS, J. F., HALVORSON, H. R. & BRENNAN, M. (1975). Consideration of the possibility that the slow step in protein denaturation reactions is due to *cis/trans* isomerism of proline residues. *Biochemistry* **14**, 4953–4963.

20 REIMER, U., ELMOKDAD, N., SCHUTKOWSKI, M. & FISCHER, G. (1997). Intramolecular assistance of *cis/trans* isomerization of the histidine-proline moiety. *Biochemistry* **36**, 13802–13808.

21 TEXTER, F. L., SPENCER, D. B., ROSENSTEIN, R. & MATTHEWS, C. R. (1992) Intramolecular Catalysis of a Proline Isomerization Reaction in the Folding of Dihydrofolate Reductase. *Biochemistry* **31**, 5687–5691.

22 CHITI, F., TADDEI, N., GIANNONI, E., VAN NULAND, N. A. J., RAMPONI, G. & DOBSON, C. M. (1999). Development of enzymatic activity during protein folding – Detection of a spectroscopically silent native-like intermediate of muscle acylphosphatase. *J. Biol. Chem.* **274**, 20151–20158.

23 YAN, S. Z., BEELER, J. A., CHEN, Y. B., SHELTON, R. K. & TANG, W. J. (2001). The regulation of type 7 adenylyl cyclase by its C1b region and Escherichia coli peptidylprolyl isomerase, SlyD. *J. Biol. Chem.* **276**, 8500–8506.

24 LU, K. P., LIOU, Y. C. & VINCENT, I. (2003). Proline-directed phosphorylation and isomerization in mitotic regulation and in Alzheimer's Disease. *Bioessays* **25**, 174–181.

25 REIERSEN, H. & REES, A. R. (2001). The hunchback and its neighbours: proline as an environmental modulator. *Trends Biochem. Sci.* **26**, 679–684.

26 CREAMER, T. P. & CAMPBELL, M. N. (2002). Determinants of the poly-

proline II helix from modeling studies. *Adv. Prot. Chem.* **62**, 263–282.

27 ONEAL, K. D., CHARI, M. V., MCDONALD, C. H., COOK, R. G., YULEE, L. Y., MORRISETT, J. D. & SHEARER, W. T. (1996). Multiple *cis/trans* conformers of the prolactin receptor proline-rich motif (prm) peptide detected by reverse-phase HPLC, CD and NMR spectroscopy. *Biochem. J.* **315**, 833–844.

28 FISCHER, G. (1994). Peptidyl-prolyl *cis/trans* isomerases and their effectors. *Angew. Chem. Intern. Ed. Engl.* **33**, 1415–1436.

29 ZHOU, X. Z., KOPS, O., WERNER, A., LU, P. J., SHEN, M. H., STOLLER, G., KÜLLERTZ, G., STARK, M., FISCHER, G. & LU, K. P. (2000). Pin1-dependent prolyl isomerization regulates dephosphorylation of Cdc25C and tau proteins. *Mol. Cell* **6**, 873–883.

30 WEIWAD, M., KÜLLERTZ, G., SCHUTKOWSKI, M. & FISCHER, G. (2000). Evidence that the substrate backbone conformation is critical to phosphorylation by p42 MAP kinase. *FEBS Lett.* **478**, 39–42.

31 BRANDSCH, M., THUNECKE, F., KÜLLERTZ, G., SCHUTKOWSKI, M., FISCHER, G. & NEUBERT, K. (1998). Evidence for the absolute conformational specificity of the intestinal H^+/peptide symporter, PEPT1. *J. Biol. Chem.* **273**, 3861–3864.

32 FENG, Y. Q., HOOD, W. F., FORGEY, R. W., ABEGG, A. L., CAPARON, M. H., THIELE, B. R., LEIMGRUBER, R. M. & MCWHERTER, C. A. (1997). Multiple conformations of a human interleukin-3 variant. *Protein. Sci.* **6**, 1777–1782.

33 KELLER, M., BOISSARD, C., PATINY, L., CHUNG, N. N., LEMIEUX, C., MUTTER, M. & SCHILLER, P. W. (2001). Pseudoproline-containing analogues of morphiceptin and endomorphin-2: Evidence for a *cis* Tyr-Pro amide bond in the bioactive conformation. *J. Med. Chem.* **44**, 3896–3903.

34 NIELSEN, K. J., WATSON, M., ADAMS, D. J., HAMMARSTROM, A. K., GAGE, P. W., HILL, J. M., CRAIK, D. J., THOMAS, L., ADAMS, D., ALEWOOD, P. F. & LEWIS, R. J. (2002). Solution structure of mu-conotoxin PIIIA, a preferential inhibitor of persistent tetrodotoxin-sensitive sodium channels. *J. Biol. Chem.* **277**, 27247–27255.

35 BRAUER, A. B. E., DOMINGO, G. J., COOKE, R. M., MATTHEWS, S. J. & LEATHERBARROW, R. J. (2002). A conserved *cis* peptide bond is necessary for the activity of Bowman-Birk inhibitor protein. *Biochemistry* **41**, 10608–10615.

36 FISCHER, G. & AUMÜLLER, T. (2003). Regulation of peptide bond *cis/trans* isomerization by enzyme catalysis and its implication in physiological processes. *Rev. Physiol Biochem. Pharmacol.* **148**, 105–150.

37 SEKERINA, E., RAHFELD, J. U., MÜLLER, J., FANGHÄNEL, J., RASCHER, C., FISCHER, G. & BAYER, P. (2000). NMR solution structure of hPar14 reveals similarity to the peptidyl-prolyl *cis/trans* isomerase domain of the mitotic regulator hPin1 but indicates a different functionality of the protein. *J. Mol. Biol.* **301**, 1003–1017.

38 SCHREIBER, S. L., LIU, J., ALBERS, M. W., ROSEN, M. K., STANDAERT, R. F., WANDLESS, T. J., SOMERS, P. K. (1992). Molecular recognition of immunophilins and immunophilin-ligand complexes. *Tetrahedron* **48**, 2545–2558.

39 HUSI, H. & ZURINI, M. G. M. (1994). Comparative binding studies of cyclophilins to cyclosporin A and derivatives by fluorescence measurements. *Anal. Biochem.* **222**, 251–255.

40 MACARTHUR, M. W. & THORNTON, J. M. (1991). Influence of proline residues on protein conformation. *J. Mol. Biol.* **218**, 397–412.

41 REIMER, U., SCHERER, G., DREWELLO, M., KRUBER, S., SCHUTKOWSKI, M. & FISCHER, G. (1998). Side-chain effects on peptidyl-prolyl *cis/trans* isomerization. *J. Mol. Biol.* **279**, 449–460.

42 LONDON, R. E., DAVIS, D. G., VAVREK, R. J., STEWART, J. M. & HANDSCHUMACHER, R. E. (1990). Bradykinin and its Gly6 analogue are substrates of cyclophilin: a fluorine-19 magnetiza-

tion transfer study. *Biochemistry* **29**, 10298–10302.
43 HÜBNER, D., DRAKENBERG, T., FORSEN, S. & FISCHER, G. (1991). Peptidyl-prolyl *cis/trans* isomerase activity as studied by dynamic proton NMR spectroscopy. *FEBS Lett.* **284**, 79–81.
44 KERN, D., KERN, G., SCHERER, G., FISCHER, G. & DRAKENBERG, T. (1995). Kinetic analysis of cyclophilin-catalyzed prolyl *cis/trans* isomerization by dynamic NMR spectroscopy. *Biochemistry* **34**, 13594–13602.
45 VIDEEN, J. S., STAMNES, M. A., HSU, V. L. & GOODMAN, M. (1994). Thermodynamics of cyclophilin catalyzed peptidyl-prolyl isomerization by NMR spectroscopy. *Biopolymers* **34**, 171–175.
46 BAINE, P. (1986). Comparison of rate constants determined by two-dimensional NMR spectroscopy with rate constants determined by other NMR techniques. *Magn. Res. Chem.* **24**, 304–307.
47 LANDRIEU, I., DE VEYLDER, L., FRUCHART, J. S., ODAERT, B., CASTEELS, P., PORTETELLE, D., VAN MONTAGU, M., INZE, D. & LIPPENS, G. (2000). The Arabidopsis thaliana PIN1At gene encodes a single-domain phosphorylation-dependent peptidyl-prolyl *cis/trans* isomerase. *J. Biol. Chem.* **275**, 10577–10581.
48 KÜLLERTZ, G., LUTHE, S. & FISCHER, G. (1998). Semiautomated microtiter plate assay for monitoring peptidyl-prolyl *cis/trans* isomerase activity in normal and pathological human sera. *Clin. Chem.* **44**, 502–508.
49 GAREL, J. R. (1980). Evidence for involvement of proline *cis/trans* isomerization in the slow unfolding reaction of RNase A. *Proc. Natl Acad. Sci. USA* **77**, 795–798.
50 GRACIA-ECHEVERRIA, C., KOFRON, J. L., KUZMIC, P., KISHORE, V. & RICH, D. H. (1992). Continuous fluorimetric direct (uncoupled) assay for peptidyl-prolyl *cis/trans*-isomerases. *J. Am. Chem. Soc.* **114**, 2758–2759.
51 JANOWSKI, B., WOLLNER, S., SCHUTKOWSKI, M. & FISCHER, G. (1997). A protease-free assay for peptidyl-prolyl *cis/trans* isomerases using standard peptide substrates. *Anal. Biochem.* **252**, 299–307.
52 GRACIA-ECHEVERRIA, C., KOFRON, J. L., KUZMIC, P. & RICH, D. H. (1993) A continuous spectrophotometric direct assay for peptidyl-prolyl *cis/trans*-isomerases. *Biochem. Biophys. Res. Commun.* **191**, 70–75.
53 ZHANG, Y. X., FÜSSEL, S., REIMER, U., SCHUTKOWSKI, M. & FISCHER, G. (2002). Substrate-based design of reversible Pin1 inhibitors. *Biochemistry* **41**, 11868–11877.
54 WEISSHOFF, H., FROST, K., BRANDT, W., HENKLEIN, P., MUGGE, C. & FROMMEL, C. (1995). Novel disulfide-constrained pentapeptides as models for beta-VIa turns in proteins. *FEBS Lett.* **372**, 203–209.
55 FISCHER, G., SCHUTKOWSKI, M., & KÜLLERTZ, G. (2001) Peptidic substrates for directly assaying enzyme activities. Patent DE 100 23743 A1.
56 LIN, L. N., HASUMI, H. & BRANDTS, J. F. (1988). Catalysis of proline isomerization during protein-folding reactions. *Biochim. Biophys. Acta* **956**, 256–266.
57 LANG, K., SCHMID, F. X. & FISCHER, G. (1987). Catalysis of protein folding by prolyl isomerase. *Nature* **329**, 268–270.
58 KIEFHABER, T., QUAAS, R., HAHN, U. & SCHMID, F. X. (1990). Folding of ribonuclease T1. 2. Kinetic models for the folding and unfolding reactions. *Biochemistry* **29**, 3061–70.
59 KIEFHABER, T., QUAAS, R., HAHN, U. & SCHMID, F. X. (1990). Folding of ribonuclease T1. 1. Existence of multiple unfolded states created by proline isomerization. *Biochemistry* **29**, 3053–61.
60 BALBACH, J., STEEGBORN, C., SCHINDLER, T. & SCHMID, F. X. (1999). A protein folding intermediate of ribonuclease T-1 characterized at high resolution by 1D and 2D real-time NMR spectroscopy. *J. Mol. Biol.* **285**, 829–842.
61 MORITZ, R., REINSTADLER, D., FABIAN,

H. & NAUMANN, D. (2002). Time-resolved FTIR difference spectroscopy as tool for investigating refolding reactions of ribonuclease T1 synchronized with *trans* to *cis* prolyl isomerization. *Biopolymers* **67**, 145–155.

62 KIEFHABER, T. & SCHMID, F. X. (1992). Kinetic coupling between protein folding and prolyl isomerization. 2. folding of ribonuclease-A and ribonuclease-T1. *J. Mol. Biol.* **224**, 231–240.

63 MAYR, L. M., ODEFEY, C., SCHUTKOWSKI, M. & SCHMID, F. X. (1996). Kinetic analysis of the unfolding and refolding of ribonuclease T1 by a stopped-flow double-mixing technique. *Biochemistry* **35**, 5550–5561.

64 MÜCKE, M. & SCHMID, F. X. (1992). Enzymatic catalysis of prolyl isomerization in an unfolding protein. *Biochemistry* **31**, 7848–7854.

65 PAPPENBERGER, G., BACHMANN, A., MULLER, R., AYGUN, H., ENGELS, J. W. & KIEFHABER, T. (2003). Kinetic mechanism and catalysis of a native-state prolyl isomerization reaction. *J. Mol. Biol.* **326**, 235–246.

66 BOSE, S., WEIKL, T., BUGL, H. & BUCHNER, J. (1996). Chaperone function of hsp90-associated proteins. *Science* **274**, 1715–1717.

67 FREEMAN, B. C., TOFT, D. O. & MORIMOTO, R. I. (1996). Molecular chaperone machines – chaperone activities of the cyclophilin cyp-40 and the steroid aporeceptor-associated protein p23. *Science* **274**, 1718–1720.

68 SCHOLZ, C., STOLLER, G., ZARNT, T., FISCHER, G. & SCHMID, F. X. (1997). Cooperation of enzymatic and chaperone functions of trigger factor in the catalysis of protein folding. *EMBO J.* **16**, 54–58.

69 SCHAFFITZEL, E., RÜDIGER, S., BUKAU, B. & DEUERLING, E. (2001). Functional dissection of Trigger factor and DnaK: Interactions with nascent polypeptides and thermally denatured proteins. *Biol. Chem.* **382**, 1235–1243.

70 SCHREIER, M. H., BAUMANN, G. & ZENKE, G. (1993). Inhibition of T-cell signaling pathways by immunophilin drug complexes – are side effects inherent to immunosuppressive properties. *Transplant. Proc.* **25**, 502–507.

71 SCHREIBER, S. L., ALBERS, M. W. & BROWN, E. J. (1993). The cell cycle, signal transduction, and immunophilin ligand complexes. *Acc. Chem. Res.* **26**, 412–420.

72 HAMILTON, G. S. & STEINER, J. P. (1998). Immunophilins: Beyond immunosuppression. *J. Med. Chem.* **41**, 5119–5143.

73 VOGEL, K. W., BRIESEWITZ, R., WANDLESS, T. J. & CRABTREE, G. R. (2001). Calcineurin inhibitors and the generalization of the presenting protein strategy. *Adv. Protein Chem.* **56**, 253–291.

74 BAUMGRASS, R., ZHANG, Y., ERDMANN, F., THIEL, A., RADBRUCH, A., WEIWAD, M. & FISCHER, G. (2004). Substitution in position 3 of cyclosporin A abolishes the cyclophilin-mediated gain-of-function mechanism but not immunosuppression. *J. Biol. Chem.*, in press.

75 STEINER, J. P., HAMILTON, G. S., ROSS, D. T., VALENTINE, H. L., GUO, H. Z., CONNOLLY, M. A., LIANG, S., RAMSEY, C., LI, J. H. J., HUANG, W., HOWORTH, P., SONI, R., FULLER, M., SAUER, H., NOWOTNIK, A. C. & SUZDAK, P. D. (1997). Neurotrophic immunophilin ligands stimulate structural and functional recovery in neurodegenerative animal models. *Proc. Natl. Acad. Sci. USA* **94**, 2019–2024.

76 BARTZ, S. R., HOHENWALTER, E., HU, M. K., RICH, D. H. & MALKOVSKY, M. (1995). Inhibition of human immunodeficiency virus replication by non-immunosuppressive analogs of cyclosporin A. *Proc. Natl. Acad. Sci. USA* **92**, 5381–5385.

77 CHRISTNER, C., HERDEGEN, T. & FISCHER, G. (2001). FKBP ligands as novel therapeutics for neurological disorders. *Mini Rev. Med. Chem.* **1**, 377–397.

78 WALDMEIER, P. C., FELDTRAUER, J. J., QIAN, T. & LEMASTERS, J. J. (2002). Inhibition of the mitochondrial permeability transition by the

nonimmunosuppressive cyclosporin derivative NIM811. *Mol. Pharmacol.* **62**, 22–29.

79 UCHIDA, T., TAKAMIYA, M., TAKAHASHI, M., MIYASHITA, H., IKEDA, H., TERADA, T., MATSUO, Y., SHIROUZU, M., YOKOYAMA, S., FUJIMORI, F. & HUNTER, T. (2003). Pin1 and Par14 peptidyl-prolyl isomerase inhibitors block cell proliferation. *Chem. Biol.* **10**, 15–24.

80 HENNIG, L., CHRISTNER, C., KIPPING, M., SCHELBERT, B., RÜCKNAGEL, K. P., GRABLEY, S., KÜLLERTZ, G. & FISCHER, G. (1998). Selective inactivation of parvulin-like peptidyl-prolyl cis/trans isomerases by juglone. *Biochemistry* **37**, 5953–5960.

81 YLI-KAUHALUOMA, J. T., ASHLEY, J. A., LO, C. H. L., COAKLEY, J., WIRSCHING, P. & JANDA, K. D. (1996). Catalytic antibodies with peptidyl-prolyl cis/trans isomerase activity. *J. Am. Chem. Soc.* **118**, 5496–5497.

82 MA, L. F., HSIEH-WILSON, L. C. & SCHULTZ, P. G. (1998). Antibody catalysis of peptidyl-prolyl cis/trans isomerization in the folding of RNase T1. *Proc. Natl. Acad. Sci. USA* **95**, 7251–7256.

83 STEIN, R. L. (1993). Mechanism of enzymic and non-enzymic prolyl cis/trans isomerization. *Adv. Protein Chem.* **44**, 1–24.

84 SCHUTKOWSKI, M., BERNHARDT, A., ZHOU, X. Z., SHEN, M. H., REIMER, U., RAHFELD, J. U., LU, K. P. & FISCHER, G. (1998). Role of phosphorylation in determining the backbone dynamics of the serine/threonine-proline motif and Pin1 substrate. *Biochemistry* **37**, 5566–5575.

85 ZYDOWSKY, L. D., ETZKORN, F. A., CHANG, H. Y., FERGUSON, S. B., STOLZ, L. A., HO, S. I. & WALSH, C. T. (1992). Active site mutants of human cyclophilin-a separate peptidyl-prolyl isomerase activity from cyclosporin-A binding and calcineurin inhibition. *Protein Sci.* **1**, 1092–1099.

86 DECENZO, M. T., PARK, S. T., JARRETT, B. P., ALDAPE, R. A., FUTER, O., MURCKO, M. A. & LIVINGSTON, D. J. (1996). FK506-binding protein mutational analysis – defining the active-site residue contributions to catalysis and the stability of ligand complexes. *Protein Eng.* **9**, 173–180.

87 TRADLER, T., STOLLER, G., RUCKNAGEL, K. P., SCHIERHORN, A., RAHFELD, J. U. & FISCHER, G. (1997). Comparative mutational analysis of peptidyl-prolyl cis/trans isomerases – active sites of *Escherichia coli* trigger factor and human FKBP12. *FEBS Lett.* **407**, 184–190.

88 KRAMER, M. L. & FISCHER, G. (1997). FKBP-like catalysis of peptidyl-prolyl bond isomerization by micelles and membranes. *Biopolymers* **42**, 49–60.

89 KERN, D., SCHUTKOWSKI, M. & DRAKENBERG, T. (1997). Rotational barriers of cis/trans isomerization of proline analogues and their catalysis by cyclophilin. *J. Am. Chem. Soc.* **119**, 8403–8408.

90 YAFFE, M. B., SCHUTKOWSKI, M., SHEN, M. H., ZHOU, X. Z., STUKENBERG, P. T., RAHFELD, J. U., XU, J., KUANG, J., KIRSCHNER, M. W., FISCHER, G., CANTLEY, L. C. & LU, K. P. (1997). Sequence-specific and phosphorylation-dependent proline isomerization – a potential mitotic regulatory mechanism. *Science* **278**, 1957–1960.

91 METZNER, M., STOLLER, G., RUCKNAGEL, K. P., LU, K. P., FISCHER, G., LUCKNER, M. & KÜLLERTZ, G. (2001). Functional replacement of the essential ESS1 in yeast by the plant parvulin DlPar13. *J. Biol. Chem.* **276**, 13524–13529.

92 LU, K. P., LIOU, Y. C. & ZHOU, X. Z. (2002). Pinning down proline-directed phosphorylation signaling. *Trends Cell Biol.* **12**, 164–172.

93 SCHIENE, C., REIMER, U., SCHUTKOWSKI, M. & FISCHER, G. (1998). Mapping the stereospecificity of peptidyl-prolyl cis/trans isomerases. *FEBS Lett.* **432**, 202–206.

94 GALAT, A. (2003). Peptidylprolyl cis/trans isomerases (immunophilins): biological diversity-targets-functions. *Curr. Top. Med. Chem.* **3**, 1315–1347.

95 BUKRINSKY, M. I. (2002). Cyclophilins: unexpected messengers in inter-

cellular communications. *Trends Immunol.* **23**, 323–325.

96 YURCHENKO, V., ZYBARTH, G., O'CONNOR, M., DAI, W. W., FRANCHIN, G., HAO, T., GUO, H. M., HUNG, H. C., TOOLE, B., GALLAY, P., SHERRY, B. & BUKRINSKY, M. (2002). Active site residues of cyclophilin A are crucial for its signaling activity via CD147. *J. Biol. Chem.* **277**, 22959–22965.

97 HOROWITZ, D. S., KOBAYASHI, R. & KRAINER, A. R. (1997). A new cyclophilin and the human homologues of yeast prp3 and prp4 form a complex associated with U4/U6 snRNPs. *RNA* **3**, 1374–1387.

98 MEZA-ZEPEDA, L. A., FORUS, A., LYGREN, B., DAHLBERG, A. B., GODAGER, L. H., SOUTH, A. P., MARENHOLZ, I., LIOUMI, M., FLORENES, V. A., MAELANDSMO, G. M., SERRA, M., MISCHKE, D., NIZETIC, D., RAGOUSSIS, J., TARKKANEN, M., NESLAND, J. M., KNUUTILA, S. & MYKLEBOST, O. (2002). Positional cloning identifies a novel cyclophilin as a candidate amplified oncogene in 1q21. *Oncogene* **21**, 2261–2269.

99 ZHOU, Z., YING, K., DAI, J., TANG, R., WANG, W., HUANG, Y., ZHAO, W., XIE, Y. & MAO, Y. (2001). Molecular cloning and characterization of a novel peptidylprolyl isomerase (cyclophilin)-like gene (PPIL3) from human fetal brain. *Cytogenet. Cell Genet.* **92**, 231–236.

100 OZAKI, K., FUJIWARA, T., KAWAI, A., SHIMIZU, F., TAKAMI, S., OKUNO, S., TAKEDA, S., SHIMADA, Y., NAGATA, W., WATANABE, T., TAKAICHI, A., TAKAHASHI, E., NAKAMURA, Y. & SHIN, S. (1996). Cloning, expression and chromosomal mapping of a novel cyclophilin-related gene (PPIL1) from human fetal brain. *Cytogenet. Cell Genet.* **72**, 242–245.

101 HOROWITZ, D. S., LEE, E. J., MABON, S. A. & MISTELI, T. (2002). A cyclophilin functions in pre-mRNA splicing. *EMBO J.* **21**, 470–480.

102 REIDT, U., WAHL, M. C., FASSHAUER, D., HOROWITZ, D. S., LUHRMANN, R. & FICNER, R. (2003). Crystal structure of a complex between human spliceosomal cyclophilin H and a U4/U6 snRNP-60K peptide. *J. Mol. Biol.* **331**, 45–56.

103 REIDT, U., REUTER, K., ACHSEL, T., INGELFINGER, D., LUHRMANN, R. & FICNER, R. (2000). Crystal structure of the human U4/U6 small nuclear ribonucleoprotein particle-specific SnuCyp-20, a nuclear cyclophilin. *J. Biol. Chem.* **275**, 7439–7442.

104 WALLACE, A. C., LASKOWSKI, R. A. & THORNTON, J. M. (1996). Derivation of 3D coordinate templates for searching structural databases: application to Ser-His-Asp catalytic triads in the serine proteinases and lipases. *Protein Sci.* **5**, 1001–1013.

105 DEMANGE, L., MOUTIEZ, M., VAUDRY, K. & DUGAVE, C. (2001). Interaction of human cyclophilin hCyp-18 with short peptides suggests the existence of two functionally independent subsites. *FEBS Lett.* **505**, 191–195.

106 MONTAGUE, J. W., HUGHES, F. M. & CIDLOWSKI, J. A. (1997). Native recombinant cyclophilins A, B, and C degrade DNA independently of peptidylprolyl *cis/trans*-isomerase activity – potential roles of cyclophilins in apoptosis. *J. Biol. Chem.* **272**, 6677–6684.

107 NICIEZA, R. G., HUERGO, J., CONNOLLY, B. A. & SANCHEZ, J. (1999). Purification, characterization, and role of nucleases and serine proteases in *Streptomyces* differentiation – Analogies with the biochemical processes described in late steps of eukaryotic apoptosis. *J. Biol. Chem.* **274**, 20366–20375.

108 SCHMIDT, B., TRADLER, T., RAHFELD, J. U., LUDWIG, B., JAIN, B., MANN, K., RÜCKNAGEL, K. P., JANOWSKI, B., SCHIERHORN, A., KÜLLERTZ, G., HACKER, J. & FISCHER, G. (1996). A cyclophilin-like peptidyl-prolyl *cis/trans* isomerase from *Legionella pneumophila* – characterization, molecular cloning and overexpression. *Mol. Microbiol.* **21**, 1147–1160.

109 RYFFEL, B., FOXWELL, B. M., GEE, A., GREINER, B., WOERLY, G. & MIHAISCH, M. J. (1988). Cyclosporine

– relationship of side effects to mode of action. *Transplantation* **46**, 90S–96S.
110 GOLDNER, F. M. & PATRICK, J. W. (1996). Neuronal localization of the cyclophilin a protein in the adult rat brain. *J. Comp. Neurol.* **372**, 283–293.
111 ARCKENS, L., VAN DER GUCHT, E., VAN DEN BERGH, G., MASSIE, A., LEYSEN, I., VANDENBUSSCHE, E., EYSEL, U. T., HUYBRECHTS, R. & VANDESANDE, F. (2003). Differential display implicates cyclophilin A in adult cortical plasticity. *Eur. J. Neurosci.* **18**, 61–75.
112 BERGSMA, D. J., EDER, C., GROSS, M., KERSTEN, H., SYLVESTER, D., APPELBAUM, E., CUSIMANO, D., LIVI, G. P., McLAUGHLIN, M. M., KASYAN, K., PORTER, T. G., SILVERMAN, C., DUNNINGTON, D., HAND, A. PRITCHETT, W. P., BOSSARD, M. J. BRANDT, M. & LEVY, M. A. (1991). The cyclophilin multigene family of peptidyl-prolyl isomerases. Characterization of three separate human isoforms. *J. Biol. Chem.* **266**, 23204–23214.
113 HARRISON, R. K. & STEIN, R. L. (1992). Mechanistic studies of enzymic and nonenzymic prolyl *cis/trans* isomerization. *J. Am. Chem. Soc.* **114**, 3464–3471.
114 HOWARD, B. R., VAJDOS, F. F., LI, S., SUNDQUIST, W. I. & HILL, C. P. (2003). Structural insights into the catalytic mechanism of cyclophilin A. *Nat. Struct. Biol.* **10**, 475–481.
115 DOLINSKI, K., MUIR, S., CARDENAS, M. & HEITMAN, J. (1997). All cyclophilins and FK506 binding proteins are, individually and collectively, dispensable for viability in *Saccharomyces cerevisiae*. *Proc. Natl. Acad. Sci. USA* **94**, 13093–13098.
116 COLGAN, J., ASMAL, M. & LUBAN, J. (2000). Isolation, characterization and targeted disruption of mouse Ppia: Cyclophilin A is not essential for mammalian cell viability. *Genomics* **68**, 167–178.
117 WANG, P., CARDENAS, M. E., COX, C. M., PERFECT, J. R. & HEITMAN, J. (2001). Two cyclophilin A homologs with shared and distinct functions important for growth and virulence of *Cryptococcus neoformans*. *EMBO Rep.* **2**, 511–518.
118 JIN, Z. G., MELARAGNO, M. G., LIAO, D. F., YAN, C., HAENDELER, J., SUH, Y. A., LAMBETH, J. D. & BERK, B. C. (2000). Cyclophilin A is a secreted growth factor induced by oxidative stress. *Circ. Res.* **87**, 789–796.
119 HELEKAR, S. A. & PATRICK, J. (1997). Peptidyl-prolyl *cis/trans* isomerase activity of cyclophilin a in functional homo-oligomeric receptor expression. *Proc. Natl. Acad. Sci. USA* **94**, 5432–5437.
120 CAPANO, M., VIRJI, S. & CROMPTON, M. (2002). Cyclophilin-A is involved in excitotoxin-induced caspase activation in rat neuronal B50 cells. *Biochem. J.* **363**, 29–36.
121 SYED, F., RYCYZYN, M. A., WESTGATE, L. & CLEVENGER, C. V. (2003). A novel and functional interaction between cyclophilin A and prolactin receptor. *Endocrine* **20**, 83–89.
122 CUI, Y., MIRKIA, K., FU, Y. H. F., ZHU, L., YOKOYAMA, K. K. & CHIU, R. (2002). Interaction of the retinoblastoma gene product, RB, with cyclophilin A negatively affects cyclosporin-inhibited NFAT signaling. *J. Cell. Biochem.* **86**, 630–641.
123 JASCHKE, A., MI, H. F. & TROPSCHUG, M. (1998). Human T cell cyclophilin18 binds to thiol-specific antioxidant protein AOP1 and stimulates its activity. *J. Mol. Biol.* **277**, 763–769.
124 LEE, J. P., PALFREY, H. C., BINDOKAS, V. P., GHADGE, G. D., MA, L., MILLER, R. J. & ROOS, R. P. (1999). The role of immunophilins in mutant superoxide dismutase-1-linked familial amyotrophic lateral sclerosis. *Proc. Natl. Acad. Sci. USA* **96**, 3251–3256.
125 BRAZIN, K. N., MALLIS, R. J., FULTON, D. B. & ANDREOTTI, A. H. (2002). Regulation of the tyrosine kinase Itk by the peptidyl-prolyl isomerase cyclophilin A. *Proc. Natl. Acad. Sci. USA* **99**, 1899–1904.
126 PORTER, R. H., BURNET, P. W., EASTWOOD, S. L. & HARRISON, P. J. (1996). Contrasting effects of electroconvulsive shock on mRNAs encoding the high affinity kainate receptor sub-

units (KA1 and KA2) and cyclophilin in the rat. *Brain Res.* **710**, 97–102.
127 SYKES, K., GETHING, M. J. & SAMBROOK, J. (1993). Proline isomerases function during heat shock. *Proc. Natl. Acad. Sci. USA* **90**, 5853–5857.
128 KÜLLERTZ, G., LIEBAU, A., RÜCKNAGEL, P., SCHIERHORN, W., DIETTRICH, B., FISCHER, G. & LUCKNER, M. (1999). Stress-induced expression of cyclophilins in proembryonic masses of *Digitalis lanata* does not protect against freezing thawing stress. *Planta* **208**, 599–605.
129 ANDREEVA, L., HEADS, R. & GREEN, C. J. (1999). Cyclophilins and their possible role in the stress response. *Int. J. Exp. Pathol.* **80**, 305–315.
130 VUADENS, F., CRETTAZ, D., SCELATTA, C., SERVIS, C., QUADRONI, M., BIENVENUT, W. V., SCHNEIDER, P., HOHLFELD, P., APPLEGATE, L. A. & TISSOT, J. D. (2003). Plasticity of protein expression during culture of fetal skin cells. *Electrophoresis* **24**, 1281–1291.
131 BORALDI, F., BINI, L., LIBERATORI, S., ARMINI, A., PALLINI, V., TIOZZO, R., PASQUALI-RONCHETTI, I. & QUAGLINO, D. (2003). Proteome analysis of dermal fibroblasts cultured in vitro from human healthy subjects of different ages. *Proteomics* **3**, 917–929.
132 ANDERSEN, H., JENSEN, O. N. & ERIKSEN, E. F. (2003). A proteome study of secreted prostatic factors affecting osteoblastic activity: identification and characterisation of cyclophilin A. *Eur. J. Cancer* **39**, 989–995.
133 CAMPA, M. J., WANG, M. Z., HOWARD, B., FITZGERALD, M. C. & PATZ, E. F. (2003). Protein expression profiling identifies macrophage migration inhibitory factor and cyclophilin A as potential molecular targets in non-small cell lung cancer. *Cancer Res.* **63**, 1652–1656.
134 LIM, S. O., PARK, S. J., KIM, W., PARK, S. G., KIM, H. J., KIM, Y. I., SOHN, T. S., NOH, J. H. & JUNG, G. (2002). Proteome analysis of hepatocellular carcinoma. *Biochem. Biophys. Res. Commun.* **291**, 1031–1037.

135 ALBERTI, J., VALENZUELA, J. G., CARRUTHERS, V. B., HIENY, S., ANDERSEN, J., CHAREST, H., SOUSA, C. R. E., FAIRLAMB, A., RIBEIRO, J. M. & SHER, A. (2003). Molecular mimicry of a CCR5 binding-domain in the microbial activation of dendritic cells. *Nat. Immunol.* **4**, 485–490.
136 DUTTA, M., DELHI, P., SINHA, K. M., BANERJEE, R. & DATTA, A. K. (2001). Lack of abundance of cytoplasmic cyclosporin A-binding protein renders free-living *Leishmania donovani* resistant to cyclosporin A. *J. Biol. Chem.* **276**, 19294–19300.
137 BROWN, C. R., CUI, D. Y., HUNG, G. G. C. & CHIANG, H. L. (2001). Cyclophilin a mediates Vid22p function in the import of fructose-1,6-bisphosphatase into vid vesicles. *J. Biol. Chem.* **276**, 48017–48026.
138 TAYLOR, P., HUSI, H., KONTOPIDIS, G. & WALKINSHAW, M. D. (1997). Structures of cyclophilin-ligand complexe. *Prog. Biophys. Mol. Biol.* **67**, 155–181.
139 HUAI, Q., KIM, H. Y., LIU, Y. D., ZHAO, Y. D., MONDRAGON, A., LIU, J. O. & KE, H. M. (2002). Crystal structure of calcineurin-cyclophilin-cyclosporin shows common but distinct recognition of immunophilin-drug complexes. *Proc. Natl. Acad. Sci. USA* **99**, 12037–12042.
140 JIN, L. & HARRISON, S. C. (2002). Crystal structure of human calcineurin complexed with cyclosporin A and human cyclophilin. *Proc. Natl. Acad. Sci. USA* **99**, 13522–13526.
141 GAMBLE, T. R., VAJDOS, F. F., YOO, S. H., WORTHYLAKE, D. K., HOUSEWEART, M., SUNDQUIST, W. I. & HILL, C. P. (1996). Crystal structure of human cyclophilin A bound to the amino-terminal domain of HIV-1 capsid. *Cell* **87**, 1285–1294.
142 ZHAO, Y. D., CHEN, Y. Q., SCHUTKOWSKI, M., FISCHER, G. & KE, H. M. (1997). Cyclophilin A complexed with a fragment of HIV-1 gag protein – insights into HIV-1 infectious activity. *Structure* **5**, 139–146.
143 FRANKE, E. K., YUAN, H. E. H. & LUBAN, J. (1994). Specific incorpora-

tion of cyclophilin A into HIV-1 virions. *Nature* **371**, 359–362.
144 OU, W. B., LUO, W., PARK, Y. D. & ZHOU, H. M. (2001). Chaperone-like activity of peptidyl-prolyl *cis/trans* isomerase during creatine kinase refolding. *Protein Sci.* **10**, 2346–2353.
145 KERN, G., KERN, D., SCHMID, F. X. & FISCHER, G. (1995). A kinetic analysis of the folding of human carbonic anhydrase II and its catalysis by cyclophilin. *J. Biol. Chem.* **270**, 740–745.
146 BOSCO, D. A., EISENMESSER, E. Z., POCHAPSKY, S., SUNDQUIST, W. I. & KERN, D. (2002). Catalysis of *cis/trans* isomerization in native HIV-1 capsid by human cyclophilin A. *Proc. Natl. Acad. Sci. USA* **99**, 5247–5252.
147 REIMER, U., DREWELLO, M., JAKOB, M., FISCHER, G. & SCHUTKOWSKI, M. (1997). Conformational state of a 25-mer peptide from the cyclophilin-binding loop of the HIV type 1 capsid protein. *Biochem. J.* **326**, 181–185.
148 TOWERS, G. J., HATZIIOANNOU, T., COWAN, S., GOFF, S. P., LUBAN, J. & BIENIASZ, P. D. (2003). Cyclophilin A modulates the sensitivity of HIV-1 to host restriction factors. *Nat. Med.* **9**, 1138–1143.
149 SCHMID, F. X. (2002). Prolyl isomerases. *Adv. Prot. Chem.* **59**, 243–282.
150 CHITI, F., MANGIONE, P., ANDREOLA, A., GIORGETTI, S., STEFANI, M., DOBSON, C. M., BELLOTTI, V. & TADDEI, N. (2001). Detection of two partially structured species in the folding process of the amyloidogenic protein beta 2-microglobulin. *J. Mol. Biol.* **307**, 379–391.
151 GOLBIK, R., FISCHER, G. & FERSHT, A. R. (1999). Folding of barstar C40A/C82A/P27A and catalysis of the peptidyl-prolyl *cis/trans* isomerization by human cytosolic cyclophilin (Cyp18). *Protein Sci.* **8**, 1505–1514.
152 VEERARAGHAVAN, S., NALL, B. T. & FINK, A. L. (1997). Effect of prolyl isomerase on the folding reactions of staphylococcal nuclease. *Biochemistry* **36**, 15134–15139.
153 READER, J. S., VAN NULAND, N. A. J., THOMPSON, G. S., FERGUSON, S. J., DOBSON, C. M. & RADFORD, S. E. (2001). A partially folded intermediate species of the beta-sheet protein apo-pseudoazurin is trapped during proline-limited folding. *Protein Sci.* **10**, 1216–1224.
154 MARTIN, A. & SCHMID, F. X. (2003). A proline switch controls folding and domain interactions in the gene-3-protein of the filamentous phage fd. *J. Mol. Biol.* **331**, 1131–1140.
155 ALTAMIRANO, M. M., GARCIA, C., POSSANI, L. D. & FERSHT, A. R. (1999). Oxidative refolding chromatography: folding of the scorpion toxin Cn5. *Nature Biotechnol.* **17**, 187–191.
156 ALTAMIRANO, M. M., WOOLSON, A., DONDA, A., SHAMSHIEV, A., BRISENO-ROA, L., FOSTER, N. W., VEPRINTSEV, D. B., DE LIBERO, G., FERSHT, A. R. & MILSTEIN, C. (2001). Ligand-independent assembly of recombinant human CD1 by using oxidative refolding chromatography. *Proc. Natl. Acad. Sci. USA* **98**, 3288–3293.
157 KARADIMITRIS, A., GADOLA, S., ALTAMIRANO, M., BROWN, D., WOOLFSON, A., KLENERMAN, P., CHEN, J. L., KOEZUKA, Y., ROBERTS, I. A. G., PRICE, D. A., DUSHEIKO, G., MILSTEIN, C., FERSHT, A., LUZZATTO, L. & CERUNDOLO, V. (2001). Human CD1d-glycolipid tetramers generated by in vitro oxidative refolding chromatography. *Proc. Natl. Acad. Sci. USA* **98**, 3294–3298.
158 SCHUTKOWSKI, M., WÖLLNER, S. & FISCHER, G. (1995). Inhibition of peptidyl-prolyl *cis/trans* isomerase activity by substrate analog structures: Thioxo tetrapeptide-4-nitroanilides. *Biochemistry* **34**, 13016–13026.
159 SANGLIER, J. J., QUESNIAUX, V., FEHR, T., HOFMANN, H., MAHNKE, M., MEMMERT, K., SCHULER, W., ZENKE, G., GSCHWIND, L., MAURER, C. & SCHILLING, W. (1999). Sanglifehrins A, B, C and D, novel cyclophilin-binding compounds isolated from *Streptomyces* sp A92-308110-I. Taxonomy, fermentation, isolation and biological activity. *J. Antibiot.* **52**, 466–473.

160 SEDRANI, R., KALLEN, J., CABREJAS, L. M. M., PAPAGEORGIOU, C. D., SENIA, F., ROHRBACH, S., WAGNER, D., THAI, B., EME, A. M. J., FRANCE, J., OBERER, L., RIHS, G., ZENKE, G. & WAGNER, J. (2003). Sanglifehrin-cyclophilin interaction: Degradation work, synthetic macrocyclic analogues, X-ray crystal structure, and binding data. *J. Am. Chem. Soc.* **125**, 3849–3859.

161 ZHANG, L. H. & LIU, J. O. (2001). Sanglifehrin A, a novel cyclophilin-binding immunosuppressant, inhibits IL-2-dependent T cell proliferation at the G(1) phase of the cell cycle. *J. Immunol.* **166**, 5611–5618.

162 WU, Y. Q., BELYAKOV, S., CHOI, C., LIMBURG, D., THOMAS, B. E., VAAL, M., WEI, L., WILKINSON, D. E., HOLMES, A., FULLER, M., MCCORMICK, J., CONNOLLY, M., MOELLER, T., STEINER, J. & HAMILTON, G. S. (2003). Synthesis and biological evaluation of non-peptidic cyclophilin ligands. *J. Med. Chem.* **46**, 1112–1115.

163 DEMANGE, L., MOUTIEZ, M. & DUGAVE, C. (2002). Synthesis and evaluation of Gly psi(PO2R-N)pro-containing pseudopeptides as novel inhibitors of the human cyclophilin hCyp-18. *J. Med. Chem.* **45**, 3928–3933.

164 WANG, H. C., KIM, K., BAKHTIAR, R. & GERMANAS, J. P. (2001). Structure-activity studies of ground- and transition-state analogue inhibitors of cyclophilin. *J. Med. Chem.* **44**, 2593–2600.

165 MARUYAMA, T. & FURUTANI, M. (2000). Archaeal peptidyl-prolyl cis/trans isomerases (PPIases). *Front. Biosci.* **5**, D821–D836.

166 AVRAMUT, M. & ACHIM, C. L. (2002). Immunophilins and their ligands: insights into survival and growth of human neurons. *Physiol. Behav.* **77**, 463–468.

167 WALENSKY, L. D., DAWSON, T. M., STEINER, J. P., SABATINI, D. M., SUAREZ, J. D., KLINEFELTER, G. R. & SNYDER, S. H. (1998). The 12 kD FK506 binding protein FKBP12 is released in the male reproductive tract and stimulates sperm motility. *Mol. Med.* **4**, 502–514.

168 LEIVA, M. C. & LYTTLE, C. R. (1992). Leukocyte chemotactic activity of FKBP and inhibition by FK506. *Biochem. Biophys. Res. Commun.* **186**, 1178–1183.

169 VEERARAGHAVAN, S., HOLZMAN, T. F. & NALL, B. T. (1996). Autocatalyzed protein folding. *Biochemistry* **35**, 10601–10607.

170 SCHOLZ, C., ZARNT, T., KERN, G., LANG, K., BURTSCHER, H., FISCHER, G. & SCHMID, F. X. (1996). Autocatalytic folding of the folding catalyst FKBP12. *J. Biol. Chem.* **271**, 12703–12707.

171 DEIVANAYAGAM, C. C. S., CARSON, M., THOTAKURA, A., NARAYANA, S. V. L. & CHODAVARAPU, R. S. (2000). Structure of FKBP12.6 in complex with rapamycin. *Acta Crystallogr. D Biol. Crystallogr.* **56**, 266–271.

172 SOMAN, K. V., HANKS, B. A., TIEN, H., CHARI, M. V., ONEAL, K. D. & MORRISETT, J. D. (1997). Template-based docking of a prolactin receptor proline-rich motif octapeptide to FKBP12 – Implications for cytokine receptor signaling. *Protein Sci.* **6**, 999–1008.

173 LAM, E., MARTIN, M. M., TIMERMAN, A. P., SABERS, C., FLEISCHER, S., LUKAS, T., ABRAHAM, R. T., OKEEFE, S. J., ONEILL, E. A. & WIEDERRECHT, G. J. (1995). A novel FK506 binding protein can mediate the immuno-suppressive effects of FK506 and is associated with the cardiac ryanodine receptor. *J. Biol. Chem.* **270**, 26511–26522.

174 PARK, S. T., ALDAPE, R. A., FUTER, O., DECENZO, M. T. & LIVINGSTON, D. J. (1992). PPIase catalysis by human FK506-binding protein proceeds through a conformational twist mechanism. *J. Biol. Chem.* **267**, 3316–3324.

175 KIRKEN, R. A. & WANG, Y. L. (2003). Molecular actions of sirolimus: Sirolimus and mTor. *Transplant. Proc.* **35**, 227S–230S.

176 YANG, W., ROZAMUS, L. W., NARULA, S., ROLLINS, C. T., YUAN, R., ANDRADE, L. J., RAM, M. K., PHILLIPS, T. B., VAN SCHRAVENDIJK, M. R., DALGARNO, D., CLACKSON, T. & HOLT,

D. A. (2000). Investigating protein-ligand interactions with a mutant FKBP possessing a designed specificity pocket. *J. Med. Chem.* **43**, 1135–1142.

177 CHRISTNER, C., WYRWA, R., MARSCH, S., KÜLLERTZ, G., THIERICKE, R., GRABLEY, S., SCHUMANN, D. & FISCHER, G. (1999). Synthesis and cytotoxic evaluation of cycloheximide derivatives as potential inhibitors of FKBP12 with neuroregenerative properties. *J. Med. Chem.* **42**, 3615–3622.

178 DRAGOVICH, P. S., BARKER, J. E., FRENCH, J., IMBACUAN, M., KALISH, V. J., KISSINGER, C. R., KNIGHTON, D. R., LEWIS, C. T., MOOMAW, E. W., PARGE, H. E. & PELLETIER, L. A. (1996). Structure-based design of novel, urea-containing FKBP12 inhibitors. *J. Med. Chem.* **39**, 1872–1884.

179 BURKHARD, P., TAYLOR, P. & WALKINSHAW, M. D. (2000). X-ray structures of small ligand-FKBP complexes provide an estimate for hydrophobic interaction energies. *J. Mol. Biol.* **295**, 953–962.

180 WEI, L., WU, Y. Q., WILKINSON, D. E., CHEN, Y., SONI, R., SCOTT, C., ROSS, D. T., GUO, H., HOWORTH, P., VALENTINE, H., LIANG, S., SPICER, D., FULLER, M., STEINER, J. & HAMILTON, G. S. (2002). Solid-phase synthesis of FKBP12 inhibitors: N-sulfonyl and N-carbamoylprolyl/pipecolyl amides. *Bioorg. Med. Chem. Lett.* **12**, 1429–1433.

181 HAMILTON, G. S., HUANG, W., CONNOLLY, M. A., ROSS, D. T., GUO, H., VALENTINE, H. L., SUZDAK, P. D. & STEINER, J. P. (1997). FKBP12-binding domain analogues of FK506 are potent, nonimmunosuppressive neurotrophic agents in vitro and promote recovery in a mouse model of Parkinsons-disease. *Bioorg. Med. Chem. Lett.* **7**, 1785–1790.

182 YAMASHITA, D. S., OH, H. J., YEN, H. K., BOSSARD, M. J., BRANDT, M., LEVY, M. A., NEWMANTARR, T., BADGER, A., LUENGO, J. I. & HOLT, D. A. (1994). Design, synthesis and evaluation of dual domain FKBP ligands. *Bioorg. Med. Chem. Lett.* **4**, 325–328.

183 HOLT, D. A., LUENGO, J. I., YAMASHITA, D. S., OH, H. J., KONIALIAN, A. L., YEN, H. K., ROZAMUS, L. W., BRANDT, M., BOSSARD, M. J., LEVY, M. A., EGGLESTON, D. S., LIANG, J., SCHULTZ, L. W., STOUT, T. J. & CLARDY, J. (1993). Design, synthesis, and kinetic evaluation of high-affinity FKBP ligands and the X-ray crystal structures of their complexes with FKBP12. *J. Am. Chem. Soc.* **115**, 9925–9938.

184 CONNELLY, P. R., ALDAPE, R. A., BRUZZESE, F. J., CHAMBERS, S. P., FITZGIBBON, M. J., FLEMING, M. A., ITOH, S., LIVINGSTON, D. J., NAVIA, M. A., THOMSON, J. A. & WILSON, K. P. (1994). Enthalpy of hydrogen bond formation in a protein-ligand binding reaction. *Proc. Natl. Acad. Sci. USA* **91**, 1964–1968.

185 AGHDASI, B., YE, K. Q., RESNICK, A., HUANG, A., HA, H. C., GUO, X., DAWSON, T. M., DAWSON, V. L. & SNYDER, S. H. (2001). FKBP12, the 12-kDa FK506-binding protein, is a physiologic regulator of the cell cycle. *Proc. Natl. Acad. Sci. USA* **98**, 2425–2430.

186 SHOU, W. N., AGHDASI, B., ARMSTRONG, D. L., GUO, Q. X., BAO, S. D., CHARNG, M. J., MATHEWS, L. M., SCHNEIDER, M. D., HAMILTON, S. L. & MATZUK, M. M. (1998). Cardiac defects and altered ryanodine receptor function in mice lacking FKBP12. *Nature* **391**, 489–492.

187 XIN, H. B., SENBONMATSU, T., CHENG, D. S., WANG, Y. X., COPELLO, J. A., JI, G. J., COLLIER, M. L., DENG, K. Y., JEYAKUMAR, L. H., MAGNUSON, M. A., INAGAMI, T., KOTLIKOFF, M. I. & FLEISCHER, S. (2002). Oestrogen protects FKBP12.6 null mice from cardiac hypertrophy. *Nature* **416**, 334–337.

188 DARGAN, S. L., LEA, E. J. A. & DAWSON, A. P. (2002). Modulation of type-1 Ins(1,4,5)P-3 receptor channels by the FK506-binding protein, FKBP12. *Biochem. J.* **361**, 401–407.

189 MARKS, A. R. (2002). Ryanodine receptors, FKBP12, and heart failure. *Front. Biosci.* **7**, D970–D977.

190 CHEN, Y. G., LIU, F. & MASSAGUE, J. (1997). Mechanism of TBF-beta receptor inhibition by FKBP12. *EMBO J.* **16**, 3866–3876.

191 LOPEZILASACA, M., SCHIENE, C., KÜLLERTZ, G., TRADLER, T., FISCHER, G. & WETZKER, R. (1998). Effects of FK506-binding protein 12 and FK506 on autophosphorylation of epidermal growth factor receptor. *J. Biol. Chem.* **273**, 9430–9434.

192 DOLINSKI, K. J. & HEITMAN, J. (1999). Hmo1p, a high mobility group 1/2 homolog, genetically and physically interacts with the yeast FKBP12 prolyl isomerase. *Genetics* **151**, 935–944.

193 ALARCON, C. M. & HEITMAN, J. (1997). FKBP12 physically and functionally interacts with aspartokinase in *Saccharomyces cerevisiae*. *Mol. Cell. Biol.* **17**, 5968–5975.

194 CHAMBRAUD, B., RADANYI, C., CAMONIS, J. H., SHAZAND, K., RAJKOWSKI, K. & BAULIEU, E. E. (1996). FAP48, a new protein that forms specific complexes both immunophilins FKBP59 and FKBP12 – prevention by the immunosuppressant drugs FK506 and rapamycin. *J. Biol. Chem.* **271**, 32923–32929.

195 BROUILLARD, P., BOON, L. M., MULLIKEN, J. B., ENJOLRAS, O., GHASSIBE, M., WARMAN, M. L., TAN, O. T., OLSEN, B. R. & VIKKULA, M. (2002). Mutations in a novel factor, glomulin, are responsible for glomuvenous malformations ("glomangiomas"). *Am. J. Hum. Genet.* **70**, 866–874.

196 YANG, W. M., INOUYE, C. J. & SETO, E. (1995). Cyclophilin A and FKBP12 interact with YY1 and alter its transcriptional activity. *J. Biol. Chem.* **270**, 15187–15193.

197 FAURE, J. D., GINGERICH, D. & HOWELL, S. H. (1998). An *Arabidopsis* immunophilin, atFKBP12, binds to atFIP37 (FKBP interacting protein) in an interaction that is disrupted by FK506. *Plant J.* **15**, 783–789.

198 JEYAKUMAR, L. H., BALLESTER, L., CHENG, D. S., MCINTYRE, J. O., CHANG, P., OLIVEY, H. E., ROLLINS-SMITH, L., BARNETT, J. V., MURRAY, K., XIN, H. B. & FLEISCHER, S. (2001). FKBP binding characteristics of cardiac microsomes from diverse vertebrates. *Biochem. Biophys. Res. Commun.* **281**, 979–986.

199 XIN, H. B., ROGERS, K., QI, Y., KANEMATSU, T. & FLEISCHER, S. (1999). Three amino acid residues determine selective binding of FK506-binding protein 12.6 to the cardiac ryanodine receptor. *J. Biol. Chem.* **274**, 15315–15319.

200 PRESTLE, J., JANSSEN, P. M. L., JANSSEN, A. P., ZEITZ, O., LEHNART, S. E., BRUCE, L., SMITH, G. L. & HASENFUSS, G. (2001). Overexpression of FK506-Binding protein FKBP12.6 in cardiomyocytes reduces ryanodine receptor-mediated Ca^{2+} leak from the sarcoplasmic reticulum and increases contractility. *Circ. Res.* **88**, 188–194.

201 GABURJAKOVA, M., GABURJAKOVA, J., REIKEN, S., HUANG, F., MARX, S. O., ROSEMBLIT, N. & MARKS, A. R. (2001). FKBP12 binding modulates ryanodine receptor channel gating. *J. Biol. Chem.* **276**, 16931–16935.

202 YANO, M., ONO, K., OHKUSA, T., SUETSUGU, M., KOHNO, M., HISAOKA, T., KOBAYASHI, S., HISAMATSU, Y., YAMAMOTO, T., NOGUCHI, N., TAKASAWA, S., OKAMOTO, H. & MATSUZAKI, M. (2000). Altered stoichiometry of FKBP12.6 versus ryanodine receptor as a cause of abnormal Ca^{2+} leak through ryanodine receptor in heart failure. *Circulation* **102**, 2131–2136.

203 GEORGE, C. H., SORATHIA, R., BERTRAND, B. M. A. & LAI, F. A. (2003). In situ modulation of the human cardiac ryanodine receptor (hRyR2) by FKBP12.6. *Biochem. J.* **370**, 579–589.

204 RAHFELD, J. U., SCHIERHORN, A., MANN, K. & FISCHER, G. (1994). A novel peptidyl-prolyl *cis/trans* isomerase from *Escherichia coli*. *FEBS Lett.* **343**, 65–69.

205 RAHFELD, J. U., RÜCKNAGEL, K. P., SCHELBERT, B., LUDWIG, B., HACKER, J., MANN, K. & FISCHER, G. (1994). Confirmation of the existence of a third family among peptidyl-prolyl

cis/trans isomerases – Amino acid sequence and recombinant production of parvulin. *FEBS Lett.* **352**, 180–184.

206 YAO, J. L., KOPS, O., LU, P. J. & LU, K. P. (2001). Functional conservation of phosphorylation-specific prolyl isomerases in plants. *J. Biol. Chem.* **276**, 13517–13523.

207 LANDRIEU, I., WIERUSZESKI, J. M., WINTJENS, R., INZE, D. & LIPPENS, G. (2002). Solution structure of the single-domain prolyl cis/trans isomerase PIN1At from *Arabidopsis thaliana*. *J. Mol. Biol.* **320**, 321–332.

208 RAY, W. C., MUNSON, R. S, & DANIELS, C. J. (2001). Tricross: using dot-plots in sequence-id space to detect uncataloged intergenic features. *Bioinformatics* **17**, 1105–1112.

209 SCHOLZ, C., RAHFELD, J., FISCHER, G. & SCHMID, F. X. (1997). Catalysis of protein folding by parvulin. *J. Mol. Biol.* **273**, 752–762.

210 ROUVIERE, P. E. & GROSS, C. A. (1996). SurA, a periplasmic protein with peptidyl-prolyl isomerase activity, participates in the assembly of outer membrane porins. *Genes Dev.* **10**, 3170–3182.

211 BITTO, E. & McKAY, D. B. (2002). Crystallographic structure of SurA, a molecular chaperone that facilitates folding of outer membrane porins. *Structure* **10**, 1489–1498.

212 TERADA, T., SHIROUZU, M., FUKUMORI, Y., FUJIMORI, F., ITO, Y., KIGAWA, T., YOKOYAMA, S. & UCHIDA, T. (2001). Solution structure of the human parvulin-like peptidyl-prolyl cis/trans isomerase, hPar14. *J. Mol. Biol.* **305**, 917–926.

213 RANGANATHAN, R., LU, K. P., HUNTER, T. & NOEL, J. P. (1997). Structural and functional analysis of the mitotic rotamase Pin1 suggests substrate recognition is phosphorylation dependent. *Cell* **89**, 875–886.

214 RYO, A., LIOU, Y. C., LU, K. P. & WULF, G. (2003). Prolyl isomerase Pin1: a catalyst for oncogenesis and a potential therapeutic target in cancer. *J. Cell Sci.* **116**, 773–783.

215 MALESZKA, R., LUPAS, A., HANES, S. D. & MIKLOS, G. L. G. (1997). The dodo gene family encodes a novel protein involved in signal transduction and protein folding. *Gene* **203**, 89–93.

216 HSIA, C, Y., GANSHAW, G., PAECH, C. & MURRAY, C, J. (1996) Active-site titration of serine proteases using a fluoride ion selective electrode and sulfonyl fluoride inhibitors. *Anal. Biochem.* **242**, 221–227.

217 FISCHER, G., WITTMANN LIEBOLD, B., LANG, K.; KIEFHABER, T. & SCHMID, F. X. (1989). Cyclophilin and peptidyl-prolyl cis/trans isomerase are probably identical proteins. *Nature* **337**, 476–478.

218 KOFRON, J. L., KUZMIC, P., KISHORE, V., COLON BONILLA, E. & RICH, D. H. (1991). Determination of kinetic constants for peptidyl-prolyl cis/trans isomerases by an improved spectrophotometric assay. *Biochemistry* **30**, 6127–6134.

11
Secondary Amide Peptide Bond *cis/trans* Isomerization in Polypeptide Backbone Restructuring: Implications for Catalysis

Cordelia Schiene-Fischer and Christian Lücke

11.1
Introduction

As a recurring element in the polypeptide backbone, peptide bonds occur at a high molar concentration in cells and tissues. The ability of the nitrogen lone electron pair to delocalize over the entire secondary amide moiety results in planarity and enhanced barriers to rotation of the C–N bond. The resonance stabilization of the planar moiety causes restrictions in the number of energy minima with respect to the torsional movements characterized by the dihedral angle ω. The *cis* conformation ($\omega \approx 0°$) is separated from the *trans* conformation ($\omega \approx 180°$) by a rotational barrier much higher than that found for neighboring covalent bonds [1]. Analyses of protein dynamics have revealed the significance of the rigidity of the peptide bond, which is flanked on either side by mobile single bonds forming a crucial folding unit of proteins.

Peptidyl-prolyl *cis/trans* isomerases (PPIases) have evolved to accelerate the interconversion between the *cis* and *trans* conformation, but catalysis is restricted to the imidic bond that is formed at the N-terminal side of prolyl residues [2]. The Hsp70 protein DnaK was identified as the first member of the novel enzyme class of secondary amide peptide bond *cis/trans* isomerases (APIases), which selectively accelerate the *cis/trans* isomerization of secondary amide peptide bonds formed by 19 out of the 20 gene-coded amino acids [3]. In contrast to prolyl bonds, these peptide linkages occur in *cis* conformation for only a minor fraction of oligopeptide molecules in solution. The close proximity of two neighboring Cα atoms in the *cis* conformation has been associated with the low percentage of the secondary amide *cis* peptide bonds, as steric strain is released in the *trans* isomer. Due to a large number of well-defined intramolecular contacts under native conditions, proteins usually exhibit each peptide bond in a defined conformational state, either *cis* or *trans*, unless the peculiarity of the tertiary structure favors native-state isomerization [4]. In a non-redundant set of 571 proteins, 5.2% of the prolyl bonds are in *cis* conformation, whereas only 0.03% secondary amide *cis* peptide bonds are found [5].

Surprisingly, the free-energy difference of the rotational barriers between imidic bonds and secondary amide peptide bonds is rather small. $\Delta\Delta G^{\neq}$ amounts to 5–

10 kJ mol^{-1} in structurally related oligopeptides at 25 °C, with the higher barriers found for the prolyl bonds [6–8]. In both cases, the low spontaneous rate of *cis/trans* isomerization indicates the potential importance of these interconversions as rate-limiting steps in protein backbone rearrangements. Hence, an enzymatic acceleration of this isomerization may be required to avoid accumulation of folding intermediates susceptible to aggregation.

Critical steps in protein folding include rotations about peptide bonds. Thus, frequent occurrence of slow kinetic folding phases with high activation enthalpies $\Delta H^{\neq}$ suggests that prolyl bond *cis/trans* isomerization may be of considerable significance [9–12]. Similarly, secondary amide *cis* peptide bonds in the native state have been found to cause slow steps in the folding kinetics for virtually all molecules of the proteins involved [13, 14]. Generally, secondary amide peptide bonds in *cis* conformation may be more common in native proteins than usually suspected, as their existence may be overlooked by the refinement algorithms [15]. These bonds often occur at positions critical to biological function. Some locations were found to be close to active sites, and others correspond to reactive linkages [16–20]. It has been hypothesized that the sterically constrained geometry associated with *cis* peptide bonds is able to provide potential energy to molecular reactions [18]. However, supporting experimental evidence is still lacking.

Most importantly, the directed bond rotation involved in the mechanism of *cis/trans* interconversion might represent a general feature of folding proteins since the barrier to rotation restricts the backbone mobility of a narrow region of the folding chain. The rotational frequency is determined by the sum of the "*cis* to *trans*" and the "*trans* to *cis*" rate constants in the unfolded and partially folded chains. Thus, a high *cis/trans* ratio in the native state, usually assumed as being critical for the contribution of *cis* secondary amide peptide bonds in refolding, is not considered essential [21, 22].

11.2
Monitoring Secondary Amide Peptide Bond *cis/trans* Isomerization

Detection of secondary amide peptide bond *cis/trans* isomerization in oligopeptides is usually hampered by the small amount of *cis* isomer in equilibrium and the halftime of re-equilibration of about 1 s at 25 °C. By using dynamic NMR spectroscopy, *cis/trans* isomerization rates of peptide bonds in the unperturbed equilibrium can be detected in water under neutral conditions, provided that the respective isomers display NMR signals at sufficiently different chemical shift values. The spontaneous interconversion of the peptide bond conformers proved to be slow on the NMR timescale. Analyses of proline-free oligopeptides (-Ala-Tyr/Phe-Ala-) showed two well-separated β-methyl proton resonances for alanyl residues located next to an aromatic amino acid, indicating conformational heterogeneity. In fact, the aromatic ring current shifts the methyl signal of the *cis* isomer upfield ($|\Delta\delta| \approx$

0.5 ppm) into an empty region of the ^{1}H spectrum. Exchange cross-peaks in two-dimensional nuclear Overhauser enhancement and exchange spectroscopy (NOESY) spectra between the minor and major alanine methyl signals are direct evidence for the existence of interconverting conformers (see the Section 11.6). High-resolution NMR techniques such as magnetization transfer experiments, line-shape analysis, and exchange spectroscopy can be utilized for the quantification of this kind of conformational dynamics [6].

Spectral differences in the UV region between the *cis* and *trans* isomers have been reported for the zwitterionic dipeptide Gly-Gly. This difference is greatest at 216 nm but exhibits an optimal signal-to-noise ratio at 220 nm. It can be utilized to study the kinetics of the *cis/trans* isomerization directly. The re-equilibration of the Gly-Gly isomers after pH jump from pH 2.0 to 7.5 is a first-order reaction and can be monitored with a stopped-flow spectrophotometer at 220 nm. The reverse signal, monitored for a pH jump from pH 7.5 to 2.0, demonstrated the reversibility of the spectral changes. The pH-jump technique can also be used for dipeptides consisting of residues other than glycine. It is thus useful to search for enzymes that catalyze the *cis/trans* isomerization of secondary amide peptide bonds. The reaction can be monitored even in the presence of crude protein extracts [7].

The O/S one-atom substitution of the carbonyl oxygen (C=O to C=S) in secondary amide peptide bonds results in isosteric peptide bonds. The $\pi \rightarrow \pi^*$(C=S) thioxo amide UV absorption is red-shifted relative to the $\pi \rightarrow \pi^*$(C=O) absorption. In zwitterionic thioxo peptide bonds (denoted as -ψ[CS-NH]-), the spectral difference between the *cis* and *trans* isomers is especially large. In addition, a light pulse will allow controlled triggering of the *cis/trans* isomerization. Large spectroscopic changes can be observed upon irradiation at 254 nm or laser irradiation at 337 nm, which is based on the absorption resulting from the n $\rightarrow \pi^*$ excitation of the thioxo carbonyl group. A large increase of the *cis* isomer content in the photostationary state, as recorded by NMR spectroscopy and capillary electrophoresis, and a slow thermal re-equilibration rate characterize the O to S substitution in a peptide unit. Thus, irradiation of a thioxo peptide bond provides a predictable conformational switch for a narrow segment of a polypeptide chain [43].

For thioxo dipeptides, the kinetics of the *cis/trans* isomerization can also be studied at 290 nm by alteration of the ionization state through pH jump. For proteins with a *cis* peptide bond in the native state, secondary amide peptide bond *cis/trans* isomerizations could be characterized both in the unfolded state and during refolding. The *trans* to *cis* isomerization could be measured directly by intrinsic fluorescence changes, which reflect the rate of slow folding [13, 14]. The kinetic parameters of the *cis* to *trans* isomerization after unfolding could be measured in double-mixing experiments (unfolding followed by refolding), since unfolded molecules that still contain the *cis* isomer refold rapidly [13]. The *cis/trans* equilibration in the unfolded form of a protein with all-*trans* peptide bonds in the native form after conformational unfolding was demonstrated for a proline-free tendamistat variant by the appearance of a slow refolding phase [21]. Again, double mixing (unfolding followed by refolding) was used in these experiments.

11.3
Kinetics and Thermodynamics of Secondary Amide Peptide Bond cis/trans Isomerization

Simple N-alkyl-substituted secondary amides such as N-methylformamide and N-methylacetamide have often been used as models to study the conformational characteristics of the peptide backbone. Equilibrium constants for the cis/trans isomerization determined by NMR methods agree with molecular mechanics calculations and Monte Carlo simulations that point to a rather low percentage of the cis isomer [23]. N-methylformamide exists as cis isomer to the extent of 8% in D_2O and 10.3% in $CDCl_3$. In N-methylacetamide, which more closely resembles a peptide unit, a free-energy difference of about 10.5 kJ mol^{-1} between the trans and cis isomers – corresponding to 1.5% cis isomer at equilibrium – has been detected in D_2O. Nonpolar solvents do not cause a significant alteration [24].

In the dipeptide Ala-Tyr, the cis content depends on the protonation states of the amino and carboxyl groups. The maximal cis fraction of 0.41% occurs in the zwitterionic state of the molecule, decreasing below the pK_a of the carboxyl group and above the pK_a of the amino group. The pH dependence of the cis content suggests that oppositely charged groups close to the isomerizing bond lead to enhanced cis populations. ^{13}C-NMR spectra revealed a cis population of 1.0% for the zwitterionic form of Gly-Gly. This value agrees with the estimate of 0.5% cis Gly-Gly detected in UV resonance Raman experiments [25].

Elongation of the peptide at both ends decreases the cis content of the central peptide bond by favoring the trans isomer with about 3 kJ mol^{-1}. The cis populations of the Ala2-Tyr3 and Tyr3-Ala4 bonds of the pentapeptide Ala-Ala-Tyr-Ala-Ala have been determined to be 0.14% and 0.11%, respectively [6]. These values approach the value of 0.1% cis Ala-Ala, as estimated by conformational energy calculations for the center of a peptide segment containing three peptide bonds [26]. Despite this rather low probability of cis isomers, the existence of at least one cis peptide bond fluctuating in an unstructured polypeptide chain of 1000 amino acid residues has to be considered.

Because the fraction of the cis isomer is so small, little is known about the isomerization kinetics. Rotational barriers of $\Delta G^{\neq} = 79$ kJ mol^{-1} (cis to trans) and $\Delta G^{\neq} = 89$ kJ mol^{-1} (trans to cis) have been determined for N-methylacetamide by NMR line-shape analysis of the N-methyl signals at 60 °C in water [27].

Peptide bond isomerization is a unimolecular process characterized by a first-order rate constant $k_{obs} = k_{cis\ to\ trans} + k_{trans\ to\ cis}$ for the reversible reaction. The equilibration of secondary amide peptide bonds, with rate constants of about 0.5 s^{-1} at 25 °C, is almost entirely determined by $k_{cis\ to\ trans}$, because the cis to trans isomerization is at least two orders of magnitude faster than the reverse reaction.

For Gly-Gly, an isomerization rate lower than 1.5 s^{-1} at 45 °C was estimated by ^{1}H-NMR. In support of this approximation, rate constants of about 0.3 s^{-1} at 25 °C were determined by directly monitoring the kinetics of the cis/trans isomerization of zwitterionic Gly-Gly following a pH jump. The Eyring activation enthalpy $\Delta H^{\neq}$ of 72 kJ mol^{-1} is in good agreement with the lower limit of 65 kJ mol^{-1} estimated from line broadening in ^{1}H-NMR spectra. For Ala-Tyr the rate constants were determined to be $k_{cis\ to\ trans} = 0.57$ s^{-1} and $k_{trans\ to\ cis} = 2.4 \times 10^{-3}$ s^{-1} by NMR magnetization transfer experiments. As exemplified by Ala-Phe, $k_{cis\ to\ trans}$ is pH-independent between pH 4.5 and pH 8.6 [6, 7]. The activation enthalpies $\Delta H^{\neq}$ of the cis/trans isomerization of secondary amide peptide bonds are in a range typical of imidic peptide bond isomerizations [8]. Increasing the peptide chain length led to an increase of the isomerization rate and to a decrease of $\Delta H^{\neq}$ for the central peptide bonds.

Experimental data for the kinetics of cis/trans isomerization in entire proteins were mainly derived from protein variants, in which (1) a cis prolyl bond was replaced by a secondary amide bond and (2) the cis conformation was retained in the native state. This substitution usually results in structural destabilization of about 8–21 kJ mol^{-1} in comparison to the wild-type proteins, as reported for the carbonic anhydrase Pro202Ala variant [28], the bovine pancreatic ribonuclease A Pro93Ala variant [29, 30], the ribonuclease T$_1$ Pro39Ala variant [31, 32], and the TEM-1 β-lactamase Pro167Thr variant [14]. This effect was attributed to the unfavorable cis/trans ratio of the secondary amide peptide bonds in the unfolded state.

The rate constant $k_{cis\ to\ trans}$ of the Tyr92-Ala93 bond of the ribonuclease A Pro93Ala variant in the unfolded state was found to be 0.702 s^{-1} at 15 °C [33]. For the ribonuclease T$_1$ Pro39Ala variant with the cis Tyr38-Ala39 bond in the native state, rate constants for the cis to trans isomerization of 0.9–1.4 s^{-1}, depending on the GdmCl concentration present in the unfolding reaction, were detected in the unfolded protein by double-mixing experiments at 25 °C. The trans to cis isomerization of the Tyr38-Ala39 bond leads to a slow refolding phase in a partially structured folding intermediate. It is characterized by a rate constant of 4×10^{-3} s^{-1} at 25 °C [13]. During the refolding of the TEM-1 β-lactamase Pro167Thr variant, the cis Glu166-Thr167 bond is formed with a rate constant between 1×10^{-3} and 4×10^{-3} s^{-1} [14]. These trans to cis isomerization rates resemble those obtained for the oligopeptides, although the cis/trans ratio differs considerably between these folded proteins and the unstructured peptide chains. Evidently, the trans to cis isomerization rates do not provide a kinetic basis for the stability of the cis conformer in the native states of the ribonuclease T$_1$ and β-lactamase variants. A large part of the thermodynamic stability of the cis conformation has to originate from a considerable deceleration of the cis to trans isomerization in native proteins.

For the folding of proteins such as *Escherichia coli* dehydrofolate reductase [34] and staphylococcal nuclease [35], kinetic complications were encountered that have been discussed assuming conformational heterogeneity of the secondary amide peptide bonds.

Experimental proof for this hypothesis came from refolding data of a proline-free variant of the α-amylase inhibitor tendamistat. A fraction of the tendamistat variant molecules, with a tryptophan fluorescence amplitude of 5%, folds in a reaction assigned to the cis to trans isomerization of secondary amide peptide bonds. The activation energy of 51 kJ mol^{-1} and the reaction enthalpy of -29 kJ mol^{-1} are both in agreement with data obtained for oligopeptides. Interestingly, the denaturation of proline-free tendamistat in LiCl/trifluoroethanol leads, dependent on the

LiCl concentration, to a linear increase of the slow refolding fraction of the molecules. Lithium chloride in anhydrous trifluoroethanol, which has already been shown to increase the *cis* content of prolyl bonds [36], evidently is able to increase the *cis* content of secondary amide peptide bonds as well. The *cis* population of the Ala-Tyr bond of Ala-Ala-Tyr in LiCl/trifluoroethanol was determined to be 0.5% compared to 0.19% in aqueous solution. These data support the assumption that the conformational heterogeneity of secondary amide peptide bonds in the unfolded state of the protein is the cause for the slow refolding phase, which can also be found in the refolding of wild-type tendamistat [21].

The observed complex folding kinetics may originate from either a particular secondary amide peptide bond exhibiting an increased propensity for the *cis* conformation in the unfolded state or a certain level of *cis* isomer at any position of the peptide chain. A secondary amide peptide bond isomerization occurring randomly throughout the unfolded chain has already been suggested as a distinct reaction during refolding of proteins lacking secondary amide *cis* peptide bonds in the native state [37, 38]. Generally, the higher number of peptide bonds of larger proteins includes a statistical advantage in favor of a higher number of fluctuating *cis* peptide bonds in the whole polypeptide chain. For larger proteins this effect might cause more complex folding kinetics.

11.4
Principles of DnaK Catalysis

Cis/trans isomerization of peptide bonds has proved to be slow on the timescale of biorecognition involving polypeptides. A general possibility to accelerate peptide bond *cis/trans* isomerization in cells is enzyme catalysis. Indeed, the chaperone DnaK displays *cis/trans* isomerase activity toward secondary amide peptide bonds, but not towards prolyl bonds, and thus represents a novel class of enzymes that have been termed secondary amide peptide bond *cis/trans* isomerases (APIases) [3]. DnaK accelerates Ala-Ala dipeptide *cis/trans* isomerization, and the time courses are in accordance with the minimal model of an enzymatically catalyzed reversible reaction with a k_{cat}/K_M value of 3.5×10^5 M^{-1} s^{-1}. The K_M and k_{cat} values are estimated at 8 mM and 3×10^3 s^{-1} (at 25 °C), respectively. Dose-dependent inhibition of the APIase activity of DnaK in the presence of the heptapeptide NRLLLTG or the undecapeptide RPKPQQFFGLM-NH$_2$ (substance P), which are known to be DnaK ligands, indicates that the catalysis occurs at or near the peptide-binding groove.

The APIase activity exhibits both regio- and stereoselectivity. DnaK catalysis displays subsite specificity for the C-terminal residue of the isomerizing peptide bond in a series of Ala-Xaa dipeptides. The best substrates were the dipeptides with Xaa = Met, Ala, and Ser, followed by Ala-Glu and Ala-Leu. Ile, Gly, and Lys in the position Xaa formed less-efficient substrates, whereas isomerizations in Asp, Tyr, Thr, Asn, Phe, Arg, Val, Gln, and Pro derivatives could not be catalyzed. Interestingly, the nature of the preferred subsites for catalysis flanking the reactive bond is at variance with the hydrophobic core affinity sites, as concluded from a library of 13mer peptides [39]. Both studies survey different properties of DnaK. According to the enzyme transition-state complementarity, the k_{cat}/K_M values describe transition-state affinities. The affinity to matrix-bound peptides of a peptide library, on the other hand, may reveal ground-state binding, which is characterized by the microscopic dissociation constant K_s. Only leucine is a preferred residue for both properties, K_s and k_{cat}/K_M. The high relevance of steric features for catalysis can be inferred from the detrimental consequences of the C_β-branched hydrophobic residues Val, Ile, and Thr. The lack of catalysis of prolyl bonds is probably due to N-alkylation. Obviously, the APIase activity is an intrinsic property of the DnaK chain, closely resembling PPIase activity in its mechanistic consequences. Neither hydrolysis of ATP nor the presence of accessory proteins is imperative to the elementary step of catalysis of *cis/trans* isomerization. However, a functional overlap between APIase and PPIase substrate specificity has been excluded. Analysis of DnaK in PPIase assays displayed no PPIase activity, and for PPIases of different families such as human Cyp18, human FKBP12, and *Escherichia coli* Par10, no APIase activity was detected in the dipeptide assay [3].

As mentioned above, two-dimensional ^{1}H-NMR NOESY spectroscopy is the most sensitive NMR method for determining the dynamics of the conformational interconversion of peptide bonds neighboring aromatic amino acids. At 5 °C, the *cis/trans* isomerization rates of the Ala2-Tyr3 and the Tyr3-Ala4 bond of the pentapeptide Ala-Ala-Tyr-Ala-Ala become sufficiently slow to suppress the exchange cross-peaks. In the presence of 20 μM DnaK, however, the Ala2-Tyr3 exchange cross-peak reappears at this temperature, indicating a DnaK-mediated dynamic exchange between the *cis* and *trans* isomers (Figure 11.1). This DnaK-mediated acceleration of the *cis/trans* isomerization found for the Ala2-Tyr3 bond represents a lower limit due to the very high substrate concentration of 25 mM required for NMR measurements. Moreover, in addition to the two bonds that flank the tyrosine residue and can be monitored by NMR in their isomeric states, the substrate contains two additional peptide bonds that compete for the catalytic site of the enzyme. Usually, DnaK complemented with ATP, as well as its co-chaperones DnaJ and GrpE, is needed to perform successful protein renaturation experiments. The observed NOESY cross-peak intensities in the presence of the complete DnaK/DnaJ/GrpE/ATP system are similar for the Ala2-Tyr3 bond, but considerably larger for the Tyr3-Ala4 linkage, in comparison to the presence of DnaK alone. These data suggest an extension of the range of productive enzyme-substrate interactions by the complete chaperone mixture, leading to a more promiscuous enzyme specificity.

DnaK also catalyzes *cis/trans* isomerization in a partially folded protein. It accelerates the *trans* to *cis* isomerization of the Tyr38-Ala39 bond in the refolding of the ribonuclease T$_1$ Pro39Ala variant under conditions where partial folding precedes Tyr-Ala *trans* to *cis* isomerization.

The interrelation between APIase function and chaperone activity was examined in the DnaK/GrpE/DnaJ-assisted refolding of GdmCl-denatured firefly luciferase. In a series of peptide ligands of DnaK, APIase inhibition parallels the inhibi-

a AAYAA

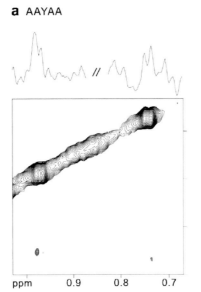

b AAYAA + DnaK

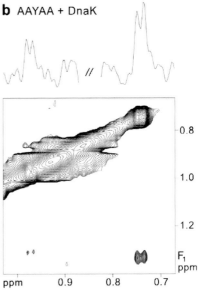

Fig. 11.1. *cis*-Alanine methyl region in the NOESY spectra of Ala-Ala-Tyr-Ala-Ala at 5 °C. The one-dimensional traces above the two-dimensional plots are taken along F_2 at the F_1 position of the correlated *trans* signals. (a) Peptide alone: the exchange cross-peaks at 0.74 ppm and 0.97 ppm in F_2 are assigned to the *cis/trans* isomerization of the Ala2-Tyr3 and the Tyr3-Ala4 peptide bond, respectively. (b) Peptide in the presence of 20 µM DnaK.

tion of chaperoning activity. This finding can be interpreted as an indication of a direct catalytic effect in the refolding of denatured proteins, since functional coupling of both folding probes is obvious. Remarkably, DnaK caused a concentration-dependent increase in the refolding yield and a concentration-independent lag phase in the refolding kinetics of luciferase. These data suggest the rapid formation of enzymatically inactive folding intermediates of luciferase, which are all captured in a metastable kinetic trap before they can reach their native conformation. The structure of most folding intermediates favors formation of aggregates. Absence of DnaK/GrpE/DnaJ/ATP is associated with a very slow rate of appearance of native luciferase via a productive folding intermediate, and only a very small fraction refolds productively. To describe the general features of the APIase-assisted refolding accelerated formation of a productive folding intermediate from aggregation-prone intermediates suffices to obtain increasing refolding yields [40].

Peptide bond *cis/trans* isomerases transiently initiate decreased torsional stiffness for a certain part of the polypeptide chain, while leaving other chain segments less flexible. As a consequence, these enzymes may mediate the reversal of local kinetic traps that can be detrimental to productive folding (Figure 11.2). DnaK may catalyze the isomerization at a large part of the secondary amide peptide bonds of a protein. However, it is not possible at present to assess the precise number of DnaK-catalyzable bond rotations. A single turnover of a substrate will affect

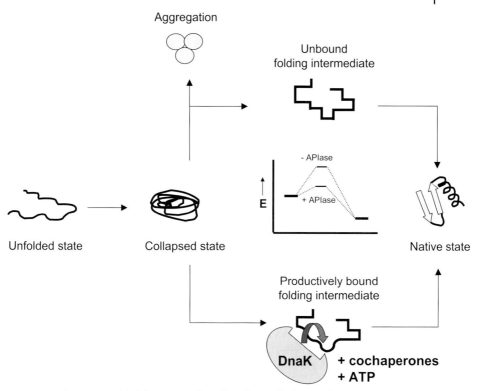

Fig. 11.2. Schematic model of the action of peptide bond *cis/trans* isomerases during refolding of a denatured protein. The action of DnaK involves scanning of the polypeptide chain by multiple attacks on peptide bonds. Thus, the torsional stiffness of a certain part of the polypeptide chain is decreased in a reaction intermediate. The function of ATP is to mediate release of the polypeptide chain in its folding-competent conformation from DnaK.

only a small polypeptide segment. Large k_{cat} values form the precondition for rapid scanning of peptide bonds throughout the entire polypeptide chain. Hence, a new catalytic function was proposed for DnaK: kinetic proofreading, which has been discussed previously for chaperone action [41], targets certain bonds in a specific context of amino acid residues. This catalytic process of bond rotation transiently abolishes the steric restrictions associated with the electronic structure of the peptide bond.

11.5
Concluding Remarks

Novel assays for the analysis of the *cis/trans* isomerization of secondary amide peptide bonds allowed the identification of DnaK as the first member of the enzyme class of APIases. APIases strongly resemble PPIases in that they confer a loss of rotational restrictions, but at different positions in the polypeptide backbone. The major difference between these two enzyme classes lies in the nature of the targeted peptide bond. PPIases are specific to imidic linkages that are formed by the imino acid proline. In contrast, the Hsp70 protein DnaK catalyzes the isomerization at the much larger contingent of secondary amide peptide bonds. Based on the above-described results, we assume that in response to the APIase activity of DnaK, the high internal rotational barrier of a particular secondary amide peptide bond in a multiply bound polypeptide substrate will be decreased. As a consequence, a highly mobile structural unit that is confined by an extended, rigid backbone on both sides of the catalyzed bond will be generated. The resulting higher backbone flexibility may impede the formation of incorrect folds. A remaining challenge will be to analyze folding intermediates of proteins susceptible to APIase activity. Future studies are likely to involve the search for additional members of the APIase family, the identification of their catalytic mechanisms, and the search for specific inhibitors of these enzymes.

11.6
Experimental Protocols

11.6.1
Stopped-flow Measurements of Peptide Bond *cis/trans* Isomerization

Peptides can be purchased from Bachem (Heidelberg, Germany). Measurements are performed with a stopped-flow spectrophotometer (Applied Photophysics, Leatherhead, UK). The path length of the observation chamber should be 2 mm. For measurement of the peptide bond *cis/trans* isomerization of dipeptides, the peptides are dissolved in 50 mM sodium phosphate buffer at the appropriate pH (pH 2.0 for the pH jump to pH 7.5, where the zwitterionic form exists) to obtain a stock solution of 20–200 mM. The required pH is adjusted with dilute NaOH or phosphoric acid. Small differences in the final concentrations of the peptide may result from the pH adjustment but are unimportant for the assay.

The reaction is started by a 26-fold dilution of the peptides into 50 mM sodium phosphate buffer at the appropriate pH (pH 7.5 for the pH jump to the zwitterionic form or pH 2.0 for the reverse jump). (Examination of the resulting pH is highly recommended.) The time course of the reaction is monitored at 220 nm and 25 °C for 10 s. The time traces should be averages of at least five measurements. Data analysis is performed by single-exponential nonlinear regression using, e.g., the SigmaPlot scientific graphing system Version 8.0 (Jandel Corp., USA) to yield first-order rate constants $k_{c/t}$ of the *cis/trans* isomerizations.

To assay APIase activity of proteins, jumps from pH 2.0 to pH 7.5 are performed with the dipeptides as described above. Either protein fractions or pure proteins are added to the dilution buffer (12.5 mM Tris·HCl pH 8.0, 50 mM KCl, 11 mM $MgCl_2$) stored in the 2.5-mL syringe. To reduce the total absorbance at 220 nm, the dipeptides are dissolved in water to obtain a concentration of 50 mM. The required pH of 2.0 is adjusted with dilute HCl. Check in controls that the final pH 7.5 is achieved.

11.6.2
Two-dimensional ^{1}H-NMR Exchange Experiments

The homonuclear, two-dimensional NOESY experiment provides a very sensitive method to detect the chemical exchange that occurs during peptide bond isomerization. The NOESY exchange cross-peaks observed between the *cis* and *trans* isomer forms are very low in intensity and can thus be clearly observed only in spectral regions that are devoid of other signals. In the case of the pentapeptide Ala-Ala-Tyr-Ala-Ala (25 mM in 12.5 mM Tris·HCl pH 7.1, 50 mM KCl, 11 mM MgCl$_2$, 90% H$_2$O/10% D$_2$O), the methyl proton resonance of Ala2 is considerably shifted upfield when the Ala2-Tyr3 peptide bond takes on the *cis* conformation, due to the ring-current effect of the neighboring Tyr3 side chain. Subsequently, the NOESY exchange cross-peak that represents the *cis/trans* interconversion appears at a resonance frequency where no other peaks occur in the spectrum of this peptide. This allows monitoring of the *cis/trans* isomerization rate under various conditions, even in the presence of substoichiometric amounts of DnaK or other proteins. To distinguish the DnaK-mediated isomer exchange from the spontaneous *cis/trans* interconversion of the pentapeptide at 20 °C, all spectra have to be collected at low temperature (5 °C), where the latter effect is too slow on the NMR timescale (300–410 ms mixing time) for the corresponding NOESY exchange cross-peaks to be observed.

Therefore, the magnetization transfer between the Ala2 methyl proton resonances of the *cis* and *trans* isomer forms directly reflects the isomerization rate [42]. In the NOESY experiment, this chemical exchange takes place almost exclusively during the mixing period, since the evolution time of the two-dimensional experiment is comparatively short relative to the mixing time (τ_m). Assuming that the rate constant (k) and the longitudinal relaxation rate (R_1) remain approximately the same in both isomer forms of the peptide, the exchange signal intensity at a constant τ_m is proportional to ($x_b - x_b^2$), where x_b represents the mole fraction of the *cis* component. Hence, the signal intensity (I_{ab}) of the NOESY exchange cross-peak between the *trans* (a) and the *cis* (b) isomer form shows a parabolic dependence on the *cis* mole fraction, whereby $I_{ab} = f(x_b)$ lies very close to the inflection point at $x_b \sim 0.15\%$. At this rather low *cis* content (0.15% of a 25-mM pentapeptide concentration equals a 37.5-µM *cis* population), the addition of 20 µM DnaK could cause a maximal increase to $x_b = 0.23\%$, if the enzyme were to raise the *cis* content at a 1:1 molar ratio. This would induce a change in I_{ab} of less than 0.5%. The experimentally observed intensity enhancement effects, however, are considerably higher, indicating an enzymatic increase of the isomerization rate.

All NMR experiments should be performed with a high-field magnet for better spectral resolution, such as a Bruker DRX500 spectrometer (operating at a proton frequency of 500.13 MHz) that is equipped with a 5-mm triple-resonance (^{1}H, ^{13}C,

^{15}N) probe with Z-gradient capability. Data acquisition and processing are achieved with special computer software (e.g., XWINNMR 2.5, Bruker, Germany). For spectra calibration, proton chemical shifts are usually referenced relative to internal DSS in the sample. All spectra should be recorded in a phase-sensitive mode, with time-proportional phase incrementation of the initial pulse and quadrature detection in both dimensions. The carrier is placed in the center of the spectrum on the water resonance for suppression of the solvent signal. Suppression of the water signal can be achieved by presaturation during the relaxation delay and with a WATERGATE pulse sequence after the mixing period. The NOESY experiments should be collected at 5 °C with a mixing time of 330 ms and a relaxation delay of 1.45 s. For optimal resolution, the time-domain data should consist of at least $512(t_1) \times 3072(t_2)$ complex points at a spectral width of 11 ppm (i.e., 5482.5 Hz) in both dimensions. Prior to Fourier transformation, the time-domain data can be zero-filled to 1024 and 4096 points in the t_1 and t_2 dimensions, respectively, and multiplied by a 90°-shifted squared sine-bell window function. Baseline correction needs to be performed in ω_2 from 1.76–1.51 ppm and from 1.06–0.56 ppm prior to peak integration with the program SPARKY (http://www.cgl.ucsf.edu/home/sparky).

During the course of the enzymatic catalysis reaction, NOESY exchange cross-peaks associated with the *cis* and *trans* isomer forms of the Ala2-Tyr3 and Tyr3-Ala4 bonds can be observed for the methyl groups of Ala2 ($\omega_1 = 1.33$ ppm, $\omega_2 = 0.74$ ppm) and Ala4 ($\omega_1 = 1.31$ ppm, $\omega_2 = 0.97$ ppm), whereby the resonances of the *cis* forms are shifted upfield in both cases. In order to provide a reliable intensity reference relative to the total peptide concentration, all signal intensities can be normalized to the NOESY exchange cross-peak between the two ^{13}C satellites of the Ala1 methyl group ($\omega_1 = 1.31$ ppm, $\omega_2 = 1.57$ ppm) after baseline correction has been performed.

References

1 FISCHER, G. (2000). Chemical aspects of peptide bond isomerisation. *Chem. Soc. Rev.*, **29**, 119–127.

2 FISCHER, G. (1994). Peptidyl-prolyl cis/trans isomerases and their effectors. *Angew. Chem. Int. Ed. Engl.*, **33**, 1415–1436.

3 SCHIENE-FISCHER, C., HABAZETTL, J., SCHMID, F. X. & FISCHER, G. (2002). The hsp70 chaperone DnaK is a secondary amide peptide bond *cis-trans* isomerase. *Nat. Struct. Biol.*, **9**, 419–424.

4 ANDREOTTI, A. H. (2003). Native state proline isomerization: an intrinsic molecular switch. *Biochemistry*, **42**, 9515–9524.

5 JABS, A., WEISS, M. S. & HILGENFELD, R. (1999). Non-proline *cis* peptide bonds in proteins. *J. Mol. Biol.*, **286**, 291–304.

6 SCHERER, G., KRAMER, M. L., SCHUTKOWSKI, M., REIMER, U. & FISCHER, G. (1998). Barriers to rotation of secondary amide peptide bonds. *J. Am. Chem. Soc.*, **120**, 5568–5574.

7 SCHIENE-FISCHER, C. & FISCHER, G. (2001). Direct measurement indicates a slow cis/trans isomerization at the secondary amide peptide bond of glycylglycine. *J. Am. Chem. Soc.*, **123**, 6227–6231.

8 STEIN, R. L. (1993). Mechanism of

enzymic and non-enzymic prolyl cis/trans isomerization. *Adv. Protein Chem.*, **44**, 1–24.

9 LIN, L. N. & BRANDTS, J. F. (1984). Involvement of prolines-114 and -117 in the slow refolding phase of ribonuclease A as determined by isomer-specific proteolysis. *Biochemistry*, **223**, 5713–5723.

10 SCHMID, F. X., GRAFL, R., WRBA, A. & BEINTEMA, J. J. (1986). Role of proline peptide bond isomerization in unfolding and refolding of ribonuclease. *Proc. Natl. Acad. Sci. USA*, **83**, 872–876.

11 SEMISOTNOV, G. V., UVERSKY, V. N., SOKOLOVSKY, I. V., GUTIN, A. M, RAZGULYAEV, O. I. & RODIONOVA, N. A. (1990). Two slow stages in refolding of bovine carbonic anhydrase B are due to proline isomerization. *J. Mol. Biol.*, **213**, 561–568.

12 THIES, M. J. W., MAYER, J., AUGUSTINE, J. G., FREDERICK, C. A., LILIE, H. & BUCHNER, J. (1999). Folding and association of the antibody domain C_H3: Prolyl isomerization preceeds dimerization. *J. Mol. Biol.*, **293**, 67–79.

13 ODEFEY, C., MAYR, L. M. & SCHMID, F. X. (1995). Non-prolyl cis-trans peptide bond isomerization as a rate-determining step in protein unfolding and refolding. *J. Mol. Biol.*, **245**, 69–78.

14 VANHOVE, M., RAQUET, X., PALZKILL, T., PAIN, R. H. & FRERE, J. M. (1996). The rate-limiting step in the folding of the cis-Pro167Thr mutant of TEM-1 β-lactamase is the trans to cis isomerization of a non-proline peptide bond. *Proteins: Struct. Funct. Genet.*, **25**, 104–111.

15 WEISS, M. S., JABS, A. & HILGENFELD, R. (1998) Peptide bonds revisited. *Nat. Struct. Biol.*, **5**, 676.

16 BOUCKAERT, J., LORIS, R., POORTMANS, F. & WYNS, L. (1995). Crystallographic structure of metal-free concanavalin A at 2.5 angstrom resolution. *Proteins: Struct. Funct. Genet.*, **23**, 510–524.

17 BOUCKAERT, J., DEWALLEF, Y., POORTMANS, F., WYNS, L. & LORIS, R. (2000). The structural features of concanavalin A governing non-proline peptide isomerization. *J. Biol. Chem.*, **275**, 19778–19787.

18 STODDARD, B. L. & PIETROKOVSKI, S. (1998). Breaking up is hard to do. *Nat. Struct. Biol.*, **5**, 3–5.

19 HEROUX, A., WHITE, E. L., ROSS, L. J., DAVIS, R. L. & BORHANI, D. W. (1999). Crystal structure of *Toxoplasma gondii* hypoxanthine-guanine phosphoribosyltransferase with XMP, pyrophosphate, and two Mg^{2+} ions bound: Insights into the catalytic mechanism. *Biochemistry*, **38**, 14495–14506.

20 BANERJEE, S., SHIGEMATSU, N., PANNELL, L. K., RUVINOV, S., ORBAN, J., SCHWARZ, F. & HERZBERG, O. (1997) Probing the non-proline cis peptide bond in β-lactamase from *Staphylococcus aureus* PC1 by the replacement Asn136 → Ala, *Biochemistry*, **36**, 10857–10866.

21 PAPPENBERGER, G., AYGUN, H., ENGELS, J. W., REIMER, U., FISCHER, G. & KIEFHABER, T. (2001). Nonprolyl cis peptide bonds in unfolded proteins cause complex folding kinetics. *Nat. Struct. Biol.*, **8**, 452–458.

22 EYLES, S. J. (2001). Proline not the only culprit? *Nat. Struct. Biol.*, **8**, 380–381.

23 JORGENSEN, W. L. & GAO, J. (1988). *Cis/trans* energy difference for the peptide bond in the gas phase and in aqueous solution. *J. Am. Chem. Soc.*, **110**, 4212–4216.

24 RADZICKA, A., PEDERSEN, L. & WOLFENDEN, R. (1988). Influences of solvent water on protein folding: free energies of solvation of cis and trans peptides are nearly identical. *Biochemistry*, **27**, 4538–4541.

25 LI, P. S., CHEN, X. G., SHULIN, E. & ASHER, S. A. (1997). UV resonance Raman ground and excited state studies of amide and peptide isomerization dynamics. *J. Am. Chem. Soc.*, **119**, 1116–1120.

26 RAMACHANDRAN, G. N. & MITRA, A. K. (1976). An explanation for the rare occurrence of cis peptide units in proteins and polypeptides. *J. Mol. Biol.*, **107**, 85–92.

27 DRAKENBERG, T. & FORSEN, S. (1971).

The barrier to internal rotation in monosubstituted amides. *J. Chem. Soc. D: Chem. Commun.*, 1404–1405.

28 TWEEDY, N. B., NAIR, S. K., PATERNO, S. A., FIERKE, C. A. & CHRISTIANSON, D. W. (1993). Structure and energetics of a non-proline *cis*-peptidyl linkage in a proline-202 → alanine carbonic anhydrase-II variant. *Biochemistry*, 32, 10944–10949.

29 SCHULTZ, D. A. & BALDWIN, R. L. (1992). *Cis* proline mutants of ribonuclease-A.I. Thermal stability. *Protein Sci.*, 1, 910–916.

30 PEARSON, M. A., KARPLUS, P. A., DODGE, R. W., LAITY, J. H. & SCHERAGA, H. A. (1998). Crystal structures of two mutants that have implications for the folding of bovine pancreatic ribonuclease A. *Protein Sci.*, 7, 1255–1258.

31 MAYR, L. M., LANDT, O., HAHN, U. & SCHMID, F. X. (1993). Stability and folding kinetics of ribonuclease T_1 are strongly altered by the replacement of *cis*-proline 39 with alanine. *J. Mol. Biol.*, 231, 897–912.

32 MAYR, L. M., WILLBOLD, D., RÖSCH, P. & SCHMID, F. X. (1994). Generation of a non-prolyl *cis* peptide bond in ribonuclease T_1. *J. Mol. Biol.*, 240, 288–293.

33 DODGE, R. W. & SCHERAGA, H. A. (1996). Folding and unfolding kinetics of the proline-to-alanine mutants of bovine pancreatic ribonuclease A. *Biochemistry*, 35, 1548–1559.

34 JENNINGS, P. A., FINN, B. E., JONES, B. E. & MATTHEWS, C. R. (1993). A reexamination of the folding mechanism of dihydrofolate reductase from *Escherichia coli*: verification and refinement of a four-channel model. *Biochemistry*, 32, 3783–3789.

35 MAKI, K., IKURA, T., HAYANO, T., TAKAHASHI, N. & KUWAJIMA, K. (1999). Effects of proline mutations on the folding of staphylococcal nuclease. *Biochemistry*, 38, 2213–2223.

36 KOFRON, J. L., KUZMIC, P., KISHORE, V., COLON-BONILLA, E. & RICH, D. H. (1991). Determination of kinetic constants for peptidyl prolyl *cis*-*trans* isomerases by an improved spectrophotometric assay. *Biochemistry*, 30, 6127–6134.

37 UDGAONKAR, J. B. & BALDWIN, R. L. (1990). Early folding intermediate of ribonuclease A. *Proc. Natl. Acad. Sci. USA*, 87, 8197–8201.

38 WALKENHORST, W. F., GREEN, S. M. & RODER, H. (1997). Kinetic evidence for folding and unfolding intermediates in staphylococcal nuclease. *Biochemistry*, 36, 5795–5805.

39 MAYER, M. P., RÜDIGER, S. & BUKAU, B. (2000). Molecular basis for interactions of the DnaK chaperone with substrates. *Biol. Chem.*, 381, 877–885.

40 SCHIENE-FISCHER, C., HABAZETTL, J., TRADLER, T. & FISCHER, G. (2002). Evaluation of similarities in the *cis/trans* isomerase function of trigger factor and DnaK. *Biol. Chem.*, 383, 1865–1873.

41 GULUKOTA, K. & WOLYNES, P. G. (1994). Statistical mechanics of kinetic proofreading in protein folding *in vivo*. *Proc. Natl. Acad. Sci. USA*, 91, 9292–9296.

42 BAINE, P. (1986). Comparison of rate constants determined by two-dimensional NMR spectroscopy with rate constants determined by other NMR techniques. *Magn. Reson. Chem.*, 24, 304–307.

43 ZHAO, J., WILDEMANN, D., JAKOB, M., VARGAS, C. & SCHIENE-FISCHER, C. (2003). Direct photomodulation of peptide backbone conformations. *Chem. Commun.*, 2810–2811.

12
Ribosome-associated Proteins Acting on Newly Synthesized Polypeptide Chains

Sabine Rospert, Matthias Gautschi, Magdalena Rakwalska, and Uta Raue

12.1
Introduction

Translation of the genetic message into a polypeptide is carried out by ribosomes. Eukaryotic ribosomes are macromolecular structures of about 4000 kDa consisting of two-thirds RNA and one-third protein [1–3]. The structures of the 30S subunit of the eubacterium *Thermus thermophilus* [4] and the 50S subunit of the archaebacterium *Haloarcula marismortui* [5] and of the eubacterium *Deinococcus radiodurans* [6] have been solved by X-ray crystallography. Although the core of the ribosome is conserved in all organisms, eukaryotes contain an additional rRNA molecule plus 20–30 more ribosomal proteins than do prokaryotes. So far neither the 60S nor the 40S eukaryotic ribosomal subunits are available in atomic detail. Our current understanding of the eukaryotic ribosome comes from a combination of cryo-electron microscopy, crystallography data on individual components, and homology modeling based on the prokaryotic ribosome [7–10].

The key event in polypeptide synthesis is the catalysis of peptide bond formation. The peptidyl transferase (PT) center is located on the large ribosomal subunit and is composed of RNA (23S rRNA) [5, 11]. A mechanism of catalysis was proposed based on the arrangement of RNA elements in the active site and their interactions with an intermediate-state analogue [11]. The newly formed polypeptide leaves the ribosome through a tunnel. First observed in the mid-1980s, the tunnel was assumed to be a passive path to exit the ribosome [12, 13]. The crystal structure of the *H. marismortui* 50S subunit revealed that the tunnel from the site of peptide synthesis to its exit is about 10 nm in length and between 1 and 2 nm in width. The tunnel can accommodate a polypeptide chain of approximately 40 amino acids [5, 11] and is formed by RNA and protein (23S rRNA, L4, L22, L39e: *H. marismortui* nomenclature; compare Table 12.1). Its surface is largely hydrophilic and includes exposed hydrogen-bonding groups from bases, phosphates, and polar amino acids. It has been suggested that this results in a Teflon-like tunnel surface, similar to the chaperonin GroEL [14] in its non-binding conformation [11]. However, recent findings suggest that the tunnel may very well interact with the nascent polypeptide chain and play a rather active role in regulating translation

Protein Folding Handbook. Part II. Edited by J. Buchner and T. Kiefhaber
Copyright © 2005 WILEY-VCH Verlag GmbH & Co. KGaA, Weinheim
ISBN: 3-527-30784-2

Tab. 12.1. Yeast ribosomal proteins with homology to ribosomal proteins of *Haloarcula marismortui* localized in close proximity to the polypeptide tunnel exit [5, 6, 11]. The name of the yeast homologue is given according to the Rp nomenclature [156].

Family of ribosomal proteins	H. marismortui	E. coli	S. cerevisiae	Phenotype of the deletion in S. cerevisiae	Attachment site for ribosome-associated proteins
Ribosomal proteins close to the exit of the polypeptide tunnel					
L24	L24	L24	Rpl26a/b	Δrpl26a: viable Δrpl26b: viable [157]	
L22	L22	L22	Rpl17a/b	Δrpl17a: lethal Δrpl17b: viable [157]	
L23	L23	L23	Rpl25	Δrpl25: lethal [158]	Binding site for TF [46] and SRP [51] in *E. coli*. Binding site for eukaryotic SRP [36]
L29	L29	L29	Rpl35a/b	Δrpl35a: viable Δrpl35b: viable [70, 157]	Binding site for eukaryotic SRP [36]
L19e	L19	–	Rpl19a/b	Δrpl19a: viable Δrpl19b: viable Δrpl19aΔrpl19b: lethal [157, 159]	
L39e	L39e	–	Rpl39	Δrpl39: viable [160]	
L31e	L31e	–	Rpl31a/b	Δrpl31a: viable Δrpl31b: viable [157]	
Ribosomal proteins lining the exit tunnel					
L4e/L4p	L4	L4	Rpl4a/b	Δrpl4a: viable Δrpl4b: viable [161]	
L22	L22	L22	Rpl17a/b	Compare above	
L39e	L39e	–	Rpl39	Compare above	

events downstream of polypeptide bond formation. An example for such an active regulatory role of the ribosomal tunnel came from mechanistic studies on cotranslational translocation into the mammalian endoplasmic reticulum (ER). Specific nascent polypeptides harbored within the ribosomal tunnel have been shown to induce structural changes in the ER translocon upon ribosome binding [15, 16]. This finding suggests that a polypeptide is able to affect the process of membrane translocation from within the tunnel. Recently, a number of studies have suggested sequence-specific interactions between the exit tunnel and nascent peptides [17]. Thus, the original idea of a passive tunnel seems to be no longer valid. Significant interaction of the tunnel surface with newly synthesized polypeptides raises the question of what drives translocation of the polypeptide.

Besides the main tunnel, there might be several additional routes for a polypeptide to leave the ribosome [18]. It has been suggested that the translation state of the ribosome influences the tunnel (system) via communication between the small and large ribosomal subunits [18]. The basis of this idea is that it might be important to hold the polypeptide chain in a defined position during peptide bond synthesis. Subsequently, opening of the tunnel would allow movement of the polypeptide and provide space for the next amino acid to be added. Indeed, the diameter of the tunnel entrance of a translating ribosome has been found to be significantly larger than that of a non-translating ribosome.

The tunnel exit is surrounded by a number of ribosomal proteins that are thought to provide the binding sites for additional proteins affecting the newly synthesized polypeptide chain upon exit from the tunnel (Figure 12.1 and Table 12.1). These ribosome-associated proteins are thought to serve a variety of functions; however, their exact cellular roles are only poorly understood. One subgroup of proteins is localized so closely to the nascent polypeptide chain that they can be covalently linked by cross-linking reagents. Some belong to the classical chaperone families of Hsp70 and Hsp40 homologues [19], while others have enzymatic functions and covalently modify nascent polypeptides [20]. Still others lack homologues in other compartments of the cell, and their role in protein biogenesis is currently

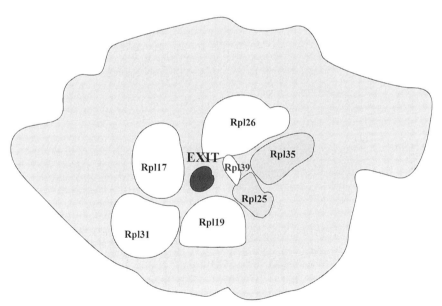

Fig. 12.1. Potential arrangement of yeast ribosomal proteins in close proximity to the polypeptide exit tunnel. The model is based on the crystal structure of the large ribosomal subunit of the archaebacterium *Haloarcula marismortui* [5, 11]. Archaebacterial ribosomes share a high degree of similarity with eukaryotic ribosomes. The mammalian homologues of Rpl25 and Rpl35 have been identified as the binding sites for SRP [36]. The eubacterial L23 (yeast homologue of Rpl25) was identified as the binding site for trigger factor and bacterial SRP [46, 51]. Proteins present in only archaebacteria and eukaryotes are lightly shaded (compare also Table 12.1).

Tab. 12.2. Major ribosome-associated proteins and complexes in yeast (in alphabetical order).

	Subunits (protein family)	Interaction with the ribosome via:	Direct binding to the nascent polypeptide chain	Selected references
NAC (Nascent polypeptide-associated complex)	α NAC = Egd2p $β_1$ NAC = Egd1p $β_2$ NAC = Btt1p	β NAC	α NAC β NAC	21, 29, 154, 162
NatA ($N^α$-acetyl-transferase A)	Nat1p Ard1p Nat5p	Nat1p	Nat1p	31, 132, 138, 139
RAC (Ribosome-associated complex)	Zuotin (Hsp40) Ssz1p (Hsp70)	Zuotin	–	63, 71, 87, 107
Ssb1p Ssb2p	– (Hsp70)	Ssb1/2p	Ssb1/2p	62, 66, 87
SRP (Signal recognition particle)	Srp72p Srp68p Sec65p Srp54p Srp21p Srp14p scR1 (RNA)	Srp54p Srp14p + part of scR1 (Alu-domain)	Srp54p	23, 24, 36, 163–165

unclear [21]. This review focuses on our current knowledge of ribosome-associated proteins affecting nascent polypeptides in the eukaryote *Saccharomyces cerevisiae* (Table 12.2) and the eubacterium *Escherichia coli*.

12.2
Signal Recognition Particle, Nascent Polypeptide–associated Complex, and Trigger Factor

12.2.1
Signal Recognition Particle

The function of signal recognition particle (SRP) is specific for a subgroup of newly synthesized proteins and is conserved throughout all kingdoms of life [22]. In eukaryotes, SRP acts on nascent polypeptide chains, which are either secreted or integrated into the membrane of the endoplasmic reticulum, and connects their translation with translocation. Eukaryotic SRP is a multi-protein complex containing an RNA molecule. In yeast, SRP is not essential; however, cells lacking functional SRP grow only very poorly and show severe translocation defects [23]. In prokaryotes, SRP-dependent export is responsible for the translocation of the majority of proteins targeted to the membrane and periplasmic space. Its structure is less complex than that of its eukaryotic counterpart. *E. coli* SRP consists of only the protein Ffh and a 4.5S RNA molecule [22, 24, 25]. In both eukaryotes and prokaryotes, SRP recognizes N-terminally localized signal sequences co-translationally. Signal sequences share no specific consensus on the amino acid level but display common structural properties: they possess the capability to adopt an α-helical conformation, plus they contain a central hydrophobic core [24, 25]. In eukaryotes, binding of SRP to nascent polypeptides leads to a translational arrest (Figure 12.2B) and is required for the productive interaction of the complex, consisting of ribosome and nascent chain (RNC), with the translocation machinery [22].

12.2.2
An Interplay between Eukaryotic SRP and Nascent Polypeptide–associated Complex?

Nascent polypeptide–associated complex (NAC) is a heterodimeric complex present in all eukaryotic cells [21]. In eubacteria, NAC homologues have not been detected thus far. Several studies based on cross-linking approaches indicate that both subunits of NAC are located in close proximity to ribosome-bound nascent polypeptide chains [26–31]. These cross-link experiments suggest that NAC does not display pronounced sequence specificity but ubiquitously binds to nascent polypeptide chains. In the mammalian system, NAC cross-links have been detected to nascent polypeptide chains as short as 17 amino acids in length [26, 32]. In yeast, the minimal length of the nascent polypeptide chain forming a cross-link to NAC has been determined to be 40–50 amino acids [31]. Based on the structure of the ribosomal tunnel (compare Section 12.1); [11], the nascent polypeptide chain is predicted to exit the ribosome at a length of approximately 40 amino acids. In yeast, NAC would thus bind just after the polypeptide emerges from the ribosome. After termination of translation, NAC does not stably interact with newly synthesized polypeptides. Binding to proteins after release from the ribosome has not been detected so far [26, 28].

It has been suggested that SRP and NAC compete for the same binding site on the ribosome [33, 34]. This competition might facilitate SRP's sampling of nascent polypeptide chains for signal sequences, which is achieved by binding and release of SRP from the ribosome during the elongation process. NAC might increase the rate at which SRP dissociates from RNCs lacking signals. However, it is presently unknown exactly how NAC influences SRP's function [33, 35]. Recently it was shown that the Srp54p subunit of SRP is positioned close to ribosomal proteins L23a (corresponding to Rpl25 in yeast) and L35 (corresponding to Rpl35a/b in yeast) at the exit site of the mammalian ribosome [36] (Figure 12.1). Upon binding of SRP to its receptor in the ER membrane, Srp54p is repositioned and loses contact with L23a [36].

12.2.3
Interplay between Bacterial SRP and Trigger Factor?

Trigger factor (TF) is a protein found in eubacteria; homologues in eukaryotes have so far not been detected. TF was initially described as a ribosome-bound protein

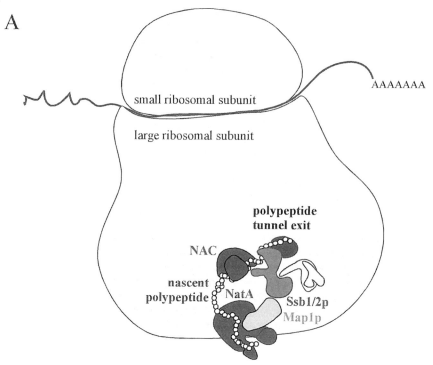

Fig. 12.2. Hypothetical arrangement of ribosome-associated proteins at the yeast polypeptide tunnel exit. (A) Ribosome-bound factors that bind to nascent polypeptide chains in a sequence-independent manner. NAC (nascent polypeptide–associated complex), the Hsp70 homologue Ssb1/2p, and the N^{α}-acetyltransferase NatA interact with nascent polypeptide chains independent of their amino acid sequence. NAC and Ssb1/2p require a minimal polypeptide length of 40–50 amino acids [31]. Binding of NAC and Ssb1/2p precedes binding of Nat1p, which is in close proximity only to nascent polypeptide chains longer than 70 amino acids [31]. Whether NAC, Ssb1/2p, and NatA simultaneously bind to one and the same ribosome (as suggested here) or independently to different molecules is currently unknown. The chaperone complex RAC and the methionine aminopeptidase Map1p bind to translating ribosomes; however, direct binding to the nascent polypeptide has so far not been observed. RAC functionally interacts with Ssb1/2p and potentially binds in its close proximity. The binding site for Map1p should enable the protein to act prior to NatA, which N^{α}-acetylates residues generated by Map1p (for details, compare text).

involved in protein export [37, 38]. Later TF was shown to possess peptidyl-prolyl isomerase activity and to localize in close proximity to a variety of nascent polypeptide chains [39–41]. TF is dispensable under normal growth conditions, and its deletion causes only a mild phenotype [42–44].

TF binds to the large ribosomal subunit [38, 45]. Recently it was shown that L23 (homologue of yeast Rpl25, Table 12.1), a ribosomal protein in close proximity to the exit tunnel, serves as the attachment site [46]. TF binds to both the ribosome and unfolded substrate proteins with an affinity in the sub- to low-micromolar range [38, 47–50]. However, the dynamics of interaction differ significantly. While

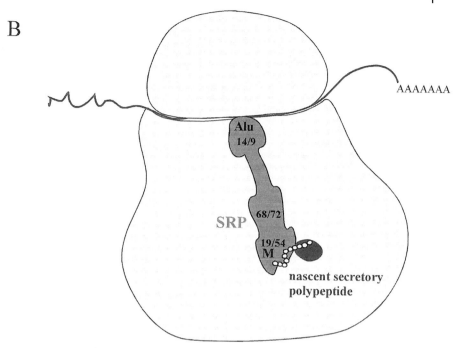

Fig. 12.2. (B) Signal recognition particle (SRP) binds to secretory nascent polypeptides. SRP binds through the M-domain of Srp54p to signal peptides of membrane or secretory proteins. This causes a transient arrest of translation, possibly through the interaction of the Alu-domain with the A-site of the peptidyl-transferase center (discussed in Ref. [155]). The numbers indicate the position of the mammalian SRP subunits on the RNA molecule (for recent reviews on structural details, see Refs. [25, 155]). Whether secretory proteins interact simultaneously with SRP plus one or more of the sequence-independent nascent-chain-binding factors shown in (A) is currently unknown (compare also Table 12.2).

the TF-ribosome complex is stable for about 30 s (2 µM concentration of TF and ribosome, 20 °C), binding to unfolded proteins is very transient (100 ms at 15 °C) [47]. As the binding sites for the ribosome and the nascent polypeptide chain reside in distinct domains, their differences in kinetic properties might enable TF to scan a growing polypeptide chain while at the same time remaining bound to the ribosome.

Like TF, SRP was localized in close proximity to ribosomal protein L23 with a cross-linking approach. In contrast to eukaryotic SRP, no cross-link to L29 (homologue of yeast Rpl35a/b, Table 12.1) was detected [51] (Table 12.1). The cross-linking pattern between L23 and SRP was unchanged in the presence of the SRP receptor FtsY; however, it was modified by the presence of a nascent signal peptide. No contacts to other ribosomal proteins were found. Given the fact that L23 exposes only a small surface at the exit of the tunnel [6], it seems unlikely that TF and SRP can bind simultaneously. Rather, they might alternate in transient binding to the ribosome until a nascent polypeptide chain emerges [52]. Upon recognition of the nascent polypeptide, binding of either SRP or TF might be stabilized. Alternatively, the ribosome may sense the nature of the nascent peptide already within its exit tunnel [15, 53–55]. In this scenario, the ribosome might actively recruit SRP or TF to polypeptides. L23 clearly plays a central role in the fate of the newly synthesized polypeptide chain.

12.2.4
Functional Redundancy: TF and the Bacterial Hsp70 Homologue DnaK

By transient association with short peptide segments, Hsp70 chaperones assist a large variety of processes, such as protein translocation, folding, and degradation (compare also below) [56]. The bacterial Hsp70 homologue DnaK is not itself directly bound to the ribosome; however, a fraction of DnaK in the cell is associated with nascent polypeptides. In *E. coli* the simultaneous deletion of both TF and DnaK is lethal and leads to the aggregation of a number of newly synthesized proteins [43, 44]. This synthetic effect suggests that TF and DnaK have overlapping functions and assist in the folding of newly synthesized polypeptides. How these structurally very different proteins cope with the newly synthesized polypeptide chain is currently only poorly understood. It is clear, however, that DnaK does not directly become associated with the ribosome in the absence of TF [57]. Recently it was shown that *Bacillus subtilis* can survive in the absence of DnaK and TF, suggesting that yet another factor can functionally complement for Dnak and TF [58].

12.3
Chaperones Bound to the Eukaryotic Ribosome: Hsp70 and Hsp40 Systems

The family of weak ATPases termed Hsp70s (70-kDa heat shock proteins) have three domains: a highly conserved N-terminal ATPase domain, a less well-conserved peptide-binding domain, and a C-terminal variable domain [56, 59, 60]. Hsp70s assist in a variety of processes in all kingdoms of life. Most importantly, they are involved in protein folding, translocation of proteins across cellular membranes, cooperation with the protein degradation machinery, and in the assembly of oligomeric complexes [56, 61]. The common principle that enables Hsp70s such broad functionality is their ability to interact with a variety of short, hydrophobic peptide segments exposed in partly folded or unfolded proteins. The interaction of Hsp70s with their substrates is regulated by an ATPase cycle controlling its affinity for peptide segments. Yeast contains a number of different Hsp70 subfamilies in the cytosol. Two subfamilies, Ssb and Ssz, are ribosome-associated [62, 63] (Figure 12.2). Published data are inconsistent regarding ribosome association of the essential Ssa subfamily (consisting of Ssa1, Ssa2, Ssa3, and Ssa4) ([64–66]; compare Section 12.3.1).

A number of co-chaperones modulate Hsp70s' function and tailor them for specific cellular processes [56, 61]. The best-studied group of Hsp70 co-chaperones are the Hsp40 homologues [56, 67–69]. Yeast contains more than 20 different Hsp40s, all containing a common domain of approximately 80 amino acids, the so-called J-domain. Two of the yeast Hsp40s have been localized on the ribosome: Sis1p and zuotin [70, 71]. The primary function of Hsp40s is to mediate ATP hydrolysis–dependent interaction of peptide segments with the peptide-binding domain of Hsp70. The J-domain is essential for this regulation and stimulates the rate of ATP hydrolysis by interacting with the ATPase domain of Hsp70s. In addition, some Hsp40s can associate with substrate proteins and transfer them to the Hsp70 partner. The mechanism of this transfer reaction is not very well understood [56, 68].

12.3.1
Sis1p and Ssa1p: an Hsp70/Hsp40 System Involved in Translation Initiation?

Sis1p is an essential Hsp40 homologue localized in the cytosol of yeast [72]. Initial experiments suggested that Sis1p was mainly bound to the small ribosomal subunit. This binding was less stable than that of other ribosome-bound chaperones and was strongly decreased at salt concentrations higher than 50 mM KCl [70]. Later it was shown that a small fraction of Sis1p is associated with polysomes even at concentrations as high as 500 mM KCl [64]. It is currently unclear how these different pools of Sis1p might relate to each other. Using a temperature-sensitive variant of Sis1p (*sis1-Δ85*), it was shown that loss of function leads to a decrease in polysomes and to the accumulation of 80S ribosomes. Consistent with these data, it was suggested that Sis1p function is required for the initiation of translation [70, 73]. Interestingly, temperature sensitivity of *sis1-Δ85* can be suppressed by deletion of either *RPL39* or *RPL35*, two proteins of the large ribosomal subunit localized close to the exit of the polypeptide tunnel [70] (Table 12.1). The unusual genetic interaction between *SIS1* and *RPL39* connects the proteins to *PAB1*, encoding poly(A)-binding protein. Deletion of *RPL39* suppresses not only temperature sensitivity of *sis1-Δ85* but also the otherwise lethal deletion of *PAB1* [74]. Pab1p binds to the poly(A)-tail of mRNA, plays a role in mRNA stabilization, and is also essential for efficient translation initiation [75, 76]. Therefore, Pab1p and Sis1p may functionally interact in translation initiation. This cooperation also seems to involve Ssa1p, a chaperone of the Hsp70 class. Ssa1p is present mainly in the soluble fraction of the cytosol and is involved in a multitude of different processes [77–81]. Like Sis1p, and also Pab1p, a small fraction of Ssa1p might be bound to polysomes in a salt-resistant manner [64]. However, in another study Ssa1p did not behave like a polysome-associated protein [65]. Two additional findings support the notion that Sis1p may act as a co-chaperone for Ssa1p. Firstly, Sis1p can stimulate the ATPase activity of Ssa1p and facilitate the refolding of denatured protein together with Ssa1p in vitro [82]. Secondly, Sis1p interacts with the C-terminal 15-amino-acid residues of Ssa1p and might be involved in the transfer of substrate proteins to Ssa1p [64, 83]. However, cross-link products indicating direct interaction of Ssa1p with nascent polypeptide chains have not been observed so far [66]. Future studies will have to clarify the role of Sis1p and Ssa1p during translation.

12.3.2
Ssb1/2p, an Hsp70 Homologue Distributed Between Ribosomes and Cytosol

The two members of the Ssb family of Hsp70 homologues, Ssb1p and Ssb2p, differ by only four amino acids and localize to the cytosol of yeast [84]. Single deletion of the respective genes does not result in a phenotype. However, cells expressing neither Ssb1p nor Ssb2p ($\Delta ssb1\Delta ssb2$) display strong growth defects. The optimal growth temperature is shifted from 30 °C to 37 °C, and $\Delta ssb1\Delta ssb2$ strains are cold-sensitive [85]. In addition, $\Delta ssb1\Delta ssb2$ strains are hypersensitive towards a number of antibiotics affecting the translational apparatus, among them, the aminoglycoside antibiotics paromomycin, hygromycin B, and G418, which increase the error rate during translation [62, 86].

Approximately 50% of Ssb1p and Ssb2p (Ssb1/2p) is associated with ribosomes [62, 63, 66]. Ssb1/2p binds both translating and non-translating ribosomes, but the salt sensitivity of the Ssb1/2p-ribosome interaction suggests two different binding modes (Table 12.3). Ssb1/2p binding to non-translating ribosomes is salt-sensitive.

Tab. 12.3. Salt sensitivity of the binding of major ribosome-bound proteins. Polysome binding was determined using sucrose-density gradients. Binding to translating ribosomes in vitro was demonstrated by cross-linking. The experimental conditions used in the different studies vary considerably. However, salt-resistant binding is always defined to be stable at concentrations of at least 500 mM salt. A question mark indicates either that salt sensitivity has not been determined or that results are inconsistent.

Eukaryotic ribosome-bound factors	Polysome (translating ribosomes)	80S ribosomes (empty ribosomes)	References
Ssb1/2p	+ Salt resistant	+ Salt sensitive	66 63, 65, 87, 88
SRP	+ Salt resistant	+ Salt sensitive	33, 166, 167
RAC (zuotin)	+ Salt sensitive	+ Salt sensitive	63, 71
NAC	+ Salt sensitive	+ Salt sensitive	29, 31, 63, 168
NatA	+ Salt sensitive	+ Salt sensitive	31
Map1p	+ ?	+ ?	137
Sis1p	+ ?	(+) ?	64, 70

Eukaryotic ribosome-bound factors	Polysome (translating ribosomes)	70S ribosomes (empty ribosomes)	References
Trigger factor	+ Salt resistant	+ Salt sensitive	41
SRP	+ ?	+ ?	51, 169

When the ribosome is involved in translation, binding of Ssb1/2p becomes salt-resistant [62, 66]. Ssb1/2p interacts directly with ribosome-associated polypeptides, suggesting that it is localized close to the exit of the ribosomal tunnel [31, 66, 87] (Figure 12.2). The ability to bind the nascent polypeptide chain is a prerequisite for the formation of the salt-resistant complex between Ssb1/2p and the ribosome [88]. Salt-resistant binding, however, is most likely not achieved exclusively via interaction of Ssb1/2p with the nascent polypeptide. Without the context of the ribosome, binding of Ssb1/2p to polypeptide chains has not been observed so far [88], suggesting that this interaction is not stable enough to resist at high salt concentrations. Conformational changes between empty and translating ribosomes, possibly induced by Ssb1/2p interaction with the nascent polypeptide chain, might contribute to the different binding modes [66]. Expression of Ssb1/2p is regulated like a core component of the ribosome [89], which is in agreement with its close relation to the translational process. Whether Ssb1/2p cycles between a ribosomal-bound and a soluble pool and whether this dual localization reflects distinct functions of this chaperone remains to be determined.

12.3.3
Function of Ssb1/2p in Degradation and Protein Folding

In eukaryotic cells, the vast majority of proteins in the cytosol and nucleus are degraded via the proteasome-ubiquitin pathway [90]. *SSB1* was isolated as a multicopy suppressor of a temperature-sensitive growth phenotype observed in a yeast strain carrying a mutation in one of the proteasome subunits. Later it was found that high levels of Sis1p lead to suppression of temperature sensitivity in the proteasome mutant and that co-overexpression of Ssb1/2p and Sis1p showed an even stronger effect [91, 92]. The findings suggest that Ssb1/2p and Sis1p together facilitate the degradation of proteins, possibly mediating the transfer of damaged proteins to the proteasome. In vitro, however, there is so far no evidence for a typical Hsp70-Hsp40-type interaction between Ssb1/2p and Sis1p [82, 93]. High levels of Ssb1/2p have also been reported to suppress a temperature-sensitive mutant displaying defects in vesicular transport [94] and a mutant with a defect in the mitochondrial import machinery [95]. Possibly, Ssb1/2p also aids the degradation of precursor proteins that fail to reach the correct destination in these mutant strains. Alternatively, high levels of Ssb1/2p might indirectly affect protein trafficking by enhancing folding or stability of factors required for the respective transport processes.

Evidence for a classical chaperone function of Ssb1/2p in protein folding comes from a study on the chaperone TRiC [96]. TRiC cooperates with GimC and assists in the folding of a subset of aggregation-sensitive polypeptides [96–98]. The function of TRiC is essential, whereas GimC function is dispensable for the life of yeast [97, 99, 100]. However, simultaneous loss of GimC and Ssb1/2p results in synthetic lethality, suggesting that GimC and Ssb1/2p are able to functionally replace each other in assisting in the folding of a subset of TRiC substrate proteins [96].

So far, no mammalian Hsp70 family member has been found to be in direct association with the ribosome. In higher eukaryotes, a number of cytosolic proteins are assisted by the cytosolic Hsp70 homologues and TRiC in co-translational folding (reviewed in Refs. [101, 102]). While TRiC might interact with the ribosome directly [103], the mammalian Hsp70 homologues studied so far seem to be attached to the ribosome–nascent chain complex via the nascent polypeptide chain only.

12.3.4
Zuotin and Ssz1p: a Stable Chaperone Complex Bound to the Yeast Ribosome

The Hsp40 homologue zuotin was found in 1998 to be a ribosome-associated protein [71]. Later it was discovered that the bulk of zuotin forms a stable complex with the Hsp70 homologue Ssz1p. The complex, termed RAC (ribosome-associated complex), is a stable dimer and is almost entirely associated with ribosomes [63]. A stable complex between an Hsp70 and an Hsp40 homologue is unusual but is not without precedence. *Thermus thermophilus* DnaK (Hsp70) and DnaJ (Hsp40) form a stable 3:3 complex that contains an additional three subunits of the 8-kDa protein Daf [104, 105]. How Ssz1p and zuotin stably interact is currently unknown. Deletion analysis of zuotin revealed that the very N-terminus up to the J-domain results in a nonfunctional protein that is still able to associate with the ribosome [71]. This N-terminus, which bears no homology to any other known J-protein, is a good candidate domain for binding to Ssz1p. Stable binding of RAC to the ribosome is mediated via the zuotin subunit: in the absence of zuotin, Ssz1p is not associated with the ribosome, while zuotin remains ribosome-associated in the absence of Ssz1p [63].

12.3.5
A Functional Chaperone Triad Consisting of Ssb1/2p, Ssz1p, and Zuotin

Deletion of either *zuo1* or *ssb1/2* results in slow growth, cold sensitivity, sensitivity towards aminoglycosides, and high osmolarity. This genetic interaction, in combination with the finding that both chaperones bind to the ribosome, strongly suggests that Ssb1/2p and zuotin have a common function [71]. Later it turned out that Ssz1p and zuotin not only form the stable RAC complex but also that deletion of *ssz1* as well as simultaneous deletion of *ssz1zuo1ssb1/2* leads to the same set of growth defects [63, 87, 106]. Overexpression of zuotin or Ssb1/2p can partially suppress the growth defect caused by deletion of *ssz1*, while overexpression of Ssz1p cannot support growth of Δ*zuo1* or Δ*ssb1/2* strains. The combination of data suggests that Ssz1p, while not obligatory for Ssb1/2p and zuotin's common function, is involved in the same cellular process [87, 106].

How RAC and Ssb1/2p work together is so far only poorly understood. In vitro studies of the purified proteins will be required to better understand their interaction on a molecular level. The first insights on how they might cooperate have come from cross-linking experiments. So far, cross-link products to neither of the RAC subunits have been detected. In contrast, Ssb1/2p can be cross-linked to a variety of nascent polypeptide chains in vitro, suggesting that it interacts with many, possibly all, newly synthesized polypeptide chains on their way through the ribosomal tunnel [66, 87]. As outlined above, high-salt treatment of ribosome–nascent chain complexes does not release Ssb1/2p but rather other factors such as RAC. Unexpectedly, on such high-salt-treated complexes, nascent polypeptide chains and Ssb1/2p no longer form cross-link products. However, cross-linking of the nascent polypeptide chain to Ssb1/2p can be restored by re-addition of purified RAC [87]. This finding suggests that RAC is required for the efficient binding of Ssb1/2p to the nascent polypeptide chain (Figure 12.3). This function of RAC depends on an intact J-domain, suggesting that zuotin acts as a J-partner for the Hsp70 Ssb1/2p while being simultaneously bound to another Hsp70, namely, Ssz1p. Interestingly, the presence of Ssz1p is required for efficient cross-linking to Ssb1/2p. Possibly the nascent polypeptide chain initially binds to Ssz1p and is subsequently

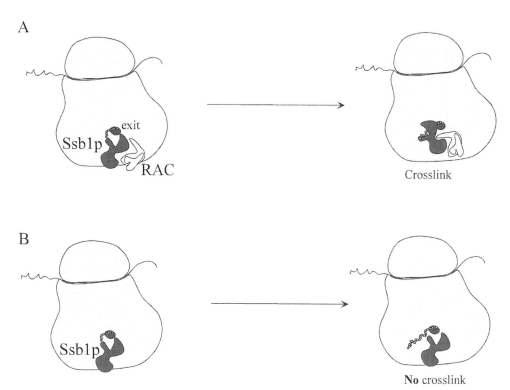

Fig. 12.3. Efficient cross-linking of Ssb1/2p depends on functional RAC. Ssb1/2p and RAC (zuotin + Ssz1p) are associated with translating ribosomes. (A) When the nascent polypeptide chain reaches a length of approximately 45 amino acids, it can be cross-linked to the Hsp70 homologue Ssb1/2p. (B) If RAC is absent from the ribosome–nascent chain complex, zuotin carries an inactivating mutation in its J-domain, or, if Ssz1p is missing, the cross-link is no longer formed, suggesting that RAC influences nascent polypeptide binding of Ssb1/2p [87] (for details, compare text).

transferred to Ssb1/2p. Fast transfer of the nascent polypeptide chain from Ssz1p to the more tightly binding Ssb1/2p might account for the lack of a direct crosslink to Ssz1p. However, sequential binding of the nascent polypeptide chain to Ssz1p and Ssb1/2p seems to be unessential for functionality of the chaperone triad. Lack of the unusually short peptide-binding domain of Ssz1p does not result in the same phenotype as observed for the deletion [106]. It is possible that Ssz1p modulates zuotin's ability to act as a partner chaperone for Ssb1/2p. Alternatively, Ssz1p might interact only with a specific, possibly small, subset of nascent polypeptide chains. Ssz1p, originally named Pdr13p, was identified as a posttranslational regulator of the transcription factor Pdr1p [107]. Whether Pdr1p or a subset of newly synthesized polypeptides directly interacts with Ssz1p and whether Ssz1p's function is influenced by zuotin's J-domain await further investigation.

Various speculations have been made on the in vivo role of ribosome-bound Ssb1/2p: folding of ribosome-bound nascent polypeptide chains, maintaining nascent polypeptide chains in an unfolded yet folding-competent state, preventing the nascent polypeptide chain from slipping backwards into the ribosome, facilitating movement of the chain through the exit tunnel or away from the ribosome, clearance of aggregated proteins from the ribosome. However, experimental evidence is scarce. Ssb1/2p cooperates with TRiC in posttranslational completion of protein folding and has a partly overlapping function with the TRiC co-chaperone GimC ([96] and compare above). It should be pointed out that this is most likely not a function of the Ssb1/2p/RAC system. While the lack of Ssb1/2p is lethal in the absence of GimC, the lack of zuotin is not [96]. This strongly suggests that soluble, and not ribosome-associated, Ssb1/2p assists in protein folding by TriC, possibly in combination with an as yet unidentified cytosolic Hsp40 homologue.

12.3.6
Effects of Ribosome-bound Chaperones on the Yeast Prion [*PSI*$^+$]

The prion-forming proteins of yeast are members of a larger class of Gln/Asn-rich proteins. Yeast prions are replicated by a protein-only mechanism resulting in a dominant, non-Mendelian mode of inheritance. The best-characterized yeast prion is [*PSI*$^+$], the polymerized form of the translation termination factor Sup35p with properties characteristic of prion proteins. The formation of [*PSI*$^+$] results in depletion of functional Sup35p and leads to a defect in translation termination that manifests itself by an increase in stop codon read-through [108–111].

Chaperones are thought to play an important role in prion biogenesis. In particular, the cytosolic chaperone Hsp104 has been implicated in yeast prion propagation [109, 112]. [*PSI*$^+$], for example, can be cured by either overproduction or inactivation of Hsp104 [113]. Hsp104, a homohexameric ATPase of the AAA protein family, functions in solubilization and refolding of aggregated proteins [114–117]. This ability provides the explanation for [*PSI*$^+$] being cured by excess Hsp104: Hsp104 overproduction leads to the release of Sup35p from [*PSI*$^+$] aggregates [118, 119]. The reason for the failure to inherit [*PSI*$^+$] in the absence of Hsp104 is most likely connected to the lack of smaller prion seeds, required for the transmission of [PSI⁺] to daughter cells. However, other models have been proposed ([120] and references therein).

In addition to Hsp104, a number of cytosolic chaperones have been implicated in yeast prion propagation, among them, the partially ribosome-associated chaperones Ssb1/2p, Ssa1p, and Sis1p. The reported effects are variable, suggesting the action of complex chaperone networks and variations between different prion strains [121, 122]. The effects are only poorly understood on a mechanistic level. High levels of Ssa have been reported to protect [PSI⁺] from being cured by an excess of Hsp104 [123]. In contrast, other studies have reported curing of [PSI⁺] due to overexpression of Ssa1p [124, 125]. Ssb1/2p increases, while lack of Ssb1/2p prevents [PSI⁺] curing by excess of Hsp104 [126, 127]. High levels of Ssa or Ssb1/2p might either prevent formation of misfolded Sup35p intermediates, which serve as the seeds for [PSI⁺], or stimulate degradation of such seeds. High levels of the Hsp40 homologues Ydj1p, Sis1p, Sti1p, and Apj1p have also been shown to influence prion propagation [124]. Sis1p might exert its function in combination with Ssa or Ssb (compare above). Sis1p is also necessary for the propagation of [RNQ⁺], another prion of yeast. As Ydj1p cannot substitute for this function of Sis1p, Hsp40s might provide specificity to the prion curing effects of Hsp70 homologues [128, 129]. Sti1p functionally interacts with Ssa and might cooperate with Ssa in prion curing [130, 131]. So far, Apj1p has been only poorly characterized, and its Hsp70 partner has not been found.

12.4
Enzymes Acting on Nascent Polypeptide Chains

Two co-translational modifications, cleavage of N-terminal methionine residues and N^{α}-terminal acetylation, are by far the most common protein modifications in eukaryotes [20, 132]. N-terminal methionine is cleaved from nascent polypeptide chains of prokaryotic and eukaryotic proteins when the residue following the initial methionine is small and uncharged. N^{α}-terminal acetylation predominantly occurs in eukaryotic cells. Most frequently modified are serine and alanine N^{α}-termini, and these residues along with methionine, glycine, and threonine account for more than 95% of the N-terminally acetylated residues [20].

12.4.1
Methionine Aminopeptidases

Methionine aminopeptidase (MetAP) catalyzes the removal of the initiator methionine from nascent polypeptides. In yeast, there are two homologous MetAPs, Map1p and Map2p. Their combined activities are essential, but their relative intracellular roles are unclear [133, 134]. The efficiency with which specific N-terminal amino acids become exposed in nascent polypeptides by the action of Map1p and Map2p correlates very well with the stability of the generated proteins according to the N-end rule [135]. As an example, methionine in front of alanine, serine, and

Tab. 12.4. Yeast N^α-acetyltransferases. NatA, NatB, NatC, and Nat4p differ in substrate specificity and subunit composition but all contain at least one catalytic subunit homologous to the GNAT family of acetyltransferases [132, 144]. For details, compare text.

	Large non-catalytic subunit	Catalytic subunit (GNAT)	Auxiliary subunit	References
NatA	Nat1p (98 kDa)	Ard1p (27 kDa)	Nat5p (20 kDa) (GNAT homologue, possibly catalytically active)	31, 138
NatB	Mdm20p (92 kDa)	Nat3p (23 kDa)		142, 143
NatC	Mak10p (84 kDa)	Mak3p (20 kDa)	Mak31p (10 kDa)	170
		Nat4p (32 kDa)		147

threonine is cleaved very efficiently, and the resulting polypeptides display high stability. On the other hand, methionine is not efficiently cleaved in front of, e.g., lysine, arginine, or glutamate. These N-terminal residues, which would destabilize the generated protein, are not commonly generated co-translationally [135, 136]. Consistent with its co-translational mode of action, it was recently demonstrated that Map1p is associated with polysomes [137] (Figure 12.2). Map2p has so far not been localized.

12.4.2
N^α-acetyltransferases

Three N^α-terminal acetyltransferase complexes termed NatA, NatB, and NatC have been identified in yeast [31, 138–143]. They differ in substrate specificity and subunit composition, but all contain at least one catalytic subunit homologous to the GNAT family of acetyltransferases [132, 144] (Table 12.4). All three NATs contain a large subunit that does not possess catalytic activity but is required for activity of the catalytic subunits in vivo. Analogous to the large, multiple tetratrico peptide repeats (TPR) containing subunits of NatA, they might anchor the catalytic subunits to the ribosome and facilitate interaction with the nascent polypeptide substrate [31, 145, 146]. Recently Nat4, a novel catalytic subunit, has been found to specifically N^α-acetylate histone H4 and H2A [147]. No potential "anchor" for Nat4 has been found so far.

NatA is the major N^α-terminal acetyltransferase in yeast, responsible for the acetylation of serine, alanine, threonine, and glycine [140, 148, 149]. NatA contains three subunits: Ard1p, the catalytically active GNAT subunit; Nat5p, a GNAT homologue for which catalytic activity has not yet been demonstrated; and the TPR domain protein Nat1p [31, 146] (Table 12.4). In accordance with its co-translational mode of action, NatA is quantitatively bound to ribosomes [31] (Figure 12.2). Binding is mediated by Nat1p, which not only anchors Ard1p and Nat5p to the ribosome but also is in close proximity to a variety of nascent polypeptide chains [31].

The lack of N^α-terminal acetylation does not cause a general defect in the stability or folding of the affected proteins [20]. In agreement with this observation, N^α-acetyltransferases are not essential for the life of yeast. However, $\Delta nat1$ and $\Delta ard1$ strains display temperature sensitivity plus some specific phenotypes, such as reduced sporulation efficiency, failure to enter G_0 under specific conditions, and defects in silencing of the silent mating-type loci [31, 139, 140, 150, 151]. The pleiotropic phenotypes are thought to reflect functional defects of various target proteins lacking proper acetylation at their N-terminus. So far there are surprisingly few examples demonstrating the biological importance of N^α-terminal protein acetylation [20]. For NatA no clear-cut example has been described. One of the few well-documented examples has been reported for Mak3p, the catalytic subunit of NatC. N^α-acetylation by Mak3p is essential for the coat protein assembly of the L-A double-stranded RNA virus in yeast [141, 152].

Ssb1/2p can partially suppress temperature sensitivity and the silencing defect caused by the lack of NatA [31]. It seems unlikely that high levels of Ssb1/2p would restore N^α-acetylation of NatA substrates. Suppression of both temperature sensitivity and the silencing defect suggests that Ssb1/2p is able to complement a general function of NatA. Whether Ssb1/2p can partially substitute for Nat1p function on the ribosome or whether it suppresses defects caused by specific proteins that lack N^α-terminal acetylation awaits further investigation.

12.5
A Complex Arrangement at the Yeast Ribosomal Tunnel Exit

Of all the nascent polypeptide chain-binding proteins discussed in this chapter, SRP is exceptional in that it interacts with nascent polypeptides bearing a signal sequence and its in vivo role precisely correlates with this specificity: SRP functions in the co-translational translocation of signal sequence bearing preproteins (compare above). The role of the other factors is less well defined; most likely they are multi-functional, as reflected by their ability to bind to a variety of nascent polypeptide chains. However, yeast Ssb1/2p and E. coli TF share an important characteristic with SRP. Consistent with a functional interaction rather than coincidental proximity, their mode of binding to empty ribosomes and ribosome-nascent polypeptide chain complexes differs. SRP, Ssb1/2p, and TF bind much more stably to ribosomes when simultaneously interacting with a nascent polypeptide chain [33, 41, 54, 66, 87] (Table 12.3). NAC and NatA also interact with the ribosome and have been found in close proximity to a variety of nascent polypeptide chains by cross-linking approaches [29, 31]. However, significant stabilization of ribosome association by nascent polypeptide chain binding has not been observed for NAC or NatA (Table 12.3). Whether proximity of NAC and NatA to nascent polypeptide chains indicates binding or rather reflects co-localization has not yet firmly been established [21, 31].

The presence of various factors at the yeast ribosomal tunnel exit implies a highly ordered arrangement and sequence of binding and release events. To date nothing is known about how proteins are arranged at the yeast exit site and to which ribosomal proteins or portions of the rRNA they bind. The first evidence for an ordered sequence of events comes from a comparison of the nascent polypeptide length required for the in vitro generation of cross-link products. Srp54p interacts with nascent secretory polypeptides of 60–70 amino acids in length [153]. Recently it has been found that longer nascent polypeptide chains also can interact with SRP [54]. Ssb1/2p and NAC can be cross-linked to nascent polypeptides – including secretory nascent polypeptides – from a minimal length of 45 amino acids, and NatA ultimately attaches to nascent polypeptides longer than 70 amino acids [31] (Figure 12.2). These results suggest that the emerging polypeptide first passes Ssb1/2p and NAC, possibly interacting with one or both. Whether NatA (and possibly other enzymes) acts on the very N-terminus while Ssb1/2p and/or NAC are still in close proximity to the nascent polypeptide is not known. Also, the interplay between SRP and the other factors on secretory nascent chains is far from being fully understood. Future studies should establish the arrangement of these factors and how they dynamically interact with the ribosome and the nascent polypeptide to perform their function.

12.6
Experimental Protocols

12.6.1
Purification of Ribosome-associated Protein Complexes from Yeast

Here we describe the protocols for the purification of wild-type NAC and RAC (Table 12.5), two abundant, ribosome-associated, heterodimeric complexes from *Saccharomyces cerevisiae*. Starting material for the purification is the pool of ribosome-associated proteins that can be released from ribosomes by high-salt treatment. The original purification of the proteins was followed using a functional assay that has been described elsewhere [63, 154]. In this protocol the first purification step is followed by Western blotting (Figure 12.4A). Later steps can be fol-

Tab. 12.5. Properties of NAC and RAC.

Complex	Oligomeric state	Gene/protein MWM, pI	References
NAC (Nascent polypeptide-associated complex)	Heterodimeric (Analytical ultracentrifugation)	*EGD2* α-NAC 18.6 kDa, pI 4.7 *EGD1* β-NAC 16.8 kDa, pI 6.3	29, 154, 162
RAC (Ribosome-associated complex)	Heterodimeric (Analytical ultracentrifugation)	*SSZ1* Ssz1p 58 kDa, pI 4.8 *ZUO1* zuotin 49 kDa, pI 8.4	63

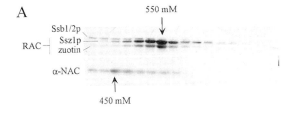

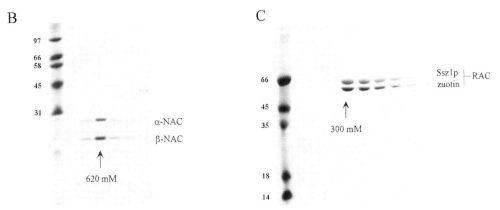

Fig. 12.4. Purification of NAC and RAC. (A) Fractions of a ribosomal salt-wash generated on ResourceQ were analyzed by Western blotting using antibodies directed against Ssb1/2p, Ssz1p, zuotin, and α-NAC. The bulk of Ssb1/2p elutes at lower K-acetate concentrations and is not shown. Note that the different intensities of the signals for Ssz1p and zuotin are caused by the different affinities of the respective antibodies. β-NAC co-elutes with α-NAC and was not analyzed in this experiment. (B) and (C) show coomassie-stained gels of the peak fractions after purification of NAC on MonoQ (B) and after purification of RAC on MonoS (C). Concentrations of K-acetate in the peak fraction of the respective proteins are given. For details, compare Appendix.

lowed by coomassie staining. Initial isolation of ribosomes, which carry the bulk of RAC and NAC, followed by high-salt release of the proteins from ribosomes provides an efficient first purification step. The purified protein complexes are able to rebind to high-salt-washed ribosomes and are functionally active [63, 87, 154].

12.6.2
Growth of Yeast and Preparation of Ribosome-associated Proteins by High-salt Treatment of Ribosomes

An initial overnight culture of wild-type yeast is used to inoculate 10 L of YPD (1% yeast extract, 2% peptone, 2% dextrose). Cells are grown in 5-L flasks, each containing 2 L of medium, to midlog phase ($OD_{600} \sim 2.0$) at 30 °C. Under these conditions wild-type yeast strains display a doubling time of approximately 1.5 h.

Cells are harvested by centrifugation for 5 min at 2700 g, resuspended in 300 mL ice-cold water, and transferred to pre-weight tubes. The cells are recollected for 5 min, the supernatant is discarded, and the mass of the cell pellet is determined.

The cell pellet is resuspended carefully in 200 mL sorbitol buffer (1.4 M sorbitol, 50 mM potassium phosphate pH 7.4, 10 mM DTT), and 2.5 mg zymolyase (20T) per gram cell pellet is added. In order to generate spheroplasts, cells are incubated in the presence of zymolyase for 30–40 min in a 30 °C water bath at 120 rpm.

Spheroplasts are collected by centrifugation for 5 min at 2700 g and are washed three times by resuspension and centrifugation with a total volume of 0.9 L sorbitol buffer (1.4 M sorbitol, 50 mM potassium phosphate pH 7.4, 5 mM DTT). After washing, spheroplasts are resuspended in 50–100 mL lysis buffer (20 mM HEPES-KOH, pH 7.4, 100 mM K-acetate, pH 7.4, 5 mM Mg-acetate, 1 mM PMSF, 1.25 $\mu g\ mL^{-1}$ leupeptin, 0.75 $\mu g\ mL^{-1}$ antipain, 0.25 $\mu g\ mL^{-1}$ chymotrypsin and elastinol, 5 $\mu g\ mL^{-1}$ pepstatin) and broken in an all-glass Dounce homogenizer (pestle-tube clearance 60 μm) with at least 20 strokes. The suspension is cleared by centrifugation for 15 min at 27 000 g.

The cleared supernatant is transferred to ultracentrifugation tubes and centrifuged for 30 min at 80 000 g. The resulting supernatant contains most of the ribosomes and the soluble cytosolic proteins and is referred to as cytosol.

In order to isolate low-salt-washed ribosomes (ribosomes and polysomes containing salt-sensitive associated proteins), the cytosol is centrifuged for 1.5–2 h at 200 000 g. The resulting supernatant (post-ribosomal supernatant) is discarded and the ribosomal pellet is resuspended in high-salt buffer (20 mM HEPES-KOH, pH 7.4, 720 mM K-acetate, 5 mM Mg-acetate, 1 mM PMSF) in a Dounce homogenizer with a Teflon piston. Resuspended ribosomes are collected by centrifugation for 1.5–2 h at 200 000 g. The resulting pellet contains high-salt-washed ribosomes, and the supernatant contains the proteins released from the ribosomes by high-salt treatment (ribosomal salt-wash) and is used for the purification of RAC and NAC.

12.6.3
Purification of NAC and RAC

The ribosomal salt-wash is diluted with six volumes of 40 mM HEPES-KOH, pH 7.4, and is subsequently loaded onto a ResourceQ (6 mL) anion-exchange column (Amersham-Pharmacia). Bound proteins are eluted with a linear 30–80 mL (100–800 mM) K-acetate gradient in 40 mM HEPES-KOH, pH 7.4. NAC is eluted at 450 mM K-acetate and RAC is eluted at 550 mM K-acetate (Figure 12.4A). Peak fractions enriched in NAC or RAC, respectively, are pooled and are used for further purification of the proteins.

The pooled fractions are diluted with 40 mM HEPES-KOH, pH 7.4, to a final concentration of 100 mM K-acetate and are loaded separately onto a MonoQ (1 mL) anion-exchange column (Amersham Pharmacia). In each case bound proteins are eluted with a linear 25–60 mL (100–1200 mM) K-acetate gradient in 40 mM HEPES-KOH pH 7.4. NAC is eluted at 620 mM and RAC at 850 M K-acetate. On MonoQ, NAC and RAC are present in two clearly separated peaks. As estimated on coomassie gels, NAC is enriched to more than 80% in the peak fractions (Figure 12.4B). NAC-containing fractions can be pooled, diluted with one volume of 40 mM HEPES-KOH pH 7.4, and concentrated in a Centricon-30 device (Millipore, Bedford, MA). Frozen in small aliquots and stored at −80 °C, the protein is stable for at least 1 year.

RAC can be further purified by cation-exchange chromatography. For this purpose RAC is diluted with six volumes of 100 mM MES, pH 6.5, and loaded onto a MonoS (1 mL) column (Amersham Pharmacia). Bound RAC is eluted with a linear 25-mL (150–600 mM) K-acetate gradient in 40 mM MES, pH 6.5. RAC is eluted at 300 mM K-acetate and is enriched to more than 95% in the peak fractions (Figure 12.4C). RAC-containing fractions are stored at −80 °C. Note that RAC at pH 6.5 cannot be concentrated to a protein concentration of higher than 1 mg mL^{-1} because of aggregation. If higher concentrations are required, the pH of the solution should be previously adjusted to pH 7.4.

References

1 GREEN, R. & NOLLER, H. F. (1997). Ribosomes and translation. *Annu. Rev. Biochem.* 66, 679–716.

2 MAGUIRE, B. A. & ZIMMERMANN, R. A. (2001). The ribosome in focus. *Cell* 104, 813–816.

3 RAMAKRISHNAN, V. (2002). Ribosome structure and the mechanism of translation. *Cell* 108, 557–572.

4 WIMBERLY, B. T., BRODERSEN, D. E., CLEMONS, W. M., JR., MORGAN-WARREN, R. J., CARTER, A. P., VONREIN, C., HARTSCH, T. & RAMAKRISHNAN, V. (2000). Structure of the 30S ribosomal subunit. *Nature* 407, 327–339.

5 BAN, N., NISSEN, P., HANSEN, J., MOORE, P. B. & STEITZ, T. A. (2000). The complete atomic structure of the large ribosomal subunit at 2.4 A resolution. *Science* 289, 905–920.

6 HARMS, J., SCHLUENZEN, F., ZARIVACH, R., BASHAN, A., GAT, S., AGMON, I., BARTELS, H., FRANCESCHI, F. & YONATH, A. (2001). High resolution structure of the large ribosomal subunit from a mesophilic eubacterium. *Cell* 107, 679–688.

7 MORGAN, D. G., MENETRET, J. F., RADERMACHER, M., NEUHOF, A., AKEY, I. V., RAPOPORT, T. A. & AKEY, C. W. (2000). A comparison of the yeast and rabbit 80S ribosome reveals the topology of the nascent chain exit tunnel, inter-subunit bridges and mammalian rRNA expansion segments. *J. Mol. Biol.* 301, 301–321.

8 AGRAWAL, R. K. & FRANK, J. (1999). Structural studies of the translational apparatus. *Curr. Opin. Struct. Biol.* 9, 215–21.

9 STARK, H. (2002). Three-dimensional electron cryomicroscopy of ribosomes. *Curr. Protein Pept. Sci.* 3, 79–91.

10 DOUDNA, J. A. & RATH, V. L. (2002). Structure and function of the eukaryotic ribosome: the next frontier. *Cell* 109, 153–156.

11 NISSEN, P., HANSEN, J., BAN, N., MOORE, P. B. & STEITZ, T. A. (2000). The structural basis of ribosome activity in peptide bond synthesis. *Science* 289, 920–930.

12 MILLIGAN, R. A. & UNWIN, P. N. (1986). Location of exit channel for nascent protein in 80S ribosome. *Nature* 319, 693–695.

13 YONATH, A., LEONARD, K. R. & WITTMANN, H. G. (1987). A tunnel in the large ribosomal subunit revealed by three-dimensional image reconstruction. *Science* 236, 813–816.

14 XU, Z., HORWICH, A. L. & SIGLER, P. B. (1997). The crystal structure of the

asymmetric GroEL-GroES-(ADP)7 chaperonin complex. *Nature* 388, 741–750.

15 LIAO, S., LIN, J., DO, H. & JOHNSON, A. E. (1997). Both lumenal and cytosolic gating of the aqueous ER translocon pore are regulated from inside the ribosome during membrane protein integration. *Cell* 90, 31–41.

16 SIEGEL, V. (1997). Recognition of a transmembrane domain: another role for the ribosome? *Cell* 90, 5–8.

17 TENSON, T. & EHRENBERG, M. (2002). Regulatory nascent peptides in the ribosomal tunnel. *Cell* 108, 591–594.

18 GABASHVILI, I. S., GREGORY, S. T., VALLE, M., GRASSUCCI, R., WORBS, M., WAHL, M. C., DAHLBERG, A. E. & FRANK, J. (2001). The polypeptide tunnel system in the ribosome and its gating in erythromycin resistance mutants of L4 and L22. *Mol. Cell* 8, 181–188.

19 CRAIG, E. A., EISENMAN, H. C. & HUNDLEY, H. A. (2003). Ribosome-tethered molecular chaperones: the first line of defense against protein misfolding? *Curr. Opin. Microbiol.* 6, 157–162.

20 POLEVODA, B. & SHERMAN, F. (2000). Nalpha-terminal acetylation of eukaryotic proteins. *J. Biol. Chem.* 275, 36479–36482.

21 ROSPERT, S., DUBAQUIÉ, Y. & GAUTSCHI, M. (2002). Nascent-polypeptide-associated complex. *Cell. Mol. Life Sci.* 59, 1632–1639.

22 KOCH, H. G., MOSER, M. & MÜLLER, M. (2003). Signal recognition particle-dependent protein targeting, universal to all kingdoms of life. *Rev. Physiol. Biochem. Pharmacol.* 146, 55–94.

23 HANN, B. C. & WALTER, P. (1991). The signal recognition particle in S. cerevisiae. *Cell* 67, 131–144.

24 KEENAN, R. J., FREYMANN, D. M., STROUD, R. M. & WALTER, P. (2001). The signal recognition particle. *Annu. Rev. Biochem.* 70, 755–775.

25 NAGAI, K., OUBRIDGE, C., KUGLSTATTER, A., MENICHELLI, E., ISEL, C. & JOVINE, L. (2003). Structure, function and evolution of the signal recognition particle. *Embo J.* 22, 3479–3485.

26 WIEDMANN, B., SAKAI, H., DAVIS, T. A. & WIEDMANN, M. (1994). A protein complex required for signal-sequence-specific sorting and translocation. *Nature* 370, 434–440.

27 RADEN, D. & GILMORE, R. (1998). Signal recognition particle-dependent targeting of ribosomes to the rough endoplasmic reticulum in the absence and presence of the nascent polypeptide-associated complex. *Mol. Biol. Cell* 9, 117–130.

28 PLATH, K. & RAPOPORT, T. A. (2000). Spontaneous release of cytosolic proteins from posttranslational substrates before their transport into the endoplasmic reticulum. *J. Cell Biol.* 151, 167–178.

29 REIMANN, B., BRADSHER, J., FRANKE, J., HARTMANN, E., WIEDMANN, M., PREHN, S. & WIEDMANN, B. (1999). Initial characterization of the nascent polypeptide-associated complex in yeast. *Yeast* 15, 397–407.

30 BEATRIX, B., SAKAI, H. & WIEDMANN, M. (2000). The alpha and beta subunit of the nascent polypeptide-associated complex have distinct functions. *J. Biol. Chem.* 275, 37838–37845.

31 GAUTSCHI, M., JUST, S., MUN, A., ROSS, S., RÜCKNAGEL, P., DUBAQUIÉ, Y., EHRENHOFER-MURRAY, A. & ROSPERT, S. (2003). The yeast N^{α}-acetyltransferase NatA is quantitatively anchored to the ribosome and interacts with nascent polypeptides. *Mol. Cell. Biol.* 23, 7403–7414.

32 WANG, S., SAKAI, H. & WIEDMANN, M. (1995). NAC covers ribosome-associated nascent chains thereby forming a protective environment for regions of nascent chains just emerging from the peptidyl transferase center. *J. Cell. Biol.* 130, 519–528.

33 POWERS, T. & WALTER, P. (1996). The nascent polypeptide-associated complex modulates interactions between the signal recognition particle and the ribosome. *Curr. Biol.* 6, 331–338.

34 MÖLLER, I., JUNG, M., BEATRIX, B., LEVY, R., KREIBICH, G., ZIMMERMANN, R., WIEDMANN, M. & LAURING, B.

(1998). A general mechanism for regulation of access to the translocon: competition for a membrane attachment site on ribosomes. *Proc. Natl. Acad. Sci. USA* 95, 13425–13430.

35 OGG, S. C. & WALTER, P. (1995). SRP samples nascent chains for the presence of signal sequences by interacting with ribosomes at a discrete step during translation elongation. *Cell* 81, 1075–1084.

36 POOL, M. R., STUMM, J., FULGA, T. A., SINNING, I. & DOBBERSTEIN, B. (2002). Distinct modes of signal recognition particle interaction with the ribosome. *Science* 297, 1345–1348.

37 CROOKE, E. & WICKNER, W. (1987). Trigger factor: a soluble protein that folds pro-OmpA into a membrane-assembly-competent form. *Proc. Natl. Acad. Sci. USA* 84, 5216–5220.

38 LILL, R., CROOKE, E., GUTHRIE, B. & WICKNER, W. (1988). The "trigger factor cycle" includes ribosomes, presecretory proteins, and the plasma membrane. *Cell* 54, 1013–1018.

39 STOLLER, G., RÜCKNAGEL, K. P., NIERHAUS, K. H., SCHMID, F. X., FISCHER, G. & RAHFELD, J. U. (1995). A ribosome-associated peptidyl-prolyl cis/trans isomerase identified as the trigger factor. *Embo J.* 14, 4939–4948.

40 VALENT, Q. A., KENDALL, D. A., HIGH, S., KUSTERS, R., OUDEGA, B. & LUIRINK, J. (1995). Early events in preprotein recognition in E. coli: interaction of SRP and trigger factor with nascent polypeptides. *Embo J.* 14, 5494–505.

41 HESTERKAMP, T., HAUSER, S., LUTCKE, H. & BUKAU, B. (1996). Escherichia coli trigger factor is a prolyl isomerase that associates with nascent polypeptide chains. *Proc. Natl. Acad. Sci. USA* 93, 4437–4441.

42 GOTHEL, S. F., SCHOLZ, C., SCHMID, F. X. & MARAHIEL, M. A. (1998). Cyclophilin and trigger factor from Bacillus subtilis catalyze in vitro protein folding and are necessary for viability under starvation conditions. *Biochemistry* 37, 13392–13399.

43 TETER, S. A., HOURY, W. A., ANG, D., TRADLER, T., ROCKABRAND, D., FISCHER, G., BLUM, P., GEORGOPOULOS, C. & HARTL, F. U. (1999). Polypeptide flux through bacterial Hsp70: DnaK cooperates with trigger factor in chaperoning nascent chains. *Cell* 97, 755–765.

44 DEUERLING, E., SCHULZE-SPECKING, A., TOMOYASU, T., MOGK, A. & BUKAU, B. (1999). Trigger factor and DnaK cooperate in folding of newly synthesized proteins. *Nature* 400, 693–696.

45 KRISTENSEN, O. & GAJHEDE, M. (2003). Chaperone Binding at the Ribosomal Exit Tunnel. *Structure* 11, 1547–1556.

46 KRAMER, G., RAUCH, T., RIST, W., VORDERWULBECKE, S., PATZELT, H., SCHULZE-SPECKING, A., BAN, N., DEUERLING, E. & BUKAU, B. (2002). L23 protein functions as a chaperone docking site on the ribosome. *Nature* 419, 171–174.

47 MAIER, R., ECKERT, B., SCHOLZ, C., LILIE, H. & SCHMID, F. X. (2003). Interaction of trigger factor with the ribosome. *J. Mol. Biol.* 326, 585–592.

48 SCHOLZ, C., STOLLER, G., ZARNT, T., FISCHER, G. & SCHMID, F. X. (1997). Cooperation of enzymatic and chaperone functions of trigger factor in the catalysis of protein folding. *Embo J.* 16, 54–58.

49 MAIER, R., SCHOLZ, C. & SCHMID, F. X. (2001). Dynamic association of trigger factor with protein substrates. *J. Mol. Biol.* 314, 1181–1190.

50 SCHOLZ, C., MÜCKE, M., RAPE, M., PECHT, A., PAHL, A., BANG, H. & SCHMID, F. X. (1998). Recognition of protein substrates by the prolyl isomerase trigger factor is independent of proline residues. *J. Mol. Biol.* 277, 723–732.

51 GU, S. Q., PESKE, F., WIEDEN, H. J., RODNINA, M. V. & WINTERMEYER, W. (2003). The signal recognition particle binds to protein L23 at the peptide exit of the Escherichia coli ribosome. *Rna* 9, 566–573.

52 ULLERS, R. S., HOUBEN, E. N., RAINE, A., TEN HAGEN-JONGMAN, C. M., EHRENBERG, M., BRUNNER, J., OUDEGA, B., HARMS, N. & LUIRINK, J.

(2003). Interplay of signal recognition particle and trigger factor at L23 near the nascent chain exit site on the Escherichia coli ribosome. *J. Cell Biol.* 161, 679–684.

53 NAKATOGAWA, H. & ITO, K. (2002). The ribosomal exit tunnel functions as a discriminating gate. *Cell* 108, 629–636.

54 FLANAGAN, J. J., CHEN, J. C., MIAO, Y., SHAO, Y., LIN, J., BOCK, P. E. & JOHNSON, A. E. (2003). Signal recognition particle binds to ribosome-bound signal sequences with fluorescence-detected subnanomolar affinity that does not diminish as the nascent chain lengthens. *J. Biol. Chem.* 278, 18628–18637.

55 GONG, F. & YANOFSKY, C. (2002). Instruction of translating ribosome by nascent peptide. *Science* 297, 1864–1867.

56 MAYER, M. P., BREHMER, D., GÄSSLER, C. S. & BUKAU, B. (2002). Hsp70 chaperone machines. *Adv. Protein Chem.* 59, 1–44.

57 KRAMER, G., RAMACHANDIRAN, V., HOROWITZ, P. M. & HARDESTY, B. (2002). The molecular chaperone DnaK is not recruited to translating ribosomes that lack trigger factor. *Arch. Biochem. Biophys.* 403, 63–70.

58 REYES, D. Y. & YOSHIKAWA, H. (2002). DnaK chaperone machine and trigger factor are only partially required for normal growth of Bacillus subtilis. *Biosci. Biotechnol. Biochem.* 66, 1583–1586.

59 FLAHERTY, K. M., DELUCA-FLAHERTY, C. & MCKAY, D. B. (1990). Three-dimensional structure of the ATPase fragment of a 70K heat-shock cognate protein. *Nature* 346, 623–628.

60 WANG, T.-F., CHANG, J.-H. & WANG, C. (1993). Identification of the peptide binding domain of hsc70. 18-kilodalton fragment located immediately after ATPase domain is sufficient for high affinity binding. *J. Biol. Chem.* 268, 26049–26051.

61 WALTER, S. & BUCHNER, J. (2002). Molecular chaperones–cellular machines for protein folding. *Angew. Chem. Int. Ed. Engl.* 41, 1098–1113.

62 NELSON, R. J., ZIEGELHOFFER, T., NICOLET, C., WERNER-WASHBURNE, M. & CRAIG, E. A. (1992). The translation machinery and 70 kd heat shock protein cooperate in protein synthesis. *Cell* 71, 97–105.

63 GAUTSCHI, M., LILIE, H., FÜNFSCHILLING, U., MUN, A., ROSS, S., LITHGOW, T., RÜCKNAGEL, P. & ROSPERT, S. (2001). RAC, a stable ribosome-associated complex in yeast formed by the DnaK-DnaJ homologs Ssz1p and zuotin. *Proc. Natl. Acad. Sci. USA* 98, 3762–3767.

64 HORTON, L. E., JAMES, P., CRAIG, E. A. & HENSOLD, J. O. (2001). The yeast hsp70 homologue Ssa is required for translation and interacts with Sis1 and Pab1 on translating ribosomes. *J. Biol. Chem.* 276, 14426–14433.

65 JAMES, P., PFUND, C. & CRAIG, E. A. (1997). Functional specificity among Hsp70 molecular chaperones. *Science* 275, 387–389.

66 PFUND, C., LOPEZ-HOYO, N., ZIEGELHOFFER, T., SCHILKE, B. A., LOPEZ-BUESA, P., WALTER, W. A., WIEDMANN, M. & CRAIG, E. A. (1998). The molecular chaperone Ssb from Saccharomyces cerevisiae is a component of the ribosome-nascent chain complex. *Embo J.* 17, 3981–3989.

67 RYAN, M. T. & PFANNER, N. (2001). Hsp70 proteins in protein translocation. *Adv. Protein Chem.* 59, 223–242.

68 BUKAU, B. & HORWICH, A. L. (1998). The Hsp70 and Hsp60 chaperone machines. *Cell* 92, 351–366.

69 KELLEY, W. L. (1998). The J-domain family and the recruitment of chaperone power. *Trends Biochem. Sci.* 23, 222–227.

70 ZHONG, T. & ARNDT, K. T. (1993). The yeast SIS1 protein, a DnaJ homolog, is required for the initiation of translation. *Cell* 73, 1175–1186.

71 YAN, W., SCHILKE, B., PFUND, C., WALTER, W., KIM, S. & CRAIG, E. A. (1998). Zuotin, a ribosome-associated DnaJ molecular chaperone. *Embo J.* 17, 4809–4817.

72 LUKE, M. M., SUTTON, A. & ARNDT, K. T. (1991). Characterization of SIS1, a Saccharomyces cerevisiae homologue

of bacterial dnaJ proteins. *J. Cell Biol.* 114, 623–638.
73 SALMON, D., MONTERO-LOMELI, M. & GOLDENBERG, S. (2001). A DnaJ-like protein homologous to the yeast co-chaperone Sis1 (TcJ6p) is involved in initiation of translation in Trypanosoma cruzi. *J. Biol. Chem.* 276, 43970–43979.
74 SACHS, A. B. & DAVIS, R. W. (1989). The poly(A) binding protein is required for poly(A) shortening and 60S ribosomal subunit-dependent translation initiation. *Cell* 58, 857–867.
75 WILUSZ, C. J., WORMINGTON, M. & PELTZ, S. W. (2001). The cap-to-tail guide to mRNA turnover. *Nat. Rev. Mol. Cell. Biol.* 2, 237–246.
76 PREISS, T. & HENTZE, M. W. (1999). From factors to mechanisms: translation and translational control in eukaryotes. *Curr. Opin. Genet. Dev.* 9, 515–521.
77 BUSH, G. L. & MEYER, D. I. (1996). The refolding activity of the yeast heat shock proteins Ssa1 and Ssa2 defines their role in protein translocation. *J. Cell Biol.* 135, 1229–1237.
78 DUTTAGUPTA, R., VASUDEVAN, S., WILUSZ, C. J. & PELTZ, S. W. (2003). A yeast homologue of Hsp70, Ssa1p, regulates turnover of the MFA2 transcript through its AU-rich 3′ untranslated region. *Mol. Cell. Biol.* 23, 2623–2632.
79 GEYMONAT, M., WANG, L., GARREAU, H. & JACQUET, M. (1998). Ssa1p chaperone interacts with the guanine nucleotide exchange factor of ras Cdc25p and controls the cAMP pathway in Saccharomyces cerevisiae. *Mol. Microbiol.* 30, 855–864.
80 KIM, S., SCHILKE, B., CRAIG, E. A. & HORWICH, A. L. (1998). Folding in vivo of a newly translated yeast cytosolic enzyme is mediated by the SSA class of cytosolic yeast Hsp70 proteins. *Proc. Natl. Acad. Sci. USA* 95, 12860–12865.
81 LIU, Y., LIANG, S. & TARTAKOFF, A. M. (1996). Heat shock disassembles the nucleolus and inhibits nuclear protein import and poly(A)+ RNA export. *Embo J.* 15, 6750–6757.
82 LU, Z. & CYR, D. M. (1998). Protein folding activity of Hsp70 is modified differentially by the hsp40 co-chaperones Sis1 and Ydj1. *J. Biol. Chem.* 273, 27824–27830.
83 QIAN, X., HOU, W., ZHENGANG, L. & SHA, B. (2002). Direct interactions between molecular chaperones heat-shock protein (Hsp)70 and Hsp40: yeast Hsp70 Ssa1 binds the extreme C-terminal region of yeast Hsp40 Sis1. *Biochem. J.* 361, 27–34.
84 SHULGA, N., JAMES, P., CRAIG, E. A. & GOLDFARB, D. S. (1999). A nuclear export signal prevents Saccharomyces cerevisiae Hsp70 Ssb1p from stimulating nuclear localization signal-directed nuclear transport. *J. Biol. Chem.* 274, 16501–16507.
85 CRAIG, E. A. & JACOBSEN, K. (1985). Mutations in cognate genes of Saccharomyces cerevisiae hsp70 result in reduced growth rates at low temperatures. *Mol. Cell. Biol.* 5, 3517–3524.
86 OGLE, J. M., CARTER, A. P. & RAMAKRISHNAN, V. (2003). Insights into the decoding mechanism from recent ribosome structures. *Trends Biochem. Sci.* 28, 259–266.
87 GAUTSCHI, M., MUN, A., ROSS, S. & ROSPERT, S. (2002). A functional chaperone triad on the yeast ribosome. *Proc. Natl. Acad. Sci. USA* 99, 4209–4214.
88 PFUND, C., HUANG, P., LOPEZ-HOYO, N. & CRAIG, E. A. (2001). Divergent functional properties of the ribosome-associated molecular chaperone Ssb compared with other Hsp70s. *Mol. Biol. Cell* 12, 3773–3782.
89 LOPEZ, N., HALLADAY, J., WALTER, W. & CRAIG, E. A. (1999). SSB, encoding a ribosome-associated chaperone, is coordinately regulated with ribosomal protein genes. *J. Bacteriol.* 181, 3136–3143.
90 HOCHSTRASSER, M., JOHNSON, P. R., ARENDT, C. S., AMERIK, A., SWAMINATHAN, S., SWANSON, R., LI, S. J., LANEY, J., PALS-RYLAARSDAM, R., NOWAK, J. & CONNERLY, P. L. (1999). The Saccharomyces cerevisiae ubiquitin-proteasome system. *Philos.*

91 OHBA, M. (1997). Modulation of intracellular protein degradation by SSB1-SIS1 chaperon system in yeast S. cerevisiae. *FEBS Lett.* 409, 307–311.

92 OHBA, M. (1994). A 70-kDa heat shock cognate protein suppresses the defects caused by a proteasome mutation in Saccharomyces cerevisiae. *FEBS Lett.* 351, 263–266.

93 LOPEZ-BUESA, P., PFUND, C. & CRAIG, E. A. (1998). The biochemical properties of the ATPase activity of a 70-kDa heat shock protein (Hsp70) are governed by the C-terminal domains. *Proc. Natl. Acad. Sci. USA* 95, 15253–15258.

94 KOSODO, Y., IMAI, K., HIRATA, A., NODA, Y., TAKATSUKI, A., ADACHI, H. & YODA, K. (2001). Multicopy suppressors of the sly1 temperature-sensitive mutation in the ER-Golgi vesicular transport in Saccharomyces cerevisiae. *Yeast* 18, 1003–1014.

95 DUNN, C. D. & JENSEN, R. E. (2003). Suppression of a defect in mitochondrial protein import identifies cytosolic proteins required for viability of yeast cells lacking mitochondrial DNA. *Genetics* 165, 35–45.

96 SIEGERS, K., BOLTER, B., SCHWARZ, J. P., BOTTCHER, U. M., GUHA, S. & HARTL, F. U. (2003). TRiC/CCT cooperates with different upstream chaperones in the folding of distinct protein classes. *Embo J.* 22, 5230–5240.

97 SIEGERS, K., WALDMANN, T., LEROUX, M. R., GREIN, K., SHEVCHENKO, A., SCHIEBEL, E. & HARTL, F. U. (1999). Compartmentation of protein folding in vivo: sequestration of non-native polypeptide by the chaperonin-GimC system. *EMBO Journal* 18, 75–84.

98 HANSEN, W. J., COWAN, N. J. & WELCH, W. J. (1999). Prefoldin-nascent chain complexes in the folding of cytoskeletal proteins. *J. Cell Biol.* 145, 265–277.

99 STOLDT, V., RADEMACHER, F., KEHREN, V., ERNST, J. F., PEARCE, D. A. & SHERMAN, F. (1996). Review: the Cct eukaryotic chaperonin subunits of Saccharomyces cerevisiae and other yeasts. *Yeast* 12, 523–529.

100 GEISSLER, S., SIEGERS, K. & SCHIEBEL, E. (1998). A novel protein complex promoting formation of functional alpha- and gamma-tubulin. *Embo J.* 17, 952–966.

101 HARTL, F. U. & HAYER-HARTL, M. (2002). Molecular chaperones in the cytosol: from nascent chain to folded protein. *Science* 295, 1852–1858.

102 FRYDMAN, J. (2001). Folding of newly translated proteins in vivo: The Role of molecular chaperones. *Annu. Rev. Biochem.* 70, 603–647.

103 MCCALLUM, C. D., DO, H., JOHNSON, A. E. & FRYDMAN, J. (2000). The interaction of the chaperonin tailless complex polypeptide 1 (TCP1) ring complex (TRiC) with ribosome-bound nascent chains examined using photo-cross-linking. *J. Cell Biol.* 149, 591–602.

104 MOTOHASHI, K., YOHDA, M., ENDO, I. & YOSHIDA, M. (1996). A novel factor required for the assembly of the DnaK and DnaJ chaperones of Thermus thermophilus. *J. Biol. Chem.* 271, 17343–17348.

105 SCHLEE, S. & REINSTEIN, J. (2002). The DnaK/ClpB chaperone system from Thermus thermophilus. *Cell. Mol. Life Sci.* 59, 1598–1606.

106 HUNDLEY, H., EISENMAN, H., WALTER, W., EVANS, T., HOTOKEZAKA, Y., WIEDMANN, M. & CRAIG, E. (2002). The in vivo function of the ribosome-associated Hsp70, Ssz1, does not require its putative peptide-binding domain. *Proc. Natl. Acad. Sci. USA* 99, 4203–4208.

107 HALLSTROM, T. C., KATZMANN, D. J., TORRES, R. J., SHARP, W. J. & MOYE-ROWLEY, W. S. (1998). Regulation of transcription factor Pdr1p function by an Hsp70 protein in Saccharomyces cerevisiae. *Mol. Cell. Biol.* 18, 1147–1155.

108 LINDQUIST, S., KROBITSCH, S., LI, L. & SONDHEIMER, N. (2001). Investigating protein conformation-based inheritance and disease in yeast. *Philos. Trans. R. Soc. Lond. B Biol. Sci.* 356, 169–76.

109 TUITE, M. F. & LINDQUIST, S. L. (1996). Maintenance and inheritance of yeast prions. *Trends Genet.* 12, 467–471.

110 OSHEROVICH, L. Z. & WEISSMAN, J. S. (2002). The utility of prions. *Dev. Cell* 2, 143–151.

111 TUITE, M. F. (2000). Yeast prions and their prion-forming domain. *Cell* 100, 289–92.

112 WICKNER, R. B., TAYLOR, K. L., EDSKES, H. K., MADDELEIN, M. L., MORIYAMA, H. & ROBERTS, B. T. (2000). Prions of yeast as heritable amyloidoses. *J. Struct. Biol.* 130, 310–322.

113 CHERNOFF, Y. O., LINDQUIST, S. L., ONO, B., INGE-VECHTOMOV, S. G. & LIEBMAN, S. W. (1995). Role of the chaperone protein Hsp104 in propagation of the yeast prion-like factor [psi+]. *Science* 268, 880–884.

114 SCHIRMER, E. C., GLOVER, J. R., SINGER, M. A. & LINDQUIST, S. (1996). HSP100/Clp proteins: a common mechanism explains diverse functions. *Trends Biochem. Sci.* 21, 289–296.

115 PARSELL, D. A., SANCHEZ, Y., STITZEL, J. D. & LINDQUIST, S. (1991). Hsp104 is a highly conserved protein with two essential nucleotide-binding sites. *Nature* 353, 270–3.

116 SANCHEZ, Y., TAULIEN, J., BORKOVICH, K. A. & LINDQUIST, S. (1992). Hsp104 is required for tolerance to many forms of stress. *Embo J.* 11, 2357–2364.

117 GLOVER, J. R. & LINDQUIST, S. (1998). Hsp104, Hsp70, and Hsp40: a novel chaperone system that rescues previously aggregated proteins. *Cell* 94, 73–82.

118 PATINO, M. M., LIU, J. J., GLOVER, J. R. & LINDQUIST, S. (1996). Support for the prion hypothesis for inheritance of a phenotypic trait in yeast. *Science* 273, 622–6.

119 PAUSHKIN, S. V., KUSHNIROV, V. V., SMIRNOV, V. N. & TER-AVANESYAN, M. D. (1996). Propagation of the yeast prion-like [psi+] determinant is mediated by oligomerization of the SUP35-encoded polypeptide chain release factor. *Embo J.* 15, 3127–3134.

120 WEGRZYN, R. D., BAPAT, K., NEWNAM, G. P., ZINK, A. D. & CHERNOFF, Y. O. (2001). Mechanism of prion loss after Hsp104 inactivation in yeast. *Mol. Cell. Biol.* 21, 4656–4669.

121 UPTAIN, S. M., SAWICKI, G. J., CAUGHEY, B., LINDQUIST, S., CHIEN, P. & WEISSMAN, J. S. (2001). Strains of [PSI(+)] are distinguished by their efficiencies of prion-mediated conformational conversion conformational diversity in a yeast prion dictates its seeding specificity. *Embo J.* 20, 6236–6245.

122 TRUE, H. L. & LINDQUIST, S. L. (2000). A yeast prion provides a mechanism for genetic variation and phenotypic diversity. *Nature* 407, 477–483.

123 NEWNAM, G. P., WEGRZYN, R. D., LINDQUIST, S. L. & CHERNOFF, Y. O. (1999). Antagonistic interactions between yeast chaperones Hsp104 and Hsp70 in prion curing. *Mol. Cell. Biol.* 19, 1325–1333.

124 KRYNDUSHKIN, D. S., SMIRNOV, V. N., TER-AVANESYAN, M. D. & KUSHNIROV, V. V. (2002). Increased expression of Hsp40 chaperones, transcriptional factors, and ribosomal protein Rpp0 can cure yeast prions. *J. Biol. Chem.* 277, 23702–23708.

125 KUSHNIROV, V. V., KRYNDUSHKIN, D. S., BOGUTA, M., SMIRNOV, V. N. & TER-AVANESYAN, M. D. (2000). Chaperones that cure yeast artificial [PSI+] and their prion-specific effects. *Curr. Biol.* 10, 1443–1446.

126 CHERNOFF, Y. O., NEWNAM, G. P., KUMAR, J., ALLEN, K. & ZINK, A. D. (1999). Evidence for a protein mutator in yeast: role of the Hsp70-related chaperone ssb in formation, stability, and toxicity of the [PSI] prion. *Mol. Cell. Biol.* 19, 8103–8112.

127 CHACINSKA, A., SZCZESNIAK, B., KOCHNEVA-PERVUKHOVA, N. V., KUSHNIROV, V. V., TER-AVANESYAN, M. D. & BOGUTA, M. (2001). Ssb1 chaperone is a [PSI+] prion-curing factor. *Curr. Genet.* 39, 62–7.

128 SONDHEIMER, N., LOPEZ, N., CRAIG, E. A. & LINDQUIST, S. (2001). The role of Sis1 in the maintenance of the [RNQ+] prion. *Embo J.* 20, 2435–42.

129 LOPEZ, N., ARON, R. & CRAIG, E. A. (2003). Specificity of Class II Hsp40 Sis1 in Maintenance of Yeast Prion [RNQ(+)]. Mol. Biol. Cell 14, 1172–1181.

130 CHANG, H. C., NATHAN, D. F. & LINDQUIST, S. (1997). In vivo analysis of the Hsp90 cochaperone Sti1 (p60). Mol. Cell. Biol. 17, 318–3125.

131 WEGELE, H., HASLBECK, M., REINSTEIN, J. & BUCHNER, J. (2003). Sti1 is a novel activator of the Ssa proteins. J. Biol. Chem. 278, 25970–25976.

132 POLEVODA, B. & SHERMAN, F. (2003). N-terminal acetyltransferases and sequence requirements for N-terminal acetylation of eukaryotic proteins. J. Mol. Biol. 325, 595–622.

133 LI, X. & CHANG, Y. H. (1995). Amino-terminal protein processing in Saccharomyces cerevisiae is an essential function that requires two distinct methionine aminopeptidases. Proc. Natl. Acad. Sci. USA 92, 12357–12361.

134 DUMMITT, B., MICKA, W. S. & CHANG, Y. H. (2003). N-terminal methionine removal and methionine metabolism in Saccharomyces cerevisiae. J. Cell Biochem. 89, 964–974.

135 GONDA, D. K., BACHMAIR, A., WUNNING, I., TOBIAS, J. W., LANE, W. S. & VARSHAVSKY, A. (1989). Universality and structure of the N-end rule. J. Biol. Chem. 264, 16700–16712.

136 MOERSCHELL, R. P., HOSOKAWA, Y., TSUNASAWA, S. & SHERMAN, F. (1990). The specificities of yeast methionine aminopeptidase and acetylation of amino-terminal methionine in vivo. Processing of altered iso-1-cytochromes c created by oligonucleotide transformation. J. Biol. Chem. 265, 19638–19643.

137 VETRO, J. A. & CHANG, Y. H. (2002). Yeast methionine aminopeptidase type 1 is ribosome-associated and requires its N-terminal zinc finger domain for normal function in vivo. J. Cell Biochem. 85, 678–688.

138 PARK, E. C. & SZOSTAK, J. W. (1992). ARD1 and NAT1 proteins form a complex that has N-terminal acetyltransferase activity. Embo J. 11, 2087–2093.

139 MULLEN, J. R., KAYNE, P. S., MOERSCHELL, R. P., TSUNASAWA, S., GRIBSKOV, M., COLAVITO-SHEPANSKI, M., GRUNSTEIN, M., SHERMAN, F. & STERNGLANZ, R. (1989). Identification and characterization of genes and mutants for an N-terminal acetyltransferase from yeast. Embo J. 8, 2067–2075.

140 POLEVODA, B., NORBECK, J., TAKAKURA, H., BLOMBERG, A. & SHERMAN, F. (1999). Identification and specificities of N-terminal acetyltransferases from Saccharomyces cerevisiae. Embo J. 18, 6155–6168.

141 TERCERO, J. C. & WICKNER, R. B. (1992). MAK3 encodes an N-acetyltransferase whose modification of the L-A gag NH2 terminus is necessary for virus particle assembly. J. Biol. Chem. 267, 20277–20281.

142 POLEVODA, B., CARDILLO, T. S., DOYLE, T. C., BEDI, G. S. & SHERMAN, F. (2003). Nat3p and Mdm20p are required for function of yeast NatB Nalpha-terminal acetyltransferase and of actin and tropomyosin. J. Biol. Chem. 3, 3.

143 SINGER, J. M. & SHAW, J. M. (2003). Mdm20 protein functions with Nat3 protein to acetylate Tpm1 protein and regulate tropomyosin-actin interactions in budding yeast. Proc. Natl. Acad. Sci. USA 100, 7644–9.

144 NEUWALD, A. F. & LANDSMAN, D. (1997). GCN5-related histone N-acetyltransferases belong to a diverse superfamily that includes the yeast SPT10 protein. Trends Biochem. Sci. 22, 154–155.

145 BLATCH, G. L. & LASSLE, M. (1999). The tetratricopeptide repeat: a structural motif mediating protein-protein interactions. Bioessays 21, 932–939.

146 POLEVODA, B. & SHERMAN, F. (2003). Composition and function of the eukaryotic N-terminal acetyltransferase subunits. Biochem. Biophys. Res. Commun. 308, 1–11.

147 SONG, O. K., WANG, X., WATERBORG,

J. H. & Sternglanz, R. (2003). An Nalpha-acetyltransferase responsible for acetylation of the N-terminal residues of histones H4 and H2A. *J. Biol. Chem.* 278, 38109–38112.

148 Takakura, H., Tsunasawa, S., Miyagi, M. & Warner, J. R. (1992). NH2-terminal acetylation of ribosomal proteins of Saccharomyces cerevisiae. *J. Biol. Chem.* 267, 5442–5445.

149 Arnold, R. J., Polevoda, B., Reilly, J. P. & Sherman, F. (1999). The action of N-terminal acetyltransferases on yeast ribosomal proteins. *J. Biol. Chem.* 274, 37035–37040.

150 Whiteway, M., Freedman, R., Van Arsdell, S., Szostak, J. W. & Thorner, J. (1987). The yeast ARD1 gene product is required for repression of cryptic mating-type information at the HML locus. *Mol. Cell. Biol.* 7, 3713–3722.

151 Whiteway, M. & Szostak, J. W. (1985). The ARD1 gene of yeast functions in the switch between the mitotic cell cycle and alternative developmental pathways. *Cell* 43, 483–492.

152 Tercero, J. C., Dinman, J. D. & Wickner, R. B. (1993). Yeast MAK3 N-acetyltransferase recognizes the N-terminal four amino acids of the major coat protein (gag) of the L-A double-stranded RNA virus. *J. Bacteriol.* 175, 3192–3194.

153 Kurzchalia, T. V., Wiedmann, M., Girshovich, A. S., Bochkareva, E. S., Bielka, H. & Rapoport, T. A. (1986). The signal sequence of nascent preprolactin interacts with the 54K polypeptide of the signal recognition particle. *Nature* 320, 634–636.

154 Fünfschilling, U. & Rospert, S. (1999). Nascent Polypeptide-associated Complex Stimulates Protein Import into Yeast Mitochondria. *Mol. Biol. Cell.* 10, 3289–3299.

155 Wild, K., Weichenrieder, O., Strub, K., Sinning, I. & Cusack, S. (2002). Towards the structure of the mammalian signal recognition particle. *Curr. Opin. Struct. Biol.* 12, 72–81.

156 Planta, R. J. & Mager, W. H. (1998). The list of cytoplasmic ribosomal proteins of Saccharomyces cerevisiae. *Yeast* 14, 471–477.

157 Winzeler, E. A., Shoemaker, D. D., Astromoff, A., Liang, H., Anderson, K., Andre, B., Bangham, R., Benito, R., Boeke, J. D., Bussey, H., Chu, A. M., Connelly, C., Davis, K., Dietrich, F., Dow, S. W., El Bakkoury, M., Foury, F., Friend, S. H., Gentalen, E., Giaever, G., Hegemann, J. H., Jones, T., Laub, M., Liao, H., Davis, R. W. & et al. (1999). Functional characterization of the S. cerevisiae genome by gene deletion and parallel analysis. *Science* 285, 901–906.

158 Rutgers, C. A., Schaap, P. J., van't Riet, J., Woldringh, C. L. & Raue, H. A. (1990). In vivo and in vitro analysis of structure-function relationships in ribosomal protein L25 from Saccharomyces cerevisiae. *Biochim. Biophys. Acta* 1050, 74–79.

159 Song, J. M., Cheung, E. & Rabinowitz, J. C. (1996). Organization and characterization of the two yeast ribosomal protein YL19 genes. *Curr. Genet.* 30, 273–278.

160 Sachs, A. B. & Davis, R. W. (1990). Translation initiation and ribosomal biogenesis: involvement of a putative rRNA helicase and RPL46. *Science* 247, 1077–1079.

161 Lucioli, A., Presutti, C., Ciafre, S., Caffarelli, E., Fragapane, P. & Bozzoni, I. (1988). Gene dosage alteration of L2 ribosomal protein genes in Saccharomyces cerevisiae: effects on ribosome synthesis. *Mol. Cell. Biol.* 8, 4792–4798.

162 George, R., Beddoe, T., Landl, K. & Lithgow, T. (1998). The yeast nascent polypeptide-associated complex initiates protein targeting to mitochondria in vivo. *Proc. Natl. Acad. Sci. USA* 95, 2296–2301.

163 Brown, J. D., Hann, B. C., Medzihradszky, K. F., Niwa, M., Burlingame, A. L. & Walter, P. (1994). Subunits of the Saccharomyces cerevisiae signal recognition particle required for its functional expression. *Embo J.* 13, 4390–4400.

164 Mason, N., Ciufo, L. F. & Brown,

J. D. (2000). Elongation arrest is a physiologically important function of signal recognition particle. *Embo J.* 19, 4164–4174.

165 WILLER, M., JERMY, A. J., STEEL, G. J., GARSIDE, H. J., CARTER, S. & STIRLING, C. J. (2003). An in vitro assay using overexpressed yeast SRP demonstrates that cotranslational translocation is dependent upon the J-domain of Sec63p. *Biochemistry* 42, 7171–7177.

166 ZOPF, D., BERNSTEIN, H. D., JOHNSON, A. E. & WALTER, P. (1990). The methionine-rich domain of the 54 kd protein subunit of the signal recognition particle contains an RNA binding site and can be crosslinked to a signal sequence. *Embo J.* 9, 4511–4517.

167 HIGH, S. & DOBBERSTEIN, B. (1991). The signal sequence interacts with the methionine-rich domain of the 54-kD protein of signal recognition particle. *J. Cell Biol.* 113, 229–233.

168 GEORGE, R., WALSH, P., BEDDOE, T. & LITHGOW, T. (2002). The nascent polypeptide-associated complex (NAC) promotes interaction of ribosomes with the mitochondrial surface in vivo. *FEBS Lett.* 516, 213–216.

169 VALENT, Q. A., DE GIER, J. W., VON HEIJNE, G., KENDALL, D. A., TEN HAGEN-JONGMAN, C. M., OUDEGA, B. & LUIRINK, J. (1997). Nascent membrane and presecretory proteins synthesized in Escherichia coli associate with signal recognition particle and trigger factor. *Mol. Microbiol.* 25, 53–64.

170 POLEVODA, B. & SHERMAN, F. (2001). NatC Nalpha-terminal acetyltransferase of yeast contains three subunits, Mak3p, Mak10p, and Mak31p. *J. Biol. Chem.* 276, 20154–20159.